Lecture Notes in Physics

Springer-Verlag Berlin Heidelberg GmbH

The Editorial Policy for Proceedings

The series Lecture Notes in Physics reports new developments in physical research and teaching – quickly, informally, and at a high level. The proceedings to be considered for publication in this series should be limited to only a few areas of research, and these should be closely related to each other. The contributions should be of a high standard and should avoid lengthy redraftings of papers already published or about to be published elsewhere. As a whole, the proceedings should aim for a balanced presentation of the theme of the conference including a description of the techniques used and enough motivation for a broad readership. It should not be assumed that the published proceedings must reflect the conference in its entirety. (A listing or abstracts of papers presented at the meeting but not included in the proceedings could be added as an appendix.)

When applying for publication in the series Lecture Notes in Physics the volume's editor(s) should submit sufficient material to enable the series editors and their referees to make a fairly accurate evaluation (e.g. a complete list of speakers and titles of papers to be presented and abstracts). If, based on this information, the proceedings are (tentatively) accepted, the volume's editor(s), whose name(s) will appear on the title pages, should select the papers suitable for publication and have them refereed (as for a journal) when appropriate. As a rule discussions will not be accepted. The series editors and Springer-Verlag will normally not interfere with the detailed editing except in fairly obvious cases or on technical matters.

Final acceptance is expressed by the series editor in charge, in consultation with Springer-Verlag only after receiving the complete manuscript. It might help to send a copy of the authors' manuscripts in advance to the editor in charge to discuss possible revisions with him. As a general rule, the series editor will confirm his tentative acceptance if the final manuscript corresponds to the original concept discussed, if the quality of the contribution meets the requirements of the series, and if the final size of the manuscript does not greatly exceed the number of pages originally agreed upon. The manuscript should be forwarded to Springer-Verlag shortly after the meeting. In cases of extreme delay (more than six months after the conference) the series editors will check once more the timeliness of the papers. Therefore, the volume's editor(s) should establish strict deadlines, or collect the articles during the conference and have them revised on the spot. If a delay is unavoidable, one should encourage the authors to update their contributions if appropriate. The editors of proceedings are strongly advised to inform contributors about these points at an early stage.

The final manuscript should contain a table of contents and an informative introduction accessible also to readers not particularly familiar with the topic of the conference. The contributions should be in English. The volume's editor(s) should check the contributions for the correct use of language. At Springer-Verlag only the prefaces will be checked by a copy-editor for language and style. Grave linguistic or technical shortcomings may lead to the rejection of contributions by the series editors. A conference report should not exceed a total of 500 pages. Keeping the size within this bound should be achieved by a stricter selection of articles and not by imposing an upper limit to the length of the individual papers. Editors receive jointly 30 complimentary copies of their book. They are entitled to purchase further copies of their book at a reduced rate. As a rule no reprints of individual contributions can be supplied. No royalty is paid on Lecture Notes in Physics volumes. Commitment to publish is made by letter of interest rather than by signing a formal contract. Springer-Verlag secures the copyright for each volume.

The Production Process

The books are hardbound, and the publisher will select quality paper appropriate to the needs of the author(s). Publication time is about ten weeks. More than twenty years of experience guarantee authors the best possible service. To reach the goal of rapid publication at a low price the technique of photographic reproduction from a camera-ready manuscript was chosen. This process shifts the main responsibility for the technical quality considerably from the publisher to the authors. We therefore urge all authors and editors of proceedings to observe very carefully the essentials for the preparation of camera-ready manuscripts, which we will supply on request. This applies especially to the quality of figures and halftones submitted for publication. In addition, it might be useful to look at some of the volumes already published. As a special service, we offer free of charge LaTeX and TeX macro packages to format the text according to Springer-Verlag's quality requirements. We strongly recommend that you make use of this offer, since the result will be a book of considerably improved technical quality. To avoid mistakes and time-consuming correspondence during the production period the conference editors should request special instructions from the publisher well before the beginning of the conference. Manuscripts not meeting the technical standard of the series will have to be returned for improvement.

For further information please contact Springer-Verlag, Physics Editorial Department II, Tiergartenstrasse 17, D-69121 Heidelberg, Germany

Paul Krée Walter Wedig (Eds.)

Probabilistic Methods in Applied Physics

Springer

Editors

Paul Krée
Université de Paris VI, Pierre et Marie Curie
Mathématiques, Tour 45-46, 5e étage
4, place Jussieu, F-75252 Paris Cedex 05, France

Walter Wedig
Institut für Technische Mechanik
Universität Karlsruhe
D-75120 Karlsruhe, Germany

Cataloging-in-Publication Data applied for

Die Deutsche Bibliothek - CIP-Einheitsaufnahme

Probabilistic methods in applied physics / Paul Krée ; Walter Wedig (ed.).
(Lecture notes in physics ; 451)
ISBN 978-3-662-14006-2 ISBN 978-3-540-44725-2 (eBook)
DOI 10.1007/978-3-540-44725-2
NE: Krée, Paul [Hrsg.]; GT

ISBN 978-3-662-14006-2

Originally published by Springer-Verlag Berlin Heidelberg New York in 1995
Softcover reprint of the hardcover 1st edition 1995

Typesetting: Camera-ready by the editors
SPIN: 10501375 55/3142-543210 - Printed on acid-free paper

Preface

Random phenomena have increasing importance in Engineering and Physics, therefore theoretical results are strongly needed. But there is a gap between the probability theory used by mathematicians and practicioners. Two very different languages have been generated in this way. Note that the objectives are also very different: mathematicians are usually concerned with abstract existence results concerning theoretical problems; and engineers or physicists are mainly interested by effective methods, i.e. methods allowing computations. Also, since these problems have been till now studied independantly in all European countries, this added difficulties.

The main goal of the research contract # STJ 01801F (CD), granted by EEC to several research centers in Europe, was to begin at the European level a scientific collaboration for the resolution of those practical problems, and to diffuse the corresponding scientific results.

Our aim is to be readable by research engineers, physicists and mathematicians, hence the necessity to use an adequate language. This has been realized by a collaboration of research engineers, mechanics, and mathematicians elaborating these new methods and by joint work for the writing of these results.

In order to have a more effective presentation of the results and to be more readable, long papers have been written in order to give simultaneously new methods and examples, and a panorama of the corresponding research fields.

As shown by the list of papers, this volume covers the following topics: simulation, stability theory, Lyapunov exponents, stochastic modelling, statistics on trajectories, parametric stochastic control, Fokker–Planck equations, and Wiener filtering.

The authors and the recipients of the contract thank Springer-Verlag for the interest in this European Scientific Project, and the help they brought through their outstanding publishing capacity in this field.

I thank Denis Talay for his help.

Paris, June 1995 Paul Krée

Table of Contents

Simulation

Stability. Lyapunov Exponents and Stochastic Bifurcation

Probability Vector Spaces and Statistics on Trajectories

Parametric Stochastic Control of Non-Linear Systems and Stochastic Equivalent Linearization

Resolution of Fokker–Planck Equation (FPE)

Wiener Filtering

Addresses of Authors

D. Ammon	Universität Karlsruhe, Institut für Technische Mechanik D-76128 Karlsruhe, Germany
S. Bellizi	Laboratoire de Mécanique et d'Acoustique-CNRS 31, chemin J. Aiguier F-13402 Marseille, France
P. Bernard	Laboratoire de Mathématiques Appliquées Université Blaise Pascal 63177 Aubière Cedex, France
R. Bouc	Laboratoire de Mécanique et d'Acoustique-CNRS 31, chemin J. Aiguier F-13402 Marseille, France
P. Boxler	Institut für Dynamische Systeme Universität Bremen Postfach 330440 D-28334 Bremen, Germany
F. Campillo	INRIA 2004, Route des Lucioles B.P. 93 F-06902 Sophia Antipolis, France
F. Casciati	Department of Structural Mechanics Universita di Pavia Pavia, Italy
P. Fayol	Service Technique des Programmes Aéronautiques (STPA) 4, avenue de la Porte d'Issy Paris F-00460 Armée, France
P. Krée	Université P. et M. Curie Département de mathématiques UFR 20. Tour 46.0 Place Jussieu F-75005 Paris, France
I. Lindeman	Institut für Dynamische Systeme, Universität Bremen Postfach 330440 D-28334 Bremen, Germany
F. Poirion	Office National d'Etudes et de Recherches Aérospatiales (ONERA) 29, Avenue de la Division Leclerc F-92230 Chatillon, France
C. Soize	Same address
D. Talay	INRIA 2004, Route des Lucioles B.P. 93 F-06902 Sophia Antipolis, France
W. Wedig	Universität Karlsruhe, Institut für Technische Mechanik D-76128 Karlsruhe, Germany

The Approximation and the Generation of Stationary Vector Processes

D. Ammon and W. Wedig

Abstract

In the stationary and gaussian case a process is completely described by its power density spectrum. W. Wedig suggested in 1985/6 the use of linear dynamical filter systems driven by White Noise for the artificial generation of the desired process.

Two different formulations of linear approximations are presented and discussed considering practical applications.

An extension of this method to approximate vector processes is described. After a discussion of some principal properties of vector processes to description of the target process (a given spectral density matrix) is "transformed" into a mathematically more convenient form. The approximation of the homogeneous parameters and the adaptation of the right hand side parameters can be preformed very similar to the scalar procedure, leading to approximations of high quality.

In the case of a piecewise constant exitation equivalent discrete time systems can be evaluated and transformed into an ARMA system or ARMA system models, respectively. The generation of the process can be performed with a minimal amount of computation time using these systems which are driven by a simple sequence of random numbers or random vectors, respectively.

The development of technical systems make it more and more necessary to take natural loads and excitations into consideration. In various cases these loads can be described by means of stochastic processes. Usually the statistical properties are given by physical theories or by measurement. One of the basic problems in this field the artificial generation of trajectories of a process, which satisfies a set of given statistical properties.

Some typical examples are turbulence phenomena in Fluid Dynamics, the shaking of a building induced by an earthquake or vehicle vibrations effected by an uneven road surface.

Because of the complex structure of a modern vehicle the main part of studies and investigations must be performed by Monte-Carlo simulations and laboratory tests. Consequently typical realizations of road processes are required in order to approximate the excitation of the considered model or component. From a practical point of view, especially with respect to computer simulations, two further conditions have to be satisfied: I. Good convergence properties (i.e. high sensibility to variations of parameters). II. Minimal amount of computation time for the generation of the process (On-line simulations).

The classical approach to the artificial generation of stochastic processes is based on a system of harmonic functions with random phase angles (Rice [6], Shinozuka [7] et. al.). The amplitudes or the frequency distribution can be adapted to the given power density spectrum. This method yields periodical or almost periodical processes respectively. Because of the special properties of the above system of functions, it may not be suitable in some cases of on-line applications.

In 1986 an alternative approach has been suggested by W. Wedig [9], the approximation of the power density spectrum by complete filter systems driven by White Noise. In a first step the parameters of this linear dynamical model are adapted to the given spectrum. Then an equivalent ARMA system, corresponding to the continuous time system, is used to generate trajectories of the desired process with a minimal amount of computation time. Both methods have been extended to approximate multi-dimensional processes [7,1,2].

1 Approximation of Scalar Processes

1.1 Target Process and System Model

A stationary Gaussian distributed process Y_t^v is completely characterized by its power density spectrum $S^v(\omega)$. The target spectrum can be defined as a smooth, piecewise linear curve in a double logarithmic scale, X_i are the spectral intensities belonging to the (cycle-) frequencies ω_i.

$$S^v(\omega) = X_i \left(\frac{\omega}{\omega_i}\right); \quad \omega_i \leq \omega < \omega_{i+1}; \ i = 0, 1, \ldots, k;$$
$$a_i = \frac{\log(X_{i+1}/X_i)}{\log(\omega_{i+1}/\omega_i}; \quad \omega_0 := \omega_1; \ \omega_{k+1} := \infty \tag{1}$$

To ensure the existence of the spectral moments of Y_t^v the slopes a_0 and a_k have to be restricted as follows: $a_0 \geq 0$ and $a_k \leq -2$. Figure 1 shows a typical road spectrum [1] and the corresponding sequence of straight lines.

The height profiles of road surfaces are stationary with respect to the spatial coordinates. Hence, frequencies correspond to wave numbers. In this case the described procedure can be applied, too, without any alteration. Only the time variables have to be exchanged by spatial ones.

$$m_\xi = E\{\xi_t\} = 0; \quad R_\xi(\tau) = E\{\xi_t \xi_{t+\tau}\} = \delta(\tau); \quad S_\xi(\omega) = 1 \tag{2}$$

The properties of the output process Y_t of the system are depending on $2n+1$

[1] The system model is a linear dynamical system of the order n (complete filter system). It is excited by a normalized stationary White Noise process ξ_t; the mean value ξ_t is assumed to be vanishing.

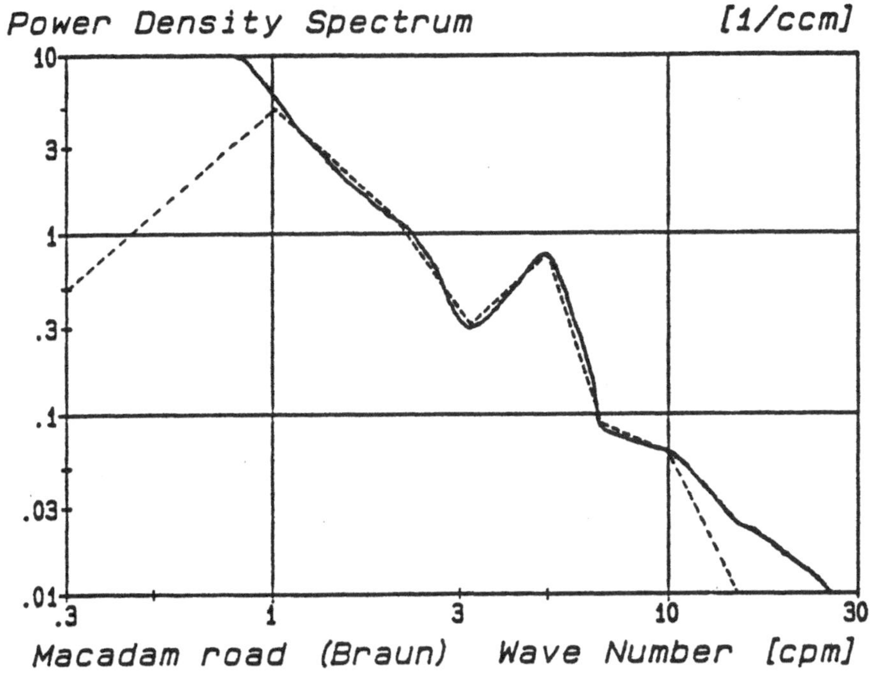

Fig. 1. The power density spectrum of a 'MacAdam' road [3]. The piecewise linear curve represents an approximative description relating to (1).

parameters a_i and b_i

$$\left.\begin{aligned} L_N\{X_t\} &= \sum_{i=0}^{n} a_i X_t^{(i)} = \xi_t; \\ Y_t = L_Z\{X_t\} &= \sum_{i=0}^{n-1} b_i X_t^{(i)} = \xi_t; \\ X_t^{(i)} &:= \frac{d^i}{dt^i} X_t \end{aligned}\right\} \tag{3}$$

Applying the Fourier-Transform to the differential operators L_N and L_Z in (3), we obtain the corresponding admittances (frequency responses) $N(j\omega)$ and $Z(j\omega)$ in the frequency domain.

$$N(j\omega) = \sum_{i=0}^{n} a_i (j\omega)^i; \qquad Z(j\omega) = \sum_{i=0}^{n-1} b_i (j\omega)^i; \tag{4}$$

$$j = \sqrt{-1}$$

With respect to the excitation properties (2) the power density spectrum $S_y(\omega)$ of the process Y_t can be expressed in terms of ω and the system parameters.

$$S_y(\omega) = \frac{Z(-j\omega)Z(j\omega)}{N(-j\omega)N(j\omega)} \tag{5}$$

The parameters a_i and b_i have to be determined in such a way, that $S_y(\omega)$ approximates the target spectrum $S^v(\omega)$ as well as possible. This is decribed in the following two sections.

In the case of a piecewise constant excitation process $\xi_t := Z_k = Z(k\Delta t)$, $k\Delta t \leq t < (k+l)\Delta t$, an ARMA system can be evaluated using the analytical solution of the continuous-time system (3).

$$Y(k+n) = \frac{1}{d_n}\left\{-\sum_{i=0}^{n-1} d_i Y(k+i) + \sum_{i=0}^{n-1} e_i Y(k+i)\right\}; \quad k = 0,1,2,\ldots \qquad (6)$$

1.2 Approximation of the Parameters

Actually we have to perform an adaptation of the spectrum of the system $S_y(\omega)$ to the given spectrum $S^v(\omega)$. The application of a global L_2-norm to the local frequency dependent error results in the quality function $J^\omega(a_i, b_i)$, which is to be minimized with respect to the parameters a_i and b_i.

$$J^\omega(a_i, b_i) = \int_0^\infty [S_y(\omega) - S^v(\omega)]^2 d\omega \equiv minimum \qquad (7)$$

The spectrum $S_y(\omega)$ of the system is a fractional rational function of a_i and b_i (4,5). Hence, for higher order systems the above formulation (7) yields a nonlinear and nonconvex optimization problem. In general, neither the existence nor the uniqueness of a solution can be proved [9].

1.3 The 'Standard' Procedure

The problems above mentioned do not appear in the linear formulation suggested by W. Wedig [8]. A L_2-approximation applied to the correlation differential equation $L_N\{R_y(\tau)\}$ yields the system parameters a_i in a linear way.

$$R_y(\tau) = E\{Y_t Y_{t+\tau}\};$$

$$L_N\{R_y(\tau)\} = L_N^0\{R_y(\tau)\} = \sum_{i=0}^{n} a_i R_y^{(i)}(\tau) = 0; \quad \tau > O \qquad (8)$$

In order to shift the 'weighting' of the approximation to the range of lower frequencies we can also use integrated forms of the operator $L_N\{R_y(\tau)\}$.

$$L^{j+1} = \frac{d}{d\tau} L^j; \qquad L^{j-1} = \int_\infty^\tau L^j d\tau;$$

$$L_N^j\{R_y(\tau)\} = \sum_{i=0}^{n} a_i R_y^{i+j}(\tau) = 0; \quad j = 0,-1,-2,\ldots \qquad (9)$$

Since the correlation operator is homogeneous for $\tau > 0$, the corresponding L_2 quality function J^{lin} only contains the correlation $R^v(\tau)$ of the target.

$$J^{\text{lin}}(a_i; j) = \int_0^{+\infty} \left[-\sum_{i=0}^{n} a_i R^{v(i+j)}(\tau) \right] d\tau \equiv min.; \qquad j = 0, -1, -2, \ldots \tag{10}$$

A necessary condition of (10) is that the partial derivatives of $J^{\text{lin}}(a_i; j)$ are vanishing. We obtain a linear system of equations to calculate the parameters a_i. The improper integrals in (10) can be solved analytically.

An adaptation of the system correlation $R_y(\tau)$ to the initial conditions of the target yields the remaining parameters b_i. In a first step we apply the Laplace-transform to the integro-differential equation (9). We obtain an algebraic equation for the positive time ($\tau > 0$) image $R^+(p)$ of the correlation function.

$$\sum_{i=0}^{n} a_i p^i R^+(p) = Q(p) := \sum_{i=0}^{n-1} q_i p^i;$$
$$q_i = \begin{cases} -\sum\limits_{k=0}^{i} a_i R^{(k-1-i)}(0); & i = 0; \ldots, -1-j; \\ \sum\limits_{k=i+1}^{n} a_i R^{(k-1-i)}(0); & i = -j; \ldots, n-1; \end{cases} \tag{11}$$

Regarding pure imaginary arguments $p \Rightarrow j\omega$, we can use the Laplace solution $R^+(j\omega)$ and the negative time image $R^-(j\omega) = R^+(-j\omega)$ to construct the corresponding power density spectrum.

$$S_y(\omega) = R^-(j\omega) + R^+(j\omega) = \frac{Q(-j\omega)}{N(-j\omega)} + \frac{Q(j\omega)}{N(j\omega)} \equiv \frac{Z(-j\omega)Z(j\omega)}{N(-j\omega)N(j\omega)} \tag{12}$$

A comparison of the above equation with the Fourier-Transform expression (5) shows that the coefficients of the numerator polynomial are wellknown. The parameters b_i can easily be evaluated using the corresponding conjugate real roots (eigenvalues) of the polynomial. With the object of minimizing the phase delay, the negative real roots should be chosen.

Figure 2 shows the linear approximation of a power density spectrum with a 7th order system.

1.4 Orthogonalization

We define a parameter vector $\mathbf{a}^T = (a_0, a_1, \ldots, a_n)$ to obtain a quadratic form for the quality function $J^{\text{lin}}(a_i; j)$ in (10). Obviously the form matrix $\mathbf{M}$ is symmetrical.

$$J^{\text{lin}}(a_i; j) = \mathbf{a}^T\mathbf{M}\mathbf{a}; \quad \mathbf{M} = (m_{k,l}) = \left(\int_{+0}^{\infty} R^{k+i}(\tau) R^{i+j}(\tau) d\tau \right) \tag{13}$$

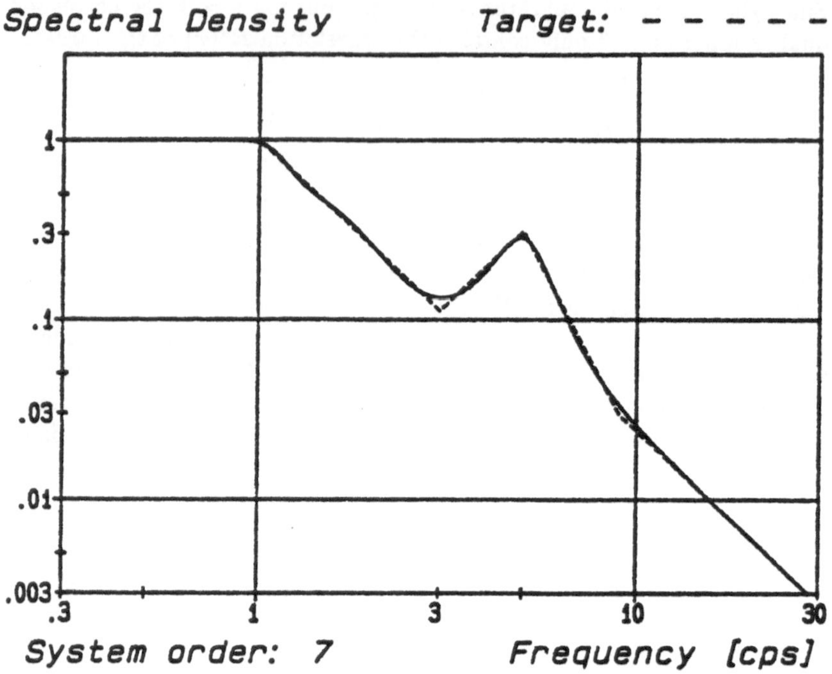

Fig. 2. Linear approximation: performed with a 7th order system.

A transformation with respect to the principal axes yields the corresponding normal form. With the eigenvalues λ_i and the eigenvectors $\mathbf{a}$, we obtain:

$$\mathbf{a} := \mathbf{C}\mathbf{y}; \qquad \mathbf{C} = (\mathbf{a}_0, \mathbf{a}_1, \ldots, \mathbf{a}_n);$$
$$J^{\text{lin}}(a_i; j) \equiv J^{\text{lin}}(y_i; j) = \sum_{i=0}^{n} \lambda_i y_i^2 > 0 \tag{14}$$

The definition (10) implies that the quality function J^{lin} is non-negative. In the considered case of a real approximation we conclude that J^{lin} is always positive, excluding the trivial solution $\mathbf{y} = \mathbf{0}$, a minimization by the parameters y_i, results in the solution y_{i*j}, corresponding to the lowest eigenvalue $\lambda_{i*} \equiv \lambda^{\text{min}}$ $(\lambda_i > 0, i = 0, 1, \ldots, n)$, giving the best approximation. Figure 3 shows an illustration of the case of a second order system.

Thus the orthogonalization procedure requires the calculation of the eigenvalues λ_i of $\mathbf{M}$. The eigenvector $\mathbf{a}_{i*}$ corresponding to the smallest eigenvalue $\lambda_{i*} = \lambda^{\text{min}}$ contains the desired set of parameters (see [8]).

From a theoretical point of view the orthogonalization seems to be more proper (or suitable respectively), because it avoids the selection of n equations (of $n+1$) which must be performed using the standard method (in order to evaluate a nontrivial solution). According to our experience the interesting eigenvalue λ^{min} of $\mathbf{M}$ is always very small and converges to zero with increasing system order. Thus the results of both formulations are in most cases almost identical.

2 Vector Processes

A set of (complex) power density spectra $S_{k,l}(\omega)$; $k,l = 1,\ldots,n$, combined in a $n \times n$ spectral density matrix $\mathbf{S}^v(\omega)$ describe the statistical properties of a stationary n-component vector process $\mathbf{Y}_t$. Applying the Fourier Transform to the spectral density matrix $\mathbf{S}^v(\omega)$ we obtain the corresponding correlation matrix $\mathbf{R}^v(\tau) = E\left\{\mathbf{Y}_t\mathbf{Y}_{t+\tau}^T\right\}$ of $\mathbf{Y}_t$. "T" stands for the transposition; $E\{\ldots\}$ is the operator of expectation. In the case of a real, stationary process $\mathbf{Y}_t$ the matrix $\mathbf{S}^v(\omega)$ is Hermitian; its real part $Re\{S(\omega)\}$ is symmetrical and its imaginary part is $Im\{S(\omega)\}$ antisymmetrical with respect to $\omega = 0$.

$$\mathbf{S}^{T*}(\omega) = \mathbf{S}(\omega)$$

$$Re\{\mathbf{S}(\omega)\} = Re\{\mathbf{S}(-\omega)\}; \qquad Im\{\mathbf{S}(\omega)\} = -Im\{\mathbf{S}(-\omega)\}; \qquad (15)$$

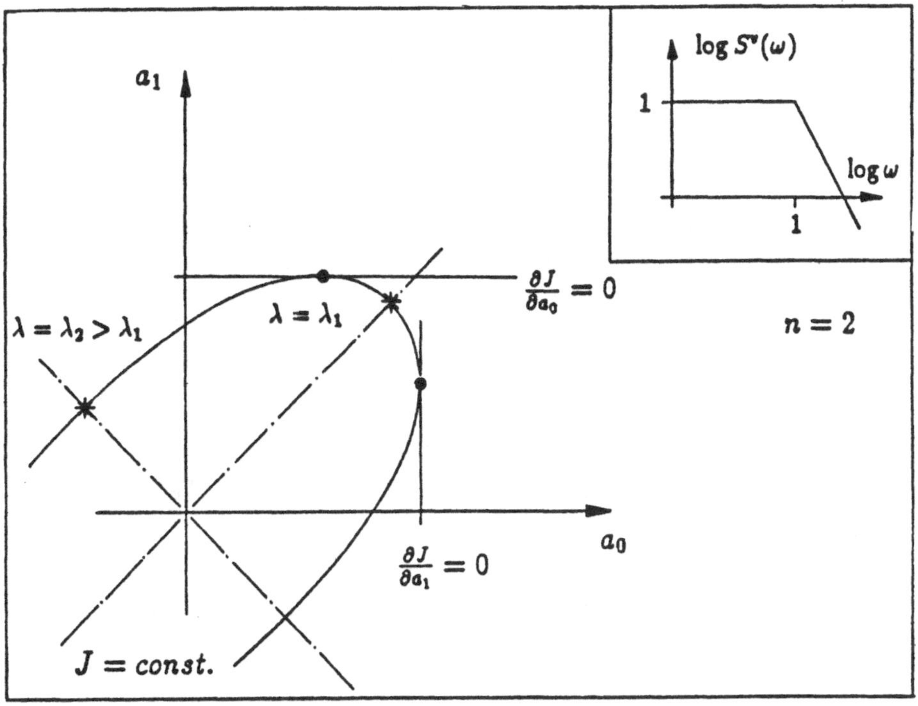

Fig. 3. Comparison of the approximations of a simple spectrum with a 2nd order system. The solutions of the standard method (•) and of the orthogonalization procedure (*) yield different results in the parameter plane (a_0, a_1).

Conjugate complex forms are denoted by "*". The above relation can be derived by a separation of the correlation matrix $\mathbf{R}(\tau)$ into a symmetrical part

$\mathbf{R}^s(\tau)$ and an antisymmetrical part $\mathbf{R}^a(\tau)$

$$\mathbf{R}^s(\tau) = \frac{1}{2}(\mathbf{R}(\tau) + \mathbf{R}(-\tau)); \qquad \mathbf{R}^a(\tau) = \frac{1}{2}(\mathbf{R}(\tau) - \mathbf{R}(-\tau)) \tag{16}$$

and the transposition property

$$\mathbf{R}^T(\tau) = E\{\mathbf{Y}_{t+\tau}\mathbf{Y}_t^T\} = E\{\mathbf{Y}_t\mathbf{Y}_{t-\tau}^T\} = \mathbf{R}(-\tau). \tag{17}$$

Thus, $\mathbf{S}^v(\omega)$ defines n real, symmetrical auto power density spectra $S_{k,k}(\omega)$ and $(1/2)n(n-1)\cdot$ complex cross power density spectra $S_{k,l}(\omega)$ (i.e. $l > k$), where the real parts are symmetrical and the imaginary parts are antisymmetrical with respect to $\omega = 0$. The symmetrical spectra $S^R(\omega) = S_{k,k}(\omega)$ or $S^R(\omega) = Re\{S_{k,l}(\omega)\}$, respectively, correspond to symmetrical correlation functions $R_{i,j}(\tau)$, $(i \leq j)$; an equivalent relation is valid for the imaginary parts $S^I(\omega) = jIm\{S_{k,l}(\omega)\}$.

The real target spectra $S^R(\omega)$ may be defined by a piecewise linear curve in a double logarithmic scale; X_i are the spectral intensities corresponding to the frequencles ω_i.

$$S^R(\omega) := S_i^R = X_i\left(\frac{\omega}{\omega_i}\right)^{a_i} \quad \text{in } \omega_i \leq \omega < \omega_{i+1};\ i = 0,\ldots,k; \tag{18}$$

$$\text{where: } a_i = \frac{\log(X_{i+1}/X_i)}{\log(\omega_{i+1}/\omega_i)}; \quad a_0 \geq 0;\ a_k \leq -2;$$

$$\omega_0 = \omega_1;\ X_0 = X_1;\ \omega_{k+1} = \infty$$

Such a power density spectrum can be approximated by a complete filter system without any loss of accuracy, as described in the previous sections. Performing this approximation we obtain another representation for the corresponding correlation function, which is more convenient for the following calculations. α_i, β_i are the eigenvalues and C_i, S_i are constants to adapt the initial conditions. The order of the filter is n; m denotes the number of conjugate complex eigenvalues.

$$R^R(\tau) = \sum_{i=1}^{m} e^{\alpha_i}(C_i \cos\beta_i\tau + S_i \sin\beta_i\tau) + \sum_{i=2m+1}^{n} C_i e^{\alpha_i}; \tag{19}$$

$$R^R(\tau) = R^R(-\tau)$$

The pure complex parts $S^I(\omega)$ of hte target spectra can be defined analogously to (18).

$$S^I(\omega) := S_i^I = jX_i\left(\frac{\omega}{\omega_i}\right)^{a_i} \quad \text{in } \omega_i \leq \omega \leq \omega_{i+1};\ i = 0,\ldots,k; \tag{20}$$

$$j = \sqrt{-1}; \quad a_0 \geq 1; \quad a_k \leq -1$$

Dividing $S^I(\omega)$ by $(j\omega)$ we define a real symmetrical 'substitute' spectrum $S^E(\omega)$.

$$S^E(\omega) := S_i^E = E_i\left(\frac{\omega}{\omega_i}\right)^{e_i} \tag{21}$$

$$E_i = \frac{X_i}{\omega_i}; \quad e_i = a_i - 1; \quad i = 0, \ldots, k$$

The above spectrum is formally equivalent to the spectra in equations (18). Thus, we can determine a corresponding symmetrical correlation function $R^E(\tau)$ of the type (19) using the filter approximation formalism. The derivative of this function with respect to τ corresponds to the target spectrum $S^I(\omega)$ and can also be expressed in the terms of exponential cosine and exponential sine functions.

$$S^I(\omega) = (j\omega)S^E(\omega) \Leftrightarrow R^I(\tau) = \frac{d}{d\tau}R^E(\tau) \tag{22}$$

Finally, we obtain an equivalent representation of the given spectral density matrix $\mathbf{S}^v(\omega)$ without any loss of accuracy, using the correlation matrix $\mathbf{R}^v(\tau)$. Its derivatives are defined at any order for $\tau > 0$. Additionally a special type of integrals $\int r_{i,j}(\tau) r_{k,l}(\tau) d\tau$ — appearing in the approximation section — can be solved analytically. In the following we assume a correlation matrix $\mathbf{R}^v(\tau)$ of the above type to be given.

3 Complete Filter Systems

A linear system of differential equations of N-th order is going to be used as the system model. The excitation process is a stationary White Noise vector process $\boldsymbol{\xi}_t$, with n independent components $\xi_{i,t} = \dot{W}_{i,t}$ and vanishing mean. The output process of the system is the stationary model process $\mathbf{Y}_t$.

$$\mathbf{m}_\xi = E\{\boldsymbol{\xi}_t\} = \mathbf{0}; \ \mathbf{R}_\xi(\tau) = E\{\boldsymbol{\xi}\boldsymbol{\xi}^T_{t+\tau}\} = \mathbf{I}\delta(\tau) \tag{23}$$

$$\sum_{i=0}^{N} \mathbf{D}_i \mathbf{X}_t^{(i)} = \boldsymbol{\xi}_t; \quad \mathbf{Y}_t = \sum_{i=0}^{N-1} \mathbf{E}_i \mathbf{X}_t^{(i)}; \quad \mathbf{D}_N = \mathbf{I} \tag{24}$$

The matrices $\mathbf{D}_i$ and $\mathbf{E}_i$ content the parameters of the system, which is formulated in the most general way (24). If the excitation processes $\mathbf{U}_t = \boldsymbol{\xi}_t$ are derivable, the system can be represented by one matrix differential equation.

$$\boldsymbol{\xi}_t := \mathbf{U}_t; \quad \sum_{i=0}^{N} \mathbf{A}_i \mathbf{Y}_t^{(i)} = \sum_{i=0}^{N-1} \mathbf{B}_i \mathbf{U}_t^{(i)}; \quad \mathbf{A}_N = \mathbf{I} \tag{25}$$

The system parameters in the matrices $\mathbf{A}_i$ and $\mathbf{B}_i$ are depending on the parameters in (24) as given in the following linear system algebraic equations.

$$\sum_{j=0}^{k} (\mathbf{A}_j \mathbf{E}_{k-j} - \mathbf{B}_{k-j} \mathbf{D}_j) = 0; \text{ for } k = 0, 1, \ldots, N-1; \tag{26}$$

$$\sum_{j=k-N+1}^{N} (\mathbf{A}_j \mathbf{E}_{k-j} - \mathbf{B}_{k-j} \mathbf{D}_j) = 0; \text{ for } k = N, N+1, \ldots, 2N-1$$

Multiplying the above differential equation by the time-shifted and transposed process and taking the expectation, we obtain the correlation differential equations of the process $\mathbf{X}_t$ and $\mathbf{Y}_t$. Because of the expectation properties (23) these equations are homogeneous for positive arguments τ.

$$\mathbf{R}_x(\tau) = E\{\mathbf{X}_t\mathbf{X}_{t+\tau}^T\}; \quad \sum_{i=0}^{N} \mathbf{D}_i\mathbf{R}_x^{(i)T}(\tau) = 0; \; \tau > 0 \tag{27}$$

$$\mathbf{R}_y(\tau) = E\{\mathbf{X}_t\mathbf{X}_{t+\tau}^T\}; \quad \sum_{i=0}^{N} \mathbf{A}_i\mathbf{R}_y^{(i)T}(\tau) = 0; \; \tau > 0 \tag{28}$$

4 Approximation of the Homogeneous Parameters

The parameters $\mathbf{A}_i$ can be determined by a L_2-approximation, applied to the correlation differential equation $D^{(j)}\{\mathbf{R}_y(\tau)\}$, $j = 0$ (28) or integrated forms ($j > 0$) of these equations, respectively.

$$D^{(j)}\{\mathbf{R}_y(\tau)\} := \sum_{i=0}^{N} \mathbf{A}_i\mathbf{R}_y^{(i+j)T}(\tau); \quad j = 0, 1, \ldots, -(N-1); \tag{29}$$

$$D^{(j+1)}\{\mathbf{R}(\tau)\} = \frac{d}{d\tau}D^{(j)}\{\mathbf{R}(\tau)\}; \quad D^{(j-1)}\{\mathbf{R}(\tau)\} = \int_{\infty}^{\tau} D^{(j)}\{\mathbf{R}(\tau)\}d\tau$$

The difference between the operator of the system correlation $D^{(j)}\{\mathbf{R}_y(\tau)\}$ and the correlation of the target $D^{(j)}\{\mathbf{R}_y^v(\tau)\}$ can be used to define the $n \times n$ error matrix ε.

$$\varepsilon = \sum_{i=0}^{N} \mathbf{A}_i\mathbf{R}_y^{(i+j)T}(\tau) - \sum_{i=0}^{N} \mathbf{A}_i\mathbf{R}_y^{v(i+j)T}(\tau); \tag{30}$$

$$\mathbf{A}_i = (a_{k,l}^i); \; \mathbf{R}^{v(i)} = (r_{k,l}^{(i)})$$

Applying a global L_2-norm to the error matrix ε we obtain the objective function $J(a_{\nu,\eta}^{\kappa})$ to be minimized with respect ot the paramters $a_{\nu,\eta}^{\kappa}$.

$$J(a_{\nu,\eta}^{\kappa}) := \int_{+0}^{\infty} \|\varepsilon\| d\tau = \int_{+0}^{\infty} \sum_{k=1}^{n}\sum_{l=1}^{n}\Big(-\sum_{i=0}^{N}\sum_{j=1}^{n} a_{k,j}^i r_{l,j}^{(i+j)}(\tau)\Big)^2 \equiv \min. \tag{31}$$

The necessary properties of a solution —vanishing partial derivatives of J with respect to $a_{\nu,\eta}^{\kappa}$— leads to a linear system of the equations to determine to desired parameters.

$$\sum_{i=0}^{N-1}\sum_{j=1}^{n} a_{\nu,\eta}^i \sum_{l=1}^{n}\int_{+0}^{\infty} r_{l,j}^{(i+j)}(\tau) r_{l,\eta}^{(\kappa+j)}(\tau)d\tau = -\sum_{l=1}^{n}\int_{+0}^{\infty} r_{l,\nu}^{(N+j)}(\tau) r_{l,\eta}^{(\kappa+j)}(\tau)d\tau \tag{32}$$

$$\nu, \eta = 1, 2, \ldots, n; \quad \kappa = 0, 1, \ldots, N-1$$

The quadratic integrals can be solved analytically, as mentioned before. The matrices $\mathbf{A}_i$, that means the system parameters, are calculated row per row for $\nu = 1, 2, \ldots, n$, where the $nN \times nN$ matrix $\mathbf{M}$ fo the system is constant. For every row ν we have to replace the inhomogenity by the corresponding vector $\mathbf{B}_\nu$.

$$\begin{aligned} \mathbf{a}_\nu^T &:= (a_{\nu,1}^0, a_{\nu,2}^0, \ldots, a_{\nu,n}^0, a_{\nu,1}^1, a_{\nu,2}^1, \ldots, \ldots, a_{\nu,n}^{N-1}); \\ \mathbf{R}_\nu^T(j) &:= (r_{\nu,1}^{(j)}, r_{\nu,2}^{(j)}, \ldots, r_{\nu,n}^{(j)}, r_{\nu,1}^{(j+1)}, r_{\nu,2}^{(j+1)}, \ldots, \ldots, r_{\nu,n}^{(j+N-1)}), \end{aligned} \tag{33}$$

$$\mathbf{M} = \sum_{l=1}^{n} \int_{+0}^{\infty} \mathbf{R}_\nu(j)\mathbf{R}_\nu^T(j) d\tau; \quad \mathbf{B}_\nu = -\sum_{l=1}^{n} \int_{+0}^{\infty} r_{l,\nu}^{v(N+j)}(\tau)\mathbf{R}_\nu^T(j) d\tau; \tag{34}$$

$$\mathbf{a}_\nu^T = \mathbf{M}^{-1}\mathbf{B}_\nu; \quad \nu = 1, 2, \ldots, n$$

5 Inhomogeneous Parameters

For a complex excitation $\boldsymbol{\xi}_t = \boldsymbol{\xi}_0$ and a corresponding response of the system $\mathbf{Y}_t = \mathbf{F}_y(j\omega)\boldsymbol{\xi}_t$ we obtain the matrix of the admittances $\mathbf{F}_y(j\omega)$ of the system (25).

$$\mathbf{H}(j\omega) = \left(\sum_{i=0}^{N} \mathbf{A}_i (j\omega)^i\right)^{-1}; \qquad \mathbf{G}(j\omega) = \sum_{i=0}^{N-1} \mathbf{B}_i (j\omega)^i; \tag{35}$$

$$\mathbf{F}_y(j\omega) = \mathbf{H}(j\omega)\mathbf{G}(j\omega)$$

Because of the special properties of the ecitation process (23), the spectral density matrix of the system $\mathbf{S}_y(\omega)$ can easily be expressed in terms of the admittances.

$$\mathbf{S}_y(\omega) = \mathbf{F}_y^*(j\omega)\mathbf{I}\mathbf{F}_y^T(j\omega) = \mathbf{H}^*(j\omega)\mathbf{G}^*(j\omega)\mathbf{G}^T(j\omega)\mathbf{H}^T(j\omega) \tag{36}$$

In order to determine the system matrices $\mathbf{B}_i$ we are going to adapt the solution of the correlation differential (29) of the system to the corresponding initial values of the target $\mathbf{R}^v(\tau)$. The Laplace transform of the system operator $D^{(j)}\{\mathbf{R}_y(\tau)\}$, $\tau > 0$ leads to an algebraic system of equations for the correlation matrix $\mathbf{R}_y^+(p)$ in the image domain.

$$\sum_{i=0}^{N} \mathbf{A}_i p^i \mathbf{R}_y^+(p) = \sum_{i=0}^{N-1} p^l \mathbf{R}_l; \quad \mathbf{R}_A(p) = \sum_{l=0}^{N-1} p^l \mathbf{R}_l; \tag{37}$$

$$\mathbf{R}_i = \begin{cases} -\sum_{i=0}^{l} \mathbf{A}_i \mathbf{R}_y^{(i-1-l)T}(0) \text{ for } l = 0, 1, \ldots, (-i-j) \\ -\sum_{i=i+1}^{N} \mathbf{A}_i \mathbf{R}_y^{(i-1-l)T}(0) \text{ for } l = (-j), \ldots, -N-1 \end{cases} \tag{38}$$

Considering pure imaginary arguments $p \rightarrow j\omega$ the complex spectrum $\mathbf{R}_y^+(j\omega)$ can be explained as the Fourier image of the $\tau \geq 0$ part of the correlation matrix $\mathbf{R}_y(\tau)$ (where $\mathbf{R}_y^+(\tau < 0) = 0$; $\mathbf{R}_y^+(\tau \geq 0) = \mathbf{R}_y(\tau)$).

$$\mathbf{R}_y^+(j\omega) = \mathbf{R}_A^T(j\omega)\mathbf{H}^T(j\omega) \tag{39}$$

A comparison of the Fourier and the Laplace Transform shows that the (entire) spectral density matrix $\mathbf{S}(\omega)$ can be presented by $\mathbf{R}_y^+(j\omega)$ and the transposed, conjugate complex form $\mathbf{R}_y^-(j\omega)$.

$$\mathbf{S}(\omega) = \mathbf{R}_y^+(j\omega) + \mathbf{R}_y^-(j\omega); \qquad \mathbf{R}_y^-(j\omega) = \mathbf{R}_y^{+*T}(j\omega) \tag{40}$$

Actually, the above equation is another description for the spectrum of the system (36). This equality leads to an expression for the Hermitian form of the $\mathbf{B}_i$ matrix polynomial, combined in the matrix polynomial $\mathbf{P}(j\omega)$.

$$\mathbf{G}^*(j\omega)\mathbf{G}^T(j\omega) = \mathbf{R}_A^*(j\omega)\mathbf{H}^{T-1}(j\omega)\mathbf{H}^{*-1}(j\omega)\mathbf{R}_A^T(j\omega) =: \tag{41}$$

$$\mathbf{P}(j\omega) = \sum_{m=0}^{2N-2} \mathbf{P}_m (j\omega)^m$$

The matrices $\mathbf{P}_m$ are determined by the homogeneous parameters $\mathbf{A}_i$ and the initial values of the Laplace Solution $\mathbf{R}_i$.

$$\mathbf{P}_m = \begin{cases} \sum_{i=0}^{m} \{(-1)^{m-i}\mathbf{R}_{m-i}\mathbf{A}_i^T + (-1)^i \mathbf{A}_i \mathbf{R}_{m-i}^T\}, & m = 0, 1, \ldots, N-1; \\ \sum_{i=m-N+1}^{N} \{(-1)^{m-i}\mathbf{R}_{m-i}\mathbf{A}_i^T + (-1)^i \mathbf{A}_i \mathbf{R}_{m-i}^T\}, & m = N, \ldots, 2N-2; \end{cases} \tag{42}$$

The calculation of the matrices $\mathbf{B}_i$, that means the solution of the nonlinear system of equations (41), can be performed within three steps. First we define a normalized polynomial $\mathbf{C}$, which is related to $\mathbf{G}$ as follows.

$$\mathbf{G}(j\omega) = \mathbf{B}\mathbf{C}(j\omega); \quad \mathbf{C}(j\omega) = \sum_{i=0}^{N-1} \mathbf{C}_i (j\omega)^i; \quad \mathbf{C}_{N-1} = \mathbf{I} \tag{43}$$

The normalization matrix $\mathbf{B}$ is assumed to be upper triangular. Therefore it can easily be determined. The above assumption is not coupled with any loss of generality, because $\mathbf{P}_{2N-2}$ belongs to a quadratic form (see Zurmühl [10]).

$$\mathbf{B}\mathbf{B}^T = (-1)^{N-1}\mathbf{P}_{2N-2} \tag{44}$$

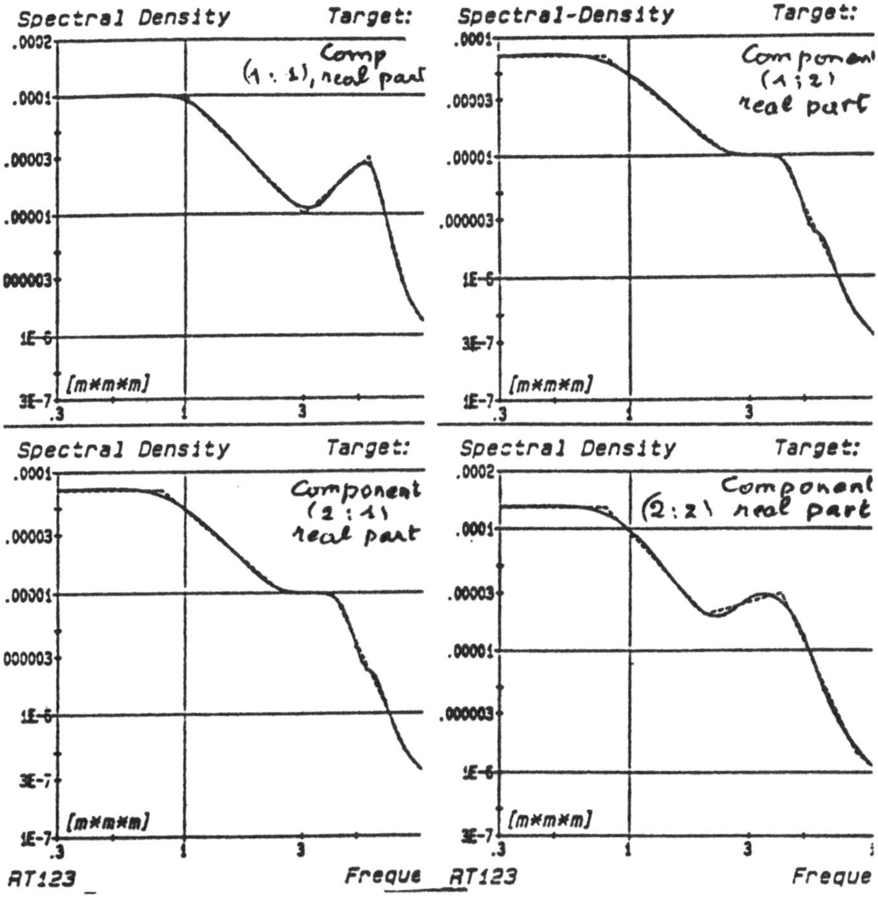

Fig. 4. Approximation of a pure real 2×2 spectral density matrix with a filter system of 7th order

The coefficients of the matrices $\mathbf{C}_i$ can be determined by a Newton iteration, applied to the nonlinear system of equations reduced by $\mathbf{B}$. The required starting vectors can be obtained, if i.e. an exact diagonal solution for $\mathbf{C}_i$ is determined first. In this case we calculate the eigenvalues of the (quadratic) diagonal polynomials. Then the corresponding elements of the $\mathbf{C}$ polynomial can be constructed using Vieta's Theorem.

At the end of the iteration procedure the matrices $\mathbf{C}_i$ and —by equation (43)— also the inhomogeneous matrices $\mathbf{B}_i$ of the system differential equation (25) are known. Using the relation (36) we can calculate the spectral density matrix of the modell system. Figure (4) shows the results of an approximation with a 7th order filter system for two state processes.

6 ARMA System Model

Using the relation (26) we obtain the parameter matrices $\mathbf{D}_i$ and $\mathbf{E}_i$ of the form (24). By the definition of a (new) state vector $\mathbf{U}_t$ we obtain first order system to describe the model system.

$$\mathbf{U}_t^T := (\mathbf{X}_t^T, \dot{\mathbf{X}}_t^T, \ldots, \mathbf{X}_t^{(N-1)T}); \qquad \mathbf{V}_t^T := (\mathbf{0}^T, \mathbf{0}^T, \ldots, \boldsymbol{\xi}_t^T);$$

$$\mathbf{A} = \begin{pmatrix} \mathbf{0} & \mathbf{I} \cdots & \mathbf{0} \\ \mathbf{0} & \mathbf{I} & \vdots \\ \vdots & \ddots \ddots & \\ \mathbf{0} \quad \cdots & \mathbf{0} & \mathbf{I} \\ -\mathbf{D}_0 \; -\mathbf{D}_1 \cdots & & -\mathbf{D}_{N-1} \end{pmatrix}; \qquad \mathbf{E} = \begin{pmatrix} \mathbf{E}_0^T \\ \mathbf{E}_1^T \\ \vdots \\ \mathbf{E}_{N-1}^T \end{pmatrix} \tag{45}$$

$$\dot{\mathbf{U}}_t = \mathbf{A}\mathbf{U}_t; \qquad \mathbf{Y}_t = \mathbf{E}^T\mathbf{U}_t \tag{46}$$

The $nN \times nN$ system matrix $\mathbf{A}$ and the $nN \times n$ excitation matrix $\mathbf{E}$ are directly depending on wellknown parameters. Replacing the noise process $\boldsymbol{\xi}_t$ by an independent, piecewise constant excitation process $\mathbf{Z}(k)$, the inhomogeneous solution of the system (46) yields a recursion formula for the generation of process $\mathbf{Y}(k)$. For a given step size Δt the transition matrix $\Phi(\Delta t)$ of the above system can be calculated by the method of Plant (see Föllinger [5]) or using the eigenvalues of the matrix $\mathbf{A}$.

$$\mathbf{M} := \Phi(\Delta t) = \exp(\mathbf{A}\Delta t); \quad \mathbf{P} = \mathbf{A}^{-1}(\mathbf{M} - \mathbf{I}) = (\mathbf{p}_1, \mathbf{p}_2, \cdots, \mathbf{p}_N) \tag{47}$$

$$\mathbf{U}(k+1) = \mathbf{M}\mathbf{U}(k) + \mathbf{q}\mathbf{Z}(k); \quad \mathbf{Y}(k) = \mathbf{E}^T\mathbf{U}(k); \quad \mathbf{q} = \frac{1}{\sqrt{\Delta t}}\mathbf{p}_N \tag{48}$$

The discretization changes the amplitudes of the filter signal. This is a consequence of the correlation properties of the discrete excitation process $\mathbf{Z}(k)$ in (48); actually it is a kind of coloured noise process. A good adaptation of the variances of the process for small step sizes Δt can be obtained taking Δt to normalize the $nN \times n$ excitation matrix $\mathbf{p}_N$ (48); in this case the energy of the excitation does not change in the interesting frequency range.

An ARMA system modell of N-th order can be used to generate the same process $\mathbf{Y}(k)$, where the excitation trajectory $\mathbf{Z}(k)$ is the same as in the above recursion formula.

$$\mathbf{Y}(k+N) = -\sum_{i=0}^{N-1} \mathbf{D}_i\mathbf{Y}(k+i) + \sum_{i=0}^{N-1} \mathbf{E}_i\mathbf{Z}(k+i); \tag{49}$$

$$\mathbf{D}_N = \mathbf{I}; \quad k = 0, 1, \ldots$$

The $2Nn^2$ elements of the matrices $\mathbf{D}_i$, and $\mathbf{E}_i$ can be determined by a comparison of the coefficients of the state sequences $\mathbf{Y}(k)$ and $\mathbf{Z}(k)$. Using the one step recursion (48) we generate $2Nn$ states k. Assuming this state sequence to correspond to the ARMA system model (49), we obtain a linear system of equations to determine the required parameter matrices $\mathbf{D}_i$ and $\mathbf{E}_i$.

As in the continuous case, the system (49) can be transformed into a form with an intermediate process, using the relation (26).

$$\sum_{i=0}^{N} \mathbf{A}_i \mathbf{X}(k+i) = \mathbf{Z}(k); \quad \mathbf{Y}(k) = \sum_{i=0}^{N-1} \mathbf{B}_i \mathbf{X}(k+i); \tag{50}$$

$$\mathbf{A}_N = \mathbf{I}; \quad k = 0, 1, \ldots$$

Applying the same operations as in the continuous case we calculate N correlation matrices $\mathbf{R}_x(k)$, $k = 0, 1, \ldots, N-1$ of the process $\mathbf{X}$ and the covariance matrix $\mathbf{R}_y(0)$ of the process $\mathbf{Y}$. A comparison with the diagonal elements of the covariance matrix of the target process $\mathbf{R}_y^v(0)$ yields correction values for the corresponding rows of the $\mathbf{B}_i$ matrices (to eliminate the effects of the discrete excitation process $\mathbf{Z}(k)$). The remaining deviations in the values of the cross covariances are comparatively small, in general. Finally, the desired parameters of the ARMA system model ($\mathbf{D}_i$, $\mathbf{E}_i$) are calculated using (26) in the inverse direction.

Figure (5) shows a trajectory of the process $\mathbf{Y}(k)$, generated by a 2 component ARMA system model od 7th order. The trajectory corresponds to the continuous approximation example presented in the previous section.

References

1. Ammon, D., Generierung von diskreten zweidimensionalen stochastischen Prozessen mit definierten Leistungsdichtespektren! Z. Angew. Math. u. Mech., 68(1988), T133-135.
2. Ammon, D., Vollstndige Filtersysteme zur Approximation und Generierung von Vektorprozessen bei vorgegebener Spektraldichtematrix to appear in Z. Angew. Math. u. Mech., 69(1989), T307-.
3. Braun, H., Untersuchungen uon Fahrbahnunebenheiten und Anwendungen der Ergebnisse Diss., University of Hannover, 1969.
4. Bronstein, I. N. and Semendjajew, K. A., Taschenbuch der Mathematik, l9th edition, Verlag Harry Deutsch, Thun, 1980.
5. Föllinger, O., Regelungstechnik, 2. Aufl., Elitera, Berlin, 1978.
6. Rice, S. O., Mathemattcal Inalysys of Random Noise in: Wax, N. (ed.), Selected Papers on Noise and Stochastic Processes, Dover Publications, New York, 1954, pp. 180-181.
7. Shinozuka, M., Simulation of Multivibrate and Multidimensional Random Prozesses, J. Acoust. Soc. Am., 49(1971), 357-368.
8. Wedig, W., Linear and Nonlinear Identification — One dimensional Road Spectra and Nonlinear Oscillators, in: Schiehlen, W., and Wedig, W. (eds.), Analysis and Estimation of Mechanical Systems, CISM Vol. No. 303, Springer, Berlin, 1987.

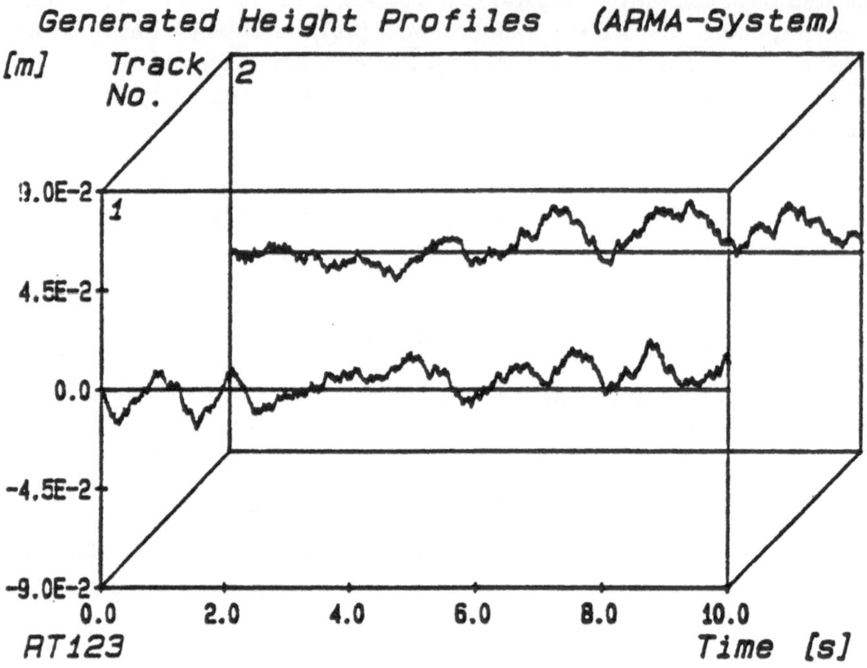

Fig. 5. Realization of a 2 component process generated by a 7th order ARMA system model

9. Wedig, W., Mean Square Stability and Spectrum Identification of Nonlinear Stochastic Systems, in: Ziegler, F. and Schuller, G. I. (eds.), Nonlinear Stochastic Dynamical Engineering Systems, IUTAM Symp., June 21-26, 1987, to be published by Springer, Berlin, in 1988, pp. 135-152.
10. Zurmühl, R. und Falk, S., Matrizen und ihre Anwendungen, Teil 1, 5. Aufl., Springer, Berlin, 1984.

Numerical Methods and Mathematical Aspects For Simulation of Homogeneous and Non Homogeneous Gaussian Vector Fields

F. Poirion and C. Soize

Abstract

In the first part, we present a method of numerical simulation of random gaussian fields, with values in a vector space, homogeneous and non homogeneous. We give a thorough description with some addenda for the general case of vector valued random fields, including some properties which seem to us essential in order to control the quality of the numerical simulation. The case of real gaussian processes, stationary and non stationary is of course a special case of the general results. The second part of the paper is dealing with the numerical simulation of the atmospheric turbulence.

1 Introduction

The numerical methods of simulation are approximate constructive methods which allow to study the solutions of certain random problems. Actually, they are in fact the only effective ones to construct a solution of various stochastic high dimensional problems for which no other methods are tractable [55, 56]. Such an approach is used for instance to study the numerical response of an aircraft flying in the continuous atmospheric turbulence when non linear effects are taken into account in the coupled system : aerodynamics - structure - airfoils (non linear effects brought by the limitation of the speed of the airfoils, by the structure itself, or by the aerodynamics in the transsonic range, ..) [45].

So we need to construct for such a type of problem numerical simulations of random fields. In the previous example, the field which has to be simulated is the atmospheric turbulence. Although the numerical methods of simulation are generally simple in their conception, they are rather tricky to use. The mathematical properties of the random field must indeed be checked, since they are the only ones to warrant the quality of the trajectories of the simulation and consequently the meaning of the simulation itself in relation to the probabilistic model. On the other hand, if the choice of the method has little effect on the cost of the simulation when one is dealing with stochastic processes (random variables indexed by a scalar parameter), it may lead to unpractical expensive schemes when dealing with random fields (variables indexed by a vectorial parameter, for instance one coordinate being the time, and three the space). There exists many numerical methods of simulation of random processes and random fields, stationary or non stationary, homogeneous or non homogeneous, gaussian or non gaussian. The main methods are described for instance in [55].

In the first part of this article, we shall present a method of numerical simulation of random gaussian fields, with values in a vector space, homogeneous and non homogeneous, this case arising very often in practice. Some basic ideas of the method are not new ones for the homogeneous case. One can find for instance a description of the method and some applications in [51, 52, 53]. However, for the reasons given previously, we shall give in this paper a thorough description for homogeneous, but also for non homogeneous cases, with some addenda for the general case of vector valued random fields, including some properties which seem to us essential in order to control the quality of the numerical simulation, and, at last, ready to use since the numerical scheme is written down.

The case of real gaussian processes, stationary and non stationary is of course a special case of the general results.

The second part of the paper is dealing with the numerical simulation of the atmospheric turbulence. First, we shall apply the previous method to the case of the isotropic atmospheric turbulence when it is modelled by an homogeneous random field[2, 9, 25, 58]. Then we shall deal with the atmospheric turbulence in the boundary layer (0 - 300 meters) : the loss of isotropy is due to the ground effect and then the vector random field is no longer homogeneous. We present a basis of construction of a model recently developped which is adapted to the numerical simulation and which might allow simulation of the response of aircrafts flying at a low altitude.

2 Numerical simulation of homogeneous and non homogeneous gaussian vector fields

2.1 Notations

Let m be a positive integer. The euclidian space $\mathbb{R}^m$ with its canonical basis is equipped with the usual scalar product $\langle x, y\rangle = \sum_{j=1}^m x_j y_j$ and with the associated norm $\|x\| = \langle x, \overline{x}\rangle^{1/2}$ where $x = (x_1, \ldots, x_m), y = (y_1, \ldots, y_m)$. For x and $y \in \mathbb{C}^m$, we denote $\langle x, \overline{y}\rangle = \sum_{j=1}^m x_j \overline{y}_j$ and $\|x\| = \langle x, \overline{x}\rangle^{1/2}$ where $\overline{y}$ means the complex conjugate of y. Let $\mathbb{K} = \mathbb{R}$ or $\mathbb{C}$, n and m two positive integers. We shall identify the vector space of the linear applications of $\mathbb{K}^m$ into $\mathbb{K}^n$ with the space $\mathrm{Mat}_{\mathbb{K}}(n, m)$ of n lines and m columns matrices which elements belong to $\mathbb{K}$. Therefore $\mathbb{K}^n \sim \mathrm{Mat}_{\mathbb{K}}(n, 1)$. Let $A \in \mathrm{Mat}_{\mathbb{C}}(n, m)$. We denote A^T the transpose of A, $\overline{A}$ the complex conjugate matrix and $A^* = \overline{A}^T$ the adjoint matrix. We also denote $\mathrm{tr}A = \sum_{j=1}^m A_{jj}$ the trace of A (when $n = m$), and we shall use the matrix norm $|A| = \sum_{j=1}^m \sum_{i=1}^m |A_{ji}|$.

Parameters and indexation sets for the fields In the following, d will be an integer ≥ 1. We shall denote $t = (t_1, \ldots, t_d)$ a point in $\mathbb{R}^d$, which will be the parameter of indexation of the fields we shall deal with. We shall also use the notations $t' = (t'_1, \ldots, t'_d)$ and $u = (u_1, \ldots, u_d)$, instead of t, when multiple points will be needed.

Let $T_1, \ldots, T_d$, d positive real numbers, $\overline{T}$ the following domain of $\mathbb{R}^d$:

$$\overline{T} = [0, T_1] \times [0, T_2] \times \ldots \times [0, T_d] \tag{1}$$

$$|T| = T_1 \times \ldots \times T_d = \int_{\overline{T}} dt \tag{2}$$

where dt is the Lebesgue measure on $\mathbb{R}^d$. We shall write:

$$T_{\inf} = \inf\{T_1, \ldots, T_d\} \tag{3}$$

The Greek letter α will be used for any d-dimensional multi-index, indexing t:

$$\alpha = (\alpha_1, \ldots, \alpha_d) \in \mathbb{N}^d \quad \alpha_j \in \mathbb{N} \tag{4}$$

For any fixed α, t_α will denote the point in $\mathbb{R}^d$:

$$t_\alpha = (t_{1,\alpha_1}, \ldots, t_{d,\alpha_d}) \tag{5}$$

where $t_{j,\alpha_j} \in \mathbb{R}$ is the value of the jth coordinate t_j of t.

Parameters and spectral domain of the fields Let $\Omega_1, \ldots, \Omega_d$, d positive fixed real numbers. We denote $\overline{\Omega}$ the domain of $\mathbb{R}^d$

$$\overline{\Omega} = [-\Omega_1, \Omega_1] \times \ldots \times [-\Omega_d, \Omega_d] \tag{6}$$

defined as the product of the intervals $[-\Omega_j, \Omega_j]$ of $\mathbb{R}$, called its edges. The compact $\overline{\Omega}$ of $\mathbb{R}^d$ will be the support of the matricial spectral measures that we shall consider and it will be called the "spectral domain". Let $\omega = (\omega_1, \ldots, \omega_d)$ a point in $\overline{\Omega} \in \mathbb{R}^d$. When another point will be needed, we shall write $\omega' = (\omega'_1, \ldots, \omega'_d)$. We shall write as we did before:

$$|\Omega| = \Omega_1 \times \ldots \times \Omega_d = \int_{\overline{\Omega}} d\omega \tag{7}$$

where $d\omega$ is the Lebesgue measure on $\mathbb{R}^d$.

The Greek letter β will be used for any d dimensional multi-index, indexing ω:

$$\beta = (\beta_1, \ldots, \beta_d) \in \mathbb{N}^d \quad \beta_j \in \mathbb{N} \tag{8}$$

For a fixed β, ω_β is the point of $\overline{\Omega}$ such that:

$$\omega_\beta = (\omega_{1,\beta_1}, \ldots, \omega_{d,\beta_d}) \tag{9}$$

where $\omega_{j,\beta_j} \in [-\Omega_j, \Omega_j]$ is the value of the jth coordinate ω_j of ω.

Subdivision of the spectral domain $\overline{\Omega}$ In order to write down the formulas for the numerical simulation of fields, we need to construct a subdivision of the spectral domain $\overline{\Omega}$. Any subdivision would work but would lead however to formulas which would not fit with the use of fast Fourier transform algorithm (FFT) and therefore the cost of the simulation would be expensive. In order to use FFT, the subdivision must be constructed with a constant step for each direction, the step beeing allowed to be different for each coordinate. Let $M_1, \ldots, M_d$, d fixed positive integers and $N_1 = 2M_1, \ldots, N_d = 2M_d$, the d even associated positive integers. Let:

$$N = N_1 \times \ldots \times N_d \tag{10}$$

$$N_{\text{inf}} = \inf\{N_1, \ldots, N_d\} \tag{11}$$

In the following we shall denote B_N the subset of $\mathbb{N}^d$ such that (cartesian product of N subsets):

$$B_N = \prod_{j=1}^{d} \{0, 1, 2, \ldots, N_j - 1\} \tag{12}$$

Therefore $\forall j \in \{1, \ldots, d\}, \beta \in B_N \Rightarrow \beta_j \in \{0, 1, 2, \ldots, N_j - 1\}$

For each $j \in \{1, \ldots, d\}$, we consider the subdivision of the edge $[-\Omega_j, \Omega_j]$, with constant step Δ_j such that:

$$\Delta_j = 2\Omega_j N_j^{-1} = \Omega_j M_j^{-1} \tag{13}$$

which meshes are the subintervals of $[-\Omega_j, \Omega_j]$, with center:

$$\omega_{j,\beta_j} = -\Omega_j + \left(\beta_j + \frac{1}{2}\right) \Delta_j; \quad \beta_j \in \{0, 1, \ldots, N_j - 1\} \tag{14}$$

The subdivision of $\overline{\Omega}$ is then defined as the finite union of subsets Q_β of $\overline{\Omega}$, with non intersecting interiors:

$$\overline{\Omega} = \bigcup_{\beta \in B_N} Q_\beta \tag{15}$$

such that

$$Q_\beta = \prod_{j=1}^{d} [\omega_{j,\beta_j} - \Delta_j/2, \omega_{j,\beta_j} + \Delta_j/2] \tag{16}$$

$$|Q_\beta| = \int_{Q_\beta} d\omega = |\Delta| \tag{17}$$

with $|\Delta|$ such that

$$|\Delta| = \Delta_1 \times \ldots \times \Delta_d \tag{18}$$

One may observe that the subset Q_β of $\overline{\Omega}$ is centered at the point ω_β and that no subset has the origin of $\overline{\Omega}$ for center, that last condition beeing a necessary condition for later on.

Let $\Delta = (\Delta_1, \dots, \Delta_d)$ the vector of $\mathbb{R}_d$ and $\|\Delta\| = \langle \Delta, \Delta \rangle^{1/2}$ its norm ($|\Delta|$ is different from $\|\Delta\|$). Then we have:

$$N_{\inf} \to +\infty \iff \|\Delta\| \to 0 \tag{19}$$

Finally, for any mapping f on $\overline{\Omega}$, we shall use the following condensed notation:

$$\sum_{\beta \in B_N} f(\omega_\beta) = \sum_{\beta_1=0}^{N_1-1} \cdots \sum_{\beta_d=0}^{N_d-1} f\left(\omega_{1,\beta_1}, \dots, \omega_{d,\beta_d}\right) \tag{20}$$

with ω_{j,β_j} defined by (14).

2.2 Hypothesis for the fields

In the following, $(\mathcal{A}, \mathcal{T}, P)$ denotes a probability space, E the mathematical expectation, d and n two integers ≥ 1. Let $X(t) = (X_1(t), \dots, X_n(t))$ be a random field defined on $(\mathcal{A}, \mathcal{T}, P)$, indexed on an open set $\mathcal{O}$ of $\mathbb{R}^d$, with values in $\mathbb{R}^n$, gaussian, of second order, centered, which auto-correlation function $t, t' \to R(t, t') = E(X(t)X(t')^T) : \mathcal{O} \times \mathcal{O} \to \mathrm{Mat}_{\mathbb{R}}(n, n)$ has the following representation:

$$R(t, t') = \int_{\mathbb{R}^d} e^{i\langle \omega, t - t' \rangle} Q(t, \omega) S(\omega) Q(t', \omega)^* d\omega \tag{21}$$

with the following assumptions:

H1 $\omega \to S(\omega)$ is a function from $\mathbb{R}^d$ into $\mathrm{Mat}_{\mathbb{C}}(n, n)$ with compact support $\overline{\Omega}$, continuous on $\overline{\Omega}$ and such that for all $\omega \in \mathbb{R}^d$, $S(\omega)$ is a positive hermitian matrix:

$$\begin{aligned} &S(\omega) = S(\omega)^* \\ &\langle S(\omega)x, \overline{x} \rangle \geq 0 \quad \forall x \in \mathbb{C}^n \end{aligned} \tag{22}$$

and such that:

$$S(-\omega) = \overline{S(\omega)} \tag{23}$$

H2 $t, \omega \to Q(t, \omega)$ is a continuous function from $\mathcal{O} \times \mathbb{R}^d$ into $\mathrm{Mat}_{\mathbb{C}}(n, n)$ such that for all $\omega \in \overline{\Omega}$:

$$Q(t, -\omega) = \overline{Q(t, \omega)} \quad \forall t \tag{24}$$

and such that for every compact set $\mathcal{K} \subset \mathcal{O}$, there exists a real, positive constant $C_{\mathcal{K}}$ independent of ω such that $\forall t, t' \in \mathcal{K}$, $\forall \omega \in \overline{\Omega}$, $\forall p, q \in \{1, 2, \dots, n\}$:

$$\begin{aligned} \big| |Q_{pq}(t, \omega)| - |Q_{pq}(t', \omega)| \big| &\leq C_{\mathcal{K}} \|t - t'\| \\ \big| |\Psi_{pq}(t, \omega)| - |\Psi_{pq}(t', \omega)| \big| &\leq C_{\mathcal{K}} \|t - t'\| \end{aligned} \tag{25}$$

where $Q_{pq}(t, \omega) = |Q_{pq}(t, \omega)| \exp(i\Psi_{pq}(t, \omega))$.

Remarks

1. Let $X^C(t) = X^R(t) + iX^I(t)$ a field with values in $\mathbb{C}^m$, $m \geq 1$, with $X^R(t)$ and $X^I(t)$ taking their values in $\mathbb{R}^m$. $\mathbb{C}$ can be identified with $\mathbb{R}^2$ and $X^C(t)$ with $X(t) = \{X^R(t), X^I(t)\}$ with values in $\mathbb{R}^n$, $n = 2m$. So the case of complex valued vector fields can be studied just as the case of real valued vector fields.
2. $X(t)$ takes its values in $\mathbb{R}^n$ so $R(t,t') \in \mathrm{Mat}_{\mathbb{R}}(n,n)$. Thus the imaginary part of the integral (21) is equal to zero and we can also write:

$$R(t,t') = \int_{\overline{\Omega}} e^{i\langle\omega,t-t'\rangle} Q(t,\omega)S(\omega)Q(t',\omega)^* d\omega \tag{26}$$

3. the hypothesis (22) implies that the function $\omega \to \mathrm{tr}S(\omega)$ on $\mathbb{R}^d$ has positive real values, is continuous on $\overline{\Omega}$ and then that the measure $\mathrm{tr}S(\omega)d\omega$ is positive, bounded on $\mathbb{R}^d$.
4. the hypothesis of continuity of Q on $\mathcal{O} \times \mathbb{R}^d$ is not limitative. Since S has a compact support $\overline{\Omega}$, it is sufficient that $t, \omega \to Q(t,\omega)$ be continuous on $\mathcal{O} \times \overline{\Omega}$. As a matter of fact, one does not need the values of Q on $\mathcal{O} \times (\mathbb{R}^d \backslash \overline{\Omega})$ and so, one can use any continuous extension of Q.

2.3 Mathematical properties of the fields

The case of an homogeneous field In all this section, we shall take $\mathcal{O} = \mathbb{R}^d$. As the field is gaussian and centered, a necessary and sufficient condition for the field to be homogeneous (it is called stationary if $d = 1$ and sometimes even if $d > 1$), is that $R(t,t') = R(t-t')$ therefore $Q(t,\omega) = I$, where I is the unity matrix of $\mathrm{Mat}_{\mathbb{R}}(n,n)$. The representation (21) can be also written:

$$R(t-t') = \mathrm{Re} \int_{\overline{\Omega}} e^{i\langle\omega,t-t'\rangle} S(\omega)d\omega \tag{27}$$

and $S(\omega)$ is the density of the matricial spectral measure of the homogeneous, second order field $X(t)$. It is well known that a necessary and sufficient condition in order to get (27) is that the field is mean square continuous and that the spectral matricial measure $M(d\omega)$ has a density $S(\omega)$ with respect to $d\omega$. The hypotheses introduced at section 2.2 imply the following properties:

P1 the function $u \to R(u)$ is continuous on $\mathbb{R}^d$ and

$$\lim_{\|u\| \to +\infty} |R(u)| = 0$$

It is infinitely differentiable on $\mathbb{R}^d$ and therefore the field $X(t)$ is mean square infinitely differentiable on $\mathbb{R}^d$. The functions S and S^2 are integrable on $\mathbb{R}^d$. Therefore the function R^2 is integrable on $\mathbb{R}^d$, but, in the general case, R itself, is not integrable on $\mathbb{R}^d$.

P2 The field $X(t)$ has almost surely continuous trajectories on $\mathbb{R}^d$. This last property will be useful to study statistics on the trajectories of the following kind: $\sup_{t\in\overline{T}} \|X(t)\|$, where $\overline{T}$ is defined by (1). The proof can be obtained by applying the proposition A1.1 of section A.1.iv of the Appendix, because $Q(t,\omega)$ being equal to I, (100) and (101) are verified.

P3 The hypotheses of Shannon's theorem for the fields are verified. For $j \in \{1,\dots,d\}$, let $\tau_j = \pi \Omega_j^{-1}$ the step of sampling for the coordinate t_j. Let $\alpha = (\alpha_1,\dots,\alpha_d) \in \mathbb{Z}^d$ the multi-index of relative integers of dimension d. Let $t_\alpha = (\alpha_1\tau_1,\dots,\alpha_d\tau_d) \in \mathbb{R}^d$ the sampling points in t, and $t = (t_1,\dots,t_d)$ any point of $\mathbb{R}^d$. Then we have:

$$X(t) = \sum_{\alpha \in \mathbb{Z}^d} X(t_\alpha) \prod_{j=1}^{d} \frac{\sin[\Omega_j(t_j - \alpha_j\tau_j)]}{\Omega_j(t_j - \alpha_j\tau_j)} \tag{28}$$

A remark on the compact support hypothesis for S

For some applications S does have not a compact support. It is the case for instance of the atmospheric isotropic turbulence. But for the physical fields, the power:

$$E\left(\|X(t)\|^2\right) = \mathrm{tr}R(0) = \int_{\mathbb{R}^d} \mathrm{tr}S(\omega)d\omega < +\infty \tag{29}$$

is finite, the field is of second order and S is integrable on $\mathbb{R}^d$. So, $\forall \varepsilon > 0$, $\exists \overline{\Omega}$ defined by (6) such that:

$$0 < \int_{\mathbb{R}^d \setminus \overline{\Omega}} \mathrm{tr}S(\omega)d\omega \leq \varepsilon E\left(\|X(t)\|^2\right) \tag{30}$$

We approximate then in a classical way S by $\mathbb{1}_{\overline{\Omega}} S$ where $\mathbb{1}_{\overline{\Omega}}$ is the characteristic function of the set $\overline{\Omega}$ on $\mathbb{R}^d$. The approximated field constructed in such a way is then mean square infinitely differentiable, when the initial field might not even be mean square differentiable. Therefore, the criterion (30) for the the determination of $\overline{\Omega}$ is adapted if no mean square differentiation on $\mathbb{R}^d$ of the field $X(t)$ is needed. However, if the initial field $X(t)$ has a spectral matricial density S which has not a compact support and which is such that:

$$\int_{\mathbb{R}^d} \|\omega\|^{2p} \mathrm{tr}S(\omega)d\omega < +\infty \tag{31}$$

for an integer $p \geq 1$, that last property going together with the existence of mean square derivatives on $\mathbb{R}^d$ of the field, and if we need the property:

$$\int_{\mathbb{R}^d} \|\omega\|^{2q} \mathrm{tr}S(\omega)d\omega < +\infty \quad \text{for } q \leq p \tag{32}$$

then the criterion (30) must necessarily be replaced by:

$$\int_{\mathbb{R}^d \setminus \overline{\Omega}} \|\omega\|^{2q} \mathrm{tr}S(\omega)d\omega \leq \varepsilon E\left(\|X(t)\|^2\right) \tag{33}$$

The non homogeneous case Here, $\mathcal{O}$ can be any open set of $\mathbb{R}^d$ and $Q(t,\omega)$ depends on t. The function S is called the density with respect to $d\omega$ of the matricial structural measure and the function $t,\omega \rightarrow S_X(t,\omega) : \mathcal{O} \times \mathbb{R}^d \rightarrow \mathrm{Mat}_{\mathbb{C}}(n,n)$ such that:

$$S_X(t,\omega) = Q(t,\omega)S(\omega)Q(t,\omega)^* \tag{34}$$

is the instantaneous spectral density. The support of $\omega \rightarrow S_X(t,\omega)$ is $\overline{\Omega}$, uniformly in t, and $t,\omega \rightarrow S_X(t,\omega)$ is continuous on $\mathcal{O} \times \overline{\Omega}$. The field is therefore of second order because

$$\forall t \in \mathcal{O} \quad E\left(\|X(t)\|^2\right) = \mathrm{tr}R(t,t) = \mathrm{Re}\int_{\overline{\Omega}} \mathrm{tr}S_X(t,\omega)d\omega < +\infty \tag{35}$$

The assumptions of section 2.2 imply the following properties:

P1 The field $X(t)$ has almost surely its trajectories continuous on $\mathcal{O}$. The proof can be obtained by applying the proposition A1.1 of section A.1.iv of the Appendix. Indeed, (100) is verified because of (25).

P2 Let $\tau = (\tau_1, \ldots, \tau_d) \in \mathbb{R}^d$ with $\tau_j = \pi\Omega_j^{-1}$. We shall say that the function $(t,\omega) \rightarrow Q(t,\omega)$ is slowly oscillating if, for every t and every $\omega \in \overline{\Omega}$, $Q(t+\tau,\omega) \sim Q(t,\omega)$. In this case the sampling in t is driven by the support $\overline{\Omega}$ of S, as in the case of homogeneous fields for which we have used the Shannon's theorem. In the following, we shall suppose this last assumption verified.

2.4 Construction of an approaching sequence of the field

Factorization of S The numerical simulation will be built for an approximation of the initial field, this approximation beeing an element of an approaching sequence which will converge towards the initial field. According to (22), for all $\omega \in \overline{\Omega}$, $S(\omega)$ is a positive hermitian matrix. Therefore there always exists a matrix $H(\omega) \in \mathrm{Mat}_{\mathbb{C}}(n,n)$ such that:

$$S(\omega) = H(\omega)H(\omega)^* \tag{36}$$

and the function $\omega \rightarrow H(\omega)$ is continuous on $\overline{\Omega}$.

Let $r = \mathrm{rank}S(\omega)$, the rank of $S(\omega)$. For $r \leq n$, we can construct $H(\omega)$ using general linear algebra methods based on the Gauss LU reduction. In the case where $r = n$, $S(\omega)$ is positive definite, and we can use the Cholesky method. For every $t \in \mathcal{O}$ and $\omega \in \overline{\Omega}$, we define the matrix $\mathbb{H}(t,\omega) \in \mathrm{Mat}_{\mathbb{C}}(n,n)$ such that:

$$\mathbb{H}(t,\omega) = Q(t,\omega)H(\omega) \tag{37}$$

The function $t,\omega \rightarrow \mathbb{H}(t,\omega)$ is continuous on $\mathrm{O} \times \overline{\Omega}$. Thus, (26) can be written:

$$R(t,t') = \mathrm{Re}\int_{\overline{\Omega}} e^{i\langle\omega,t-t'\rangle}\mathbb{H}(t,\omega)\mathbb{H}(t',\omega)^* d\omega \tag{38}$$

If the field is homogeneous, $\mathbb{H}(t,\omega) = \mathbb{H}(\omega)$ is independent of t.

Definition of the approaching sequence All the notations of section 2.1 will be used. For fixed N, we define the random field $X^N(t) = \left(X_1^N(t), \ldots, X_n^N(t)\right)$ on $(\mathcal{A}, \mathcal{T}, P)$, indexed on $\mathcal{O}$, with values in $\mathbb{R}^n$, such that for every $p \in \{1, \ldots, n\}$,

$$X_p^N(t) = \sqrt{2|\Delta|}\text{Re}\left\{\sum_{q=1}^{n} \sum_{\beta \in B_N} \mathbb{H}_{pq}(t, \omega_\beta) Z_{q,\beta} \exp\left(i\Phi_{q,\beta} + i\langle t, \omega_\beta\rangle\right)\right\} \tag{39}$$

with

a $\{\Phi_{q,\beta}\}_{q,\beta}$, $K = n \times N$ random variables defined on $(\mathcal{A}, \mathcal{T}, P)$, uniform on $[0, 2\pi]$, independent.

b $\{Z_{q,\beta}\}_{q,\beta}$ corresponds to the two following choices:

b.1 choice number 1 the $Z_{q,\beta}$ are constants equal to 1;

$$\forall p \in \{1, \ldots, n\},\ \forall \beta \in B_N,\ Z_{q,\beta} = 1 \tag{40}$$

b.2 choice number 2 the $Z_{q,\beta}$ are random variables defined by:

$$Z_{q,\beta} = \sqrt{-\log \Psi_{q,\beta}} \tag{41}$$

where $\{\Psi_{q,\beta}\}$ are K random variables defined on $(\mathcal{A}, \mathcal{T}, P)$, uniform on $[0, 1]$, independent, and independent of the random variables $\{\Phi_{q,\beta}\}_{q,\beta}$.

Remarks The two choices lead to the same second order characteristics for the field X^N. However, the first choice gives a field X^N which is not gaussian, but which will be asymptotically gaussian for $N_{\text{inf}} \to +\infty$, when the second choice gives a gaussian field X^N for every fixed N.

2.5 Properties of the fields in the approaching sequence

Second order properties For each fixed N, the random field $X^N(t)$ has, whatever the choice (40) or (41) for $Z_{q,\beta}$ is, the following properties:

1. it is a second order field and for each $t \in \mathcal{O}$:

$$E\left(\|X^N(t)\|^2\right) = |\Delta| \sum_{\beta \in B_N} \text{tr}S_x(t, \omega_\beta) < +\infty \tag{42}$$

2. it is centered and its autocorrelation function

$$t, t' \to R^N(t, t') = E(X^N(t) X^N(t')^T) : \mathcal{O} \times \mathcal{O} \to \text{Mat}_{\mathbb{R}}(n, n)$$

can be written:

$$R^N(t, t') = \text{Re} \int_{\mathbb{R}^d} e^{i\langle\omega, t-t'\rangle} Q(t, \omega) M^N(d\omega) Q(t', \omega)^* \tag{43}$$

with $M^N(d\omega)$ a measure on $\mathbb{R}^N$, with values $\text{Mat}_{\mathbb{C}}(n, n)$ such that:

$$M^N(d\omega) = |\Delta| \sum_{\beta \in B_N} s(\omega_\beta) \delta_{\omega_\beta}(\omega) \text{ and} \tag{44}$$

$$\delta_{\omega_\beta} = \otimes_{j=1}^{d} \delta_{\omega_{j,\beta_j}} \tag{45}$$

where $\delta_{\omega_{j,\beta_j}}$ is the Dirac measure on $\mathbb{R}$ at point ω_{j,β_j} , for the coordinate $\omega_j \in \mathbb{R}$ of ω.

3. the field $X^N(t)$ is mean square continuous.
4. if $\mathcal{O} = \mathbb{R}^d$ and $Q(t,\omega) = I$, the field $X^N(t)$ is weakly stationary.

Proof. For every y and y' in $\mathbb{R}$, we have

$$E(\cos(y + \Phi_{q,\beta})) = 0 \tag{46}$$

$$E(\cos(y + \Phi_{q,\beta})\cos(y' + \Phi_{q',\beta'})) = \delta_{qq'}\delta_{\beta\beta'}\frac{1}{2}\cos(y - y') \tag{47}$$

- For the choice number 1,

$$E\left(Z_{q,\beta}^2\right) = 1 \tag{48}$$

- For the choice number 2,

$$E\left(Z_{q,\beta}^2\right) = \int_0^1 \left(-\sqrt{\log \Psi}\right)^2 d\Psi = 1 \tag{49}$$

Writing $\mathbb{H}_{pq}(t,\omega_\beta) = |\mathbb{H}_{pq}(t,\omega_\beta)| \exp(i\theta_{pq}(t,\omega_\beta))$, relation (39) can be written:

$$X_p^N(t) = \sqrt{2|\Delta|}\sum_{q=1}^{n}\sum_{\beta\in B_N} |\mathbb{H}_{pq}(t,\omega_\beta)|\, Z_{q,\beta}\cos(\langle t,\omega_\beta\rangle + \theta_{pq}(t,\omega_\beta) + \Phi_{q,\beta}) \tag{50}$$

According to (46) up to (49) and since the $\Phi_{q,\beta}$ are independent of the $\Psi_{q,\beta}$, we see that $E\left(X_p^N(t)\right) = 0$, so the field is centered and we get:

$$E\left(X_p^N(t)X_{p'}^N(t')\right) =$$

$$|\Delta|\sum_{q=1}^{n}\sum_{\beta\in B_N} |\mathbb{H}_{pq}(t,\omega_\beta)\mathbb{H}_{p'q}(t',\omega_\beta)| \cos(\langle t-t',\omega_\beta\rangle + \theta_{pq}(t,\omega_\beta) - \theta_{p'q}(t,\omega_\beta))$$

$$= \mathrm{Re}\left\{|\Delta|\sum_{q=1}^{n}\sum_{\beta\in B_N} \mathbb{H}_{pq}(t,\omega_\beta)\overline{\mathbb{H}_{p'q}(t',\omega_\beta)}\exp\left(i\langle\omega_\beta, t-t'\rangle\right)\right\}$$

As $Q(t,\omega)S(\omega)Q(t',\omega)^* = \mathbb{H}(t,\omega)\mathbb{H}(t',\omega)^*$, we obtain (43), (44). We deduce (42) from (43)-(44). As B_N has a finite cardinal, R^N is continuous on $\mathcal{O}\times\mathcal{O}$, whence the third property. Finally, since $X^N(t)$ is a centered field and that, in the case 4, $R^N(t,t') = R^N(t-t')$, the field is homogeneous in the weak sense, which proves the fourth property.

Properties in law

1. **choice number 1** If the $Z_{q,\beta}$ are defined by (40), for every fixed N, the random field $X^N(t)$ is not gaussian.
2. **choice number 2** If the $Z_{q,\beta}$ are defined by (41), then, for every fixed N ,
   - the random field $X^N(t)$ is a gaussian field
   - moreover, if $\mathcal{O} = \mathbb{R}^d$ and $Q(t,\omega) = I$, the field $X^N(t)$ is homogeneous.

Proof. Let us begin by proving the point 2. For every $y_{q,\beta} \in \mathbb{R}$ we set

$$U_{q,\beta} = \sqrt{-\log \Psi_{q,\beta}} \cos(\Phi_{q,\beta} + y_{q,\beta}) \tag{51}$$

Then the random variables $\{U_{q,\beta}\}_{q,\beta}$ are independent and for every fixed q and β, the random variable $U_{q,\beta}$, which is centered and of covariance 1/2, following (46), (47) and (49), is a gaussian random variable. According to (50), it follows that X^N is a gaussian field. (We can show that $U_{q,\beta}$ is gaussian by calculating for instance its characteristic function). In the case $\mathcal{O} = \mathbb{R}^d$ and $Q(t,\omega) = I$, following 2.5.4, X^N is weakly homogeneous. But, beeing gaussian, it is homogeneous in the strict sense. The proof of point 1 can be deduced from point 2.

Properties of the trajectories

i choice number 1 For every $m \geq 1$ and every compact set $\mathcal{K}$ of $\mathcal{O}$, there exists a positive constant $C_{\mathcal{K}}$ independent of N such that $\forall t, t' \in \mathcal{K}$ and $\forall N$:

$$\sum_{p=1}^{n} E\left(\left|X_p^N(t) - X_p^N(t')\right|^{2m}\right) \leq C_{\mathcal{K}} \|t - t'\|^{2m} \tag{52}$$

and the field $X^N(t)$ which is not gaussian, has almost surely its trajectories continuous on $\mathcal{O}$.

ii choice number 2 For every compact $\mathcal{K} \subset \mathcal{O}$, there exists a positive constant $C_{\mathcal{K}}$ independent of N such that $\forall t, t' \in \mathcal{K}$ and $\forall N$:

$$E(\|X(t) - X(t')\|^2) \leq C_{\mathcal{K}} \|t - t'\|^2 \tag{53}$$

and the gaussian field $X^N(t)$ has almost surely continuous trajectories on $\mathcal{O}$.

iii homogeneous case If $\mathcal{O} = \mathbb{R}^d$ and $Q(t,\omega) = I$, and whatever choice is considered, the trajectories $t_j \to X^N(t,a)$, $a \in \mathcal{A}$, are periodic functions with period:

$$T_j^0 = 2\pi \left(\frac{1}{2}\Delta_j\right)^{-1} \quad j \in \{1, \ldots, d\} \tag{54}$$

Writing $T^0 = (T_1^0, \ldots, T_d^0)$, $t \to X^N(t,a)$ is a periodic function on $\mathbb{R}^d$, with period T^0.

Proof. **i** The result proceeds directly from proposition A1.2 of section A.1.vi for the non gaussian case. The hypothesis (105) is verified since $t, \omega \to \mathbb{H}(t,\omega)$ is continuous on $\mathcal{O} \times \overline{\Omega}$, and the assumption (106) results from (25)-(37).

ii The result proceeds from proposition A1.1 of section A.1.iv for the gaussian case, knowing that (99) is given by (43). Relation (53) is nothing else that (102).

iii The proof is trivial.

2.6 Study of the convergence of the approaching sequence

We have constructed an approaching sequence X^N of the field X. The numerical simulation will be made for an element X^N, N fixed, of this sequence. So it is necessary to study the convergence of X^N towards X. We shall establish three types of results.

- The first one is concerning second order quantities.
- The second one is related to the convergence in law (weak convergence) of the sequence X^N toward the gaussian field X.
- The last one, more precise than the previous, shows the weak convergence on the continuous functions with compact support, result which will allow to get the convergence of statistics on the trajectories.

Convergence of second order quantities For each N, the field $X^N(t)$ is centered as the field $X(t)$. We have the following results for second order quantities (valid for the two choices (40) and (41)). For every t and t' in $\mathcal{O}$:

i the sequence $\{Q(t,\omega)M^N(d\omega)Q(t',\omega)^*\}_N$ of measures converges narrowly for $N_{\text{inf}} \to +\infty$ towards the measure $Q(t,\omega)S(\omega)Q(t',\omega)^*d\omega$
ii the sequence $R^N(t,t')$ converges towards $R(t,t')$
iii if for any p and $q \in \{1,\ldots,n\}$ the functions

$$\omega \to s(\omega) = [Q(t,\omega)S(\omega)Q(t',\omega)^*]_{pq}$$

belong to $C^p(\overline{\Omega},\mathbb{C})$ with $p = 1$ or 2, the speeds of convergence are of order N_{inf}^{-p}.

Proof. For every t and t' fixed in O and p, q fixed in $1,\ldots,n$, the function $s \in C^0(\overline{\Omega},\mathbb{C})$. We use lemma A2.2 of section A.2.(ii). Point **i** proceeds directly from it ; and also do the points **ii** and **iii** by taking $s'(\omega) = s(\omega)\exp(i\langle\omega, t-t'\rangle)$.

Convergence in law of the approaching sequence

i choice number 1

a the sequence of non gaussian random fields X^N converges in law (weakly) towards the gaussian field X when $N_{\text{inf}} \to +\infty$.
b if $\mathcal{O} = \mathbb{R}^d$ and $Q(t,\omega) = I$, the approaching sequence X^N, weakly homogeneous, converges in law towards the gaussian homogeneous field X.

ii choice number 2

a the sequence of gaussian field X^N converges in law towards the gaussian field X when $N_{\text{inf}} \to +\infty$.
b if $\mathcal{O} = \mathbb{R}^d$ and $Q(t,\omega) = I$, the sequence X^N of homogeneous gaussian fields converges towards the homogeneous gaussian field X.

Proof. **i** the proof of (a) is given by section A.3. It is the central limit theorem which gives the result. The proof of point (b) results of (a) and of the fact that a weakly homogeneous gaussian field is strictly homogeneous.
ii the proof is trivial since the sequence X^N is gaussian, centered and that $R^N(t,t') \to R(t,t')$ by 2.6:*Convergence of second order quantities (ii).*

Weak convergence on continuous functions Let $\mathcal{K}$ be any compact set of $\mathcal{O}$, for instance the set $\overline{T}$ defined by (1). Let $C^0(\mathcal{K})$ the space of continuous functions on $\mathcal{K}$ with values in $\mathbb{R}$, equipped with the uniform convergence norm:

$$\forall f \in C^0(\mathcal{K}) \quad |||f||| = \sup_{t\in\mathcal{K}} |f(t)| \tag{55}$$

Let $p \in \{1,\ldots,n\}$ fixed. We would like to know for instance if the random variable $\sup_{t\in\mathcal{K}} |X_p^N(t)|$ converges in law towards $\sup_{t\in\mathcal{K}} |X_p(t)|$, or if the second order characterisitcs converge. We cannot get this type of result from just the weak convergence result of the sequence X^N towards X that we have established in section 2.6:*Convergence in law of the approaching sequence.* We need the following result: let g a continuous function of $C^0(\mathcal{K})$ into $\mathbb{R}$. Then:

$$\mathcal{L}(g(X^N)) \underset{N_{\inf}\to+\infty}{\overset{\text{narrowly}}{\longrightarrow}} \mathcal{L}(g(X)) \tag{56}$$

Proof. We have shown in section 2.3 that the field $X(t)$ has almost surely its trajectories continuous on $\mathcal{O}$. We have also shown in section 2.5:*Properties of the trajectories* that for the two choices and for each fixed N, the field $X^N(t)$ has almost surely its trajectories continuous on $\mathcal{O}$. Moreover, from results of 2.5:*Properties in law* we know that X^N converges weakly towards X. But since the inequalities (52) (for choice 1) and (53) (for choice 2), one uniform in N, we get the weak convergence on $C^0(\mathcal{K})$.

Application
The functional $g(f) = \sup_{t\in\mathcal{K}} |f(t)|$ is the norm $|||f|||$ of f on $C^0(\mathcal{K})$ and thus it is a continuous functional. So we have:

$$\mathcal{L}\left(\sup_{t\in\mathcal{K}} |X_p^N(t)|\right) \underset{N_{\inf}\to+\infty}{\overset{\text{narrowly}}{\longrightarrow}} \mathcal{L}\left(\sup_{t\in\mathcal{K}} |X_p(t)|\right) \tag{57}$$

2.7 Numerical simulation formulas

Homogeneous case. Fast formula using FFT (i) General case
We consider the case $\mathcal{O} = \mathbb{R}^d$, $Q(t,\omega) = I$ and thus (39) can be written for $p \in \{1,\ldots,n\}$ and fixed N:

$$X_p^N(t) = \sqrt{2\Delta}\,\mathrm{Re}\left\{\sum_{q=1}^{n} \sum_{\beta\in B_N} H_{pq}(\omega) Z_{q,\beta} \exp(i\Phi_{q,\beta} + i\langle t, \omega_\beta\rangle)\right\} \tag{58}$$

where $H(\omega)$ is given in section 2.4:*factorization of S.*

Using Shannon's theorem, we sample the coordinate t_j of t at the frequency $f_{s,j} = 2f_{\max,j}$ with $f_{\max,j} = \Omega_j/2\pi$. So $f_{s,j} = \tau_j^{-1}$.

More generally, we introduce the parameters $\nu_j = 2m_j$, m_j integer ≥ 1 and the coordinate t_j is sampled at $f_{s,j} = \nu_j f_{\max,j} = \nu_j \Omega_j/2\pi$. The sampling step δ_j can be written $\delta_j = f_{s,j}^{-1}$, or using (13):

$$\delta_j = 2\pi(\hat{N}_j \Delta_j)^{-1} \quad j \in \{1, \ldots, d\} \tag{59}$$

$$\hat{N}_j = \frac{1}{2}\nu_j N_j \quad \nu_j = 2m_j \quad m_j \geq 1 \quad j \in \{1, \ldots, d\} \tag{60}$$

Let $A_{\hat{N}}$ the subset of $\mathbb{N}^d$ such that

$$A_{\hat{N}} = \prod_{j=1}^{d} \{0, 1, \ldots, \hat{N}_j - 1\} \tag{61}$$

If the multi-index $\alpha \in A_{\hat{N}}$, $\alpha_j \in 0, 1, \ldots, \hat{N}_j - 1$ for every $j \in 1, \ldots, d$. The sampling points for the coordinate t_j are:

$$t_{j,\alpha_j} = \alpha_j \delta_j \quad \alpha_j \in \{0, 1, \ldots, \hat{N}_j - 1\} \tag{62}$$

and, according to (5), the set of sampling points for t is:

$$t_\alpha = (t_{1,\alpha_1}, \ldots, t_{d,\alpha_d}) \quad \alpha \in A_{\hat{N}} \tag{63}$$

The set of sampling points for ω is:

$$\omega_\beta = (\omega_{1,\beta_1}, \ldots, \omega_{d,\beta_d}) \quad \beta \in B_N \tag{64}$$

with (12) and (14). Note that according to (54), the period T_j^0 for the coordinate t_j can be written $T_j^0 = 4\pi/\Delta_j$ and since $(\hat{N}_j - 1)\delta_j \sim 2\pi/\Delta_j$, we see that the choice (62) is correct since only half a period is used. A simple calculation gives, for $\alpha \in A_{\hat{N}}$ and $b \in B_N$:

$$\langle \omega_\beta, t_\alpha \rangle = -2\pi \sum_{j=1}^{d} \alpha_j (\nu_j^{-1} - (2\hat{N}_j)^{-1}) + 2\pi \sum_{j=1}^{d} \alpha_j \beta_j \hat{N}_j^{-1} \tag{65}$$

Since $\hat{N}_j \geq N_j$, $A_{\hat{N}} \supseteq B_N$, and for $\beta \in A_{\hat{N}}$, we put:

$$\begin{cases} x_\beta^{(p)} = 0 & \text{if } \beta \notin B_N \\ x_\beta^{(p)} = \sum_{q=1}^{n} H_{pq}(\omega_\beta) Z_{q,\beta} \exp(i\Phi_{q,\beta}) & \text{if } \beta \in B_N \end{cases} \tag{66}$$

Then, the relation (58) can be written for every $a \in A_{\hat{N}}$:

$$X_p^N(t_\alpha) = \sqrt{2\Delta}\,\mathrm{Re}\left\{ \tilde{X}_p^N(\alpha) \exp\left[-2i\pi \sum_{j=1}^{d} \alpha_j (\nu_j^{-1} - (2\hat{N}_j^{-1})) \right] \right\} \tag{67}$$

with:

$$\tilde{X}_p^N(\alpha) = \sum_{\beta \in A_{\hat{N}}} x_\beta^{(p)} \exp\left[\sum_{j=1}^{d} 2i\pi\alpha_j\beta_j\hat{N}_j^{-1}\right] \tag{68}$$

where $\tilde{X}_p^N$, given by (68) can be calculated by FFT.

(ii) **Particular case: folded formula**

We introduce the following assumption concerning the field symmetry: $\forall j \in \{1,\ldots,d\}$, $\forall p$, $\forall q \in \{1,\ldots,n\}$, we have:

$$H_{pq}(\omega_1,\ldots,-\omega_j,\ldots,\omega_d) = H_{pq}(\omega_1,\ldots,\omega_j,\ldots,\omega_d) \tag{69}$$

Then, if M_j is such that $N_j = 2M_j$, we put:

$$\hat{M}_j = \nu_j M_j \quad \nu_j = 2m_j \quad m_j \geq 1$$

$$A_{\hat{M}} = \prod_{j=1}^{d}\{0,1,\ldots,\hat{M}_J - 1\}$$

$$B_M = \prod_{j=1}^{d}\{1,2,\ldots,M_j\}$$

$$\delta_j = 2\pi(\hat{M}_j\Delta_j)^{-1};\ t_{j,\alpha_j} = \alpha_j\delta_j;\ \Delta_j = \Omega_j M_j^{-1};\ \omega_{j,\beta_j} = \beta_j\Delta_j$$

and, as in (66), we put for $\beta \in A_{\hat{M}}$:

$$\begin{cases} x_\beta^{(p)} = 0 & \text{if } \beta \notin B_N \\ x_\beta^{(p)} = \sum_{q=1}^{n} H_{pq}(\omega_\beta)Z_{q,\beta}\exp(i\Phi_{q,\beta}) & \text{if } \beta \in B_N \end{cases}$$

Then, for every $\alpha \in A_{\hat{M}}$, we have:

$$X_p^N(t_\alpha) = \sqrt{2\Delta}\,\mathrm{Re}\left\{\tilde{X}_p^N(\alpha)\right\} \tag{70}$$

$$\tilde{X}_p^N(\alpha) = \sum_{\beta \in A_{\hat{N}}} x_\beta^{(p)} \exp\left[\sum_{j=1}^{d} 2i\pi\alpha_j\beta_j\hat{M}_j^{-1}\right] \tag{71}$$

where $\tilde{X}_p^N$, given by (71), can be calculated by FFT.

(iii) **Remarks**

The introduction of the parameters $\nu_1,\ldots,\nu_d$ allows to use FFT algorithmes, even if we need a higher resolution for the coordinate t_j than for the coordinate ω_j. For $m_j = 1$, we have $\nu_j = 2$ and $\hat{N}_j = N_j$, so we find the classical case.

Non homogeneous case We have to use (39) with $H_{pq}(t,\omega)$ depending of t. We can develop the same formulas than in the homogeneous case. But $x_\alpha^{(p)}$ defined by (66) depends of α: $x_\beta^{(p)}(\alpha)$ and (71) cannot be any longer calculated by FFT. However if the field stays homogeneous in some coordinates, non homogeneous for the others, the approach (70)-(71) is useful since it permits to do a part of the calculation using partial FFT for the homogeneous coordinates.

3 Model and simulation of the atmospheric turbulence

3.1 Introduction

A simple model of turbulence, called cylindrical (the turbulence is supposed constant spanwise), is often used [9,58] to calculate the responses of an aircraft. Recently, after that experimental on board measures were made on an aircraft, it was shown that this simple model gave results which were further of the experimental ones than those given by the isotropic turbulence model. However, the isotropic model which gives good results when the plane is flying at a constant altitude is not any more fitted for the lowest altitude [0 - 300] meters since we have to take into account the loss of isotropy and of homogeneity due to the closeness of the ground. In this section, we shall illustrate the previous methods by constructing simulations of the atmospheric turbulence when it is first modelled by an homogeneous isotropic field then when it is modelled by a non homogeneous field. In the following, the three coordinates of the space (x,y,z) will be denoted (t_1, t_2, t_3), the dual vector, (wave vector) will be denoted $(\omega_1, \omega_2, \omega_3)$.

3.2 Model for the isotropic turbulence and simulation

The atmospheric turbulence is modelled by a random field

$$X(t) = (X_1(t_1, t_2, t_3), X_2(t_1, t_2, t_3), X_3(t_1, t_2, t_3))$$

indexed on $\mathbb{R}^3$ with values in $\mathbb{R}^3$, gaussian, homogeneous, independent of time (frozen turbulence), of second order, centered and whose matricial spectral measure has a density given by [2,25]

$$S(\omega) = -\frac{f(\|\omega\|)}{4\pi\|\omega\|^2}\left(\frac{\omega\omega^T}{\|\omega\|^2} - I\right) \tag{72}$$

where I is the unit matrix of $\mathrm{Mat}_{\mathbb{R}}(3,3)$ and,

$$f(\|\omega\|) = \frac{55}{9}\frac{\Gamma(5/6)}{\sqrt{\pi}\Gamma(1/3)}\frac{\sigma^2}{\omega_e}\frac{(\|\omega\|/\omega_e)^4}{(1+(\|\omega\|/\omega_e)^2)^{17/6}}$$

$\sigma^2 = E(\|X(t)\|^2)$ and ω_e a reference wave number.

The density $S(\omega)$ does not have a compact support so we are going to simulate in fact the field whose spectral density is $\mathbb{1}_{\overline{\Omega}} S(\omega)$.

We can use directly the formula (58). In order to control the results, we construct an estimate of the spectral density the mean of this estimate beeing :

$$\hat{S}_{pq}(\omega) = (2\pi)^{-3}\mathcal{N}^{-1}\sum_{k=1}^{\mathcal{N}} \hat{X}_p^{(k)}(\omega)\overline{\hat{X}_q^{(k)}(\omega)} \quad 1 \leq p, q \leq 3$$

with

$$\hat{X}_p^{(k)}(\omega) = (T_1T_2T_3)^{-1}\int_0^{T_1}\int_0^{T_2}\int_0^{T_3} X_p^{(k)}(t_1, t_2, t_3)e^{-i(\omega_1t_1+\omega_2t_2+\omega_3t_3)}dt_1dt_2dt_3$$

where $X_p^{(k)}(t)$ is the result of the kth simulation of the field X, $\mathcal{N}$ beeing the total number of simulation made.

Figure 1 represents the comparison of the power spectral measure given by (72) or by the estimate, the power spectral measure beeing equal to the trace of $S(\omega)$ and which depends, since it is isotropic, only of the wave number $\|w\|$.

We have used 16 points per coordinate in order to construct the simulation, the mean of the estimate beeing calculate over $\mathcal{N} = 200$ simulations, the spectral domain $\overline{\Omega}$ beeing troncated in order to obtain a good spectral resolution for the estimate.

3.3 Construction of an anisotropic model for the atmospheric turbulence in the [0 - 300m] boundary layer

The model that we propose builds, from experimental datas, an instantaneous spectral density

$$S_X(t,\omega) = Q(t,\omega)S(\omega)Q(t,\omega)^*$$

for the turbulent field. Such an approach has been presented in a much more simple context, using a different simulation technic [20,21]. Once we have the instantaneous matricial spectral density, we can write the relation (21):

$$R(t,t') = \iiint_{\mathbb{R}^3} e^{i\langle\omega,t-t'\rangle} Q(t,\omega)S(\omega)Q(t',\omega)^* d\omega$$

and use the methods described in the first part in order to simulate the field. First we are going to set up the mathematical frame for the construction of the model.

Assumptions for the field The turbulent atmospheric field is modelled by a gaussian field, indexed on $\mathbb{R}^3$, with values in $\mathbb{R}^3$, centered, mean square continuous, horizontally homogeneous, that means for the first two coordinates of the point $t = (t_1, t_2, t_3)$ of the space. The autocorrelation function verifies the property :

$$R(t,t') = R(t_1 - t_1', t_2 - t_2', t_3, t_3') \tag{73}$$

We denote

$$\Phi(\omega_1, t_2 - t_2', t_3, t_3') \in \mathrm{Mat}_{\mathbb{C}}(3,3)$$

the one dimensional spectral density defined by:

$$R(t_1 - t_1', t_2 - t_2', t_3, t_3') = \int_{\mathbb{R}} e^{i\omega_1(t_1-t_1')} \Phi(\omega_1, t_2 - t_2', t_3, t_3') d\omega_1 \tag{74}$$

In the same way, we can define the matricial transversal spectral densities

$$\Psi(\omega_1, \omega_2, t_3, t_3') \in \mathrm{Mat}_{\mathbb{C}}(3,3)$$

by:

$$R(t_1 - t_1', t_2 - t_2', t_3, t_3') = \iint_{\mathbb{R}^2} e^{i\omega_1(t_1-t_1') + i\omega_2(t_2-t_2')} \Psi(\omega_1, \omega_2, t_3, t_3') d\omega_1 d\omega_2 \tag{75}$$

The functions Ψ and Φ are then related to each other by the relation :

$$\Phi(\omega_1, t_2 - t_2', t_3, t_3') = \int_{\mathbb{R}^3} e^{i\omega_2(t_2 - t_2')}\Psi(\omega_1, \omega_2, t_3, t_3')d\omega_2 \tag{76}$$

We associate to these spectral densities the transversal correlation matrices ρ and γ with values in $\mathrm{Mat}_{\mathbb{R}}(3,3)$ and the phase angles α and θ with values in $\mathrm{Mat}_{\mathbb{R}}(3,3)$ such that for j and $k \in \{1,2,3\}$:

$$\begin{cases} \rho_{kj}(\omega_1, t_2 - t_2', t_3)\Phi_{kk}(\omega_1, 0, t_3, t_3)\Phi_{jj}(\omega_1, 0, t_3, t_3) = |\Phi_{kj}(\omega_1, t_2 - t_2', t_3, t_3)|^2 \\ \Phi_{kj}(\omega_1, t_2 - t_2', t_3, t_3) = \exp(i\alpha_{kj}(\omega_1, t_2 - t_2', t_3))|\Phi_{kj}(\omega_1, t_2 - t_2', t_3, t_3)| \end{cases} \tag{77}$$

$$\begin{cases} \gamma_{kj}(\omega_1, \omega_2, t_3, t_3')\Psi_{kk}(\omega_1, \omega_2, t_3, t_3)\Psi_{jj}(\omega_1, \omega_2, t_3', t_3') = |\Psi_{kj}(\omega_1, \omega_2, t_3, t_3')|^2 \\ \Psi_{kj}(\omega_1, \omega_2, t_3, t_3') = \exp(i\theta_{kj}(\omega_1, \omega_2, t_3, t_3'))|\Psi_{kj}(\omega_1, \omega_2, t_3, t_3')| \end{cases} \tag{78}$$

Existence of a spectral measure We introduce the next assumption that we shall suppose verified later on :

Hypothesis H: We assume that the functions $t_3, t_3' \rightarrow \gamma(\omega_1, \omega_2, t_3, t_3')$ and $t_3, t_3' \rightarrow \theta(\omega_1, \omega_2, t_3, t_3')$ depend only of $\tau_3 = t_3 - t_3'$. We denote them:

$$\tau_3 \rightarrow \gamma(\omega_1, \omega_2, \tau_3) \quad \tau_3 \rightarrow \theta(\omega_1, \omega_2, \tau_3) \tag{79}$$

Let Γ be the function with values in $\mathrm{Mat}_{\mathbb{C}}(3,3)$ such that:

$$\Gamma_{kj}(\omega_1, \omega_2, \tau_3) = (\gamma_{kj}(\omega_1, \omega_2, \tau_3))^{1/2} \exp(i\theta_{kj}(\omega_1, \omega_2, \tau_3)) \tag{80}$$

Then, according to the assumption H of (79), Γ appears to be the complex coherence function related to Ψ and we have for k and $j \in \{1,2,3\}$

$$\Psi_{kj}(\omega_1, \omega_2, t_3, t_3') = \Gamma_{kj}(\omega_1, \omega_2, t_3, t_3')[\Psi_{kk}(\omega_1, \omega_2, t_3, t_3)\Psi_{jj}(\omega_1, \omega_2, t_3', t_3')]^{1/2} \tag{81}$$

If, for every ω_1 and ω_2 in $\mathbb{R}$ the functions $\tau_3 \rightarrow \Gamma(\omega_1, \omega_2, t_3)$ and $\tau_3 \rightarrow \hat{\Gamma}(\omega_1, \omega_2, \omega_3)$ are integrable on $\mathbb{R}$, with

$$\hat{\Gamma}(\omega_1, \omega_2, \omega_3) = \frac{1}{2}\pi \int_{\mathbb{R}} e^{-i\omega_3\tau_3}\Gamma(\omega_1, \omega_2, \tau_3)d\tau_3 \tag{82}$$

there exists a function

$$(t_3, \omega) \rightarrow Q(t_3, \omega) : \mathbb{R} \times \mathbb{R}^3 \rightarrow \mathrm{Mat}_{\mathbb{R}}(3,3)$$

and a function

$$\omega \rightarrow S(\omega) : \mathbb{R}^3 \rightarrow \mathrm{Mat}_{\mathbb{C}}(3,3)$$

with the properties (22) and (23) for all $\omega \in \mathbb{R}^3$ and such that:

$$R(t,t') = \iiint_{\mathbb{R}^3} e^{i\langle\omega, t-t'\rangle}Q(t_3, \omega)S(\omega)Q(t_3', \omega)d\omega \tag{83}$$

with $Q(t_3,\omega)$ the diagonal matrix such that:

$$[Q(t_3,\omega)]_{kj} = \delta_{kj}[\Psi_{kk}(\omega_1,\omega_2,t_3,t_3)]^{1/2} \tag{84}$$

where δ_{kj} is the Kroenecker function and

$$S(\omega) = \hat{\Gamma}(\omega_1,\omega_2,\omega_3) \tag{85}$$

As a matter of fact:

$$\Psi_{kj}(\omega_1,\omega_2,t_3,t_3') =$$

$$\Psi_{kk}(\omega_1,\omega_2,t_3,t_3)^{1/2}\left[\int_{\mathbb{R}} e^{i\omega_3(t_3,t_3')}\hat{\Gamma}_{kj}(\omega_1,\omega_2,\omega_3)d\omega_3\right]\Phi_{jj}(\omega_1,\omega_2,t_3',t_3')^{1/2}$$

Using this last relation in (75), we obtain (83) to (85). The properties (22) is general and (23) results from the fact that the field has real values.

Remark: In our problem, the datas are the functions γ and θ. The matrix $S(\omega)$ is then constructed from (85), (82) and (80), that is, for j and $k \in \{1,2,3\}$:

$$[S(\omega)]_{kj} = \frac{1}{2\pi}\int_{\mathbb{R}} e^{i\omega_3\tau_3}\gamma_{kj}(\omega_1,\omega_2,\tau_3)^{1/2}\exp(i\theta_{kj}(\omega_1,\omega_2,\tau_3))d\tau_3 \tag{86}$$

Application to the low altitude [0 - 300 m] turbulence The experimental datas that we can get on the low altitude atmospheric turbulence are essentially given by the one dimensional spectral densities $\Phi_{kk}(\omega_1,0,t_3)$, $k \in \{1,2,3\}$ (refer, for instance, to [15,16,33,41,42,50,57,62]). We can find, besides, [10,14,20,21], empirical models which describe the transversal correlations $\rho_{kk}(\omega_1,t_2-t_2',t_3)$ and phase angles $\alpha_{kk}(\omega_1,t_2-t_2',t_3)$, $k \in \{1,2,3\}$:

$$\rho_{kk}(\omega_1,t_2-t_2',t_3) = \exp\left(-\frac{1}{2\pi}A_k|\omega_1||t_2-t_2'|\right) \tag{87}$$

$$\alpha_{kk}(\omega_1,t_2-t_2',t_3) = -B_k\omega_1(t_2-t_2') \tag{88}$$

where the quantities A_k and B_k are functions of the altitude t_3, that can be assumed constant at low altitude, *and what we shall assume from now on.*

From relations (76), (77), (87) and (88) we get:

$$\Psi_{kk}(\omega_1,\omega_2,t_3,t_3) = A_k|\omega_1|[\frac{1}{4}A_k^2\omega_1^2 + 4\pi^2(\omega_1 B_k+\omega_2)^2]^{-1}\Phi_{kk}(\omega_1,0,t_3,t_3) \tag{89}$$

Actually, we don't have any data concerning the transversal correlations γ_{kj} and phase angle θ_{kj}. Therefore, by analogy with the previous models, we have introduced the hypothesis H and we have put:

$$\gamma_{kj}(\omega_1,\omega_2,t_3-t_3') = \exp[-\frac{1}{2\pi}C_{kj}|\omega_1+\omega_2||t_3-t_3'|] \tag{90}$$

$$\theta_{kj}(\omega_1,\omega_2,t_3-t_3') = -D_{kj}(\omega_1+\omega_2)(t_3-t_3') \tag{91}$$

where C_{kj} and D_{kj} are real constants independent of the altitude t_3.

The results of the previous section can be applied since the functions $\tau_3 \to \Gamma_{kj}(\omega_1, \omega_2, t_3)$ and $\omega_3 \to \hat{\Gamma}_{kj}(\omega_1, \omega_2, \omega_3)$ are integrable functions. In this case, the matrix $S(\omega)$ is obtained by computing (86) and we get, for $k, j \in \{1, 2, 3\}$:

$$[S(\omega)]_{kj} = C_{kj}|\omega_1 + \omega_2|[\frac{1}{4}C_{kj}^2(\omega_1 + \omega_2)^2 + 4\pi^2(D_{kj}(\omega_1 + \omega_2) + \omega_3)^2]^{-1} \quad (92)$$

The matrix $Q(t_3, \omega)$ is given by (84) and (89). The diagonal terms, which are the only ones not null, are, for $k \in \{1, 2, 3\}$:

$$Q_{kk}(t_3, \omega) = A_k|\omega_1|[\frac{1}{4}A_k^2\omega_1^2 + 4\pi^2(\omega_1 B_k + \omega_2)^2]^{-1}\Phi_{kk}(\omega_1, 0, t_3, t_3) \quad (93)$$

Practical considerations (i) From relations (93) and (92) we see that the functions $Q(t_3, \omega)$ and $S(\omega)$ are not defined for the null wave vector $\omega = (0, 0, 0)$. It comes from the fact that the experimental spectra $\Phi_{kk}(\omega_1, 0, t_3)$ are known only for $|\omega_1| > \omega_{1,0} (\approx 10^{-2}\text{rad/m})$. So we extend $S(\omega)$ and $Q(t_3, \omega)$ by continuity on the set $[-\omega_{1,0}, \omega_{1,0}]^3$. It is clear then that the function Q checks the conditions (24) and (25). We can anyway check that these functions are integrable in the neighborhood of the origin.

(ii) The construction of this model does not depend on the aspect of the experimental spectra and so it can be used for different meteorological situations. It depends however on the form (87), (88), (90) and (91) of the correlations and phase angles. Moreover, one must check that with the choice of A_k, B_k, C_{kj} and D_{kj} the matrix $S(\omega)$ is positive and hermitian.

(iii) Example of experimental datas

We find in [16,10,35,62,41] empirical models of one dimensional spectra describing the turbulence in a neutral meteorological situation. We have, for instance :

$$\Phi_{11}(\omega_1, 0, t_3) = \frac{105}{4\pi}u_*^2 t_3 \left(.44 + \frac{33}{2\pi}t_3|\omega_1|\right)^{-5/3}$$

$$\Phi_{22}(\omega_1, 0, t_3) = \frac{17}{4\pi}u_*^2 t_3 \left(.38 + \frac{9.5}{2\pi}t_3|\omega_1|\right)^{-5/3}$$

$$\Phi_{33}(\omega_1, 0, t_3) = \frac{2}{4\pi}u_*^2 t_3 \left(.44 + 5.3\left(\frac{t_3}{2\pi}|\omega_1|\right)^{5/3}\right)^{-1} \quad (94)$$

where u_* is the surface velocity.

In [20,21,50] we find some values for the constants:

$$C_{12} = C_{21} = C_{23} = C_{32} = 0$$

$$C_{11} = 19, C_{22} = 13, C_{33} = 13, C_{13} = C_{31} = 16$$

$$D_{11} = 1, D_{22} = 2, D_{33} = 2, D_{13} = D_{31} = 1; A_1 = 19$$

For the other constants A_k and B_k for which we have no datas, we put:

$$A_k = C_{kk}; k = 2, 3$$

$$B_k = D_{kk}; k = 1,2,3.$$

Figures 2-5 represent the mean of an estimate of the second coordinate, one dimensional spectrum of the turbulence, which was simulated numerically for a fixed altitude t_3, compared with the empirical spectrum $\Phi_{22}(\omega_1, 0, t_3)$ given by (94), and this, for different number of simulations.

Conclusion

We have presented the general results on simulation of gaussian fields. The applications of the numerical simulation algorithms to the case of homogeneous and non homogeneous turbulence show how effective are the developped methods. We have proposed, besides, a consistant model of spectrum for the atmospheric turbulence in the boundary layer [0 - 300m]. What is left to do is to find datas for every constant related to the transversal correlation and phase angle.

A ADDENDUM

A.1 Stochastic field with a.s. continuous trajectories

Let d and n be two positive integers ≥ 1. The vector space $\mathbb{R}^d$ and $\mathbb{R}^n$ are equipped with the euclidian scalar product denoted $\langle .,. \rangle$ and with the associated norm $\|.\|$. Let $(\mathcal{A}, \mathcal{T}, P)$ a probability space and E the mathematical expectation. Let $X(t) = (X_1(t), \ldots, X_n(t))$ a random field defined on $(\mathcal{A}, \mathcal{T}, P)$, indexed on some open $\mathcal{O}$ of $\mathbb{R}^d$ with values in $\mathbb{R}^n$. The field X(t) has a.s. continuous trajectories if for P almost all $a \in \mathcal{A}$, the trajectories $t = (t_1, \ldots, t_d) \to X(t, a)$ are continuous functions from $\mathcal{O}$ to $\mathbb{R}^n$. The next result, with is a generalization of Kolmogorov's lemma for the fields case $(d > 1)$, gives a sufficient condition which is very useful.

(i) Sufficient condition in the general case [63]

If there exists two positive constants η and λ, and if for every compact set $\mathcal{K}$ of $\mathcal{O}$ there exists a third constant $C_{\mathcal{K}}$ such that:

$$E(\|X(t) - X(t')\|^{\lambda}) \leq Ck\|t - t'\|^{d+\eta}, \forall t, t' \in \mathcal{K} \tag{95}$$

the random process X(t) has a.s. continuous trajectories on $\mathcal{O}$.

(ii) Remarks

1. On $\mathbb{R}^n$ all the norms are equivalent. We can take the norm:

$$\|x\| = \left(\sum_{j=1}^{n} |x_j|^p \right)^{1/p} \qquad 1 \leq p \leq +\infty \tag{96}$$

which gives the euclidian norm for p = 2.

The condition (95) can thus be replaced by the following condition with $\lambda \geq 1$:

$$\sum_{q=1}^{n} E\left(|X_q(t) - X_q(t')|^{\lambda}\right) \leq C_{\mathcal{K}}\|t - t'\|^{d+\eta} \quad t, t' \in \mathcal{K} \tag{97}$$

(A1.3)

2. The condition (95) or (97) is sufficient. It is not necessary, but no assumptions had been made on the field law. The random field can be gaussian or not, homogeneous or not. If X(t) is gaussian, we can use (95) or (97), but it may be possible in this case to weaken the condition if we take into account the gaussian character. We obtain then the classical following result:

(iii) Sufficient condition in the gaussian case [35]

If X(t) is a gaussian field, of second order, whose mean-function $t \to m_X(t) = E(X(t))$ is continuous from $\mathcal{O}$ into $\mathbb{R}^n$, if there exists a positive constant η, and if for every compact $\mathcal{K}$ of $\mathcal{O}$ there exists another positive constant $C_{\mathcal{K}}$, such that:

$$E(\|X(t) - X(t')\|^2) \leq C_{\mathcal{K}} |\log(\|t - t'\|)|^{-(1+\eta)} \tag{98}$$

for all t and t' in $\mathcal{K}$, with $\|t - t'\| \to 0$, then X(t) has a.s. continuous trajectories on $\mathcal{O}$.

(iv) Proposition A1.1

Let $X(t) = (X_1(t), \ldots, X_n(t))$ a stochastic field defined on $(\mathcal{A}, \mathcal{T}, P)$, indexed on an open set $\mathcal{O}$ of $\mathbb{R}^d$, with values in $\mathbb{R}^n$, gaussian, of second order, whose function of autocorrelation $t, t' \to R(t,t') = E(X(t)X(t')^T) : \mathcal{O} \times \mathcal{O} \to \mathrm{Mat}_{\mathbb{R}}(n,n)$ has the following representation:

$$R(t,t') = \mathrm{Re} \int_{\mathbb{R}^d} e^{i\langle \omega, t-t' \rangle} Q(t,\omega) M(d\omega) Q(t',\omega)^* \tag{99}$$

with the following assumptions:

H1 $M(d\omega)$ is a measure on $\mathbb{R}^d$ with values in $\mathrm{Mat}_{\mathbb{C}}(n,n)$ such that for every borel set of $\mathbb{R}^d$, $M(B) = \int_B M(d\omega)$ is an hermitian positive matrix and $\|M(\mathbb{R}^d)\| < +\infty$.

H2 $t, \omega \to Q(t,\omega)$ is a continuous application of $\mathcal{O} \times \mathbb{R}^d$ into $\mathrm{Mat}_{\mathbb{C}}(n,n)$ such that for every compact $\mathcal{K}$ of $\mathcal{O}$:

$$\sup_{t \in \mathcal{K}} \int_{\mathbb{R}^d} \|\omega\|^2 \|Q(t,\omega)\|^2 \|M(d\omega)\| = C_{\mathcal{K}} < +\infty \tag{100}$$

and for all t and t' in $\mathcal{K}$:

$$\int_{\mathbb{R}^d} \|Q(t,\omega) - Q(t',\omega)\|^2 \|M(d\omega)\| \leq C''_{\mathcal{K}} \|t - t'\|^2 \tag{101}$$

where $C'_{\mathcal{K}}$ and $C''_{\mathcal{K}}$ are two real positive finite constants depending of $\mathcal{K}$. Under these hypothesises the field X(t) has a.s. continuous trajectories on $\mathcal{O}$.

Proof.

$$\begin{aligned}
E(\|X(t) - X(t')\|^2) &= \mathrm{tr}\{R(t,t) + R(t',t') - R(t,t') - R(t',t)\} \\
&= \mathrm{Re} \int_{\mathbb{R}} \mathrm{tr}\{A(t,t',\omega) M(d\omega) A(t,t',\omega)^*\} \\
&\leq \|A(t,t',\omega)\|^2 \|M(d\omega)\|
\end{aligned}$$

with

$$A(t,t',\omega) = Q(t,\omega)e^{i\langle\omega,t-t'\rangle} - Q(t',\omega)$$
$$= [Q(t,\omega - Q(t',\omega)]e^{i\langle\omega,t-t'\rangle} + (e^{i\langle\omega,t-t'\rangle} - 1)Q(t',\omega)$$

but since

$$|e^{i\langle\omega,t-t'\rangle} - 1|^2 = 2(1-\cos\langle\omega,u\rangle) \leq 2\langle\omega,u\rangle^2 \leq 2\|\omega\|^2\|u\|^2$$

we have

$$\|A(t,t',\omega)\| \leq \|Q(t,\omega) - Q(t',\omega)\| + \sqrt{2}\|t-t'\|\,\|\omega\|\,\|Q(t',\omega)\|$$

for a and b positive real numbers, $(a+b)^2 \leq 2a^2 + 2b^2$; thus, following (100) and (101), for all t and t' in $\mathcal{K}$:

$$E(\|X(t) - X(t')\|^2) \leq C_{\mathcal{K}}\|t-t'\|^2 \tag{102}$$

with $C_{\mathcal{K}} = 4C'_{\mathcal{K}} + 2C''_{\mathcal{K}}$; for $\|u\| \to 0$, $\|u\|^2|\log(\|u\|)|^2 \to 0 < 1$.

The use of (98) gives the results.

(v) Remark

If, in proposition A1.1, we take $\mathcal{O} = \mathbb{R}^d$ and if we assume the field to be homogeneous, then we have (99) with $Q(t,\omega) = I$ the unit matrix, and M the matricial spectral measure.

In this case (101) is automatically satisfied and a sufficient condition for X(t) to have its trajectories a.s. continuous on $\mathbb{R}^d$ is that we have (100), which is equivalent to:

$$\int_{\mathbb{R}^d} \|\omega\|^2 \mathrm{tr} M d\omega = C < +\infty \tag{103}$$

according to the properties of the matricial spectral measure.

(vi) Proposition A1.2

Let $\overline{\Omega}$ be the compact of $\mathbb{R}^d$ defined by (6), N defined by (10), B_N by (12), ω_β by (9) - (14), $|\Delta|$ by (18). We use notation (20).

Let $X^N(t) = (X_1^N(t),\ldots,X_n^N(t))$ the stochastic field defined on $(\mathcal{A},\mathcal{T},P)$, indexed on an open set $\mathcal{O}$ of $\mathbb{R}^d$ with values in $\mathbb{R}^n$, such that for any $p \in \{1,\ldots,n\}$

$$X_p^N(t) = \sqrt{2|\Delta|}\mathrm{Re}\left\{\sum_{q=1}^{n}\sum_{\beta\in B_N} \mathbb{H}_{pq}(t,\omega_b)\exp(i\Phi_{q,\beta} + i\langle t,\omega_\beta\rangle)\right\} \tag{104}$$

with the following hypothesis:

H1 the $\Phi_{q,\beta}$ are $K = n \times N$ random variables, uniform on $[0,2\pi]$, independent.

H2 for any p and q in $\{1,\dots,n\}$, the functions $(t,\omega)\to \mathbb{H}_{pq}(t,\omega)$ are continuous from $\mathcal{O}\times\overline{\Omega}$ into $\mathbb{C}$ and for any compact $\mathcal{K}$ of $\mathcal{O}$:

$$\sup_{t\in\mathcal{K},\omega\in\overline{\Omega}} |\mathbb{H}_{pq}(t,\omega)| \le C'_{\mathcal{K}} \tag{105}$$

which means that the $\mathbb{H}_{pq}$ are bounded on $\mathcal{K}\times\overline{\Omega}$. For every t and t' in $\mathcal{K}$, there exists $C''_{\mathcal{K}}>0$, independent of ω such that:

$$\begin{aligned} \big||\mathbb{H}_{pq}(t,\omega)| - |\mathbb{H}_{pq}(t',\omega)|\big| &\le C''_{\mathcal{K}}\|t-t'\| \\ \big|\theta_{pq}(t,\omega)-\theta_{pq}(t',\omega)\big| &\le C''_{\mathcal{K}}\|t-t'\| \end{aligned} \tag{106}$$

where

$$|\mathbb{H}_{pq}(t,\omega)| = |\mathbb{H}_{pq}(t,\omega)|\exp(i\theta_{pq}(t,\omega)) \tag{107}$$

with these assumptions:

1. For every integer $m\ge 1, \exists C_{\mathcal{K}}>0$, such that $\forall t,t'\in\mathcal{K},\forall N$:

$$\sum_{p=1}^{n} E\left(\left|X_p^N(t)-X_p^N(t')\right|^{2m}\right) \le C_{\mathcal{K}}\|t-t'\|^{2m} \tag{108}$$

which means that the inequality (108) is uniform in N;

2. The field $X^N(t)$ has a.s. continuous trajectories on $\mathcal{O}$.

Proof. The criterion (97) allows to consider, in a first step, only one coordinate $X_p(t)$ of $X(t)$. According to (107), for every $t,t'\in\mathcal{O}$, we get:

$$Z_p^N = X_p^N(t)-X_p^N(t') = \sum_{k=1}^{K}\{A_k\cos(a_k+\Phi_k)+B_k\sin(b_k+\Phi_k)\}$$

where we have introduced the condensed index $k=(q,\beta), k\in\{1,\dots,K\}, K = nN$, and where

$$\Phi_k = \Phi_{q,\beta} \quad A_k = \sqrt{2|\Delta|}(|\mathbb{H}_{pq}(t,\omega_\beta)| - |\mathbb{H}_{pq}(t',\omega_\beta)|)$$

$$B_k = -2\sqrt{2|\Delta|}|\mathbb{H}_{pq}(t',\omega_\beta)|\sin\left(\frac{1}{2}\left[\langle t-t',\omega_\beta\rangle+\theta_{pq}(t,\omega_\beta)-\theta_{pq}(t',\omega_\beta)\right]\right)$$

$$a_k = \langle t,\omega_\beta\rangle + \theta_{pq}(t,\omega_\beta)$$

$$b_k = \frac{1}{2}\left[\langle t+t',\omega_\beta\rangle + \theta_{pq}(t,\omega_\beta)+\theta_{pq}(t',\omega_\beta)\right]$$

For $m\ge 1$, we have:

$$(Z_p^N)^{2m} = \sum_{i_1+\dots+i_k=2m} \frac{(2m)!}{i_1!\dots i_K!}\prod_{K}^{k=1} J_k^{i_k} \tag{109}$$

with $J_k = A_k\cos(a_k+\Phi_k)+B_k sin(b_k+\Phi_k)$.

As the random variables $\Phi_1,\dots,\Phi_K$ are independent, so are $J_1,\dots,J_K$ and we can write $E\prod_k = \prod_k E$. On the other hand $E(J_k^{i_k})$ if $i_k = 2r+1$. So, in the sum (109), only the terms with even indexes i_k will bring a non null contribution. Hence:

$$E((Z_p^N)^{2m}) = \sum_{i_1+\ldots+i_K=2m} \frac{(2m)!}{(2i_1)!\ldots(2i_K)!} \prod_{k=1}^{K} E(J_k^{2i_k})$$

But $E(J_k^{2i_k}) \leq (|A_k| + |B_k|)^{2i_k}$, $\forall k$. So, according to (105), (106)

$$E(J_k^{2i_k}) \leq (\sqrt{2|\Delta|}\tilde{C}_{\mathcal{K}}\|t - t'\|)^{2(i_1+\ldots+i_K)}, \ \forall k$$

with

$$\tilde{C}_{\mathcal{K}} = C''_{\mathcal{K}} + C'_{\mathcal{K}}(C''_{\mathcal{K}} + \sup_{\omega \in \overline{\Omega}} \|\omega\|)$$

In another hand, since $(2i_k)! \geq i_k!$, we have

$$E((Z_p^N)^{2m}) \leq \frac{(2m)!}{m!} \sum_{i_1+\ldots+i_K=m} \frac{m!}{i_1!\ldots i_K!}(\sqrt{2|\Delta|})\tilde{C}_{\mathcal{K}}\|t - t'\|^{2(i_1+\ldots+i_K)}$$

and

$$C_N = \sum_{i_1+\ldots+i_K=m} \frac{m!}{i_1!\ldots i_K!} = \left(\sum_{k=1}^{K} 1\right)^m = K^m$$

we obtain

$$\sum_{p=1}^{n} E((Z_{2p}^N)^{2m}) \leq C_{\mathcal{K}}\|t - t'\|^{2m} \quad \forall t, t' \in \mathcal{K}$$

with $C_{\mathcal{K}}$ equal to:

$$C_{\mathcal{K}} = (m!)^{-1}(2m)!2^m m^{m+1}|\Omega|^m \tilde{C}_{\mathcal{K}}^{2m}$$

since $|\Delta|K = |\Delta|Nn = |\Omega|n$ which proves the point (1) of the proposition. The point 2 then results from (108) and (97).

(vii) Remark

Since the functions $\mathbb{H}_{pq}$ are continuous on $\mathcal{O} \times \overline{\Omega}$, if they are independent of t the conditions (105) and (106) are automatically verified.

A.2 Convergence lemmas

We establish two convergence lemmas which allow to prove second order quantities convergence for the approaching sequence. We use all the notations of section 2.1.

(i) Lemma A2.1 for the rate of convergence

Let $\omega \rightarrow f(\omega)$ a function defined on $\overline{\Omega}$ with values in $\mathbb{C}$, p times continuously differentiable, with p = 1 or 2: $f \in C^p(\overline{\Omega}, \mathbb{C})$. There exists then a finite positive real constant $C_0 > 0$ such that, $\forall N$ we have:

$$\left| \int_{\overline{\Omega}} f(\omega)d\omega - |\Delta| \sum_{\beta \in B_N} f(\omega_\beta) \right| \leq C_0\|\Delta\|^p \tag{110}$$

That means that $|\Delta|\sum_{\beta\in B_N} f(\omega_\beta)$ converges toward $\int_{\overline{\Omega}} f(\omega)d\omega$ when $N_{\text{inf}} \to +\infty$, (i.e $\|\Delta\| \to 0$ by (19)) with the convergence rate N_{inf}^{-p} since:

$$\|\Delta\|^p = \left[\sum_{j=1}^{d} \frac{(2\Omega_j)^2}{N_j^2}\right]^{p/2} \tag{111}$$

Proof. We show the result first when p = 2. Let

$$(\nabla f)(\omega) = (\partial_1 f(\omega), \ldots, \partial_d f(\omega)) \in \mathbb{C}^d$$

the gradient of f in ω and $h(\omega) \in \text{Mat}_{\mathbb{C}}(d,d)$ the hessian matrix of f in ω:

$$[h(\omega)]_{ij} = \partial_i \partial_j f(\omega) \tag{112}$$

where ∂_j stands for the partial derivative relative to the jth coordinate ω_j of ω: let $R(\omega, \omega_\beta) \in \mathbb{C}$ such that

$$R(\omega,\omega_\beta) = f(\omega) - f(\omega_\beta) - \langle(\nabla f)(\omega_\beta), \omega - \omega_\beta\rangle - \frac{1}{2}\langle h(\omega_\beta)(\omega,\omega_\beta), \omega - \omega_\beta\rangle \tag{113}$$

Using Taylor's formula, $\forall\varepsilon > 0, \exists r > 0$ such that for $\|\omega - \omega_\beta\| \leq r$,

$$|R(\omega, \omega_\beta)| \leq \varepsilon\|\omega - \omega_\beta\|^2 \tag{114}$$

In another hand, we can write (using (15) - (18)):

$$\int_{\overline{\Omega}} f(\omega)d\omega = |\Delta| \sum_{\beta\in B_N} f(\omega_\beta) + \sum_{\beta\in B_N} \int_{Q_\beta} (f(\omega) - f(\omega_\beta))d\omega \tag{115}$$

Using (113), we deduce from (115):

$$Int = \int_{\overline{\Omega}} f(\omega)d\omega - |\Delta| \sum_{\beta\in B_N} f(\omega_\beta) \tag{116}$$

$Int = Int_1 + Int_2 + Int_3$ with

$$\begin{aligned} Int_1 &= \sum_{\beta\in B_N} \int_{Q_\beta} R(\omega,\omega_\beta)d\omega \\ Int_2 &= \sum_{\beta\in B_N} \int_{Q_\beta} \langle(\nabla f)(\omega_\beta), \omega - \omega_\beta\rangle d\omega \\ Int_3 &= \frac{1}{2} \sum_{\beta\in B_N} \int_{Q_\beta} \langle h(\omega_\beta)(\omega - \omega_\beta), (\omega - \omega_\beta)\rangle d\omega \end{aligned} \tag{117}$$

We introduce the following change of variable:

$$\omega = \omega_\beta + \omega' \tag{118}$$

and we put

$$Q_0 = \prod_{j=1}^{d} \left[-\frac{\Delta_j}{2}; \frac{\Delta_j}{2} \right]$$

$$R_0(\omega_\beta, \omega') = R(\omega_\beta + \omega', \omega_\beta) \tag{119}$$

According to (114), $\exists C_1 > 0$ such that:

$$|R_0(\omega_\beta, \omega')| \leq C_1 \|\omega'\|^2, \ \forall \beta \in B_N, \ \forall \omega' \in Q_0 \tag{120}$$

Thus

$$|Int_1| \leq C_1 \sum_{\beta \in B_N} \int_{Q_0} \|\omega'\|^2 d\omega' = N C_1 \int_{Q_0} \|\omega'\|^2 d\omega'$$

and whence:

$$|Int_1| \leq \frac{C_1}{12} 2^d |\Omega| \|\Delta\|^2 \tag{121}$$

In another hand, for each fixed N,

$$Int_2 = \sum_{\beta \in B_N} \int_{Q_0} \langle (\nabla f)(\omega_\beta), \omega' \rangle d\omega' = 0 \tag{122}$$

since $\int_{Q_0} \omega'_j d\omega' = 0$, $\forall j \in \{1, \ldots, d\}$. In the same way $\int_{Q_0} \omega'_i \omega'_j d\omega' = 0$ for $i \neq j$, thus:

$$\begin{aligned} Int_3 &= \frac{1}{2} \sum_{\beta \in B_N} \sum_{j=1}^{d} [h(\omega_\beta)]_{ij} \int_{Q_0} (\omega'_j)^2 d\omega' \\ &= \frac{1}{24} |\Delta| \sum_{j=1}^{d} \Delta_j^2 \sum_{\beta \in B_N} [h(\omega_\beta)]_{ij} \end{aligned}$$

Since $\omega \rightarrow h(\omega)$ is continuous on the compact $\overline{\Omega}$, $\exists C_2 > 0$ such that, $\forall j \in \{1, \ldots, d\}$ and $\forall \beta \in B_N$, we have:

$$|[h(\omega_\beta)]_{ij}| \leq C_2 \tag{123}$$

we deduce from this:

$$|Int_3| \leq \frac{C_2}{24} 2^d |\Omega| \|\Delta\|^2 \tag{124}$$

Putting $C_0 = 1/12\, 2^d |\Omega| \sup(C_1, C_2/2)$, and according to (117), (121), (122) and (124) we get (110) for p = 2. For p = 1 the proof is the same, stopping the Taylor's expansion at the first order.

(ii) Lemma A2.2 of convergence of a measures sequence

Let $\omega \rightarrow s(\omega)$ a continuous function of $\overline{\Omega}$ into $\mathbb{C}$ and $\mu(d\omega) = s(\omega)d\omega$ the measure on $\overline{\Omega}$ with values in $\mathbb{C}$ defined by the density s. We have:

$$\mu(\overline{\Omega}) = \int_{\overline{\Omega}} \mu(d\omega) = r \quad |r| < +\infty \tag{125}$$

Let $\{\mu^N(d\omega)\}_N$ be the sequence of measures on $\overline{\Omega}$ with complex values, such that:

$$\mu^N(d\omega) = |\Delta| \sum_{\beta \in B_N} s(\omega_\beta)\delta_{\omega_\beta} \tag{126}$$

where δ_{ω_β} is the measure defined in (43), and such that:

$$\mu^N(\overline{\Omega}) = \int_{\overline{\Omega}} \mu^N(d\omega) = |\Delta| \sum_{\beta \in B_N} s(\omega_\beta) = r^N \tag{127}$$

with $|r^N| < +\infty$. We have:

1. the measures sequences $\{\mu^N\}_N$ converges narrowly towards the measure μ on $\overline{\Omega}$ when $N_{\text{inf}} \rightarrow +\infty$ (narrowly convergence implies weak convergence), and thus the sequence $\{r^N\}_N$ converges towards r,
2. if $s \in C^p(\overline{\Omega}, \mathbb{C})$, for p = 1 or 2, the rate of convergence is N_{inf}^{-p}.

Proof. 1. The sequence $\{\mu^N\}_N$ converges narrowly towards μ if $\forall j \in C^0(\overline{\Omega}, \mathbb{C})$, the complex sequence:

$$\mu^N(\varphi) = \int_{\overline{\Omega}} \varphi(\omega)\mu^N(d\omega) \tag{128}$$

converges towards:

$$\mu(\varphi) = \int_{\overline{\Omega}} \varphi(\omega)\mu(d\omega) \tag{129}$$

But

$$\mu(\varphi) = \int_{\overline{\Omega}} \varphi(\omega)s(\omega)d\omega \quad \mu^N(\varphi) = |\Delta| \sum_{\beta \in B_N} \varphi(\omega_\beta)s(\omega_\beta)$$

Since the function $\omega \rightarrow f(\omega) = \varphi(\omega)s(\omega)$ is continuous on $\overline{\Omega}$, we know that for $N_{\text{inf}} \rightarrow +\infty$, we have $|\Delta| \sum_{\beta \in B_N} f(\omega_\beta) \rightarrow \int_{\overline{\Omega}} f(\omega)d\omega$. Taking $\varphi(\omega) = 1, \forall \omega \in \overline{\Omega}$, we obtain $r^N \rightarrow r$.

2. The proof comes directly from lemma A2.1.

A.3 Convergence in law of the non gaussian approaching sequence

We use all the notations of section 2. In a first time, we recall some known facts useful to prove the results.

(i) Definition of the convergence in law of a sequence

Let $\alpha(1),\dots,\alpha(L)$, L multi-indexes of $\mathbb{N}^d$, L any positive integer, with

$$\alpha(\ell) = (\alpha_1(\ell),\dots,\alpha_d(\ell)) \in \mathbb{N}^d, \alpha_j(\ell) \in \mathbb{N}$$

We say that the sequence of random fields $\{X^N(t)\}_N$ indexed on $\mathcal{O}$ with values in $\mathbb{R}^N$, converges in law, (or weakly) towards the random field X(t), indexed on $\mathcal{O}$ with values in $\mathbb{R}^N$ for $N_{\text{inf}} \longrightarrow +\infty$ if the system of finite-dimensional distributions of the field $X^N(t)$ converges narrowly towards the system of finite-dimensional distributions of $X(t)$, i.e., if for every $L \in \mathbb{N}$ and every finite family $(t_{\alpha(1)},\dots,t_{\alpha(L)})$ of points in $\mathbb{R}^d$,

$$\mathcal{L}(X^N(t_{\alpha(1)}),\dots,X^N(t_{\alpha(L)})) \underset{N_{\text{inf}}\to+\infty}{\overset{\text{narrowly}}{\longrightarrow}} \mathcal{L}(X(t_{\alpha(1)}),\dots,X(t_{\alpha(L)})) \tag{130}$$

where $\mathcal{L}(\,)$ stands for the law of the random variable.

(ii) Utilisation of Paul Levy's theorem

Let $v_\ell = (v_{\ell,1},\dots,v_{\ell,n}) \in \mathbb{R}^n$ and $v = (v_1,\dots,v_L) \in \mathbb{R}^{nL}$. We write $V \to \hat{F}_N(v)$ the characteristic function on $\mathbb{R}^{nL}$ with values in $\mathbb{C}$ of the random variable $\mathcal{U}^N = (X^N(t_{\alpha(1)}),\dots,X^N(t_{\alpha(L)}))$ with values in $\mathbb{R}^{nL}$:

$$\hat{F}_N(v) = E\{\exp(i\langle\mathcal{U}^N, v\rangle\} \tag{131}$$

Then, from Paul Levy's theorem, if the sequence of characteristic functions $\{\hat{F}_N\}_N$ converges simply towards a complex function $\hat{F}$ defined on $\mathbb{R}^{nL}$, whose real part Re$\hat{F}$ is continuous at the origin, then $\hat{F}$ is the characteristic function of a random variable with values in $\mathbb{R}^{nL}$:

$$\hat{F}(v) = E\{\exp(i\langle\mathcal{U}, v\rangle\} \tag{132}$$

and we have:

$$\mathcal{L}(\mathcal{U}^N) \underset{N_{\text{inf}}\to+\infty}{\overset{\text{narrowly}}{\longrightarrow}} \mathcal{L}(\mathcal{U}) \tag{133}$$

(iii) the central limit theorem [49]

Let $Y_1,\dots,Y_K$, K random variables on $(\mathcal{A},\mathcal{T},P)$, with values in $\mathbb{R}$ of second order, independent. For $k \in \{1,\dots,K\}$, let $m_k = E(Y_k)$ the mean of Y_k, $\sigma_k^2 = E[(Y_k - m_k)^2]$ its variance, $P_k(dy_k)$ the probability law of the centered random variable $Y_k - m_k$. The laws $P_1,\dots,P_K$ may be different. Let S such that

$$S_K^2 = \sum_{k=1}^{K} \sigma_K^2 \tag{134}$$

Then if $\forall\varepsilon > 0$ we have:

$$\lim_{K\to+\infty} \frac{1}{S_K^2} \sum_{k=1}^{K} \int_{|y_k|>\varepsilon S_K} y_k^2 P_K(dy_k) = 0 \tag{135}$$

the law of the random variable

$$\frac{1}{S_K}\sum_{k=1}^{K}(Y_k - m_k) \tag{136}$$

converges narrowly for $K \to +\infty$ towards the canonical gaussian law LG(0,1) on $\mathbb{R}$.

(iv) Convergence in law of the sequence for the choice 1

The approaching sequence X^N is defined by (39) with the choice 1, which means that, according to (40), $Z_{q,\beta} = 1, \forall q$ and $\forall \beta$.

Let $K = n \times N$ and $k = (q, \beta)$ the condensed index. We put:

$$A_{k,\ell_p} = |\mathbb{H}_{pq}(t_{\alpha(\ell)}, \omega_\beta)|$$
$$b_{k,\ell_p} = \langle t_{\alpha(\ell)}, \omega_\beta\rangle + \theta_{pq}(t_{\alpha(\ell)}, \omega_\beta)$$
$$\Phi_{q,\beta} = \Phi_k$$
$$Y_k = \sqrt{2\Delta}\sum_{\ell=1}^{L}\sum_{p=1}^{n} v_{\ell,p} A_{k,\ell_p}\cos(\Phi_k + b_{k,\ell_p}) \tag{137}$$

Then, according to (50) and since $Z_{q,\beta} = 1$, we have, with the notation of the previous section (ii):

$$\langle \mathcal{U}^N, v\rangle = \sum_{k=1}^{K} Y_K \tag{138}$$

Since the random variables $\{\Phi_k\}_k$ are independent, so are the random variables $(Y_k)_k$. In another hand, for every $k \in \{1, \dots, K\}$, $m_k = E(Y_k) = 0$ according to (46), and a simple calculation gives, by (47):

$$E(\langle \mathcal{U}_N, v\rangle^2) = \sum_{k=1}^{K}\sigma_k^2 = S_k^2 \tag{139}$$

where the notations of the central limit theorem were used.

This last relation can be written:

$$S_k^2 = \sum_{\ell=1}^{L}\sum_{\ell'=1}^{L}\langle R_N(t_{\alpha(\ell')}, t_{\alpha(\ell)})v_\ell, v_{\ell'}\rangle \tag{140}$$

When $N_{\text{inf}} \to +\infty$, $K \to +\infty$ and according to 2.6: *Convergence of second order quantities (ii)*, $R^N(t,t') \to R(t,t')$. Then if

$$\mathcal{U} = (X(t_{\alpha(1)}), \dots, X(t_{\alpha(L)}))$$

is the $\mathbb{R}^{nL}$-valued random variable such that:

$$S_\infty^2 = E(\langle \mathcal{U}^N, v\rangle^2) = \sum_{\ell=1}^{L}\sum_{\ell'=1}^{L}\langle R^N(t_{\alpha(\ell')}, t_{\alpha(\ell)}v_\ell, v_{\ell'}\rangle \tag{141}$$

we have:

$$\lim_{K\to+\infty} S_K^2 = S_\infty^2 < +\infty \tag{142}$$

Following (142), (135) will be checked if:

$$\lim_{K\to+\infty} \sum_{k=1}^{K} Int_k \to 0 \text{ with } Int_k = \int_{|y_k|>\varepsilon S_K} y_k^2 P_k(dy_k)$$

Writing

$$y_k(\varphi) = \sqrt{2\Delta} \sum_{\ell=1}^{L} \sum_{p=1}^{n} v_{\ell,p} A_{k,\ell_p} \cos(\varphi + b_{k,\ell_p})$$

we have,

$$Int_k = \frac{1}{2\pi} \int_0^{2\pi} \mathbb{1}_{\{|y_k(\varphi)|>\varepsilon S_K\}}(y(\varphi)) \times y(\varphi)^2 d\varphi$$

The inequality $|y_k(\varphi)| > \varepsilon S_K$ can also be written:

$$\left| \sum_{\ell=1}^{L} \sum_{p=1}^{n} v_{\ell,p} A_{k,\ell_p} \cos(\varphi + b_{k,\ell_p}) \right| > \frac{\varepsilon}{\sqrt{2\Delta}} S_K$$

But we have $|\Delta| = \frac{2^d|\Omega|}{N}$, so, $\forall \varepsilon > 0, \exists N(\varepsilon)$ such that $\forall N > N(\varepsilon)$: $\frac{\varepsilon}{\sqrt{2|\Delta|}} S_K > 2\pi$. Thus, $\forall k, Int_k = 0$ and we can apply the central limit theorem (iii), which, together with Paul Lévy's theorem, show that the law $\mathcal{L}(\mathcal{U}^N)$ converges narrowly towards $\mathcal{L}(\mathcal{U})$, where $\mathcal{U}$ is a gaussian random variable, which shows the narrow convergence of the sequence X^N towards the gaussian field X.

References

1. BATCHELOR G., The theory of axisymmetric turbulence. Proc. Roy. Soc. A, 186, p. 480, 1945.
2. BATCHELOR G., The theory of Homogeneous Turbulence, Cambridge University Press, 1953.
3. BENDAT J.S., PIERSOL A.G., Engineering applications of correlation and spectral analysis, John Wiley, New York, 1980.
4. BELLANGER M., Traitement Numérique du Signal, Masson, 1981.
5. BORKOWSKI J., Spectra of anisotropic turbulence in the atmosphere, Radio Science, vol. 4, number 12, 1969.
6. BUSCH N., On the mechanics of atmospheric turbulence, Workshop on micrometeorology, American Meteo Society, Boston 1973, Haugen Edition.
7. CAMPBELL W., Monte Carlo turbulence simulation using rational approximations to Von Karman spectrum, AIAA Journal, vol. 24, number 1 p. 62, 1986.
8. CAMPBELL W., SANBORN V.A., A spatial model of wind shear and turbulence, J. of Aircraft, vol. 21, number 12 p. 929, 1984.
9. COUPRY G., Problème du vol d'un avion en turbulence, Progress in aerospace vol. 11, Pergamon, 1970.

10. COUNIHAN J., Adiabatic atmospheric boundary layers, A review and analysis of data 1880-1972, Atmospheric Environment, vol. 9, p. 871, 1975.
11. CRAMER H., LEADBETTER M.R., Stationnary and Related Stochastic Processes, John Wiley, New York, 1967.
12. DEARDORFF J., The dimensional numerical modeling of the planetary boundary layer, Workshop on micrometeorology, American Meteo Society, Boston 1973, Haugen Edition.
13. DOOB J.L., Stochastic Processes, John Wiley, New York, 1967.
14. DUCHÊNE-MARULLAZ P., Full scale measurement of atmospheric turbulence in suburban area, Fourth international conference on wind effects on buildings and structures, Heathrow 1975.
15. DUCHÊNE-MARULLAZ P., Effect of high roughness on the characteristics of turbulence in case of strong winds, Wind Engineering, Pergamon Press, Cerdak Editor, 1980.
16. DUCHÊNE-MARULLAZ P., Le spectre du vent en zone urbaine, Construire le vent, Colloque 15-19 juin 1981 C.S.T.B. Nantes.
17. DUTTON J., DEAVEN D., A self similar view of atmospheric turbulence, Radio Science, vol. 4, number 12, 1969.
18. ESTOQUE M.A., Numerical modeling of the planetary boundary layer, Workshop on micrometeorology, American Meteo Society, Boston 1973, Haugen Edition.
19. ETKIN B., Theory of the flight of airplanes in isotropic turbulence. Review and extension, Report 372, NATO 1961.
20. FICHTL G.H., PERLMUTTER M., Non stationary atmospheric boundary layer turbulence modeling, J. of Aircraft, vol. 12, p. 639, 1975.
21. FICHTL G.H., FROST W., PERLMUTTER M., Three velocity component non homogeneous atmospheric boundary layer turbulence modeling, AIAA Journal, Vol. 15, number 10, p. 1444, 1977.
22. GRANT A.L., Observations of boundary layer structure made during the 1981 KONTUR experiment, Q. J. R. Meteo-Soc., Vol. 112, p. 825, 1986.
23. GUIKHMAN L., SKOROKHOD A.V., The Theory of Stochastic Processes, Springer Verlag, Berlin, 1979.
24. HARRIS C.J., Simulation of multivariate non linear stochastic system, Int. J. for Num. Meth. in Eng., Vol. 14, p. 37-50, 1979.
25. HINZE J.O., Turbulence, Mc Graw Hill, New York 1959.
26. HOGSTROM U., Turbulence characteristics in a near neutrally stratified urban atmosphere, Boundary Layer Meteo., 23, 1982.
27. JENKINS G.M., WATT D.G., Spectral Analysis and its Applications, Holden Day, San Francisco, 1968.
28. HOUBOLD J.C., Atmospheric turbulence, AIAA J., Vol. 11, number 4, p. 421, 1973.
29. KAIMAL J.C., Turbulence spectra, length scale and structure parameter in the stable surface layer, Boundary Layer Meteo., 4, 1973.
30. KAIMAL J.C., Estimating velocity spectra in an unstable surface layer, J. Appl. Meteo., Vol. 21, number 8, 1982.
31. KAIMAL J.C., Horizontal velocity in an unstable surface layer, J. Atm. Sci., Vol. 35, p. 18, 1978.
32. KAIMAL J.C. et al., Spectral characteristics of the convective boundary layer over uneven terrain, J. Atm. Sci., Vol. 39, p. 1098, 1982.
33. KAIMAL J.C., WYNGAARD J.C., IZUMI Y., COTE O.C., Spectral characteristics of surface layer turbulence, Q.J.R. Meteo, Soc., number 98, p. 563, 1972.

34. KAIMAL J.C., WYNGAARD J.C., HAUGEN M., Turbulence structure in the convective boundary layer, J. Atm. Sci., Vol. 33, p. 2152, 1976.
35. KRÉE P., SOIZE C., Mathematics of Random Phenomena, Reidel publishing company, Dordrecht, Holland, 1986.
36. MARK W., Characterization of non gaussian atmospheric turbulence for prediction of aircraft response statistics, NASA CR-2745.
37. METIVIER M., Notions Fondamentales de la Théorie des Probabilités, Dunod, Paris, 1972.
38. NEVEU J., Bases Mathématiques de la Théorie des Probabilités, Masson, Paris, 1969.
39. NICHOLLS S., READING C.S., Spectral characteristics of surface layer turbulence over the sea, Q.J.R. Meteo. Soc., Vol. 107, p. 591, 1981.
40. OPPENHEIM A., SCHAFER R., Digital Signal Processing, Prentice Hall, N.J., 1975.
41. PANOFSKY H.A. et al., Spectra of velocity components over complex terain, Q.J.R. Meteo. Soc., number 108, p. 215, 1982.
42. PANOFSKY H.A., LEVI Z., Wind fluctuation in stable air at the Boulder tower, Boundary Layer Meteo., number 25, p. 353, 1983.
43. PICINBONO B., Eléments de la Théorie du Signal, Dunod, Paris, 1977.
44. POIRION F., Simulation des forces généralisées en turbulence isotrope et simulation de la couche limite atmosphérique anisotrope
45. mètres, Rapport technique number 5108RH073, ONERA, 1987.
46. POIRION F., Bounded random oscillations. Model and numerical resolution for an airfoil, Nonlinear Stochastic Dynamic Engineering Systems, IUTAM symposium Innsbruck, Austria 1987.
47. PRIESTLEY M.B., Power spectral analysis of non stationary random process, J. of Sound and Vibration, 6, p. 86-97, 1967.
48. RABINER L.R., GOLD B., Theory and Application of Digital Signal Processing, Prentice Hall, N.J., 1975.
49. REEVES P., JOPPA R., A non gaussian model of continuous atmospheric turbulence for use in aircraft design, NASA CR 2639, 1976.
50. RENYI A., Calcul des Probabilités, Dunod, Paris 1966.
51. SACRE C., La couche limite atmosphérique, REEF, CSTB, 1974.
52. SHINOZUKA M., Simulations of multivariate and multidimensional random processes, J. Acoust. Soc. Amer., Vol. 49, number 1, part 2, pp. 357-367, 1971.
53. SHINOZUKA M., JAN C.M., Digital simulation of random processes and its applications, J. of Sound and Vibration, 25, pp. 111-128, 1972.
54. SHINOZUKA M., WEN Y.K., Monte carlo solution of non linear vibration, AIAA J., vol. 10, number 1, pp. 37-40, 1972.
55. SOIZE C., Eléments mathématiques de la théorie déterministe et aléatoire du signal, ENSTA, Paris, 1985.
56. SOIZE C., Processus stochastiques et méthodes de résolution des problèmes aléatoires, Cours de l'Ecole Centrale des Arts et Manufactures, 1986.
57. SOIZE C., Méthode d'études des problèmes classiques de vibrations aléatoires, Edition TI, série Mathématiques Appliquées, 1988 (à paraitre).
58. STEYN D.G., Turbulence in an unstable surface layer over suburban terrain, Boundary Layer Meteo, 22, p. 183, 1982.
59. TAYLOR J., Manual on Aircraft Loads, Agardograph 83, Pergammon Press 1965.
60. TATOM F., SMITH S., Advanced space shuttle simulation model, NASA CR 354, 1982.

61. TATOM F., SMITH S., Simulation of atmospheric turbulent gusts and gust gradients, AIAA J. Aircraft, Vol. 19, 1982.
62. TENNEKES H., Similarity laws and scale relations in planetary boundary layer, Workshop on micrometeorology, American Meteo Society, Boston 1973, Haugen Edition.
63. TEUNISSEN H.W., Structure of near winds and turbulence in the planetary boundary layer over rural terrain, Boundary Layer Meteo., 19, 1980.
64. TOTOKI H., A method of construction of measures on function spaces and its applications to stochastic processes, Mem. Fac. Sci. Kyushu Uni. Ser. A15-178-190, 1962.
65. TREVINO G., Turbulence for flight simulation, J.of Aircraft Eng. Notes, Vol. 23, number 4, p. 348, 1986.
66. TREVINO G., Time invariant structure of non stationary turbulence, J. of Aircraft, Vol. 22, number 9, p. 827, 1985.
67. WANG S.T., FROST W., Atmospheric turbulence simulation analysis with application to flight analysis, NASA CR 3309, 1980.
68. WEBSTER I., BURLING R.W., A test of isotropy and Taylor's hypothesis in the atmospheric boundary layer, Boundary Layer Meteo. 20, 1981.
69. WYNGAARD J., On surface layer turbulence, Workshop on micrometeorology, American Meteo Society, Boston 1973, Haugen Edition.

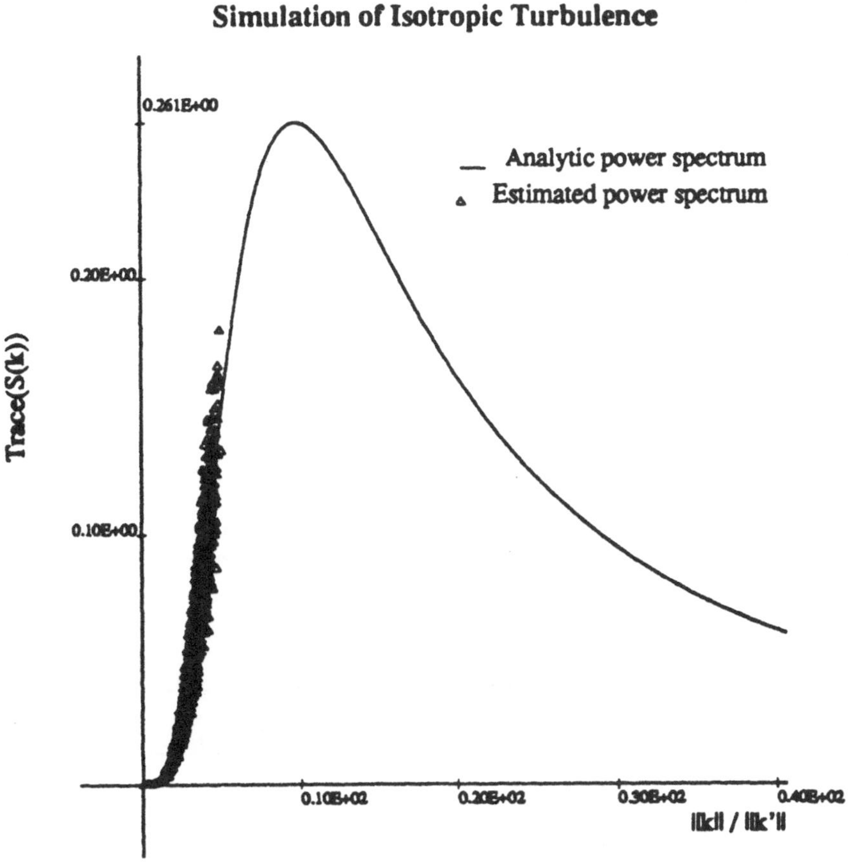

Fig. 1. Estimation built with 200 trajectories - 16 × 16 × 16 points.

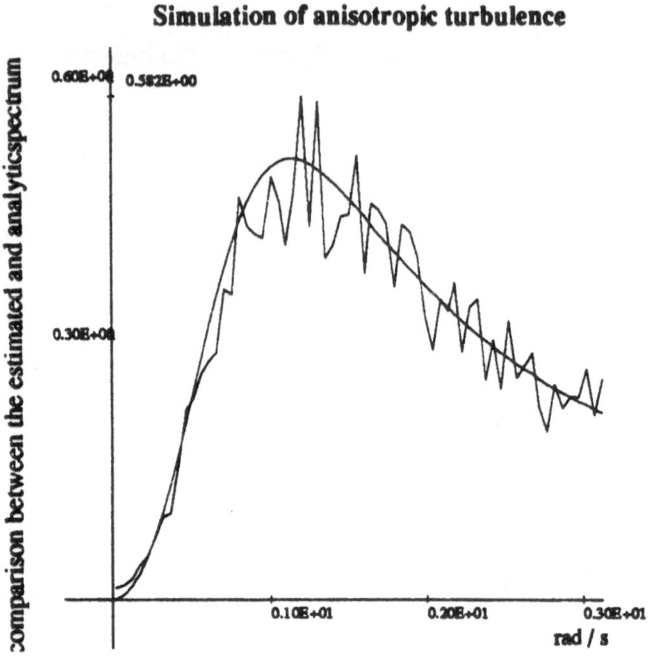

Fig. 2. Estimation built with 100 trajectories - 128 × 32 × 32 points.

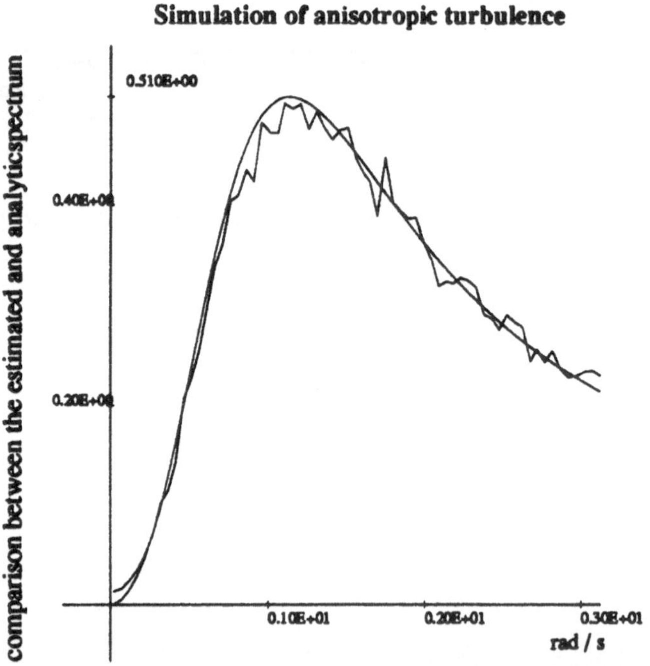

Fig. 3. Estimation built with 500 trajectories - 128 × 32 × 32 points.

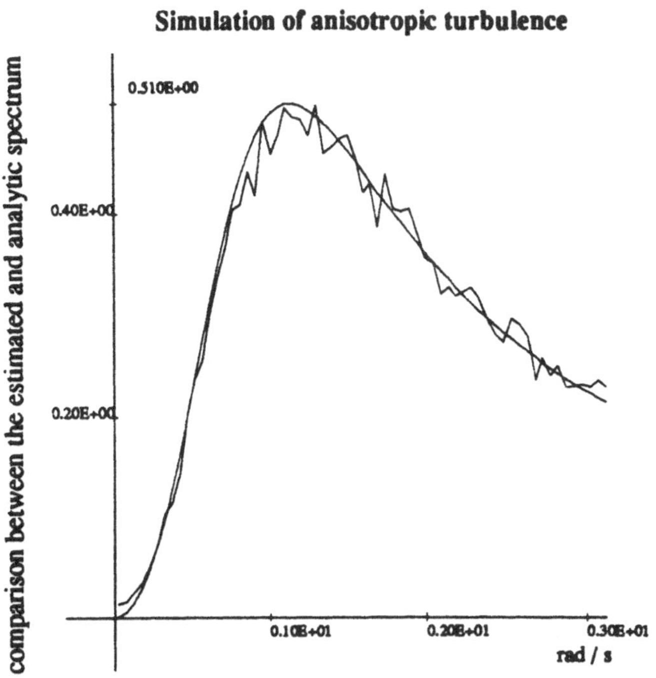

Fig. 4. Estimation built with 500 trajectories - 128 × 32 × 32 points.

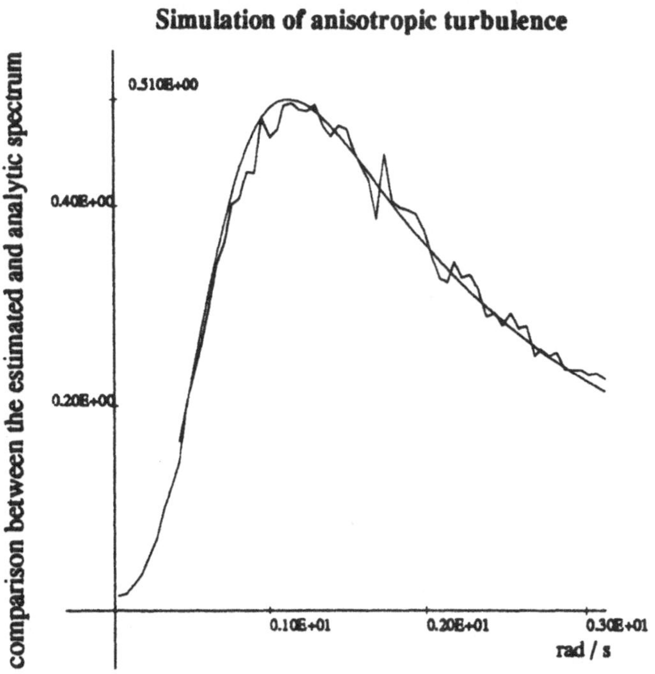

Fig. 5. Estimation built with 700 trajectories - 128 × 32 × 32 points.

Simulation of Stochastic Differential Systems

Denis Talay

Abstract

We present approximation methods for quantities related to solutions of stochastic differential systems, based on the simulation of time-discrete Markov chains. The motivations come from Random Mechanics and the numerical integration of certain deterministic P.D.E.'s by probabilistic algorithms.

We state theoretical results concerning the rates of convergence of these methods. We give results of numerical tests, and we describe an application of this approach to an engineering problem (the study of stability of the motion of a helicopter blade).

1 Introduction

Let us consider a differential system in $\mathbb{R}^d$ excited by a r-dimensional multiplicative random noise $(\xi(t,\omega))$:

$$\frac{d}{dt}X(t,\omega) = b_0(X(t,\omega)) + \sigma(X(t,\omega))\xi(t,\omega) \tag{1}$$

where $b_0(\cdot)$ is an application from $\mathbb{R}^d$ to $\mathbb{R}^d$, $\sigma(\cdot)$ is an application from $\mathbb{R}^d$ to the space of $d \times r$-matrices, and ω denotes the random parameter (in the sequel, we will omit it).

The characteristics of the random noise (bandwidth, energy, law, ...) depend on the modelled physical problem.

Here we are essentially interested in the white-noise case, which is a limit case of systems with "physically realizable" random perturbations (cf. Kushner [32] e.g.); in the section devoted to the computation of Lyapunov exponents, we will examine also systems with coloured noises.

The aim of this paper is to present efficient numerical methods to compute certain quantities depending on the unknown process $(X(t))$, with algorithms based on simulations on a computer of other processes.

We will try to justify this approach (at least, to explain why it may be interesting), we also will underline its limitations; we will estimate the theoretical errors of our approximations, and we will give the results of some illustrative numerical experiments; we will describe an application to an engineering problem (a study of stability for the motion of a helicopter blade) and, finally, we will describe PRESTO, a system of automatic generation of Fortran programs corresponding to the different problems and methods presented here.

Before going on, it must be emphasized that the numerical analysis of stochastic differential systems is at its very beginning: at our knowledge, at the present time only a few algorithms have been proposed (some of them irrealistic ...), and only a few systematic numerical investigations have been pursued. Besides, the theoretical results are not very numerous. Nevertheless, it already appears that this field is not at all a direct continuation of what has been done for the numerical solving of ordinary differential equations. For example, we will underline that **it is often unuseful and even clumsy to try to approximate the diffusion process on the space of trajectories, when one wants to compute a quantity which depends on its law**: approximate processes efficient for simulations may not converge almost surely to the considered diffusion process.

In any case, this paper treats a very few approximation problems, and we have chosen to present only algorithms which have been studied from a theoretical and numerical point of view. Therefore, this paper must be read as a subjective description of the present state of a new art, and also as a hope that numerical problems which can be efficiently solved by probabilistic algorithms will justify and cause new developments; in particular, recent results related to Random Mechanics (Arnold & Kloeden [4], Schenk [49] e.g.) or related to the numerical integration of certain deterministic nonlinear P.D.E.'s by stochastic particles methods (see Bossy & Talay ([10], [11]), Bossy [9], Bernard & Talay & Tubaro [7] e.g.), show that the mathematical or numerical techniques developed to establish some of the results stated below are useful in various contexts.

For complements and variations on the themes of this paper, one can also read the contributions to the volume [14], and consult the extended list of references in Kloeden & Platen's book [30].

The book by Bouleau & Lepingle [13] presents the various mathematical tools necessary to construct and analyse the numerical methods of approximation of a wide class of stochastic processes.

2 Examples of applications and objectives

2.1 Preliminaries

For the theoretical prerequisites, we refer to the basic book of Arnold [1], or the books of Ikeda & Watanabe [28] and Karatzas & Shreve [29] for example. Here we will just briefly introduce some very elementary concepts, which explain the construction of the discretization schemes.

From a mathematical point of view, one must first give a sense to the limit system of systems of type (1) when $(\xi(t))$ tends to a white noise. The answer is provided by the stochastic calculus; the limit system (in a sense we do not precise here) is a stochastic differential system in the Stratonovich sense (cf. Kushner [32]).

Let us consider r independent Wiener processes, $(W^i(t))$, i.e of Gaussian processes with almost surely continuous trajectories, such that

$$E(W^i(t)) = 0 \ ,$$

$$E(W^i(s)W^j(t)) = \delta_{ij}\inf(s,t) \ .$$

The limit system is written in the Stratonovich sense as follows:

$$dX(t) = b_0(X(t))dt + \sigma(X(t)) \circ dW(t) \ , \tag{2}$$

equivalent to the "integral" formulation:

$$X(t) = X(0) + \int_0^t b_0(X(s))ds + \int_0^t \sigma(X(s)) \circ dW(s) \ .$$

Almost surely, the trajectories of the Wiener process have unbounded variations on each finite time interval (this implies that they are nowhere differentiable), therefore the integral $\int_0^t \sigma(X(s)) \circ dW(s)$ cannot be defined as a Stieljes integral. Let us give indications on its construction.

A process $(Y(t))$ is said adapted to the filtration generated by the Wiener process $(W(t))$ (we will also simply say "adapted") if, for each t, $Y(t)$ is measurable w.r.t. the σ-field generated by $(W(s), s \leq t)$; in particular, $Y(t)$ is independent of all the

$$(W(t_1) - W(t), W(t_2) - W(t_1), \ldots, W(t_n) - W(t_{n-1})) \ ,$$

for any n and $t < t_1 < t_2 < \ldots < t_n$.

Let us consider a one-dimensional Wiener process $(B(t))$. For the class Q of real continuous adapted (to the filtration generated by $(B(t))$) processes $(Y(t))$ which can be represented as $Y(t) = Y(0) + M(t) + A(t)$, where $Y(0)$ is a r.v., $(M(t))$ is a continuous locally square integrable martingale relative to the previous filtration, and $(A(t))$ is a continuous adapted process of bounded variation on every finite time interval, one can show that for any $T > 0$, the following limit in probability exists:

$$\lim_{|\Delta|\to 0} \sum_{i=1}^{n} \frac{Y(t_i) + Y(t_{i-1})}{2}(B(t_i) - B(t_{i-1})) \ , \tag{3}$$

where Δ denotes a partition $0 = t_0 < t_1 < \ldots < t_n = T$, and $|\Delta|$ denotes $\max_{1\leq i\leq n}(t_i - t_{i-1})$.

This limit is called the Stratonovich integral of $(Y(t))$ w.r.t. $(B(t))$ on $[0,T]$ and denoted $\int_0^T Y(s) \circ dB(s)$. If $(Y(t))$ is a matrix-valued process, and the Wiener process multi-dimensional, the integral is defined coordinate by coordinate.

If b_0 and σ are continuous functions such that each component is twice continuously differentiable with bounded derivatives of first and second orders, then for each Borel probability measure μ on $\mathbb{R}^d$, there exists a process $(X(t))$ in Q, satisfying (2) and such that the law of $X(0)$ coincides with μ, unique in the sense that, if $(Y(t))$ is another solution with $Y(0) = X(0)$ *a.s.*, then for each t, $X(t) = Y(t)$ *a.s.*

We will see that the discretization of a Stratonovich integral leads to some difficulties, which do not exist for another stochastic integral, the Itô stochastic integral.

To simplify, we restrict ourselves to consider those of the previous processes $(Y(t))$ which also satisfy:

$$\forall t > 0 \ , \quad E\left[\int_0^t |Y(s)|^2 ds\right] < \infty \ .$$

One can show that the following limit exists in the space of the square integrable random variables:

$$\lim_{|\Delta| \to 0} \sum_{i=1}^{n} Y(t_i)(B(t_i) - B(t_{i-1})) \ ,$$

where Δ denotes a partition $0 = t_0 < t_1 < \ldots < t_n = t$, and $|\Delta|$ denotes $\max_{1 \le i \le n}(t_i - t_{i-1})$.

This limit is called the Itô integral of $(Y(t))$ w.r.t. $(B(t))$, and is denoted by $\int_0^t Y(s)dB(s)$. Under our hypothesis on $(Y(t))$, the process $(\int_0^t Y(s)dB(s))$ is a square integrable martingale, which is often used in the proofs of most approximation theorems stated below.

Moreover, under the above assumptions on $b_0(\cdot)$ and $\sigma(\cdot)$, one can also show that the solution of the Stratonovich system (2) is the unique solution of:

$$X(t) = X(0) + \int_0^t b(X(s))ds + \int_0^t \sigma(X(s))dW(s)$$

where, if σ_j denotes the jth column of σ and $\partial\sigma_j$ the matrix whose element of the ith row and k^{th} column is $\partial_k \sigma_j^i$:

$$b(\cdot) = b_0(\cdot) + \frac{1}{2}\sum_{j=1}^{r} \partial\sigma_j(\cdot)\sigma_j(\cdot) \ .$$

In differential notations, the previous equation is written under the form of an Itô stochastic differential system:

$$dX(t) = b(X(t))dt + \sigma(X(t))dW(t) \ . \tag{4}$$

As a consequence of the definition of the Itô integral, it appears that the differential chain rule is different from the deterministic case: for any real function of class C^2, we have the following formula (the Itô formula):

$$df(X(t)) = Lf(X(t))dt + \sigma(X(t))\nabla f(X(t)) \cdot dW(t) \ , \tag{5}$$

where, if the matrix a is $a = \sigma\sigma^*$, L is the following differential operator:

$$L = \sum_{i=1}^{d} b^i(x)\partial_i + \frac{1}{2}\sum_{i,j=1}^{d} a_j^i(x)\partial_{ij} \ . \tag{6}$$

Let us now briefly present the situations that we will treat.

2.2 Simulation of trajectories

We are supposed to have at our disposal exact or approximate trajectories of the Wiener process $(W(t))$, and we want to "see" the corresponding approximate trajectories of $(X(t))$. As we will show later on, when the dimension of the noise is larger than 1 and only approximate trajectories of $(W(t))$ are available, this problem has a signification under a stringent condition which must be fulfilled by $\sigma(\cdot)$ and will be called in the sequel the "commutativity condition" (its precise formulation will be given below).

Let us give 2 examples of situations where one may wish to get trajectories of the solution of a S.D.E.

First, let us suppose that the process $(X(t))$ depends on a parameter θ, that one wants to estimate from a unique observation of $(X(t))$ during a time interval $[0,T]$.

In order to test the quality of different estimators, one may choose a particular value for θ, simulate a few trajectories of the corresponding process $(X(t))$, and then apply the estimators on these simulated trajectories. For applications to financial models, see Fournié and Talay [23] and Fournié [22], e.g.

A less elementary example is a filtering situation, where $(X(t))$ is a non observed process solution of (2), whereas one observes realizations of a process $(Y(t))$ satisfying:

$$dY(t) = g(X(t))dt + \alpha dW(t) + \beta dV(t) \ ,$$

where $(V(t))$ is a Wiener process independent of $(W(t))$; one wants to get the conditional law of $(X(t))$, given the observations $(Y(s), 0 \leq s \leq t)$. In the non-linear case, under some regularity assumptions on the functions b_0, σ, g, the answer is given by the so-called Zakai equation satisfied by $p(t, x_1, \ldots, x_d)$, the unnormalized density of this conditional law:

$$dq(t) = Aq(t)dt + B_0 q(t)dt + B_1 q(t)dt + C_0 q(t) \circ dY(t) + C_1 q(t) \circ dY(t) \ ,$$

where A is a second-order operator, B_1 and C_1 are first-order operators, B_0 and C_0 are zero order operators.

The previous equation is a stochastic partial differential equation. To solve it numerically, Florchinger and Le Gland [19] propose the following algorithm.

Let (t_p) be a dicretization of the time interval $[0,t]$, and $\overline{q}$ be the approximate density. On each time interval, $[t_p, t_{p+1}]$, one first numerically solves the deterministic P.D.E.

$$\begin{aligned} \frac{d}{dt}u(t) &= Au(t) \ , \\ u(t_p) &= \overline{q}(t_p) \ , \end{aligned}$$

and then one considers the stochastic P.D.E.

$$\begin{aligned} dv(t) &= B_0 v(t)dt + B_1 v(t)dt + C_0 v(t) \circ dY(t) + C_1 v(t) \circ dY(t) \ , \\ v(t_p) &= u(t_{p+1}) \ . \end{aligned}$$

Let us write the operators B_1 and C_1 under the form: $B_1 = b_1(x)\nabla$ and $C_1 = c_1(x)\nabla$.

Let $(Z(t;s,x))$ the flow associated to the stochastic differential equation:

$$dZ(t) = -b_1(Z(t))dt - c_1(Z(t)) \circ dY(t) \ . \tag{7}$$

Then one computes the value of $v(t,x)$ at points $Z(t;t_p,z)$ according to the formula (d_1 and d_2 being appropriate functions):

$$v(t,Z(t;t_p,z)) = v(t_p,z) \exp\left\{ \int_{t_p}^{t} d_1(Z(s;t_p,z))ds + \int_{t_p}^{t} d_2(Z(s;t_p,z)) \circ dY(s) \right\} \ ,$$

and $\overline{q}(t_{p+1})$ is given by $\overline{q}(t_{p+1}) = v(t_{p+1})$.

This procedure requires to solve (7) in a pathwise sense: one wants to get the path of $(Z(t))$ corresponding to the particular observed path of $(Y(t))$.

2.3 Computation of statistics of $(X(t))$ on a finite time interval

For example, one wants to compute the first moments of the response of the dynamical system $(X(t))$, or, more generally, $Ef(X(t))$, $f(\cdot)$ being an explicitly given function.

Another motivation is to construct Monte Carlo methods to solve parabolic P.D.E.'s in $\mathbb{R}^d$

$$\begin{cases} \frac{d}{dt}u(t,x) = Lu(t,x) \ , \\ u(0,x) \ = f(x) \ , \end{cases}$$

in some situations where deterministic methods are not efficient: the theoretical accuracy and the numerical behaviour of the probabilistic algorithms are not affected by the possible non coercivity of the second-order elliptic differential operator L, and the computational cost growths only linearly w.r.t. the dimension d of the state space.

Therefore these methods and the stochastic particles methods (random vortex methods for the integration of certain non-linear P.D.E.'s in Fluid Mechanics e.g.) which also require the simulation of stochastic processes (see the references given at the end of the Introduction) can be useful in degenerate situations or when the state space has a large dimension; in Random Mechanics, often $(X(t))$ is a vector (position, speed), and therefore both degeneracy and a high dimensional state space occur. Other examples are the situations where $u(t,x)$ needs to be computed only at a small number of points, for example in order to separate the integration space in subdomains where deterministic methods become efficient.

To compute $Ef(X(t))$, if we could simulate the process $(X(t))$ itself, we would simulate several independent paths of $(X(t))$, denoted by $(X(t,\omega_1),\ldots,X(t,\omega_N))$ and then we would compute the average

$$\frac{1}{N}\sum_{i=1}^{N} f(X(t,\omega_i)) \ .$$

Instead of $(X(t))$, we propose to simulate another process $(\overline{X}(t))$; as we are interested in the approximation of the law of $(X(t))$, it is unnecessary that $(\overline{X}(t))$ is a trajectorial approximation of $(X(t))$, and the "commutativity condition" will not be required (better, some efficient processes $(\overline{X}(t))$ in that context are not at all approximations of $(X(t))$ in the pathwise sense, and even do not converge almost surely to $(X(t))$.

2.4 Asymptotic behaviour of $(X(t))$, Lyapunov exponents

In the section 6.1, we present an industrial problem leading to the study of a bilinear system for which it can be shown that, $(X(t,x))$ denoting the solution of (2) with initial condition x, the almost-sure limit

$$\lambda = \lim_{t \longrightarrow +\infty} \frac{1}{t} \log |X(t,x)|$$

exists and is independent of x (it is the upper Lyapunov exponent of the system); the problem is to determine the sign of that limit: if it is strictly negative, almost surely $(X(t))$ tends to 0 exponentially fast for any initial condition x, the system (2) is then said to be stable.

The proposed algorithm consists in simulating one particular path of a process $(\overline{X}(t))$ over a long time $[0,T]$ and in computing

$$\overline{\lambda}_T = \frac{1}{T} \log |\overline{X}(T,x)| \ .$$

We will classify different processes $(\overline{X}(t))$ according to the following criterium: how large is

$$|\lambda - \overline{\lambda}|$$

where $\overline{\lambda}$ is the Lyapunov exponent of the process $(\overline{X}(t))$, defined by

$$\begin{aligned} \overline{\lambda} &= \lim_{T \to \infty} \overline{\lambda}_T \\ &= \lim_{T \to \infty} \frac{1}{T} \log |\overline{X}(T,x)| \ . \end{aligned}$$

Remark: this criterium does not take into account the error due to the necessary approximation of $\overline{\lambda}$ by $\overline{\lambda}_T$, which only depends on the choice of the integration time T. We will see that, from a practical point of view, this choice may be very difficult.

An extension of the method has been developed and analysed for nonlinear systems.

2.5 Computation of the stationary law

Under some conditions on the coefficients $b_0(\cdot)$ and $\sigma(\cdot)$, one a priori knows that the process $(X(t))$ is ergodic. Let us denote by μ its unique invariant probability law.

One may be interested in computing $\int f(x)d\mu(x)$ for a given function f (for example, in order to get the asymptotic value of $Ef(X(t))$ when t goes to infinity, i.e to describe the stationary distribution of the response of the system, which often is of prime importance in Random Mechanics). For reasons already underlined, the numerical solving of the stationary Fokker-Planck equation may be extremely difficult.

Here, we propose to choose T "large enough" and to compute

$$\frac{1}{T}\int_0^T f(\overline{X}(s,x))ds \ .$$

2.6 Remark

As it has been mentioned above, it will appear that the choice of the convenient process $(\overline{X}(t))$ must be related to the final purpose of the simulation.

Our basic tool to construct this process is the time discretization of the system (4).

3 Discretization methods

3.1 Introduction to the Milshtein scheme

Let us consider the expression

$$X(t) = X(0) + \int_0^t b_0(X(s))ds + \int_0^t \sigma(X(s)) \circ dW(s) \ .$$

From the definition (3) of the Stratonovich integral, for small t the integral $\int_0^t \sigma(X(s)) \circ dW(s)$ can be approximated by

$$\frac{1}{2}(\sigma(X(t)) + \sigma(X(0)))(W(t) - W(0)) \ ,$$

and therefore this procedure would lead to an implicit discretization scheme.

According to the definition of an Itô integral, a rough approximation of $(X(t))$ (for a small t) can be:

$$X(t) \simeq X(0) + b(X(0))t + \sigma(X(0))(W(t) - W(0)) \ .$$

Let h be a discretization step.

The above remark justifies the Euler scheme for (4):

$$\overline{X}_{p+1}^h = \overline{X}_p^h + b(\overline{X}_p^h)h + \sigma(\overline{X}_p^h)(W((p+1)h) - W(ph)) \ . \tag{8}$$

This scheme can easily be simulated on a computer: at each step p, one has just to simulate the vector $W((p+1)h) - W(ph)$, whose law is Gaussian.

As we will see, nevertheless this scheme may be unsatisfying: for example, it is divergent for the pathwise approximation of $(X(t))$.

Let us introduce a new scheme, and first consider the case $d = r = 1$.

If we perform a Taylor expansion of $\sigma(X(t))$, we easily get:

$$X(t) \simeq X(0) + b(X(0))t + \sigma(X(0))(W(t) - W(0)) \\ + \sigma(X(0))\sigma'(X(0)) \int_0^t (W(s) - W(0))dW(s) \ .$$

At a first glance, the situation is more complex than previously, because of the presence of the stochastic integral $\int_0^t (W(s) - W(0))dW(s)$. But the Itô formula shows:

$$\int_0^t (W(s) - W(0))dW(s) = \frac{1}{2}(W(t)^2 - t) \ .$$

Then, again only Gaussian laws are involved in the previous scheme, due to Milshtein [37] who introcuded it in 1974 for the mean-square approximation of $(X(t))$.

3.2 The multi-dimensional Milshtein scheme

Let us now examine the general case.

Let us introduce the notation

$$\Delta_{p+1}^h W := W((p+1)h) - W(ph) \ .$$

The same procedure as before leads to the multi-dimensional Milshtein scheme:

$$\overline{X}_{p+1}^h = \overline{X}_p^h + \sum_{j=1}^r \sigma_j(\overline{X}_p^h)\Delta_{p+1}^h W^j + b(\overline{X}_p^h)h \\ + \sum_{j,k=1}^r \partial\sigma_j(\overline{X}_p^h)\sigma_k(\overline{X}_p^h) \int_{ph}^{(p+1)h} (W^k(s) - W^k(ph))dW^j(s) \ . \quad (9)$$

Now, the situation is really complex, because of the presence of the multiple stochastic integrals $\int_{ph}^{(p+1)h} (W^k(s) - W^k(ph))dW^j(s)$: these integrals do not depend continuously on the trajectories of $(W(t))$ (therefore are annoying for the trajectorial approximation), and the joint law of these integrals and the increments $\Delta_{p+1}^h W$ seems difficult to simulate: in particular, it cannot be seen as the law of a simple transformation of a Gaussian vector (see the work by Gaines & Lyons [25]).

How to get rid of this difficulty is one of the main features of the numerical analysis of stochastic differential systems.

3.3 Mean-square approximation and Taylor formula

Milshtein [37] proved the following result:

Theorem 1. *Let us suppose that the functions b and σ are of class C^2, with bounded derivatives of first and second orders.*

Then the Euler scheme satisfies: for any integration time T, there exists a positive constant $C(T)$ such that, for any step-size h of type $\frac{T}{n}$, $n \in \mathbb{N}$:

$$\left[E|X(T) - \overline{X}_n^h)|^2\right]^{\frac{1}{2}} \leq C(T)\sqrt{h} \ .$$

For the Milshtein scheme, we can substitute the following bound for the error:

$$\left[E|X(T) - \overline{X}_n^h)|^2\right]^{\frac{1}{2}} \leq C(T)h \ .$$

It can be shown that the Milshtein scheme is not "asymptotically efficient" in the sense that the leading coefficient in the expansion of the mean square error as power series in h is not the smallest possible. Clark [17] and Newton ([40], [41] and [42]) have introduced new schemes which are asymptotically efficient; these schemes may be seen as versions of the Milshtein scheme with additional terms of order $h\Delta_{p+1}^h W$ and $(\Delta_{p+1}^h W)^3$ (when the Wiener process is scalar). In the two last references, efficient schemes based upon first passage times of the Wiener samples through given points, and efficient Runge-Kutta schemes are presented. Another very interesting approach can be found in Castell & Gaines [16], based upon the representation of diffusion processes in terms of the solutions of ordinary differential equations.

Let us now examine the question of the order of convergence.

Let us say that a random variable is of order k if its variance is upper bounded by $Constant \times h^{2k}$.

The Milshtein scheme involves only random variables of order less than 1.

To get a better rate of convergence in the mean-square sense than this scheme, one must involve multiple stochastic integrals of order strictly larger than 1, for example:

$$\int_{ph}^{(p+1)h} (W^k(s) - W^k(ph))(W^l(s) - W^l(ph))dW^j(s) \ ,$$

$$\int_{ph}^{(p+1)h} (s - ph)dW^j(s) \ , \quad \int_{ph}^{(p+1)h} (W^k(s) - W^k(ph))ds \ ,$$

in order to get an error of order $h^{\frac{3}{2}}$.

The coefficients of these integrals in the schemes are given by a Taylor formula (see Platen & Wagner [47]).

Of course, most of these integrals, as those involved in the multi-dimensional Milshtein scheme, have probability laws difficult to simulate (see Gaines [24]). Therefore, in the general case, the Euler scheme is the only efficient scheme for the mean-square approximation.

Nevertheless, there exists a situation where the multi-dimensional Milshtein scheme involves only the increments of the Wiener process $\Delta_{p+1}^h W$.

3.4 The commutativity condition

Suppose that the column vectors of the matrix σ satisfy the following condition:

$$\forall j, \forall k \ : \ \partial\sigma_j(\cdot)\sigma_k(\cdot) = \partial\sigma_k(\cdot)\sigma_j(\cdot) \ . \tag{10}$$

That condition means that the vector fields defined by the column vectors of σ commute. It is obviously satisfied when the noise is one-dimensional, or when the function σ is constant.

The Itô formula and this hypothesis imply:

$$\begin{aligned}&\partial\sigma_k(\cdot)\sigma_j(\cdot)\int_{ph}^{(p+1)h}(W^k(s)-W^k(ph))dW^j(s)\\&+\partial\sigma_j(\cdot)\sigma_k(\cdot)\int_{ph}^{(p+1)h}(W^j(s)-W^j(ph))dW^k(s)\\&=\partial\sigma_k(\cdot)\sigma_j(\cdot)(W^k((p+1)h)-W^k(ph))(W^j((p+1)h)-W^j(ph)) \ ,\end{aligned} \tag{11}$$

and therefore the Milshtein scheme can be rewritten:

$$\begin{aligned}\overline{X}_{p+1}^h &= \overline{X}_p^h + \sum_{j=1}^{r}\sigma_j(\overline{X}_p^h)\Delta_{p+1}^h W^j + b(\overline{X}_p^h)h\\&+\sum_{k=2}^{r}\sum_{j<k}\partial\sigma_j(\overline{X}_p^h)\sigma_k(\overline{X}_p^h)\Delta_{p+1}^h W^j\Delta_{p+1}^h W^k\\&+\frac{1}{2}\sum_{j=1}^{r}\partial\sigma_j(\overline{X}_p^h)\sigma_j(\overline{X}_p^h)\left[(\Delta_{p+1}^h W^j)^2-h\right] \ .\end{aligned} \tag{12}$$

A very nice result due to Clark & Cameron [18] shows that, under the commutativity condition, the Milshtein scheme leads to the best possible rate of convergence for the mean-square approximation (i.e h) among all the discretization schemes involving only values of the process $(W(t))$ at times $(ph, 0 \leq p \leq n = \frac{T}{h})$.

4 Almost Sure and Pathwise Approximation

4.1 Statement of the problems

First, let us suppose that we observe or simulate increments of the Wiener process during time intervals of length h; then we construct a continous time process $(\overline{X}(t))$ by using the Euler scheme and by interpolating linearly between the times ph. Does this process converge almost surely to $(X(t))$ on finite time intervals when h goes to 0 ?

Second, let us now suppose that we dispose of a deterministic function $t \rightarrow u(t)$ which approximates a given trajectory of $(W(t))$ in the sense of the topology of uniform convergence on the space of continuous functions on $[0, T]$.

We hope to get an approximation of the trajectory of $(X(t))$ on $[0,T]$ corresponding to this particular trajectory of $(W(t))$.

To give a sense to this new problem, a natural condition is that there exists a continuous mapping F from $\mathbb{R}^d \times \mathcal{C}_0(\mathbb{R}_+;\mathbb{R}^r)$ [2] to $\mathcal{C}(\mathbb{R}_+;\mathbb{R}^r)$ such that $X(t) = F(X(0), W)(t)$, a.s.

A result due to McShane [36], Doss [20] and Sussman [51] shows that this mapping exists if the above "commutativity condition" is satisfied.

Now, the problem is to build a scheme defined by functionals ϕ_p^h on the space $\mathbb{R}^d \times \mathcal{C}([0,ph])$:

$$\overline{X}_0^{h,u} = X(0) \quad , \quad \overline{X}_{p+1}^{h,u} = \phi_p^h(\overline{X}_p^{h,u}, (u(t), 0 \le t \le ph))$$

such that: for any entry $(u(t))$ belonging to a large set of functions (including the trajectories of $(W(t))$ and their reasonable approximations), if $x_u(t)$ denotes $F(X(0),u)(t)$, and if $t \to \overline{X}_u^h(t)$ is the function defined by

$$\overline{X}_u^h(t) = \overline{X}_p^{h,u} \quad , \quad ph \le t < (p+1)h \ ,$$

then

$$\lim_{h\to 0} \sup_{0\le t\le T} |x_u(t) - \overline{X}_u^h(t)| = 0 \ .$$

When this property is fulfilled, the scheme is robust w.r.t. small pertubations of the trajectory $u(\cdot)$; we say that the scheme converges in the pathwise (or trajectorial) sense.

4.2 Example

Let $(B(t))$ be a one-dimensional Wiener process. From the Itô formula (5), the process

$$X(t) = \exp(t + B(t))$$

solves the one-dimensional stochastic differential system:

$$dX(t) = \frac{3}{2}X(t)dt + X(t)dB(t) \ .$$

It is easy to see that the Euler scheme (8) converges almost surely, and converges in the above pathwise sense only if the function $(u(t))$ has the same quadratic variation as the trajectories of the Wiener process.

The situation is different with the Milshtein scheme.

[2] $\mathcal{C}_0(\mathbb{R}_+;\mathbb{R}^r)$ denotes the set of continuous functions f from $\mathbb{R}_+$ to $\mathbb{R}^d$ such that $f(0) = 0$.

4.3 Main results

For the almost sure convergence problem of the Euler problem, a first result appears in Newton ([40] and [42]). More precise statements appear in Faure ([21]), for example:

Theorem 2. *Let us suppose that the coefficients $b(\cdot)$ and $\sigma(\cdot)$ are Lipschitz.*

(i) If for some integer $K > 1$ the initial condition X_0 satisfies $E|X_0|^{2K} < \infty$, then the interpolated Euler scheme with step-size $\frac{T}{n}$, $(\overline{X}^h(t))$, converges almost surely to $(X(t))$ on $[0,T]$ when n goes to infinity.
(ii) If the initial condition has moments of any order, then the order of convergence is given by

$$\forall \alpha < \frac{1}{2} \quad , \quad n^{\alpha} \sup_{t\in[0,T]} |X(t) - \overline{X}(t)| \overset{n\to+\infty}{\longrightarrow} 0 \quad , \quad a.s.$$

Let us now turn to the trajectorial problem.

In the multidimensional case, the remarkable point is that the commutativity condition, which is necessary to have a well-posed problem, is also sufficient to make the Milshtein scheme (9) depend only on the values of $(W(t))$ at the discretization points (formula (11)). This leads us to introduce the trajectorial Milshtein scheme defined by

$$\begin{aligned} \overline{X}^{h,u}_{p+1} = \overline{X}^{h,u}_{p} &+ \sum_{j=1}^{r} \sigma_j(\overline{X}^{h,u}_{p}) \Delta^h_{p+1} u^j + b(\overline{X}^{h,u}_{p}) h \\ &+ \sum_{k=2}^{r} \sum_{j<k} \partial\sigma_j(\overline{X}^{h,u}_{p}) \sigma_k(\overline{X}^{h,u}_{p}) \Delta^h_{p+1} u^k \Delta^h_{p+1} u^j \\ &+ \frac{1}{2} \sum_{j=1}^{r} \partial\sigma_j(\overline{X}^{h,u}_{p}) \sigma_k(\overline{X}^{h,u}_{p}) \left[(\Delta^h_{p+1} u^j)^2 - h \right] \ . \end{aligned} \tag{13}$$

In Talay [52] the following result is proven:

Theorem 3. *Let us suppose that b and σ are bounded, of class C^3 with bounded derivatives up to the order 3, and that the function $(u(t))$ satisfies:*

$$\lim_{|\Delta|\to 0} \sum_i |u(t_i) - u(t_{i-1})|^3 = 0 \ , \tag{14}$$

where Δ denotes a partition $0 = t_0 < t_1 < \ldots < t_m = t$, and $|\Delta|$ denotes $\max_{1\leq i\leq m}(t_i - t_{i-1})$.

Then, if $(\overline{X}^h_u(t))$ is defined as in the previous section, under the commutativity condition:

$$\lim_{h\to 0} \sup_{0\leq t\leq T} |x_u(t) - \overline{X}^h_u(t)| = 0 \ . \tag{15}$$

Remarks

- The condition (14) is satisfied by the paths of the Wiener process, but also by a much larger class of functions (for example, the differentiable functions).
- The proof of the theorem (3) is based on an analytical expression of the mapping F.
- It is also shown that the Milshtein scheme has the best possible rate of convergence for the criterium (15).
- The commutativity condition is a strong limitation to the trajectorial approximation of the solution of an Itô differential system. But, in some sense, this problem forgets the fact that $(X(t))$ is a stochastic process, whose statistics may be more interesting than some particular paths. We will not need this condition to approximate quantities depending on the law of $(X(t))$.
- The asymptotic distribution of the normalized Euler scheme error is analysed in Kurtz & Protter [31].

4.4 Numerical example

The following numerical test illustrates the divergent behaviour of the Euler scheme for the pathwise approximation.

Let $(X(t))$ the 2-dimensional process defined by

$$X(t) = (sin(W(t)), cos(W(t))) \ ,$$

where $(W(t))$ is a one-dimensional Wiener process.

This process solves a system with the function $b(x_1, x_2)$ defined by

$$\begin{aligned} b^1 &= -\frac{x^1}{2} \ , \\ b^2 &= -\frac{x^2}{2} \ , \end{aligned}$$

and the matrix σ is defined by

$$\begin{aligned} \sigma_{11} &= x^2 \ , \\ \sigma_{21} &= -x^1 \ . \end{aligned}$$

We have simulated a trajectory of $(W(t))$, and a perturbation of it: we have simulated a second Brownian trajectory (denoted by $t \to V(t)$), and, for each t, we have added $\varepsilon V(t)$ to $W(t)$.

The figure 1 shows the "exact" path of $(X(t))$ corresponding to the simulated path of $(W(t))$. The figure 2 compares the evolution in time of the errors (in the trajectorial sense) due to the Milshtein scheme (thick line) and the Euler scheme (thin line), corresponding to $\varepsilon = 0.001$ and $h = 0.01$.

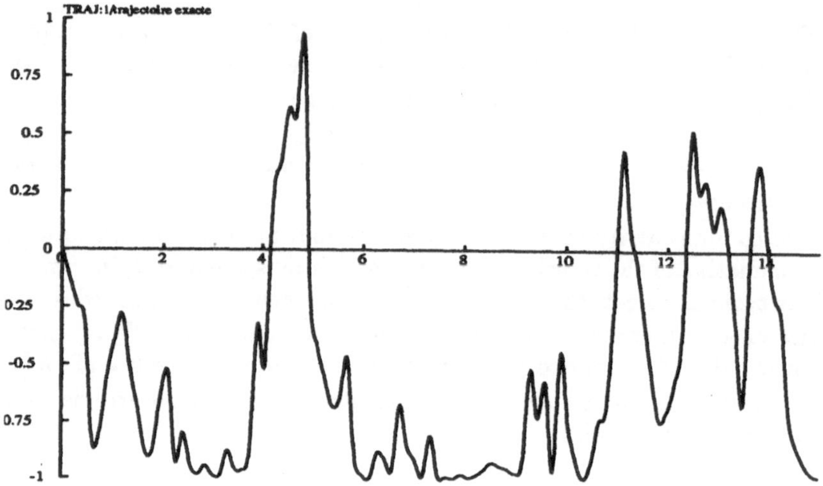

Fig. 1. Exact path.

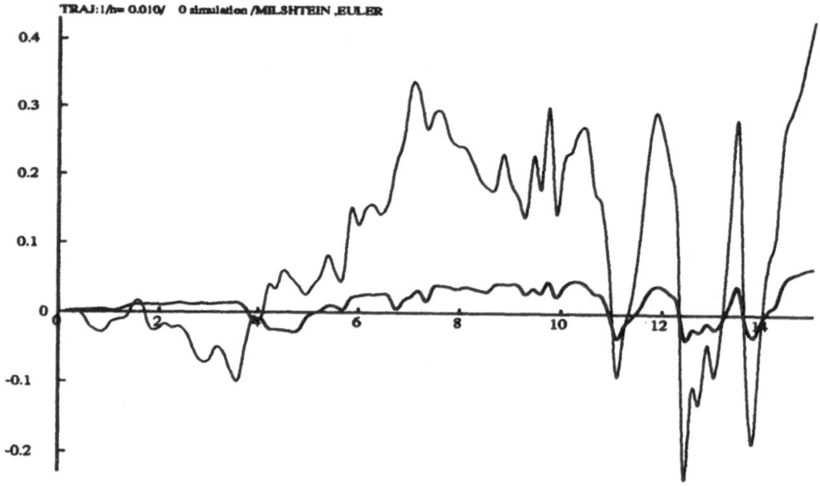

Fig. 2. Milshtein and Euler schemes.

5 Computation of $Ef(X(t))$

5.1 Methodology

Now, we are interested in the approximation of $Ef(X(t))$ on a fixed finite time interval $[0,T]$.

Suppose that the coefficients b and σ are smooth enough, and that the operator L defined in (6) is hypoelliptic; if the law of the initial condition $X(0)$ has a density $p_0(\cdot)$, then for any $t > 0$ the law of $X(t)$ has a density $p(t,\cdot)$ solution of the Fokker-Planck equation:

$$\begin{cases} \frac{d}{dt}p(t,x) = L^*p(t,x) \ , \\ p(0,x) \ = p_0(x) \ , \end{cases}$$

where L^* is the adjoint of the differential operator L (see [28] e.g.).

Thus a first method to compute $Ef(X(t))$ consists in integrating the previous P.D.E. But the numerical solving of this P.D.E. can be difficult, for example when the dimension d of the process $(X(t))$ is very large, or when the differential operator L is degenerate (*it very often is the case in problems coming from Mechanics, in particular each time* $(X(t))$ *is a vector (position,velocity)*).

A second method consists in using a Monte Carlo method. Let us begin by choosing the Euler scheme (8), and let us simulate (if possible, in parallel) a large number N of independent realizations of the Gaussian sequence $(\Delta_{p+1}^h W, p \in \mathbb{N})$. Then, for each discretization step, we get N independent realizations of $\overline{X}_p^h$, denoted by $\overline{X}_p^h(\omega_i)$, and we can compute

$$\frac{1}{N}\sum_{i=1}^{N} f(\overline{X}_p^h(\omega_i)) \ . \tag{16}$$

By the strong law of large numbers, this gives us an approximate value of $Ef(\overline{X}_p^h)$. The quality of this approximation depends only on the choice of N.

It remains to estimate the error $|Ef(X(ph))-Ef(\overline{X}_p^h)|$. It can be shown that, under some smoothness assumptions on b, σ, if the law of $X(0)$ has moments of any order, then, for any time T, there exists a positive constant $C(T)$ such that, for any discretization step h of type $h = \frac{T}{n}$, $n \in \mathbb{N}$:

$$|Ef(X(T)) - Ef(\overline{X}_n^h)| \leq C(T)h \ .$$

It can be also shown that, even under the commutativity condition, the Milshtein scheme has the same rate of convergence.

This is illustrated by the following example: choose $d = r = 1$, $b(x) = \frac{1}{2}x$, $\sigma(x) = x$ and $f(x) = x^4$. Then, for the Euler or Milshtein scheme, there exist constants C_1, C_2 such that:

$$|Ef(X(T)) - Ef(\overline{X}_n^h)| = C_1 T\exp(C_2T)h + O(h^2) \ .$$

Therefore, the Milshtein scheme which is "optimal" in the mean-square sense and in the trajectorial sense is so poor as the Euler scheme for an approximation of the law of $(X(t))$.

The technique introduced in Talay [53] or [54], Milshtein [38] permits to analyse the error on $Ef(X(T))$ without using estimates in L^p of $X(T)-\overline{X}^h_n$. It also permits to construct second-order schemes. A refinement of the analysis leads to a very efficient procedure (see the section 5.6 below).

5.2 Second-order schemes

Let $\mathcal{P}$ be the set of numerical functions of $\mathbb{R}^d$, of class C^6, such that f and its partial derivatives up to order 6 have a growth at most polynomial at infinity.

A scheme is said of second-order if it satisfies for any system whose coefficients b, σ are smooth and have bounded derivatives of any order: for any function f in $\mathcal{P}$, for any time T, there exists a positive constant $C(T)$ such that, for any discretization step h of type $h=\frac{T}{n}$, $n\in\mathbb{N}$:

$$|Ef(X(T))-Ef(\overline{X}^h_n)|\leq C(T)h^2 \ . \tag{17}$$

Let $\mathcal{F}_p$ be the σ-algebra generated by $(\overline{X}^h_0,\ldots,\overline{X}^h_p)$.

In Talay [54], it is shown that a sufficient condition for a scheme to satisfy (17) is the set of hypotheses (C1), (C2), (C3):

(C1) $\overline{X}^h_0=X(0)$;
(C2) $\forall n\in\mathbb{N}\ ,\ \forall N\in\mathbb{N}\ ,\ \exists C>0\ ,\ \forall p\leq N\ ,\quad E|\overline{X}^h_p|^n\leq C$;
(C3) the following properties are satisfied for all $p\in\mathbb{N}$, where all the right-side terms of the equalities must be understood evaluated at $\overline{X}^h_p$:

$$\begin{aligned}
E\left(\Delta^h_{p+1}\overline{X}|\mathcal{F}_p\right) &= bh+\frac{1}{2}(Lb)h^2+\xi_{p+1}\ ,\quad E|\xi_{p+1}|\leq Ch^3\ ,\\
E\left((\Delta^h_{p+1}\overline{X})^{i_1}(\Delta^h_{p+1}\overline{X})^{i_2}|\mathcal{F}_p\right) &= \sigma^{i_1}_j\sigma^{i_2}_j h+(b^{i_1}b^{i_2}+\frac{1}{2}\partial_{k_1}\sigma^{i_1}_j\partial_{k_2}\sigma^{i_2}_j\sigma^{k_1}_l\sigma^{k_2}_l\\
&\quad+\frac{1}{2}\partial_k b^{i_2}\sigma^{i_1}_j\sigma^k_j+\frac{1}{2}\partial_k b^{i_1}\sigma^{i_2}_j\sigma^k_j\\
&\quad+\frac{1}{2}\sigma^{i_1}_j\partial_k\sigma^{i_2}_j b^k+\frac{1}{2}\sigma^{i_2}_j\partial_k\sigma^{i_1}_j b^k\\
&\quad+\frac{1}{4}\sigma^{i_1}_j\partial_{kl}\sigma^{i_2}_j\sigma^k_n\sigma^l_n+\frac{1}{4}\sigma^{i_2}_j\partial_{kl}\sigma^{i_1}_j\sigma^k_n\sigma^l_n)h^2\\
&\quad+\xi^{i_1 i_2}_{p+1}\ ,\quad E|\xi^{i_1 i_2}_{p+1}|\leq Ch^3\ ,\\
E\left((\Delta^h_{p+1}\overline{X})^{i_1}\ldots(\Delta^h_{p+1}\overline{X})^{i_3}|\mathcal{F}_p\right) &= (b^{i_1}\sigma^{i_2}_j\sigma^{i_3}_j+b^{i_2}\sigma^{i_3}_j\sigma^{i_1}_j+b^{i_3}\sigma^{i_1}_j\sigma^{i_2}_j\\
&\quad+\frac{1}{2}\sigma^{i_2}_l\partial_k\sigma^{i_3}_l\sigma^{i_1}_j\sigma^k_j+\frac{1}{2}\sigma^{i_3}_l\partial_k\sigma^{i_2}_l\sigma^{i_1}_j\sigma^k_j\\
&\quad+\frac{1}{2}\sigma^{i_3}_l\partial_k\sigma^{i_1}_l\sigma^{i_2}_j\sigma^k_j+\frac{1}{2}\sigma^{i_1}_l\partial_k\sigma^{i_3}_l\sigma^{i_2}_j\sigma^k_j
\end{aligned}$$

$$+\frac{1}{2}\sigma_l^{i_1}\partial_k\sigma_l^{i_2}\sigma_j^{i_3}\sigma_j^k+\frac{1}{2}\sigma_l^{i_2}\partial_k\sigma_l^{i_1}\sigma_j^{i_3}\sigma_j^k)h^2$$

$$+\xi_{p+1}^{i_1i_2i_3} \quad , \quad E|\xi_{p+1}^{i_1i_2i_3}|\leq Ch^3 \ ,$$

$$E\left((\Delta_{p+1}^h\overline{X})^{i_1}\dots(\Delta_{p+1}^h\overline{X})^{i_4}|\mathcal{F}_p\right)=(\sigma_j^{i_1}\sigma_j^{i_2}\sigma_l^{i_3}\sigma_l^{i_4}+\sigma_j^{i_1}\sigma_j^{i_3}\sigma_l^{i_2}\sigma_l^{i_4}$$

$$+\sigma_j^{i_1}\sigma_j^{i_4}\sigma_l^{i_2}\sigma_l^{i_3})h^2+\xi_{p+1}^{i_1\dots i_4} \quad ,$$

$$E|\xi_{p+1}^{i_1\dots i_4}|\leq Ch^3 \ ,$$

$$E\left((\Delta_{p+1}^h\overline{X})^{i_1}\dots(\Delta_{p+1}^h\overline{X})^{i_5}|\mathcal{F}_p\right)=\xi_{p+1}^{i_1\dots i_5} \quad , \quad E|\xi_{p+1}^{i_1\dots i_5}|\leq Ch^3 \ ,$$

$$E\left((\Delta_{p+1}^h\overline{X})^{i_1}\dots(\Delta_{p+1}^h\overline{X})^{i_6}|\mathcal{F}_p\right)=\xi_{p+1}^{i_1\dots i_6} \quad , \quad E|\xi_{p+1}^{i_1\dots i_6}|\leq Ch^3 \ .$$

An interesting fact happens.

To satisfy the condition $(C3)$, a scheme does not need to involve any stochastic integral (even $\Delta_{p+1}^h W=\int_{ph}^{(p+1)h}dW(s)$!). Very simple random variables may be used, the only requirement is that the law of these random variables must have the same small number of statistics (for example, the same expectation vector, the same correlation matrix, and so on) as a certain finite family of stochastic integrals. Let us see 2 examples.

5.3 Two examples of efficient second-order schemes

Let us introduce the random variables that will be involved in our schemes.

– The sequence

$$(U_{p+1}^j,\tilde{U}_{p+1}^{kj},j,k=1,\dots,r;p\in\mathbb{N})$$

is a family of independent random variables; the (U_{p+1}^j) are i.i.d. and satisfy the following conditions:

$$E[U_{p+1}^j]=E[U_{p+1}^j]^3=E[U_{p+1}^j]^5=0 \ , \tag{18}$$

$$E[U_{p+1}^j]^2=1 \ , \tag{19}$$

$$E[U_{p+1}^j]^4=3 \ , \tag{20}$$

$$E[U_{p+1}^j]^6<+\infty \ ; \tag{21}$$

the $(\tilde{U}_{p+1}^{kj})$ are i.i.d., their common law being defined by

$$P(\tilde{U}_p^{kj}=\frac{1}{2})=P(\tilde{U}_p^{kj}=-\frac{1}{2})=\frac{1}{2} \ .$$

For example, one can choose

$$U_{p+1}^j=\frac{1}{\sqrt{h}}\Delta_{p+1}^h W^j$$

and as well one can choose for U_{p+1}^j the discrete law of mass $\frac{2}{3}$ at 0 and of mass $\frac{1}{6}$ at the points $+\sqrt{3}$ and $-\sqrt{3}$;

– the family (Z_p^{kj}) is defined by

$$Z_{p+1}^{kj} = \frac{1}{2} U_{p+1}^k U_{p+1}^j + \tilde{U}_{p+1}^{kj} \ , \quad k < j \ ,$$
$$Z_{p+1}^{kj} = \frac{1}{2} U_{p+1}^k U_{p+1}^j - \tilde{U}_{p+1}^{jk} \ , \quad k > j \ ,$$
$$Z_{p+1}^{jj} = \frac{1}{2} \left((U_{p+1}^j)^2 - 1 \right) \ .$$

Now, a being $a = \sigma\sigma^*$, we define the vectors A_j by

$$A_j = \frac{1}{2} \sum_{k,l=1}^{d} a_l^k \partial_{kl} \sigma_j \ .$$

We recall that we denote by L the infinitesimal generator of the process $(X(t))$:

$$L = \sum_{i=1}^{d} b^i(x) \partial_i + \frac{1}{2} \sum_{i,j=1}^{d} a_j^i(x) \partial_{ij} \ .$$

We consider the scheme defined by

$$\begin{aligned} \overline{X}_{p+1}^h = \overline{X}_p^h &+ \sum_{j=1}^{r} \sigma_j(\overline{X}_p^h) U_{p+1}^j \sqrt{h} + b(\overline{X}_p^h) h + \sum_{j,k=1}^{r} \partial\sigma_j(\overline{X}_p^h) \sigma_k(\overline{X}_p^h) Z_{p+1}^{kj} h \\ &+ \frac{1}{2} \sum_{j=1}^{r} \left\{ \partial b(\overline{X}_p^h) \sigma_j(\overline{X}_p^h) + \partial\sigma_j(\overline{X}_p^h) b(\overline{X}_p^h) + A_j(\overline{X}_p^h) \right\} U_{p+1}^j h^{\frac{3}{2}} \\ &+ \frac{1}{2} L b(\overline{X}_p^h) h^2 \ . \end{aligned} \tag{22}$$

It has been shown that this scheme is of second order (see Talay [53] or [54], Milshtein [38]). The Taylor formula given in Platen & Wagner [47] helps to understand how it has been constructed.

Another example of second-order scheme is the following "MCRK"scheme of Talay [54], which is of Runge-Kutta type, and therefore may be more interesting than the previous scheme from a numerical point of view, since most derivatives of the coefficients are avoided.

We define the new family

$$(V_{p+1}^j, \tilde{V}_{p+1}^{kj}, j, k = 1, \ldots, r; p \in \mathbb{N})$$

so that the family

$$(U_{p+1}^j, V_{p+1}^j, \tilde{U}_{p+1}^{kl}, \tilde{V}_{p+1}^{mn})_{j,k,l,m,n,p}$$

is a sequence of independent variables, the V_{p+1}^j's having the same distribution as the U_{p+1}^j's, and the $\tilde{V}_{p+1}^{mn}$'s having the same distribution as the $\tilde{U}_{p+1}^{kl}$'s.

Now we define (S_p^{kj}), (T_p^{kj}) and (Z_p^{kj}) by

$$\begin{aligned}
S_{p+1}^{kj} &= \frac{1}{4}(U_{p+1}^k U_{p+1}^j + \tilde{U}_{p+1}^{kj}) \ , \quad k<j \ , \\
S_{p+1}^{kj} &= \frac{1}{4}(U_{p+1}^k U_{p+1}^j - \tilde{U}_{p+1}^{jk}) \ , \quad k>j \ , \\
S_{p+1}^{jj} &= \frac{1}{4}\left((U_{p+1}^j)^2 - 1\right) \ , \\
T_{p+1}^{kj} &= \frac{1}{4}(V_{p+1}^k V_{p+1}^j + \tilde{V}_{p+1}^{kj}) \ , \quad k<j \ , \\
T_{p+1}^{kj} &= \frac{1}{4}(V_{p+1}^k V_{p+1}^j - \tilde{V}_{p+1}^{jk}) \ , \quad k>j \ , \\
T_{p+1}^{jj} &= \frac{1}{4}\left((V_{p+1}^j)^2 - 1\right) \ , \\
Z_{p+1}^{kj} &= S_{p+1}^{kj} + T_{p+1}^{kj} + \frac{1}{2}U_{p+1}^k V_{p+1}^j \ , \quad k \neq j \ , \\
Z_{p+1}^{jj} &= \frac{1}{4}\left((U_{p+1}^j + V_{p+1}^j)^2 - 2\right) \ .
\end{aligned}$$

The MCRK scheme proceeds in 2 steps.
From $\overline{X}_p^h$, one first computes:

$$\begin{aligned}
\overline{X}_{p+\frac{1}{2}}^h = \overline{X}_p^h &+ \frac{\sqrt{2}}{2}\sigma(\overline{X}_p^h)U_{p+1}^j\sqrt{h} + \frac{1}{2}b(\overline{X}_p^h)h \\
&+ \sum_{j,k=1}^r \partial\sigma_j(\overline{X}_p^h)\sigma_k(\overline{X}_p^h)S_{p+1}^{kj}h \ .
\end{aligned}$$

Then the new value $\overline{X}_{p+1}^h$ is obtained according to the formula:

$$\begin{aligned}
\overline{X}_{p+1}^h = \overline{X}_p^h &+ \left[\sigma(\overline{X}_p^h)U_{p+1} + \sigma(\overline{X}_{p+\frac{1}{2}}^h)V_{p+1} - \frac{1}{2}\sigma(\overline{X}_p^h)(U_{p+1} + V_{p+1})\right]\sqrt{2h} \\
&+ b(\overline{X}_{p+\frac{1}{2}}^h)h \\
&+ \sum_{j,k=1}^r \left[2\partial\sigma_j(\overline{X}_p^h)\sigma_k(\overline{X}_p^h)S_{p+1}^{kj}h + 2\partial\sigma_j(\overline{X}_{p+\frac{1}{2}}^h)\sigma_k(\overline{X}_{p+\frac{1}{2}}^h)T_{p+1}^{kj}h \right. \\
&\left. -\partial\sigma_j(\overline{X}_p^h)\sigma_k(\overline{X}_p^h)Z_{p+1}^{kj}h\right] \ . \qquad (23)
\end{aligned}$$

Other examples of Runge-Kutta schemes are studied for the quadratic mean error, in Rumelin [48] and in Newton [42].

5.4 Remarks

- The law of the family $(\sqrt{h}U_{p+1}^j, hS_p^{kj}, hT_p^{kj}, hZ_p^{kl})$ can be chosen in various ways. A precise formulation of families such that the above schemes are of second order is given by the notion of "Monte Carlo equivalence" in

Talay [54]. The idea is that the law of the family must only have a small number of properties; in particular, these properties imply that the expectation vector and the correlation matrix of

$$(\sqrt{h}U^j_{p+1}, hS^{kj}_{p+1})$$

and

$$\left(\Delta^h_{p+1}W^j, \int_{ph}^{(p+1)h}(W^k(s) - W^k(ph))dW^j(s)\right)$$

differ only by terms of order h^3.

- As already mentioned, the above schemes may diverge for the almost sure and pathwise approximations (even under the commutativity condition), in particular if one chooses for the U^j_{p+1}'s a discrete law.
- The choice of the number of the independent realizations to simulate in order to perform the Monte Carlo computation depends on the wished accuracy. In practice, one may first roughly estimate the maximal value of the variance of $f(\overline{X}^h_p)$ on the considered time interval, by simulating a small number N_0 of sample paths of $(\overline{X}^h_p)$; then, one chooses N according to the central-limit theorem, and finally one simulates $N - N_0$ other samples to approximate $E[f(\overline{X}^h_p)]$ with a better accuracy.
  Of course, this is the critical point of the procedure: the Monte Carlo algorithms converge slowly, and in practice N must be large, especially when the variance of $f(X(t))$ increases with t. If a particular discretization method permits to obtain a variance reduction is an extremely difficult question, deeply examined in a recent work by N. Newton [43], based on Haussmann's integral representation for functional of Itô processes (see also, for a different approach, Wagner [59]).

5.5 Numerical experiments

We now show that the numerical performances of the above different schemes may be extremely surprizing: the Euler scheme may be more efficient than second-order schemes (as for the ordinary differential equations, this will be explained by an expansion of the discretization error as a power series with respect to the discretization step).

First, let us consider an example where the second-order schemes are much better than the Euler and Milshtein schemes.

The processes $(X(t))$ and $(W(t))$ are one-dimensional, and $X(t) = atan(Z(t))$, where $(Z(t))$ is a stationary $\mathcal{N}(0,1)$ Ornstein-Uhlenbeck process solution of

$$dZ(t) = -Z(t)dt + \sqrt{2}dW(t)$$

so that $(X(t))$ solves a stochastic differential equation whose coefficients are:

$$b(x) = -\frac{1}{4}\sin(4x) - \sin(2x) \quad , \quad \sigma(x) = \sqrt{2}\cos^2(x) \ .$$

We compute $E(\cosh(1.3X(t)+2)) \sim 5.36168895$.

The figure 3 compares the evolution in time of the errors due to the Euler scheme (thin line) and the second-order scheme (22) (thick line). The figure 4 compares the evolution in time of the errors due to the Milshtein scheme (thin line) and the second-order scheme (22) (thick line).

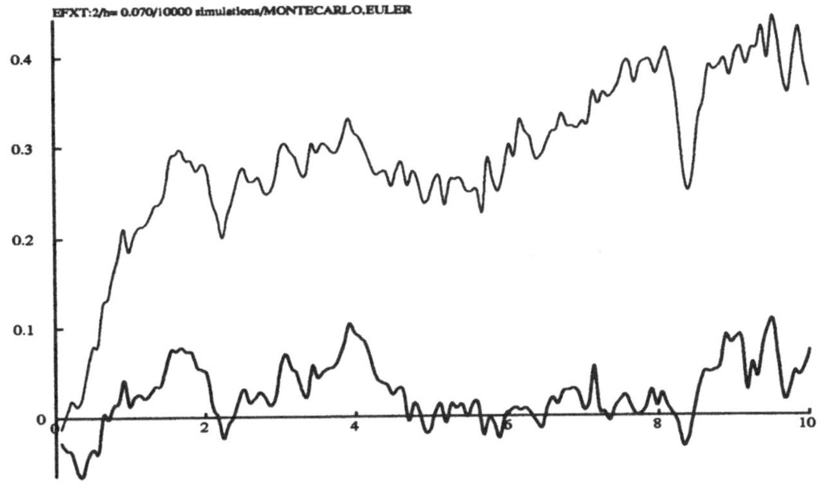

Fig. 3. Euler and second-order schemes.

But, in the next example, a strange fact occurs: the Euler scheme gives as good results as the second-order schemes, whereas the Milshtein scheme may give very bad results.

The function $b(\cdot)$ is defined by

$$b^1 = (3\sqrt{2}x^1 + 6\sqrt{2}x^2 - 2\sin(\Omega t)\Omega x^1 - 12x^1 - 6x^2)/(4(\cos(\Omega t)+2)) \ ,$$
$$b^2 = (3\sqrt{2}x^2 + 6\sqrt{2}x^1 - 2\sin(\Omega t)\Omega x^2 - 12x^2 - 6x^1)/(4(\cos(\Omega t)+2)) \ ,$$

and the matrix σ is defined by

$$\sigma_1^1 = \sin\left(\nu\left(x^1+x^2\right)\right) \ ,$$
$$\sigma_1^2 = \cos\left(\nu\left(x^1+x^2\right)\right) \ ,$$
$$\sigma_2^1 = \sin\left(\frac{\pi + 3\nu x^1 + 3\nu x^2}{3}\right) \ ,$$
$$\sigma_2^2 = \cos\left(\frac{\pi + 3\nu x^1 + 3\nu x^2}{3}\right) \ .$$

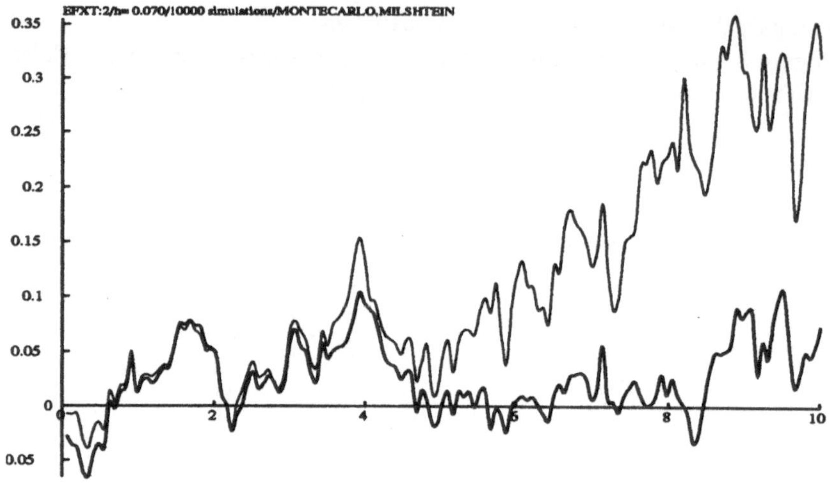

Fig. 4. Milshtein and second-order schemes.

One can check that, if the initial law is Gaussian with zero mean and a covariance matrix equal to

$$C = \begin{bmatrix} 1 & \sqrt{2}/2 \\ \sqrt{2}/2 & 1 \end{bmatrix}$$

then the law of $X(t)$ is also Gaussian with zero mean and a covariance matrix equal to $\frac{2+cos(\Omega t)}{3} C$.

Let us fix $\nu = 2$, $\Omega = 5$, and the number of simulations $N = 10,000$.

The figure 5 shows the time evolution of the true value (thick line) of $E|X^1(t)|^2$ and of the approximate value corresponding to the Euler scheme (thin line): the approximation error is weak.

The figure 6 compares the time evolution of the errors due to the Euler scheme (thin line) and the second-order scheme (22) (thick line): these errors are similar.

The figure 7 compares the time evolution of the errors due to the Euler scheme (thin line) and the Milshtein scheme (thick line). These two schemes have identical theoretical convergence rates, but, in that particular situation, the Milshtein scheme leads to absurd results.

Numerical experiments have shown that this bad behaviour of the Milshtein scheme is not avoided by an (even large) increase of the number of simulations. The only remedy is a choice of a smaller discretization step. One also observes that the Milshtein scheme has a better behaviour for small ν. Let us explain why, and present a new algorithm.

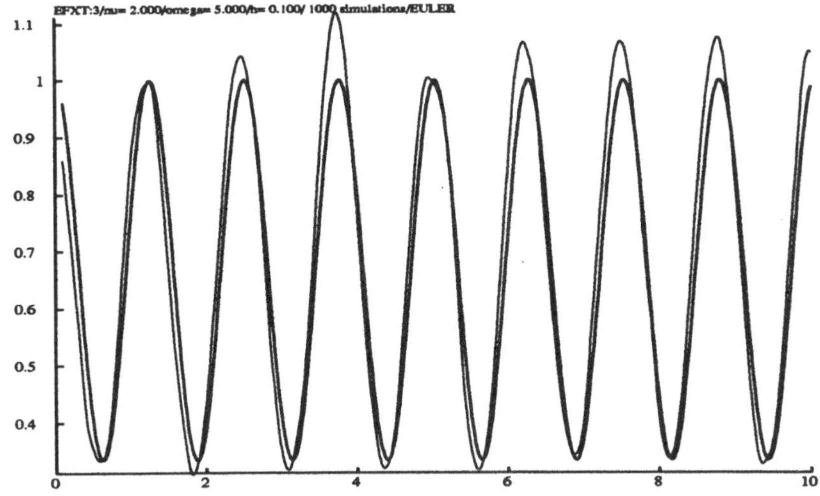

Fig. 5. Exact value and Euler scheme.

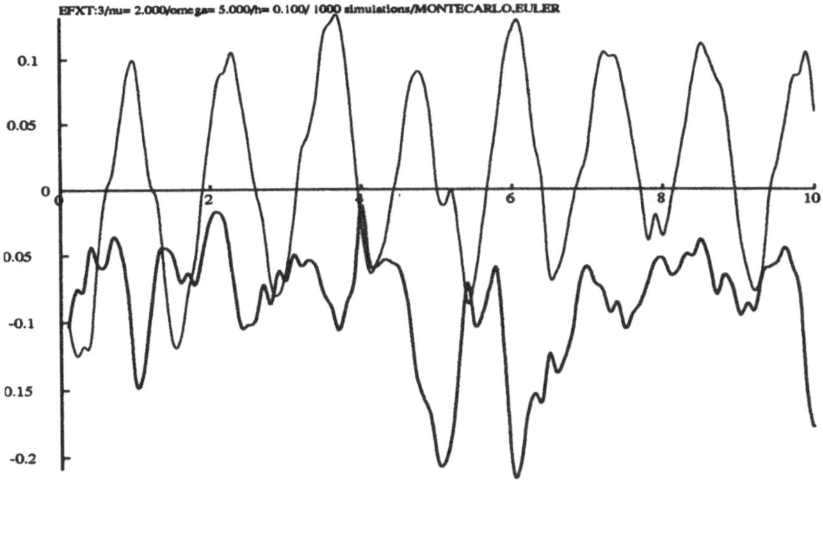

Fig. 6. Second-order and Euler schemes.

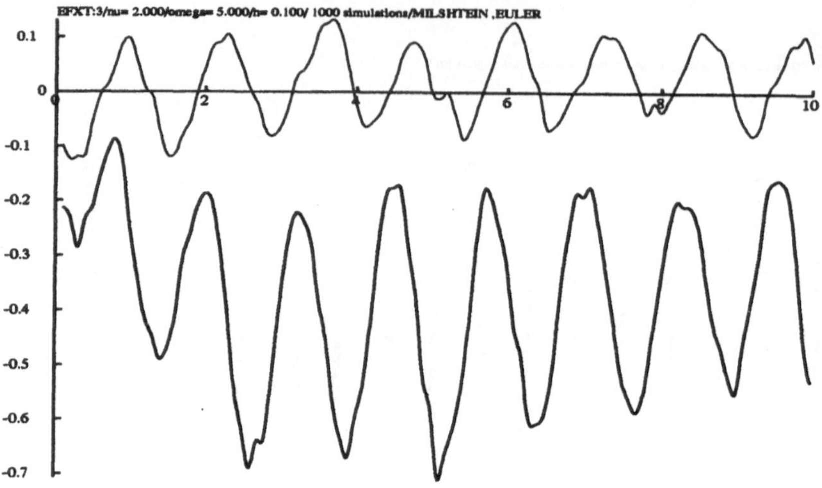

Fig. 7. Milshtein and Euler schemes.

5.6 Expansion of the error

The following theorem (Talay & Tubaro [58]) explains the numerical results of the previous section.

Theorem 4. *Let us suppose that the functions $b(\cdot)$, $\sigma(\cdot)$ and $f(\cdot)$ are C^∞; the derivatives of all orders of b and σ are supposed bounded, those of f are supposed to have a growth at most polynomial at infinity*[3]*. Then, for any step-size h of the form $\frac{T}{n}$:*

(i) For the Euler scheme, the error at time T is given by

$$Err_e(T,h) = Ef(X(T)) - Ef\left(\overline{X}_n^h\right) = -h\int_0^T E\psi_e(s,X_s)ds + \mathcal{O}(h^2)\ , \tag{24}$$

where, if $u(t,x) := Ef(X(t,x))$:

$$\begin{aligned}\psi_e(t,x) = {} & \frac{1}{2}\sum_{i,j=1}^{d} b^i(t,x)b^j(t,x)\partial_{ij}u(t,x) \\ & +\frac{1}{2}\sum_{i,j,k=1}^{d} b^i(t,x)a_k^j(t,x)\partial_{ijk}u(t,x)\end{aligned}$$

[3] See the remark thereafter for a generalization of the result when f is not a smooth function.

$$+\frac{1}{8}\sum_{i,j,k,l=1}^{d} a_j^i(t,x)a_l^k(t,x)\partial_{ijkl}u(t,x)+\frac{1}{2}\frac{\partial^2}{\partial t^2}u(t,x)$$

$$+\sum_{i=1}^{d} b^i(t,x)\frac{\partial}{\partial t}\partial_i u(t,x)+\frac{1}{2}\sum_{i,j=1}^{d} a_j^i(t,x)\frac{\partial}{\partial t}\partial_{ij}u(t,x) \quad . \quad (25)$$

(ii) The same result extends to the Milshtein scheme:

$$Err_m(T,h) = -h\int_0^T E\psi_m(s,X_s)ds + \mathcal{O}(h^2)$$

where $\psi_m(\cdot)$ is defined by

$$\psi_m(t,x) = \psi_e(t,x) + \frac{1}{4}\sum_{i_1,i_2,j,k,l} a_k^l(t,x)\partial_l\sigma_j^{i_1}(t,x)\partial_k\sigma_j^{i_2}(t,x)\partial_{i_1 i_2}u(t,x)$$

$$+\frac{1}{2}\sum_{\substack{i_1,i_2,i_3\\ j_1,j_2,k}} \sigma_{j_1}^{i_1}(t,x)\sigma_{j_2}^{i_2}(t,x)\sigma_{j_1}^{k}(t,x)\partial_k\sigma_{j_2}^{i_3}(t,x)\partial_{i_1 i_2 i_3}u(t,x). (26)$$

(iii) The same result also extends to the schemes (22) and MCRK; besides, for these 2 schemes, as well as for the Euler and Milshtein schemes, an expansion of the error as power series in h exists: for any integer n, there exist constants $C_1,\ldots,C_n$ independent of h (but depending on the scheme) such that:

$$Ef(X(T)) - Ef\left(\overline{X}_n^h\right) = C_1h + C_2h^2 + \ldots + C_nh^n + \mathcal{O}(h^{n+1}) \quad .$$

Remarks. In the preceding statement, f is supposed smooth. In Bally & Talay [5], this hypothesis is relaxed (f is only supposed measurable with a polynomial growth at infinity) under a condition on L which is slightly more than hypoellipticity; the technique of the proof uses the Malliavin calculus.

For the example illustrated by the figure 7, tedious computations show (cf [58]) that the difference between the Euler and Milshtein schemes errors behaves (for large T) like

$$\frac{\nu^2}{2}(2+\cos(\Omega T))h \quad .$$

5.7 Romberg extrapolations

An interesting consequence of the previous theorem is the justification of a Romberg extrapolation between values corresponding to two different step-sizes. More precisely, let us consider a scheme such that ($h = \frac{T}{n}$):

$$Err(T,h) = Ef(X(T)) - Ef\left(\overline{X}_n^h\right) = e_1(T)h + \mathcal{O}(h^2) \quad ,$$

and consider the following new approximation (the Romberg extrapolation):

$$Z_T^h = 2Ef(\bar{X}_{2n}^{h/2}) - Ef(\bar{X}_n^h) \ , \tag{27}$$

then:

$$Ef(X_T) - Z_T^h = \mathcal{O}(h^2) \ .$$

That is, it is possible to get a result of precision of order h^2 from results given by a first-order scheme.

This procedure seems to be very robust w.r.t the choice of the discretization step. Let us give an illustration of this remark: let us compare the time evolution of the errors due to the extrapolation based on the Milshtein scheme with the 2 step-sizes $h = 0.05$ and $h = 0.1$ (thick line), and to the Milshtein scheme itself with the step-size $h = 0.05$ (thin line): the extrapolation has extremely improved the accuracy (see figure 8).

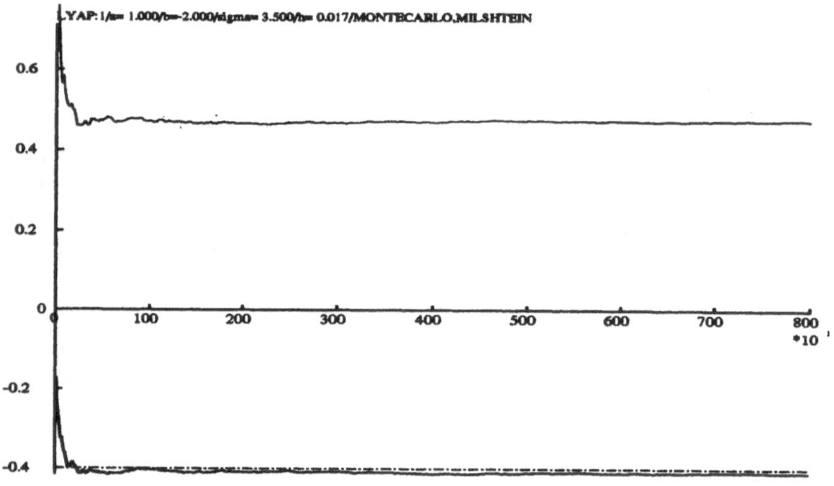

Fig. 8. Romberg extrapolation.

5.8 High order schemes

Of course, from a theoretical point of view it is possible to construct schemes with a convergence rate of an arbitrary order, or to get a precision of an arbitrary order by linearly combining results corresponding to a given scheme (Euler, MCRK, etc) and an appropriate number of different step-sizes.

We are not sure that high order procedures can be useful. Generally, it is impossible to choose a step-size too large without losing too much information on the law of the increments of $(X(t))$. For small values of h (0.05 for example), it is difficult to see the gain in accuracy due to a 3rd order method, compared to a 2nd order one: the error due to the approximation of $Ef(\overline{X}_p^h)$ by the average (16) cannot be reduced enough (one cannot choose N so large as we would like!). Besides, the coefficients C_i in the expansion of the error depend on the successive derivatives of $b(\cdot)$, $\sigma(\cdot)$ and $f(\cdot)$. Often, C_i is rapidly increasing w.r.t. i.

6 Computation of Lyapunov exponents, study of stability

A summary of the theory of Lyapunov exponents of dynamical stochastic systems can be found in Arnold [2]; for bilinear systems, complements can be found in Pardoux & Talay [45].

To compute the upper Lyapunov exponent of 2-dimensional bilinear systems, W. Wedig [60] proposes an ingenious deterministic algorithm.

This section is a summary of results concerning the approximation of Lyapunov exponents of (non necessarily 2-dimensional) bilinear stochastic differential systems and the application to a helicopter blade problem, based on simulations and presented in Talay [56].

To be complete, we mention that an algorithm of computation of the Lyapunov spectrum for nonlinear systems in $\mathbb{R}^d$ or on compact manifolds has been developed, and its convergence rate given (see Grorud & Talay [27]).

6.1 An engineering stability problem

Let us study the stability of the motion of the movement of a rotor blade with 2 freedom degrees in terms of various physical parameters: velocity of the helicopter, geometric characteristics of the blade, statistical characteristics of the process modelizing the turbulency around the blade.

In first approximation, the stability of the movement of the blade is equivalent to the stability of the solution of a linearized ordinary differential equation in $\mathbb{R}^4$

$$\frac{dX(t)}{dt} = A(t)X(t) + F(t) \ ,$$

where the matrix-valued function $A(t)$ and the vector-valued function $F(t)$ are periodic of same period (the period of rotation of the blade).

When one takes into account the turbulent flow around the blade, one may consider the following linearized model

$$\frac{dX(t)}{dt} = A(t)X(t) + F(t) + [B(t)X(t) + G(t)]\sigma(t)\xi^{\varepsilon}(t) \ , \tag{28}$$

where $b(t)$ (resp. $G(t)$) has the same property as $A(t)$ (resp. $F(t)$), and $(\xi^{\varepsilon}(t))$ is a one-dimensional noise. The intensity of the noise, $\sigma(t)$ is also a periodic function of the azimuth angle Ωt, where Ω is the angular velocity of the blade.

All the coefficients $A(t)$, $b(t)$, $F(t)$, $G(t)$ are explicitly known in terms of different physical parameters of the blade.

Here the "stability" we are interested in, is the following: the system is stable when it admits a unique periodic in law solution $(Y(t))$ and when, for each initial deterministic condition, the corresponding process $(X(t))$ satisfies:

$$\lim_{t\to+\infty} |Y(t) - X(t)| = 0 \ \ a.s. \tag{29}$$

First we will consider the white-noise case. The system (28) becomes:

$$dX(t) = [A(t)X(t) + F(t)]dt + [B(t)X(t) + G(t)]\sigma(t) \circ dW(t) \ , \tag{30}$$

where $(W(t))$ is a standard one-dimensional Wiener process.

One can show (cf Pardoux[44]):

Theorem 5. *Let us suppose there exists $\lambda_0 < 0$ such that the solution of the system*

$$dX(t) = A(t)X(t)dt + B(t)X(t)\sigma(t) \circ dW(t) \tag{31}$$

satisfies, for any deterministic initial condition x :

$$\limsup_{t\to+\infty} \frac{1}{t} \log |X(t)| \leq \lambda_0 \ \ a.s. \tag{32}$$

Then the system (30) is stable in the sense (29).

In her thesis, M. Pignol ([46]) has proven the existence of the Lyapunov exponent $\lim_{t\to+\infty} \frac{1}{t} \log |X(t)|$ for the blade system. Let us see how we can compute it (in order to know its sign!).

6.2 Numerical tests

Let us consider an example of Baxendale for which there exists an explicit formula giving the Lyapunov exponent. More precisely, let us consider a one-dimensional Wiener process $(W(t)$ and the system:

$$dX(t) = AX(t)dt + \sigma BX(t) \circ dW(t) \ , \tag{33}$$

with

$$A = \begin{bmatrix} a & 0 \\ 0 & b \end{bmatrix} \ , \ B = \begin{bmatrix} 0 & -1 \\ 1 & 0 \end{bmatrix} \ .$$

Then:

$$\lambda = \frac{1}{2}(a+b) + \frac{1}{2}(a-b)\frac{\int_0^{2\pi} \cos(2\theta)\exp(\frac{a-b}{2\sigma^2}\cos(2\theta))d\theta}{\int_0^{2\pi} \exp(\frac{a-b}{2\sigma^2}\cos(2\theta))d\theta} \ .$$

We discretize (33). For each Markov chain defined by one of our schemes, and for any h small enough, one can show that there exists $\overline{\lambda}^h$ such that, for any deterministic initial condition:

$$\overline{\lambda}^h = a.s \lim_{p\to+\infty} \frac{1}{ph} \log |\overline{X}_p^h| \ .$$

We want to compare $\overline{\lambda}^h$ and λ.

It is important to note that we cannot use the formula $\overline{\lambda}^h \sim \frac{1}{ph} \log |\overline{X}_p^h|$ in practice, because it leads to numerical instabilities, the process $(|\overline{X}_p^h|)$ decreasing to 0 or increasing to infinity exponentially fast. This is avoided by a projection at each step technique, described in the next section.

We have tested the Milshtein method and the second-order method (22).

By example, let us choose $a = 1$, $b = -2$, $\sigma = 3.5$. Then an accurate numerical computation gives $\lambda = -0.4$. In this example, the second-order schemes are extremely accurate, whereas the Euler scheme and the Milshtein scheme lead to completely wrong results, as illustrated by the 2 figures below, which represent the evolution (in terms of time ph) of our estimator of $\overline{\lambda}^h$.

First, we compare the scheme (22) (thick line) and the Euler scheme (thin line): figure 9.

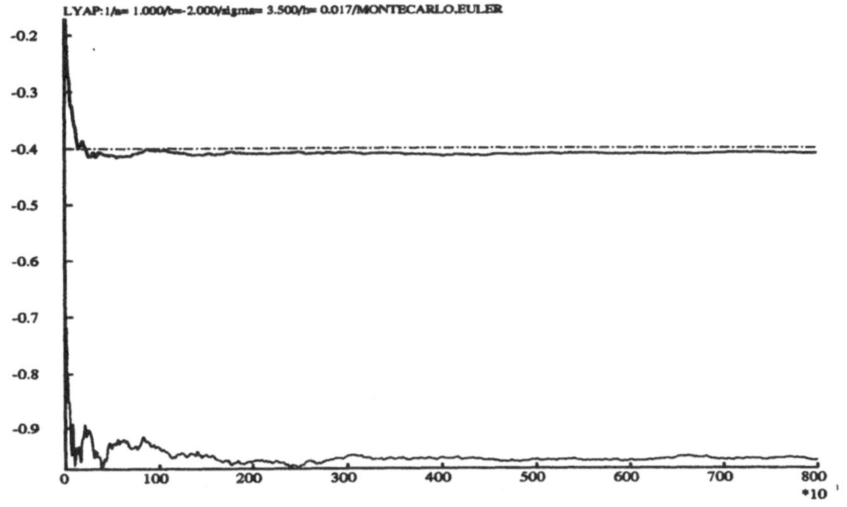

Fig. 9. Euler and second-order schemes.

Next, we compare the scheme (22) (thick line) and the Milshtein scheme (thin line): figure 10.

6.3 Algorithm for the helicopter problem

In the deterministic case (corresponding to $\sigma(t) \equiv 0$), only the Runge-Kutta methods of order larger than 4 have given good results (because of the numerical instability of the system, due to the large coefficients of the matrix $A(t)$ and their very short period).

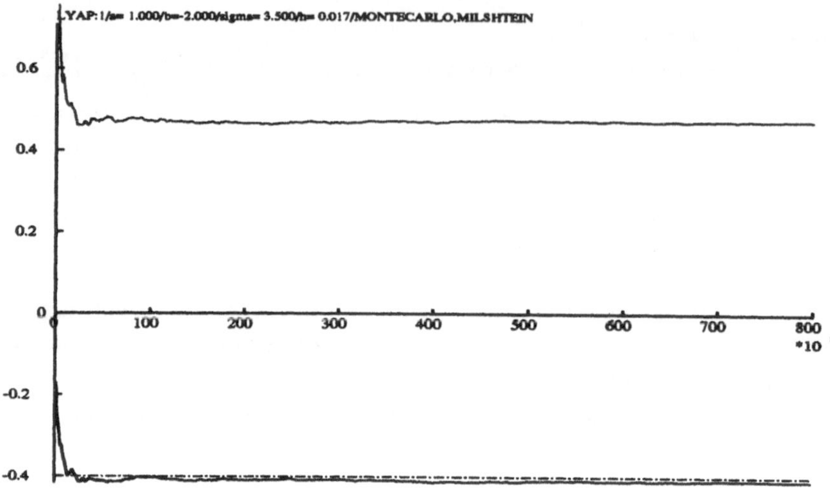

Fig. 10. Milshtein and second-order schemes.

Moreover, the discretization step had to be choosen smaller than 10^{-4}.

For the stochastic case, the use of a second-order scheme has also been necessary. We have observed the need of a large final integration time, therefore too small steps would have led to too large CPU times. Morevover, the second-order schemes have been less sensitive than the first-order ones to the above underlined strong instability of the system.

Besides, the above study of the deterministic case shows the necessity to improve the scheme (22), in order that it reduces to the Runge-Kutta scheme of order 4 when the intensity of the noise is nought.

Finally, our algorithm has been the following:

1. One chooses an initial condition on the unit sphere $\mathcal{S}^3$;
2. At step $(p+1)$, one proceeds in two stages:
   - One applies to a single step the Runge-Kutta method of order 4 in order to integrate the system

$$y(0) = \overline{X}_p^h$$
$$\dot{y}(t) = \tilde{A}(ph+t)y(t)$$

   (the presence of $\tilde{A} = A + \frac{1}{2}B^2$ is due to the discretization of the system written in the Itô sense);
   - Then one computes:

$$\overline{X}_{p+1}^{h\,'} = y(h) + \left\{ \sigma(ph)B\Delta_{p+1}^h W + \frac{1}{2}\sigma(ph)^2 B^2((\Delta_{p+1}^h W)^2 - h) \right.$$

$$+ \frac{1}{2}(\sigma(ph)(\tilde{A}B + B\tilde{A} + B')$$
$$+\sigma'(ph)B)h\Delta_{p+1}^h W\}\overline{X}_p^h \ , \tag{34}$$

where we have used the following notation:
$(\Delta_{p+1}^h W)$:= sequence of mutually independent Gaussian random variables $\mathcal{N}(0, \sqrt{h})$;
B' : derivative of the matrix $B(\cdot)$;
all the matrices are computed at time $t = ph$;

3. One computes the new approximate value λ_{p+1} from the previous one, λ_p by

$$\lambda_p \left(1 - \frac{1}{p+1}\right) + \frac{\log(|\overline{X}_{p+1}^h|)}{(p+1)h} \ ; \tag{35}$$

4. One projects $\overline{X}_{p+1}^h$ on the unit sphere.

The theorem 8 below shows that this algorithm is of second order, in the sense that, for any deterministic initial condition x:

$$\left|\lambda - \lim_{p \longrightarrow +\infty} \frac{1}{ph} \log|\overline{X}_p^h(x)|\right| = O(h^2) \ .$$

For the given models of blades, the deterministic system was extremely stable for admissible velocities.

Let us suppose that $\sigma(t)$ is a constant function $\sigma(t) \equiv \sigma_0$ (i.e the effects of the turbulency are independent of the azimuth angle).

For a velocity equal to $100m/s$, we obtain the figure 11.

The destabilization of the system could occur only for intensities of the noise larger than 0.3; such intensities are not realistic for the turbulent winds around the blade.

Moreover, let us consider a more precise modelization of the noise. We suppose that its intensity is a periodic function of the azimuth angle $\Psi = \Omega t$, reaching its maximum for $\Psi = \frac{3\pi}{4}$, and defined by the function:

$$\sigma_0 \exp(\delta(cos(0.75\pi - 0.5\Omega t)^2 - 1)) \ .$$

We have observed a strong dependency of the Lyapunov exponent upon δ but in that case also the instability could not appear for realistic intensities of the noise.

6.4 Choice of the integration time

This choice is much more complex that the choice of the number of simulations to perform a Monte Carlo approximation: the error due to the fact that one integrates during a finite time asymptotically has a Gaussian law, but the variance of this law is given by the solution of a P.D.E., and moreover this P.D.E. depends on the Lyapunov exponent that one wants to compute!

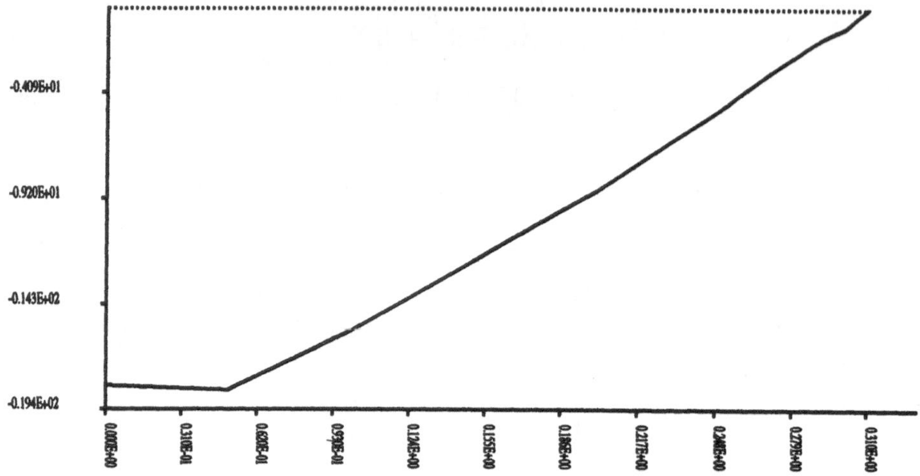

Fig. 11. Variations of the Lyapunov exponent in terms of σ_0.

More precisely, considering again the case of matrices A and B independent of t for the sake of simplicity, let $\mathcal{L}$ define the infinitesimal generator of the process $(s(t))$ defined by $s(t) = \frac{X(t)}{|X(t)|}$; under technical assumptions on the matrices A and B (satisfied by the blade system and described below), this process has a unique invariant probability measure μ on the projective space P^{d-1} (i.e the set obtained by identifying s and $-s$ on the sphere).

We have the following central-limit theorem (Bhattacharya [8]):

Theorem 6. *Let the function Q be defined by*

$$Q(s) := (As, s) + \frac{1}{2}\left[(B^2 s, s) + |Bs|^2 - 2(Bs, s)^2\right] \quad .$$

For t tending to $+\infty$:

$$\frac{1}{\sqrt{t}} \int_0^t (Q(s(\theta)) - \lambda) d\theta \to \mathcal{N}(0, \nu^2) \ \ in \ \ distribution \quad ,$$

where the constant ν^2 depends only on the coefficients of the system and is given by ($< \cdot, \cdot >$ denoting the inner product in $\mathcal{L}^2(P^{d-1}, \mu)$):

$$\nu^2 = -2 < Q - \lambda, \mathcal{L}^{-1}(Q - \lambda) > \quad .$$

An estimation of the integration time can be got by a first estimation of λ, and by a numerical resolution of the Poisson P.D.E. on the projective space

$$\mathcal{L}u = Q - \lambda \ .$$

Of course, this phase of the algorithm is very critical. Further researches in that direction are necessary.

6.5 Algorithm for the wideband noise case

Let us consider the system

$$\frac{dX(t)}{dt} = A(t)X(t) + B(t)X(t)\sigma(t)\xi^{\varepsilon}(t) \ , \tag{36}$$

with a wide-band noise of the form

$$\xi^{\varepsilon}(t) = \frac{1}{\sqrt{\varepsilon}} Z(\frac{t}{\varepsilon}) \ ,$$

where $(Z(t))$ is a stationary$\mathcal{N}(0,1)$ Ornstein-Uhlenbeck process.

The above system is an ordinary differential system, so that the pathwise simulation of the solution can be achieved using, by example, the Euler scheme:

$$\overline{X}^h_{p+1} = \overline{X}^h_p + (A(ph)\overline{X}^h_p + \sigma(ph)B(ph)\xi^{\varepsilon}(ph))\overline{X}^h_p h \ .$$

Pardoux [44] and Kushner [33] have shown that the Lyapunov exponent of (36) converges to the Lyapunov exponent of (31).

But it appears that, even for small h, the Lyapunov exponent of that dicrete-time process defined by the Euler scheme does not converge, when ε goes to 0, towards the Lyapunov exponent of the system (31).

One reason is that the process $\{\xi^{\varepsilon}(ph)\}$ does not converge in law, so that the scheme must rather involve the sequence

$$\Delta^h_{p+1}\xi = \int_{ph}^{(p+1)h} \xi^{\varepsilon}(s)ds$$

which converges in law to the sequence $(\Delta^h_{p+1}W)$.

But one has to be careful: the new scheme

$$\overline{X}^h_{p+1} = \overline{X}^h_p + (A(ph)h + \sigma(ph)B(ph)\Delta^h_{p+1}\xi)\overline{X}^h_p$$

does not converge to a discretization scheme of (31) $(A \neq \tilde{A})$.

Finally, we introduce a convenient second-order scheme, similar to (34).

First, one applies to a single step the Runge-Kutta method of order 4 in order to integrate the system

$$y(0) = \overline{X}^h_p$$
$$\dot{y}(t) = A(ph+t)y(t)$$

and then

$$\overline{X}^h_{p+1} = y(h) + \Big\{\sigma(ph)B\Delta^h_{p+1}\xi + \frac{1}{2}\sigma(ph)^2B^2(\Delta^h_{p+1}\xi)^2$$
$$+ \frac{1}{2}(\sigma(ph)(AB + BA + B') + \sigma'(ph)B)h\Delta^h_{p+1}\xi\Big\}\overline{X}^h_p \ . \tag{37}$$

It is interesting to note that the limit of the above scheme when ε goes to 0 is not the scheme (34), the difference including only terms of order $h\Delta^h_{p+1}W$

and h^2. We have not succeeded to build a second-order scheme of the wideband system converging to a second-order scheme of the white noise system.

Let us describe our simulation of the integrals $\Delta_{p+1}^h \xi$.

Let $(V(t))$ is a Wiener process independent of $(W(t))$ such that:

$$dZ(t) = -Z(t)dt + \sqrt{2}dV(t) \ .$$

Then we have the formula:

$$\int_{ph}^{(p+1)h} \xi^\varepsilon(s)ds = \sqrt{\varepsilon}\left[\frac{1-e^{-\frac{2h}{\varepsilon}}}{2}\xi^\varepsilon(ph) + \Delta_{p+1}^{\frac{1}{\varepsilon}} V - e^{-\frac{2(p+1)h}{\varepsilon}} \int_{ph}^{(p+1)h} e^{2s} dV_s\right] \ .$$

Therefore, it is possible to simulate the vector

$$(\xi^\varepsilon(ph), \Delta_{p+1}^h \xi)$$

by the simulation of the Gaussian vector

$$(\Delta_{p+1}^{\frac{1}{\varepsilon}} V, \int_{ph}^{(p+1)h} e^s dV_s, \int_{ph}^{(p+1)h} e^{2s} dV_s) \ .$$

For our models, we have observed that the coloration of the noise tended to stabilize the system (the limit case being the case of very large ε, equivalent to the deterministic case).

6.6 Remarks

The models for the blade and the noise were simplified; in particular, only physical experiments during a flight could permit to improve the modelling of the noise, and overall a more realistic model should be nonlinear.

In this simplified context, the conclusion is that the turbulency around the blade has small effects on the stability of the blade.

6.7 Convergence rate

Let us consider a bilinear system

$$dX(t) = AX(t)dt + \sum_{i=1}^{r} B_i X(t) \circ dW_i(t) \ . \tag{38}$$

Let $(s(t))$ be the process on the projective space of $\mathbb{R}^d$, P^{d-1}, defined as the equivalence class of $\frac{X(t)}{|X(t)|}$ with respect to the equivalence relation: $x \sim y$ iff $x = -y$ or $x = y$.

The process $(s(t))$ is the solution of the following Stratonovich stochastic differential equation, describing a diffusion process in P^{d-1}:

$$ds(t) = g(A, s(t))dt + \sum_{i=1}^{r} g(B_i, s(t)) \circ dW_i(t) \ , \tag{39}$$

where

$$g(C,s) := Cs - (Cs,s)s \ .$$

Now, let us introduce the Lie algebra $\Lambda = L.A.\{g(A,\cdot), g(B_1,\cdot), \ldots, g(B_k,\cdot)\}$, i.e. the smallest vector space of differential operators containing the operators

$$\sum_i g^i(A,\cdot)\partial_i \ , \quad \sum_i g^i(B_j,\cdot)\partial_i \quad (j = 1,\ldots r)$$

and closed under the brackett operation $[P_1, P_2] = P_1 \circ P_2 - P_2 \circ P_1$.

For s in the projective space P^{d-1}, $\Lambda(s)$ denotes the space obtained by considering all the elements of Λ with all the coefficients of the operators frozen at their value in s.

In Arnold, Oeljeklaus and Pardoux [3] is proven the following theorem:

Theorem 7. *Let us suppose:*

(H) $dim\Lambda(s) = d - 1 \ , \ \forall s \in P^{d-1}$.

Then the process $(s(t))$ on P^{d-1} has a unique invariant probability measure μ, and there exists a real number λ such that, for any x in $\mathbb{R}^d - \{0\}$:

$$\lambda = \lim_{t \longrightarrow +\infty} \frac{1}{t} \log |X(t,x)| \ , \quad a.s.$$

In Talay [56], is proven the

Theorem 8. *Let us suppose that the system (39) satisfies the hypothesis:*

(H0) *The infinitesimal generator $\mathcal{L}$ of the process $(s(t))$ on S^{d-1} is uniformly elliptic, i.e there exists a strictly positive constant α such that, for any x in S^{d-1} and any vector ξ in the tangent space $T_{S^{d-1}}(x)$:*

$$\sum_{i=1}^{r} (h(B_i,x),\xi)^2 \geq \alpha |\xi|^2 \ .$$

(HU) **(i)** *The (U^j_{p+1})'s are i.i.d., and the following conditions on the moments are fulfilled:*

$$\begin{aligned} E[U^j_{p+1}] = E[U^j_{p+1}]^3 = E[U^j_{p+1}]^5 &= 0 \ , \\ E[U^j_{p+1}]^2 &= 1 \ , \\ E[U^j_{p+1}]^4 &= 3 \ , \\ E[U^j_{p+1}]^n &< +\infty \ , \ \forall n > 5 \ . \end{aligned} \tag{40}$$

(ii) *The common law of the (U^j_{p+1})'s has a continuous density w.r.t. the Lebesgue measure; the support of this density contains an open interval including 0 and is compact.*

Let $(\overline{X}_p^h, p \in \mathbb{N})$ be defined by the Euler scheme, the Milshtein scheme or the scheme (22).

Then, if λ is the upper Lyapunov exponent of (38):

(i) *for $(\overline{X}_p^h)$ defined by the Euler or Milshtein scheme, $|\lambda - \overline{\lambda}^h| = O(h)$;*

(ii) *for $(\overline{X}_p^h)$ defined by the scheme (22), $|\lambda - \overline{\lambda}^h| = O(h^2)$.*

Remark: the hypothesis (HU) is not limitative from a practical point of view, but (ii) was unnecessary to obtain the results concerning the Monte Carlo type approximation.

7 Computation of the invariant law

7.1 Position of the problem

We again consider the general system (4).

Under the hypotheses below, the system has a unique invariant measure μ, which has a smooth density, p. One way to compute $\int f(x)d\mu(x)$ for a given function f is to solve the stationary Fokker-Plank equation $L^*p = 0$, where L^* is the adjoint of the infinitesimal generator of the process $(X(t))$.

This stationary Fokker-Plank equation is a P.D.E., and its numerical resolution could be extremely difficult or impossible, especially when the dimension of the state-space, d, is large, or when L is degenerate (remember the Remark of the section 5.1).

In [26], Gerardi, Marchetti & Rosa propose to approximate $(X(t))$ by a sequence of ergodic pure jump processes which converges in law.

Since for any μ-integrable function f we have:

$$\int f(x)d\mu(x) = \lim_{t\to+\infty} \frac{1}{t}\int_0^t f(X(s))ds \ ,$$

we propose to simulate one long trajectory of a process $(\overline{X}_p^h)$, and to approximate $\int f(x)d\mu(x)$ by

$$\frac{1}{N}\sum_{p=1}^{N} f(\overline{X}_p^h) \ .$$

As in the preceding section, the critical point is the choice of N: again the random variable

$$\frac{1}{\sqrt{t}}\int_0^t [f(X(s)) - \int f(\theta)d\mu(\theta)]ds$$

is asymptotically Gaussian, but the variance of the limit law depends on the solution of a P.D.E. which itself depends on the unknown $\int f(x)d\mu(x)$.

At the present time, we do not know what could be a good procedure to estimate this variance.

7.2 Second-order schemes

As for the approximation of Lyapunov exponents, these schemes seem to have a better long-time behaviour than simpler ones.

For example, one can show (cf Talay [55]):

Theorem 9. *Suppose that the hypotheses (H1), (H2), (H3) hold:*

(H1) *the functions b, σ are of class C^∞ with bounded derivatives of any order; the function σ is bounded;*

(H2) *the operator L is uniformly elliptic: there exists a positive constant α such that*

$$\forall x, \xi \in \mathbb{R}^d \ , \ \sum_{i,j} a^i_j(\xi) x^i x^j \geq \alpha |x|^2 \ ;$$

(H3) *there exists a strictly positive constant β and a compact set K such that:*

$$\forall x \in \mathbb{R}^d - K \ , \ x \cdot b(x) \leq -\beta |x|^2 \ .$$

Consider the scheme (22) and the MCRK scheme (23), with the law of the involved random variables defined as in the section (5.3), with the additional hypothesis: the law of the U^j_{p+1}'s and of the U^j_{p+1}'s has a continuous density w.r.t. the Lebesgue measure.

The schemes (22) and MCRK define ergodic Markov chains and for any function f of $\mathbb{R}^d$ of class C^∞, having the property that f, as well as all its derivatives, have an at most polynomial growth at infinity:

$$\forall x \in \mathbb{R}^d \ : \ \lim_{N \to \infty} \frac{1}{N} \sum_{p=1}^{N} f(\overline{X}^h_p(x)) = \int f(\theta) d\mu(\theta) + O(h^2) \ , \ a.s.$$

For the Euler and the Milshtein schemes, the convergence rate is of order h.

We underline that now we do not require anymore that the law of the U^j_{p+1}'s has a compact support, therefore the Gaussian law of the increments of the Wiener process is allowed.

Again, another good (and probably usually better) algorithm is to perform a Romberg extrapolation (see the subsection 5.7); indeed, as for the problem of computation of $Ef(X(t))$ on a finite time interval, the errors due to the different schemes introduced above can be expanded as power series in the discretization step h (see Talay & Tubaro [58]):

Theorem 10. *Suppose that the hypotheses of the preceding theorem hold. Let ψ_e and ψ_m be defined by (25) and (26) respectively and set*

$$\lambda_e := \int_0^{+\infty} \int_{\mathbb{R}^d} \phi_e(t, y) \mu(dy) dt \ ,$$

$$\lambda_m := \int_0^{+\infty} \int_{\mathbb{R}^d} \phi_m(t, y) \mu(dy) dt \ .$$

Then the Euler scheme error satisfies: for any deterministic initial condition $\xi = \overline{X}_0^h$,

$$\int f(y)\mu(dy) - a.s. \lim_{N\to+\infty} \frac{1}{N}\sum_{p=1}^{N} f(\overline{X}_p^h(\xi)) = -\lambda_e h + O(h^2) \ . \qquad (41)$$

For the Milshtein scheme, an analogous result can be written, substituting λ_m *to* λ_e.

7.3 Numerical experiments

The example is the same as in the section (5.5), with $\Omega = 0$, so that the invariant law is Gaussian of zero mean and of covariance matrix C.

We have observed a strong numerical instability of all the schemes (the results are very different when $h = 0.01$ and $h = 0.015$), but the second-order schemes lead to good results.

Below (figure 12), we compare $\frac{1}{N}\sum_{p=1}^{N} f(\overline{X}_p^h(x))$ for the Milshtein scheme (thin line) and the scheme (22) (thick line), for

$$f(x^1, x^2) = |x^1|^4$$

(the correct value is 3.0; in x-axis : Nh).

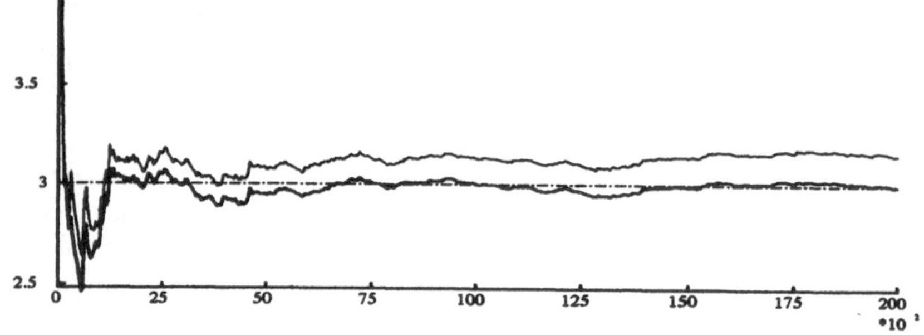

Fig. 12. Milshtein and second-order schemes.

Remark : for the two schemes, $\frac{1}{N}\sum_{p=1}^{N} f(\overline{X}_p^h(x))$ converges to $\int f(\theta)d\mu^h(\theta)$, where μ^h is the invariant law of the Markov chain $(\overline{X}_p^h)$.

8 PRESTO : a generator of Fortran programs

PRESTO is a system which generates Fortran programs solving Stochastic Differential Systems.

The user describes his problem using a bitmap environment; then PRESTO treats the data, performs the transformation Stratonovich/Itô of the system if necessary, uses its knowledge base in order to decide what particular scheme can be used in the context described by the user, what random variables must be involved and how they must be simulated, and finally writes a commented complete Fortran program ready to be run.

Internally, the analytical expressions of the coefficients of the Itô system and of the chosen scheme are computed by procedures written in a Computer Algebra Programming System Language (Reduce in the first version, Maple in the current one).

A complete description can be found in Talay [57].

9 Conclusion

We have proposed some discretization methods of stochastic differential systems, which seem efficient when one wants to simulate trajectories of diffusion processes, or when one wants to compute certain quantities depending on the law of a diffusion process, by techniques involving simulations.

A lot of open problems still remain, some of them are being studied: as examples, we could quote the discretization of reflected diffusions processes (which are studied in extremely recent interesting papers, see e.g Calzolari & Costantini & Marchetti [15] on the simulation of obliquely reflecting Brownian motions, Liu [35] who uses a penalization technique, Lépingle [34] for reflections at the boundary of a half–space or an orthant, and in a more abstract way Slominski [50]), the approximation of stopped diffusions and the numerical approximation of elliptic P.D.E.'s in bounded domains (see Milshtein [39]), the estimation of the necessary simulation time corresponding to a given accuracy for ergodic computations (computation of the stationary law, Lyapunov exponents, ...), etc.

From the numerical implementation point of view, Bouleau [12] and Ben Alaya [6] have just opened new perspectives by the use of the shift method to generate Brownian paths with few calls to a random number generator, and by their mathematical analysis of their algorithm (which is a nice application of the ergodic theory) in particular when the objective is to compute expectations of functionals of diffusion processes.

All these works show that the numerical analysis of diffusion processes is a field which is developing so fast that a new review paper will be necessary in a next future.

References

1. L. ARNOLD. *Stochastic Differential Equations.* Wiley, New-York, 1974.

2. L. ARNOLD. Lyapunov exponents of nonlinear stochastic systems. In *Nonlinear Stochastic Dynamic Engineering Systems (G.I. Schueller & F. Ziegler (Eds.)*. Springer-Verlag, 1988.
3. L. ARNOLD, E. OELJEKLAUS, E. PARDOUX. Almost sure and moment stability for linear Itô equations. In *Lyapunov Exponents, L. Arnold & V. Wihstutz (Eds.)*, volume 1186 of *Lecture Notes in Mathematics*. Springer, 1986.
4. L. ARNOLD, P. KLOEDEN. Discretization of a random system near a hyperbolic point. Report 302, Institut für Dynamische Systeme, Universität Bremen, 1994.
5. V. BALLY, D. TALAY. The law of the Euler scheme for stochastic differential equations (I) : convergence rate of the distribution function. *Probability Theory and Related Fields*, to appear.
6. M. BEN ALAYA. *Les Théorèmes Ergodiques en Simulation*. PhD thesis, Ecole Nationale des Ponts et Chaussées, 1993.
7. P. BERNARD, D. TALAY, L.TUBARO. Rate of convergence of a stochastic particle method for the Kolmogorov equation with variable coefficients. *Mathematics of Computation*, 63, 1994.
8. R.N. BHATTACHARYA. On the functional central limit theorem and the law of the iterared logarithm for Markov processes. *Z. Wahrscheinlichkeitstheorie verw. Gebiete*, 60:185–201, 1982.
9. M. BOSSY. *Vitesse de Convergence d'Algorithmes Particulaires Stochastiques et Application à l'Equation de Burgers*. PhD thesis, Université de Provence, 1995.
10. M. BOSSY, D. TALAY. A stochastic particle method for McKean-Vlasov PDE's and the Burgers equation. *Math. of Computation*, to appear.
11. M. BOSSY, D. TALAY. Convergence rate for the approximation of the limit law of weakly interacting particles : application to the Burgers equation. Rapport de Recherche 2410, Inria, novembre 1994. Submitted for publication.
12. N. BOULEAU. On numerical integration by the shift and application to Wiener space. *Acta Applicandae Mathematicae*, 3(25):201–220, 1991.
13. N. BOULEAU, D. LEPINGLE. *Numerical Methods for Stochastic Processes*. J. Wiley, 1993.
14. N. BOULEAU, D. TALAY (Eds.). *Probabilités Numériques*. Collection Didactique. INRIA, 1992.
15. A. CALZOLARI, C. COSTANTINI, F. MARCHETTI. A confidence interval for Monte Carlo methods with an application to simulation of obliquely reflecting Brownian motion. *Stochastic Processes and their Applications*, 29(2), 1988.
16. F. CASTELL, J.G. GAINES. The ordinary differential equation approach to asymptotically efficient schemes for solutions of stochastic differential equations. To appear in "Annales de l'Institut H. Poincaré".
17. J.M.C. CLARK. An efficient approximation for a class of stochastic differential equations. In W. Fleming et L. Gorostiza, editor, *Advances in Filtering and Optimal Stochastic Control*, volume 42 of *Lecture Notes in Control and Information Sciences*. Proceedings of the IFIP Working Conference, Cocoyoc, Mexico, 1982, Springer-Verlag, 1982.
18. J.M.C. CLARK, R.J. CAMERON. The maximum rate of convergence of discrete approximations for stochastic differential equations. In B. Grigelionis, editor, *Stochastic Differential Systems – Filtering and Control*, volume 25 of *Lecture Notes in Control and Information Sciences*. Proceedings of the IFIP Working Conference, Vilnius, Lithuanie, 1978, Springer-Verlag, 1980.
19. D. FLORENS-ZMIROU. Estimation de la variance d'une diffusion à partir d'une observation discrétisée. *Note au Compte-Rendu de l'Académie des Sciences*,

309(I):195–200, 1989.

20. H. DOSS. Liens entre équations différentielles stochastiques et ordinaires. *Ann. Inst. H. Poincaré*, XIII(2):99–125, 1977.
21. O. FAURE. Simulation du Mouvement Brownien et des Diffusions. *Thèse de Doctorat, Ecole des Ponts et Chaussées*, 1992.
22. E. FOURNIE. *Statistiques des Diffusions Ergodiques avec Applications en Finance*. PhD thesis, Université de Nice-Sophia-Antipolis, 1993.
23. E. FOURNIE, D. TALAY. Application de la statistique des diffusions à un modèle de taux d'intérêt. *Finance*, 12(2), 1991.
24. J.G. GAINES. The algebra of iterated stochastic integrals. *Stochastics and Stochastic Reports*, 49:169–179, 1994.
25. J.G. GAINES, T.J. LYONS. Random generation of stochastic area integrals. *SIAM Journal of Applied Mathematics*, 54(4):1132–1146, 1994.
26. A. GERARDI, F. MARCHETTI, A.M. ROSA. Simulation of diffusions with boundary conditions. *Systems & Control Letters*, 4, 1984.
27. A. GRORUD, D. TALAY. Approximation of Lyapunov exponents of nonlinear stochastic differential systems. *SIAM J. Applied Math.*, to appear.
28. N. IKEDA, S. WATANABE. *Stochastic Differential Equations and Diffusion Processes*. North Holland, 1981.
29. I. KARATZAS, S.E. SHREVE. *Brownian Motion and Stochastic Calculus*. Springer-Verlag, 1988.
30. P.E. KLOEDEN, E.PLATEN. *Numerical Solution of Stochastic Differential Equations*. Springer–Verlag, 1992.
31. T. KURTZ, P. PROTTER. Wong–Zakai corrections, random evolutions and numerical schemes for S.D.E.'s. In *Stochastic Analysis: Liber Amicorum for Moshe Zakai*, pages 331–346, 1991.
32. H.J. KUSHNER. *Approximation and Weak Convergence Methods for Random Processes*. M.I.T. Press, 1984.
33. H.J. KUSHNER. Approximations and optimal control for the pathwise average cost per unit time and discounted problems for wideband noise driven systems. *SIAM J. on Control and Optimization*, 27(3):546–562, 1987.
34. D. LEPINGLE. Un schéma d'Euler pour équations différentielles réfléchies. *Note aux Comptes-Rendus de l'Académie des Sciences*, 316(I):601–605, 1993.
35. Y. LIU. Numerical approaches to reflected diffusion processes. Submitted for publication, 1992.
36. E.J. Mc SHANE. *Stochastic Calculus and Stochastic Models*. Acad. Press, 1974.
37. G.N. MILSHTEIN. Approximate integration of stochastic differential equations. *Theory of Probability and Applications*, 19:557–562, 1974.
38. G.N. MILSHTEIN. Weak approximation of solutions of systems of stochastic differential equations. *Theory of Probability and Applications*, 30:750–766, 1985.
39. G.N. MILSHTEIN. The solving of the boundary value problem for parabolic equation by the numerical integration of stochastic equations. To appear in Theory of Probability and Applications.
40. N.J. NEWTON. An asymptotically efficient difference formula for solving stochastic differential equations. *Stochastics*, 19:175–206, 1986.
41. N.J. NEWTON. An efficient approximation for stochastic differential equations on the partition of symmetrical first passage times. *Stochastics and Stochastic Reports*, 29:227–258, 1990.

42. N.J. NEWTON. Asymptotically efficient Runge-Kutta methods for a class of Ito and Stratonovich equations. *SIAM Journal of Applied Mathematics*, 51(2):542–567, 1991.
43. N.J. NEWTON. Variance reduction for simulated diffusions. *SIAM Journal of Applied Mathematics*, to appear.
44. E. PARDOUX. Stabilité du mouvement des pales d'hélicoptères dans le cas d'un écoulement de l'air turbulent. In *Actes du Colloque "L'Automatique pour l'Aéronautique"*, Paris, 1986.
45. E. PARDOUX, D. TALAY. Stability of linear differential systems with parametric excitation. In *Nonlinear Stochastic Dynamic Engineering Systems (G.I.Schueller & F.Ziegler (Eds.)*. Springer-Verlag, 1988.
46. M. PIGNOL. *Stabilité Stochastique des Pales d'Hélicoptère*. PhD thesis, Université de Provence, 1985.
47. E. PLATEN, W. WAGNER. On a Taylor formula for a class of Ito processes. *Probability and Mathematical Statistics*, 3:37–51, 1982.
48. W. RUMELIN. Numerical treatment of stochastic differential equations. *SIAM Journal on Numerical Analysis*, 19:604–613, 1982.
49. K.R. SCHENK-HOPPE. Bifurcation scenarios of the noisy Duffing-van der Pol oscillator. Report 295, Institut für Dynamische Systeme, Universität Bremen, 1993.
50. L. SLOMINSKI. On approximation of solutions of multidimensional s.d.e.'s with reflecting boundary conditions. *Stochastic Processes and their Applications*, 50(2):197–219, 1994.
51. H.J. SUSSMANN. On the gap between deterministic and stochastic differential equations. *Ann. Prob.*, 6:19–44, 1978.
52. D. TALAY. Résolution trajectorielle et analyse numérique des équations différentielles stochastiques. *Stochastics*, 9:275–306, 1983.
53. D. TALAY. Efficient numerical schemes for the approximation of expectations of functionals of S.D.E. In J. Szpirglas H. Korezlioglu, G. Mazziotto, editor, *Filtering and Control of Random Processes*, volume 61 of *Lecture Notes in Control and Information Sciences*. Proceedings of the ENST-CNET Colloquium, Paris, 1983, Springer-Verlag, 1984.
54. D. TALAY. Discrétisation d'une E.D.S. et calcul approché d' espérances de fonctionnelles de la solution. *Mathematical Modelling and Numerical Analysis*, 20(1), 1986.
55. D. TALAY. Second order discretization schemes of stochastic differential systems for the computation of the invariant law. *Stochastics and Stochastic Reports*, 29(1):13–36, 1990.
56. D. TALAY. Approximation of upper Lyapunov exponents of bilinear stochastic differential systems. *SIAM Journal on Numerical Analysis*, 28(4):1141–1164, 1991.
57. D. TALAY. Presto: a software package for the simulation of diffusion processes. *Statistics and Computing Journal*, 4(4), 1994.
58. D. TALAY, L. TUBARO. Expansion of the global error for numerical schemes solving stochastic differential equations. *Stochastic Analysis and Applications*, 8(4):94–120, 1990.
59. W. WAGNER. MonteCarlo evaluation of functionals of stochastic differential equations—variance reduction and numerical examples. *Stochastic Analysis and Applications*, 6:447–468, 1988.
60. W. WEDIG. Contribution in this volume. 1995.

Lyapunov Exponents Indicate Stability and Detect Stochastic Bifurcations

Petra Boxler

1 Introduction

Assume that you are interested in the behaviour of a linear oscillator $\ddot{y} + (\gamma + \sigma\xi_t)y = 0$ with negative restoring force (i.e. $\gamma < 0$) which is disturbed by white noise of intensity σ. This equation may be modelled as a stochastic differential equation, and if you simulate its solution for different values of σ and γ, as described e.g. in Talay [29], then you will see that all in a sudden your solution will "disappear". If you ask a physicist or an engineer for an explanation of such a phenomenon he will answer that the solution must have become unstable and hence invisible.

The aim of the present paper is to describe a device, the so-called Lyapunov exponents, which enables you to decide whether a stochastic system is stable or not. Furtherrnore, if you deal with nonlinear systems the Lyapunov exponents will be able to tell you whether a "bifurcation" may occur in your system, i.e. whether it is possible that the original solution loses its stability at a certain parameter value and new stable solutions appear.

To achieve this aim we will proceed as follows:

1. • In section 2 we will restrict ourselves to *linear* systems. The Lyapunov exponents will be the stochastic counterparts to the real parts of the eigenvalues. For this reason we will start with a description of the deterministic situation. After having understood the role of the eigenvalues for stability questions the notion of a Lyapunov exponent should no longer be mysterious.
2. • In section 3 it will be explained how *nonlinear* systems fit into the frame developed so far. By looking at the deterministic case once again we will understand in the sequel what bifurcations are and why Lyapunov exponents may help to detect them.
3. • At that point you will hopefully be convinced that Lyapunov exponents can be a powerful tool both in stability and bifurcation theory. In this case you might be interested in further aspects of a stochastic bifurcation theory: just have a look at the two subsequent papers in this volume !

2 Lyapunov Exponents of Linear Stochastic Systems

2.1 Stability of Deterministic Systems

As we have pointed out in the introduction we will try to learn from the deterministic case how Lyapunov exponents ought to be defined. For this reason let us consider a linear ordinary differential equation

$$\dot{x} = Ax, x \in \mathbb{R}^d \tag{1}$$

where A is a d x d-matrix with constant coefficients. Then the solution starting at x_0 at time $t = 0$ is given by $\phi(t, x_0) := \Psi(t)x_0 = e^{At}x_0$ for all $t \in \mathbb{R}$ (see, e.g., Coddington and Levinson [12], Chapter 3, Theorem 4.1.).

It is also possible to study the solution of (2.1) for different initial values. This means that we consider $\phi(t,.)$ as a map from $\mathbb{R}^d$ to $\mathbb{R}^d$, which assigns to each initial value $x \in \mathbb{R}^d$ the value $\phi(t, x)$ reached by the solution at time t. It is obvious that for any value of t this map is a linear isomorphism and thus in particular a diffeomorphism. Furthermore, it has the following property:

$$\phi(t + s, x) = \phi(t,.) \circ \phi(s, x) \text{ for all } t, s \in \mathbb{R} \text{ and all } x \in \mathbb{R}^d \tag{2}$$

This is the so-called *flow property* and it simply means that instead of calculating the solution at time t + s starting at x one may first determine the solution at time s starting at x and then calculate the solution at time t with this value as an initial value. For this reason the map $\phi(t,.)$, which is generated by the solution of (1), is called a *flow of diffeomorphisms.*

Of course $\phi(t, 0) \equiv 0$ is a solution of (1). But would an engineer consider this zero solution as being relevant for practical purposes? He will only do so if a solution starting very close to zero will finally approach zero; otherwise he has no chance to "find" the zero solution numerically. Thus, the engineer will ask whether the zero solution is stable, and this is made precise by the following definition:

Definition 1. The zero solution of (1) is called stable if for any $\varepsilon > 0$ there exists a $\delta > 0$ such that for all $x \in \mathbb{R}^d$ satisfying $\|x\| < \delta$ we have: $\|\phi(t.x)\| < \varepsilon$ for all $t \geq 0$.

If, in addition, $\|\phi(t.x)\| \to 0$ for $t \to \infty$ then the zero solution is called asymptotically stable.

Here, $\|.\|$ describes the usual Euclidean norm in $\mathbb{R}^d$,

i.e. $\|x\| := \sqrt{x_1^2 + \cdots x_d^2}$.

Being only interested in asymptotic stability in the sequel we will usually drop the word "asymptotic" although we will go on asking whether the zero solution shows the behaviour Fig. 1.

Is it possible to decide about (asymptotic) stability without checking the condition of Definition 1 explicitly ?

This is in fact the case; the following theorem tells us that it is sufficient to examine the real parts of the eigenvalues of the matrix A:

Theorem 2. *The zero solution of (1) is asymptotically stable if and only if all the eigenvalues of the matrix A have negative real parts.*

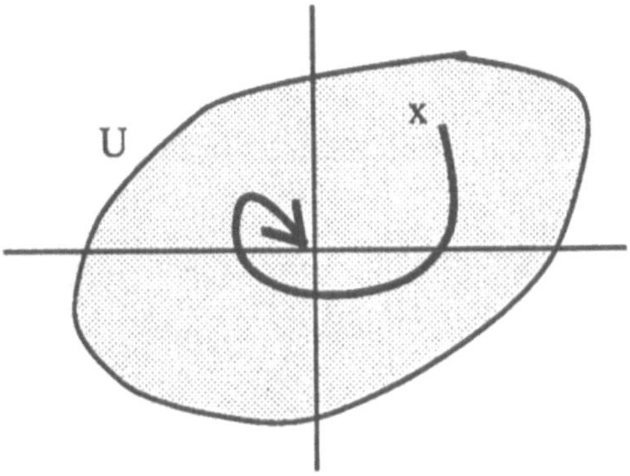

Fig. 1. Asymptotic behaviour

Proof. See, e.g., Knobloch and Kappel[18], ChapterIII, Theorem 7.5.

It is easily seen that negative real parts imply stability because if all the real parts are negative there will be positive constants K and γ such that

$$\|\phi(t,x)\| = \|\Psi(t)x\| = \|e^{At}x\| \leq Ke^{-\gamma t}\|x\| \tag{3}$$

This indicates that the real parts of the eigenvalues determine the exponential growth rate of the solution starting in the direction of the corresponding eigenspace. More precisely the following theorem yields a complete description of the asymptotic behaviour of a linear system of the form (1):

Theorem 3. *Suppose that the eigenvalues of the matrix A have $r \leq d$ different real parts $\lambda_1 > \ldots > \lambda_r$ and let $E_1, \ldots, E_r$ be the corresponding (generalized) eigenspaces. Then we obtain:*

$$\mathbb{R}^d = E_1 \oplus \cdots \oplus E_r, \tag{4}$$

$$\Psi(t)E_i = E^{At}E_i = E_i \text{ for all } t \in \mathbb{R}, \tag{5}$$

$$\lim_{t\to\pm\infty} \frac{1}{t} \log \|\Psi(t)x_0\| = \lambda_i \Leftrightarrow x_0 \in E_i. \tag{6}$$

Proof. This follows from the classical theory of ordinary differential equations as it may be found e.g. in Coddington and Levinson [12] or in Knobloch and Kappel [18]. See also the original work of Lyapunov [22].

The first statement says that the generalized eigenspaces fix a coordinate system in $\mathbb{R}^d$ which is invariant with respect to the flow (property (5)). Finally, the third property tells us which exponential growth rate we will have to expect if we start in the eigenspace directions. If our initial vector has components in more than one coordinate direction, for example in E_i and $E_j, i > j$, then the solution will be attracted by the eigenspace E_i, i.e. by the eigenspace corresponding to the biggest real part.

A two-dimensional example might look as in Fig. 2.

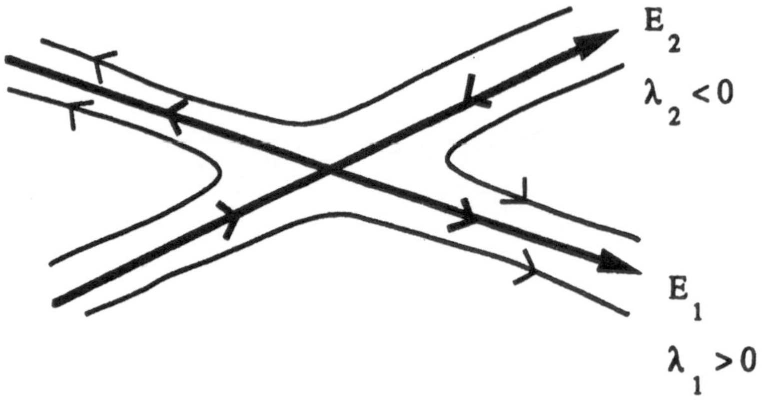

Fig. 2. A two-dimensional example

2.2 Stability of Linear Stochastic Systems

Stochastic Models for Systems Perturbed by Noise. After having investigated linear deterministic systems we are now going to ask what happens if the deterministic system is perturbed by noise. How can we model the influence of "noise" on an ordinary differential equation ? We suggest two different possibilities which will cover most of the practically relevant cases. Hence, starting from equation (1) we will consider

$$\dot{x}_t = A(\xi_t(\omega))x_t, \; t \in \mathbb{R}, \tag{7}$$

$$dx_t = A_0 x_t dt + \sum_{i=1}^{m} A_i x_t \circ dW_i(t), t \in \mathbb{R}, \tag{8}$$

In (7) $\{\xi_t | \; t \in \mathbb{R}\}$ denotes a measurable stationary stochastic process on a probability space $(\Omega, \mathcal{F}, \mathcal{P})$ with values in a measurable space $(Y, \mathcal{Y}), A(.)$ takes values in the set of d x d-matrices, and $\xi \to A(\xi)$ is supposed to depend measurably on $\xi \in Y$.

For the probability space Ω we may take the space of trajectories of the process ξ_t.

In (8) $A_0, \dots, A_m$ are d x d-matrices and $W_i(t), i = 1, \dots, m$, denote independent Brownian motions on the whole time axis. They may be constructed from two independent copies of canonical Wiener spaces and two canonical Brownian motions for $t \geq 0$. Hence, the probability space Ω may be identified with

$\{\omega \in \mathcal{C}(\mathbb{R}, \mathbb{R}^m) | \ \omega(0) = 0\}$, i.e. with the space of continuous functions from $\mathbb{R} \to \mathbb{R}^m$ which are 0 at 0. For further details see Boxler [10], Section 3.1.

This identification enables us to consider the elements of Ω as functions instead of working with "abstract" $\omega \in \Omega$. Hence, the elements of our probability space might look as follows in Fig. 3

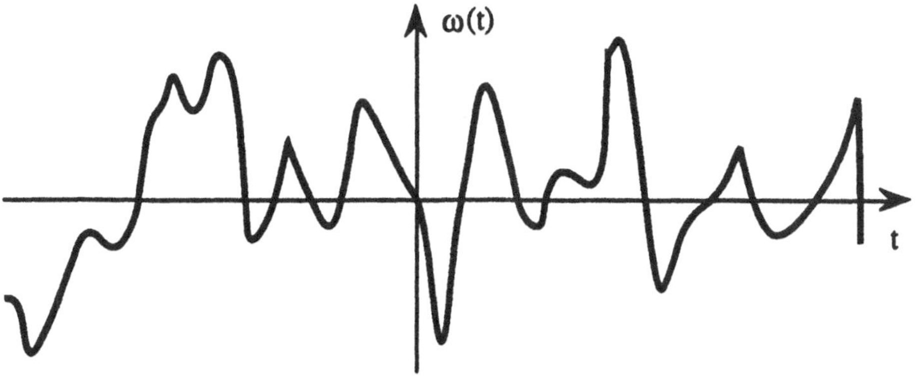

Fig. 3. An element of our probability space

It is of no importance that we have interpreted equation (8) in the sense of Stratonovich. The approach we are going to present here works equally well for Ito equations.

Of course it will depend on the concrete problem whether one chooses (7) or (8) as a model for the noise-influenced deterministic system. In the first case we speak of *real noise systems* whereas (8) describes *white noise systems*.

Real noise equations may be treated as ordinary differential equations with random coefficients; thus, in contrast to white noise equations, no Ito calculus is necessary to handle them.

Next we shall need a so-called *shift* on Ω which gives us the possibility of "switching" from one element of the probability space to another one. Such a shift will be described by a family of bimeasurable bijections $\mathcal{V}_t : \Omega \to \Omega, t \in \mathbb{R}$, which satisfy the group property $\mathcal{V}_{t+s} = \mathcal{V}_t \circ \mathcal{V}_s$ for all $t, s \in \mathbb{R}$ and preserve the measure P.

In (7) we obtain such a shift by putting

$$\mathcal{V}_t\omega(s) := \omega(s+t).$$

In equation (8) we have

$$\mathcal{V}_t\omega(s) := \omega(s+t) - \omega(t).$$

The subtraction of $\omega(t)$ in the second case is necessary to stay in the space of noise trajectories starting at 0. We see that the shift provides us with the possibility of evaluating the noise trajectory at a different time argument.

The Wiener measure P in the white noise case is not only invariant with respect to this choice of the shift but it is even ergodic, i.e. the only $\mathcal{V}_t$-invariant sets have P-measure 0 or 1 (see Boxler [10], Section 3.1., for a proof). This will also hold true in the real noise case if we assume an ergodic process ξ_t what we shall do in the sequel.

Remark: Instead of considering two different systems (7) and (8) it is possible to deal with both equations simultaneously. This is the so-called *unified treatment*, an approach that may be found in Arnold and Kliemann [2]. There one considers systems of the form

$$dx_t = A_0(\xi_t(\omega))x_t dt + \sum_{i=1}^{m} A_i(\xi_t(\omega))x_t \circ dW_i(t).$$

Cocycles of Linear Diffeomorphisms. Up to now we have been working with two different models (7) and (8). In this section we ask whether it is possible to analyze them simultaneously. Thus we will have to look for their common feature and it will turn out that it consists in something comparable to the flow generated by the solution of an ordinary differential equation like (1). In fact, let us consider the map

$$\phi : \mathbb{R} \times \Omega \times \mathbb{R}^d \to \mathbb{R}^d$$
$$(t, \omega, x) \to \phi(t, \omega, x) = \Psi(t, \omega)x,$$

which assigns to each initial value $x \in \mathbb{R}^d$ the value reached by the solution of the linear systems (7, 8 resp.) at time $t \in \mathbb{R}$ under the influence of $\omega \in \Omega$. Then we can prove the following theorem:

Theorem 4. *The map ϕ defines a cocycle of diffeomorphisms w.r.t. $\mathcal{V}_t$, i.e. ϕ is measurable in ω, continuous in (t,x), and for all $t \in \mathbb{R}$ and almost all ω $\phi(t,\omega,.) = \Psi(t,\omega) : \mathbb{R}^d \to \mathbb{R}^d$ is a (linear) diffeomorphism which satisfies the cocycle property:*

$$\Psi(t+s,\omega) = \Psi(t, \mathcal{V}_s\omega) \circ \Psi(s,\omega) \text{ for all } t,s \in \mathbb{R} \quad P\text{-a.s.}, \tag{9}$$

Proof. a) In the case of equation (7) this follows immediately because (7) may be interpreted as an ordinary differential equation with random coefficients.

b) In the white noise case the theorem follows e.g. from Arnold [1] (where the cocycle property is, however, not stated explicitly) or from Kunita [21], page 227 and 241, if one takes into account that Kunita works with two time arguments instead of using the shift.

It is heuristically clear why the shift is involved in the cocycle property:

If we first calculate the solution up to time s under the influence of ω this means that at time s we will have evaluated the noise trajectory up to time s. In order to calculate the solution at time t with the new initial value $\Psi(s,\omega)x$ we will thus start evaluating the noise trajectory at time arguments shifted by s.

(7) and (8) generating a cocycle, it will be reasonable to take this cocycle, which replaces the deterministic flow, as the basic object for our further investigations.

Lyapunov Exponents and Oseledec's Theorem. As in the deterministic case we want to decide whether the zero solution of the equations (7, 8) is (asymptotically) stable. This is understood in the sense that the cocycle has to satisfy the condition given in Definition 1 *for almost all* $\omega \in \Omega$.

In the deterministic case it was sufficient to look at the real parts of the eigenvalues to determine whether the zero solution was stable or not. In the stochastic situation, however, the eigenvalues completely lose their meaning because the systems we have to deal with do no longer have constant, i.e. time independent, coefficients. Thus, we will have to replace the real parts of the eigenvalues by new objects with properties similar to those of the eigenvalues. The following definition is inspired by property (6) of Theorem 3:

Definition 5. The Lyapunov exponent of the linear cocycle $\phi(t, \omega, .) = \Psi(t, \omega)$ in the direction of the vector $x \neq 0$ under the influence of $\omega \in \Omega$ is defined by:

$$\lambda^{\pm}(\omega, x) := \lim_{t \to \pm\infty} \frac{1}{t} \log \|\Psi(t, \omega)x\|.$$

As in the deterministic situation we may interpret this quantity as the exponential growth rate of the linear cocyle in the direction x. But up to now we do not know yet whether the limit above exists. This question is settled by the following theorem:

Theorem 6 (Oseledec's multiplicative ergodic theorem). *Assume the following integrability condition in the real noise case (there is not any condition in the white noise case):*

$$E\|A(\xi_0(.))\| < \infty.$$

Then there are r *real numbers* $\lambda_1 > \ldots > \lambda_r$ *with multiplicities* $d_i, \sum_{i=1}^r d_i = d$, *such that for all* ω *belonging to a* $\mathcal{V}_t$*-invariant set* $\Gamma \subset \Omega, P(\Gamma) = 1$, *the following properties hold:*

(i) There are r *random linear subspaces* $E_i(\omega) \subset \mathbb{R}^d$ *such that:*

$$\mathbb{R}^d = E_1(\omega) \oplus \ldots \oplus E_r(\omega), \;\; dim\ E_i(\omega) = d_i,$$

$$\Psi(t, \omega)E_i(\omega) = E_i(\mathcal{V}_t\omega) \;\; (stochastic\ invariance)$$

(ii) For any $x \neq 0$ *the limit* $\lambda^{\pm}(\omega, x)$ *exists and*

$$\lambda^{\pm}(\omega, x) = \lim_{t \to \pm\infty} \frac{1}{t} \log \|\Psi(t, \omega)x\| = \lambda_i \text{ if and only if } x \in E_i(\omega).$$

Proof. By now, there are several proofs available which deal with situations similar to that described in the theorem. See e.g. Oseledec [24], Ruelle [27], Carverhill [11], Crauel [13] or Mañé [23]. In Boxler [10] it is indicated how these proofs may be adapted to establish the theorem stated above.

The numbers λ_i are called *Lyapunov exponents* and the random subspaces $E_i(\omega)$ are called *Oseledec spaces.*

It is important to notice that the Lyapunov exponents do not depend on chance (this is due to the fact that we work with an ergodic measure; otherwise one would have to decompose P into its ergodic components). The Oseledec spaces, however, do in general depend on chance.

The first property of the theorem provides us with a random coordinate system in $\mathbb{R}^d$, and this coordinate system moves in time according to the property of stochastic invariance. The intuitive reason for the shift in the expression of stochastic invariance is the same that explained the shift in the cocycle property (6).

To obtain this random coordinate system we have to work on the whole time axis although a statement similar to Oseledec' s theorem still holds true if we restrict ourselves to positive time. But in this case we will not obtain a direct sum decomposition of $\mathbb{R}^d$ but only a nested sequence of random subspaces $\mathbb{R}^d$.

As in the deterministic case property (5) tells us that the Lyapunov exponent λ_i reflects the exponential growth rate of the solution starting in the direction of the random Oseledec space $E_i(\omega)$.

It is easy to see that if we switch off the noise the Lyapunov exponents will coincide with the real parts of the eigenvalues and the Oseledec spaces will become the generalized eigenspaces.

The following corollary summarizes in which way the Lyapunov exponents determine the stability of system (7, 8):

Corollary 7. *If all the Lyapunov exponents of the linear cocycle are negative then the system will be asymptotically stable. If at least one Lyapunov exponent is positive then the system will be unstable.*

Proof. This follows immediately from property (5) of Oseledec's theorem.

Let us end this section with an illustrating example. We consider the *harmonic oscillator* with random restoring force, which is described by the following equation:

$$\ddot{y} + 2\beta\dot{y} + (\gamma^2 + \xi(t))y = 0, \quad \beta, \gamma \in \mathbb{R},$$

where $\xi(t)$ is supposed to be a real-valued Omstein-Uhlenbeck process with variance σ^2. Putting $x := \begin{pmatrix} y \\ \dot{y} \end{pmatrix}$, this equation is equivalent to:

$$\dot{x} = \begin{pmatrix} 0 & 1 \\ -\gamma^2 - \xi(t) & -2\beta \end{pmatrix} x.$$

The level curves of the Lyapunov exponent $\lambda_1 = \lambda(\beta, \sigma^2)$ (here depicted for $\gamma^2 = 1$) yield the following refinement of the so-called *stability diagram* which may thus be considered as an improvement of the corresponding result of Kozin [20] (see Fig. 4).

We see that for $\beta > 1$ the stability is first being improved by putting noise on the system. Only beyond a certain critical noise intensity the noise acts as a destabilizing effect.

Thus, noise can either stabilize or destabilize a system. For the example above and many further examples see Arnold and Kliemann [3].

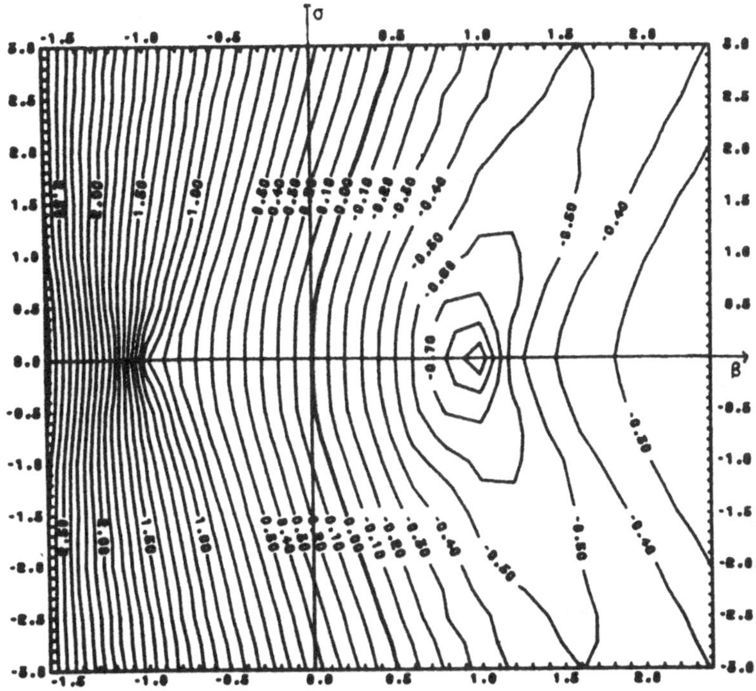

Fig. 4. Stability diagram

How to Calculate Lyapunov Exponents. In general it is not easy and often impossible to obtain explicit expressions for the Lyapunov exponents. The formula for the biggest Lyapunov exponent derived below may, however, be used as a starting point both for asymptotic expansions and for numerical calculations:

a) The white noise case: In the definition of the Lyapunov exponent only the *norm* of the solution, i.e. the distance from the origin reached by a point at time t, is involved. If we think in terms of polar coordinates this simply means that it should be sufficient to study the radial component to capture the evolution of the norm of the solution and hence to determine the biggest Lyapunov exponent. For this reason we would like to obtain decoupled equations describing the behaviour of the radial and angular part of the solution. This is in fact possible, by means of the following procedure, which is originally due to Has'minskii [16]:

The idea consists in projecting the solution $\phi(t, \omega, x_0)$ onto the unit sphere in $\mathbb{R}^d$, i.e. onto $\mathbf{S}^{d-1} := \{X \in \mathbb{R}^d | \; \|x\| = 1\}$. For this we put

$$s(t, \omega, s_0) := \frac{\phi(t, \omega, x_0)}{\|\phi(t, \omega, x_0)\|}$$

$$s_0 := \frac{x_0}{\|x_0\|}$$

Thus: see Fig. 5

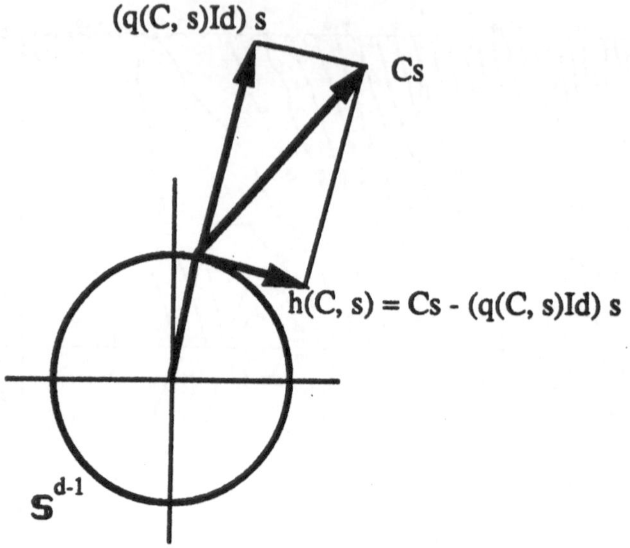

Fig. 5. System on the sphere

Here C is a d x d-matrix, $s \in \mathbf{S}^{d-1}$, and we have denoted:

$$q(C,s) := s'Cs, \quad h(C,s) = Cs - [q(C,s)\mathrm{Id}]s. \tag{10}$$

The usual rules of stochastic calculus yield the following stochastic differential equation for s:

$$ds = h(A_0, s)dt + \sum_{i=1}^{m} h(A_i, s) \circ dW_i(t), \;\; s(0) = s_0. \tag{11}$$

For the radial part $\|\phi(t,\omega,x_0)\|$ we obtain:

$$\|\phi(t,\omega,x_0)\| = \|x_0\| \exp \left[\int_0^t q(A_0, s(\tau,\omega,s_0))d\tau \right. \\ \left. + \sum_{i=1}^{m} \int_0^t q(A_i, s(\tau,\omega,s_0)) \circ dW_i(\tau) \right] \tag{12}$$

To deduce a formula for the biggest Lyapunov exponent from this equality we have to wnte it in Ito form because we want to make use of the fact that

$$M_t := \int_0^t q(A_i, s(\tau,\omega,s_0))dW_i(\tau) \tag{13}$$

is a martingale (recall that, in contrast to the Ito integral, the Stratonovich integral is not a martingale) because then we may use that $\lim_{t\to\infty} \frac{1}{t} M_t = 0$ a. s. (see Has'minskii [16]).

Due to the Stratonovich-Ito conversion we thus obtain:

$$\lambda(x_0) := \lim_{t\to\infty} \frac{1}{t} \log \|\Psi(t,\omega)x_0\| = \lim_{t\to\infty} \frac{1}{t} \int_0^t Q(s(\tau,\omega,s_0))d\tau, \tag{14}$$

where

$$Q(s) := q(A_0,s) + \sum_{i=1}^{m} \left[\frac{1}{2} q((A_i + A_i')A_i, s) - q(A_i,s)^2\right] \tag{15}$$

Of course we would like to replace this time average by a space average by means of Birkhoff's ergodic theorem. But this is only possible if there is a *unique* invariant measure on $\mathbb{S}^{d-1}$. The sphere being compact there is at least one invariant measure but if there is more than one it is intuitively clear that the trajectories will not "know" in which part of the space they will have to cluster. Taking this into account we obtain:

Theorem 8. *Under conditions which ensure the existence of a unique invariant measure μ for the process s on $\mathbb{P}^{d-1}$ the biggest Lyapunov exponent is given by:*

$$\lambda_{\max} = \lambda_1 = \int_P Q(s)d\mu(s).$$

Furthermore, for almost all $x_0 \neq 0$ we have: $\lambda(x_0) = \lambda_1$

Remark: We have unserstood the differential equation for s as an equation on the projective space $P := \mathbb{P}^{d-1}$, which is possible by identifying s and $-$s, to avoid trivial non-uniqueness of the invariant mesure.

Proof. See Amold, Oeljeklaus and Pardoux [6], Theorem 1.2. There one may also find conditions ensuring the existence of a unique invariant measure.

b) The special case of two-dimensional systems: While dealing with stochastic systems in $\mathbb{R}^2$ we can write the formula for the biggest Lyapunov exponent in a more explicit way by introducing polar coordinates, i.e. by putting

$$s_t := s(t) := \begin{pmatrix} \cos\theta_t \\ \sin\theta_t \end{pmatrix}, \qquad \theta_t \in [0,\pi[.$$

If for each $s \in \mathbb{P}^1$ at least one of the vectors $h(Ai,s)$, $i = 0,\ldots,m$, does not vanish there exists a unique invariant probability measure for s_t and the process it will have a unique invariant measure on $[0,\pi[$ which has a smooth density p that may be obtained as a solution of the Fokker-Planck equation. λ_1 may thus be written as

$$\lambda_1 = \int_0^\pi \bar{Q}(\theta)p(\theta)d\theta,$$

where the explicit expression for $\bar{Q}(\theta)$ may be found in Pardoux and Wihstutz [25].

Putting $A_i := \begin{pmatrix} a_{11}^i & a_{12}^i \\ a_{21}^i & a_{22}^i \end{pmatrix}$ we can write: $\bar{Q}(\theta) = \bar{Q}_0(\theta) + \frac{1}{2}\sum_{i=1}^m \bar{Q}_i(\theta)$, where

Putting $A_i := \begin{pmatrix} a_{11}^i & a_{12}^i \\ a_{21}^i & a_{22}^i \end{pmatrix}$ we can write: $\bar{Q}(\theta) = \bar{Q}_0(\theta) + \frac{1}{2}\sum_{i=1}^m \bar{Q}_i(\theta)$, where

$$\bar{Q}_0(\theta) = a_{11}^0 \cos^2\theta + (a_{12}^0 + a_{21}^0)\cos\theta\sin\theta + a_{22}^0\sin^2\theta$$

$$\begin{aligned}\bar{Q}_i(\theta) = {} & [2(a_{11}^i)^2 + (a_{21}^i)^2]\cos^2\theta + [2(a_{22}^i)^2 + (a_{12}^i)^2]\sin^2\theta + a_{21}^i a_{12}^i \\ & +(3a_{11}^i a_{12}^i + 3a_{21}^i a_{22}^i + a_{11}^i a_{21}^i + a_{22}^i a_{12}^i)\sin\theta\cos\theta \\ & -2[(a_{11}^i)^2\cos^4\theta + (a_{22}^i)^2\sin^4\theta] \\ & -2[(a_{12}^i)^2 + 2a_{12}^i a_{21}^i + 2a_{11}^i a_{22}^i + (a_{21}^i)^2]\sin^2\theta\cos^2\theta \\ & -4(a_{12}^i + a_{21}^i)\sin\theta\cos\theta[a_{22}^i\sin^2\theta + a_{11}^i\cos^2\theta].\end{aligned}$$

An application of this method may be found in the paper of Wedig [30] in this volume.

Up to now we have only investigated the behaviour of the radial part of the solution. In two dimensions, however, we can also characterize its angular behaviour:

Definition 9. The random variable ρ defined by

$$\rho := \limsup_{t\to\infty} \frac{\theta_t}{t}$$

which describes the average angle velocity is called the rotation number.

Assume the same condition as above to ensure the existence of a unique invariant measure. Then we will obtain:

$$\rho = \int_0^\pi \bar{h}(\theta)p(\theta)d\theta$$

where $\bar{h}(\theta)$ is explicitly given by the following expression (see Pardoux and Wihstutz [25]):

$$\bar{h}(\theta) = \bar{h}(A_0,\theta) + \frac{1}{2}\sum_{i=1}^m \bar{h}(A_i,\theta)\frac{d\bar{h}(A_i,\theta)}{d\theta} \qquad .$$

where

$$\bar{h}(A_i,\theta) = (a_{22}^i - a_{11}^i)\cos\theta\sin\theta - a_{12}^i\sin^2\theta + a_{21}^i\cos^2\theta \quad i = 0,\dots,m.$$

If the noise is switched off then one easily checks that the rotation numbers will coincide with the imaginary parts of the eigenvalues (up to signs). Hence, in the two-dimensional case we have found stochastic analogues to both, the real and the imaginary parts of the eigenvalues.

In higher dimensions, however, it is more difficult to introduce the notion of a rotation number because it is not obvious at all by which tool one might replace the polar coordinate representation we have used above. Despite these difficulties Arnold and San Martin [8] have recently been successful in generalizing the concept of rotation numbers to higher dimensions.

c) The real noise case: In this case a completely analogous projection procedure is possible if the noise is a diffusion process. Hence we may omit the details which can be found, together with the real noise counterpart of Theorem 8, in Arnold, Kliemann and Oeljeklaus [4], Theorem 4.1.

d) Asymptotic expansions: The possibility of expressing the biggest Lyapunov exponent by means of the invariant measure can be used to derive asymptotic expansions of λ_1 and ρ in the case of small noise for two-dimensional systems because the density of the invariant measure, which may be obtained as a solution of the Fokker-Planck equation, can be expanded. The real noise case is treated in Arnold, Papanicolaou and Wihstutz [7], the white noise case in Pardoux and Wihstutz [25] and in Pinsky and Wihstutz [26], and the case of dichotomic noise (i.e. which can only take two different values) in Arnold and Kloeden [5].

d) Numerical calculation of Lyapunov exponents: The biggest Lyapunov exponent may be determined numerically either by starting directly from its definition as a time average or by using the formula described in Theorem 8. For further details, the reader is refered to the paper of Talay [29] in this volume and the references therein.

3 Lyapunov Exponents of Nonlinear Stochastic Systems

3.1 Cocycles of Nonlinear Diffeomorphisms

Numerous examples in several contributions in this volume should already have convinced you that many problems in physics and engineering have to be treated as nonlinear problems. This means that instead of systems like (7, 8) we have to consider equations of one of the following types:

$$\dot{x}_t = f(x_t, \xi_t(\omega)), \quad x_0 = x \in \mathbb{R}^d,\ t \in \mathbb{R} \tag{16}$$

$$dx_t = f_0(x_t)dt + \sum_{i=1}^{m} f_i(x_t) \circ dW_i(t), \quad x_0 = x \in \mathbb{R}^d, \quad t \in \mathbb{R} \tag{17}$$

where f and $f_i, i = 0, \ldots, m$, have to be sufficiently smooth functions on $\mathbb{R}^d$. Furthermore we assume global Lipschitz conditions which ensure in particular the existence of unique solutions of (16, 17) (for this and for weaker assumptions see, e. g., Has'minskĭi [16], Theorem 11 for the real noise case and Kunita [21], page 227 and 241 for the white noise case).

In complete analogy to Theorem 4 we find that the solution $\phi(t, \omega, x)$ of the systems under consideration generates a *cocycle of (nonlinear) diffeomorphisms* with respect to the shift $\mathcal{V}_t$. This implies in particular:

$$\phi(t + s, \omega, x) = \phi(t, \mathcal{V}_s\omega, .) \circ \phi(s, \omega, x) \text{ for all } t, s \in \mathbb{R},\ x \in \mathbb{R}^d \text{ a.s}$$

In the deterministic situation, i.e. while dealing with an equation of the form $\dot{x} = f(x)$, one usually starts with the assumption that $f(x_0) = 0$ for a certain x_0, and this means that the system is supposed to have an equilibrium point (= steady state, fixed point) at x_0. The stochastic counterpart to such an equilibrium point would be a stationary solution, which means in particular that the probability distribution of the solu tion is the same at time t and at time t+s for any real s. Although this seems to be much more general than the assumption of a fixed point we may (and do) assume without loss of generality (see Boxler [10], Theorem 4.1. for a proof) that:

$$\phi(t,\omega,0) = 0 \text{ for all } t \in \mathbb{R} \text{ a.s}$$

This is possible because we can introduce a new coordinate system which moves with the stationary solution. Looking at our original system from this moving frame we will always see a fixed point at 0.

3.2 Stability of the Linearized System

As in the linear case we would like to know whether the zero solution of (16, 17) is (asymptotically) stable. We have seen in section 2 that for *linear* systems the Lyapunov exponents may be a powerful tool to decide about this question. How can we use this insight in the present situation ?

For this one has to take into account that the map $\phi(t,\omega,.) : \mathbb{R}^d \to \mathbb{R}^d$ can be linearized (= differentiated) at 0 because it is a diffeomorphism. Hence we obtain a *linear* map

$$\begin{aligned} \Psi(t,\omega) &: \mathbb{R}^d \to \mathbb{R}^d, \\ &x \to \Psi(t,\omega)x =: v(t,\omega,x). \end{aligned}$$

This map may also be interpreted as the solution of the linearization of (16, 17):

$$\dot{v}_t = A(\xi_t(\omega))v_t, \quad v_0(x) = x, \quad t \in \mathbb{R} \tag{18}$$

$$dv_t = A_0 v_t dt + \sum_{i=1}^{m} A_i v_t \circ dW_i(t), \quad v_0(x) = x, \quad t \in \mathbb{R} \tag{19}$$

where $A(\xi) := \left(\frac{\partial f}{\partial x}\right)(0,\xi)$, $A_i := \left(\frac{\partial f_i}{\partial x}\right)(0)$, $i = 0,\ldots,m$ denote the corresponding Jacobian matrices.

But (18, 19) are linear systems of the form (2.3 a, b) which we can handle. Thus, the concept of Lyapunov exponents makes sense for these equations, and we are ready to define:

Definition 10. The Lyapunov exponent at 0 of a nonlinear cocycle ϕ in the direction of the vector $x \neq 0$ under the influence of $\omega \in \Omega$ is defined by

$$\lambda^{\pm}(\omega,x) = \lim_{t\to\pm\infty} \frac{1}{t} \log \|\Psi(t,\omega)x\|$$

where $\Psi(t,\omega)$ denotes the linearization of ϕ.

Hence all we have learnt about Lyapunov exponents and Oseledec's theorem will carry over to nonlinear systems if we simply think of the linear map $\Psi(t,\omega)$ appearing in the various statements as being obtained from the nonlinear map ϕ by differentiation at 0, i. e.

$$\Psi(t,\omega) = \left(\frac{\partial}{\partial x}\phi(t,\omega,.)\right)\Big|_{x=0}$$

Up to now we are able to get information about the stability of the system (18, 19) obtained from the original nonlinear system (16, 17) by linearization. If we know for example that all the Lyapunov exponents of the *linearized* system are negative then we may conclude that this system is asymptotically stable.

But what does this tell us about the stability of the *original* system ? Is it possible that the stability of the linearized equation does not carry over to the nonlinear system ? Let us mention the following result:

Theorem 11. *If all the Lyapunov exponents of a nonlinear cocycle are negative then the nonlinear stochastic system generating this cocycle will be asymptotically stable. If at least one Lyapunov exponent is positive then the nonlinear system will be unstable.*

Proof. This is an obvious consequence of the so-called stable manifold theorem a proof of which may be found e.g. in Ruelle [27] or Carverhill [11].

But this is only partially satisfactory because you will see in the next section that there are many important cases where the theorem above is not applicable because one Lyapunov exponent vanishes. In this situation we do not know a priori in which way the nonlinear system and its linearization are related. New concepts have to be developped: for more details see the paper of Boxler [9] in this volume.

4 Lyapunov Exponents as Indicators of Stochastic Bifurcation

In this section we are going to talk about a new type of phenomena, the so-called *bifurcations*, which may occur in nonlinear parameter dependent systems.

Our aim is two-fold:

First, we want to explain why the linearization of a nonlinear equation does only contain part of the infornation about the dynamics of the nonlinear system and what one may discover by taking the nonlinearities into account.

Second, we will show that despite these facts the Lyapunov exponents and hence the linearization of the system are able to detect the points where this new bifurcation behaviour is possible.

For these reasons the "message" concerning the treatment of nonlinear stochastic systems will be as follows: of course one should first examine the linearized system and its Lyapunov exponents. But in case Theorem 11 does not apply it will then be necessary to consider the nonlinearities and to see what happens.

4.1 Deterministic Bifurcations and How to Detect Them

Once again we will start by looking at deterministic systems; later on we will answer the question how to cope with systems perturbed by noise.

Let us begin with an elementary one-dimensional example. We consider the equation

$$\dot{x} = \alpha x - x^3, \quad \alpha \in \mathbb{R}, \quad x \in \mathbb{R}, \tag{20}$$

whose equilibrium states may be obtained by solving $\alpha x - x^3 = 0$. We find

$$\begin{cases} x = 0 & \text{if } \alpha < 0 \\ \left.\begin{array}{l} x = 0 \\ x = \pm\sqrt{\alpha} \end{array}\right\} & \text{if } \alpha > 0 \end{cases}$$

If we draw a so-called *bifurcation diagram* where we plot the steady states versus the different parameter values we will get:

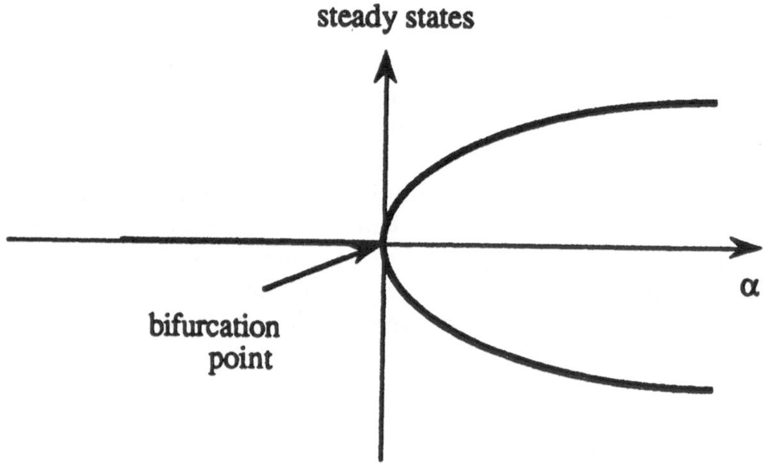

Fig. 6. Bifurcation diagram

We did not draw the zero solution for positive values of α because this equilibrium state will be "invisible" for an engineer or a physicist. In section 2 we have learnt that this means in mathematical terms that the solution has become unstable. We have also seen that the eigenvalues are suitable tools to tackle stability questions.

Hence let us calculate the linearization of (20) at $x = 0$. This yields the equation $\dot{v} = \alpha v$, whose only eigenvalue is α, and thus the zero solution is stable for $\alpha < 0$ and unstable for $\alpha > 0$. Of course we could have seen this directly because the solution of the linearized equation is $\Psi(t)v_0 = e^{\alpha t}v_0$.

We deduce from the bifurcation diagram that the point $\alpha = 0$ where the zero solution loses its stability coincides with the point where new steady states appear. These new solutions are stable, as we verify by calculating the linearization of (20) at $\pm\sqrt{\alpha}$: we obtain $-2\alpha < 0$ for $\alpha > 0$. Thus, we might call $\alpha = 0$ a bifurcation point.

From this simply example we learn that a bifurcation can only occur if the original solution loses its stability, i. e. if the eigenvalue vanishes for a certain parameter value. If this is not the case, which means that the eigenvalue remains negative, then nothing new will happen and we will not speak about a bifurcation.

After these preparations we are ready to attack the general situation. We consider the following nonlinear parameter depending equation (for f_α sufficiently smooth):

$$\dot{x} = f_\alpha(x), \quad f_\alpha(0) = 0, \quad x \in \mathbb{R}^d, \quad \alpha \in \mathbb{R} \tag{21}$$

Hence we suppose 0 to be an equilibrium point. To decide whether it is stable we calculate the linearization at 0 and obtain

$$\dot{v} = A_\alpha v, \text{ where } A_\alpha = \left.\frac{\partial f_\alpha}{\partial x}\right|_{x=0}$$

Let us start in a parameter region where the zero solution is stable, i. e. where the real parts of all the eigenvalues of A_α are negative. Then it is reasonable to define:

Definition 12. If at $\alpha = \alpha_0$ the zero solution becomes unstable and at least one non-trivial solution appears then α_0 is called a (deterministic) bifurcation point.

Of course this definition is not the most general one but it is sufficient for our purposes here. A more general definition and an overview of deterrninistic bifurcation theory may be found, e. g., in Guckenheimer and Holmes [15] .

Our previous discussion yields the following necessary condition for a bifurcation:

Theorem 13. *If α_0 is a bifurcation point and if we denote by $\lambda_i(\alpha)$ the real parts of the eigenvalues of A_α then we have:*

1. *For all $i = 1, \ldots, r$: $\lambda_i(\alpha) < 0$ either for $\alpha < \alpha_0$ or for $\alpha > \alpha_0$.*
2. *There is at least one index j such that: $\lambda_j(\alpha_0) = 0$.*

Hence a bifurcation point is characterized by a vanishing real part of at least one eigenvalue. However, this condition is not sufficient because for example in the case of a linear system an eigenvalue crossing the imaginary axis will simply announce that the zero solution has become unstable. We would not call this a bifurcation because no non-trivial solutions appear.

Thus, it is due to the nonlinearities in the system that non-trivial solutions may bifurcate for a certain pararneter value. This means t hat the eigenvalues of the system linearized at 0 are able to indicate where a bifurcation might occur. But in the next step the entire nonlinear equation has to be considered to decide whether there is really a bifurcation. For this, new methods have to be used (if one is not able to calculate directly the bifurcating solutions) because Theorem 2, which allowed to carry over the behaviour of the linear system to the nonlinear system, is not applicable at the bifurcation point.

4.2 Lyapunov Exponents Detect Stochastic Bifurcations

A Necessary Condition for a Stochastic Bifurcation. Once again we will consider systems of the form (16) or (17) but this time the right-hand side is supposed to depend on a real parameter α, the so-called *bifurcation parameter.* The cocyle generated by the solution will then depend on a, too, as well as the linearized cocycle $\Psi_\alpha(t,\omega)$ and the Lyapunov exponents $\lambda_i(\alpha)$.

Then a *stochastic bifurcation point* may be defined in complete analogy to Definition 12, and a necessary condition for a stochastic bifurcation is obtained if we reformulate Theorem 13 in terms of Lyapunov exponents. This enables us to conclude:

If the biggest Lyapunov exponent vanishes at α_0 then a stochastic bifurcation may occur at this point.

Stochastic Bifurcations in One-dimensional Systems. For simplicity we will restrict ourselves to the white noise case here; the real noise case is to be treated exactly the same way.

Before examining the bifurcation behaviour of equations of the type

$$dx_t = f_\alpha(x_t)dt + \sigma g(x_t) \circ dW(t), \quad x_0 = x \in \mathbb{R} \tag{22}$$

where $\sigma > 0$ denotes the noise intensity, we will make use of Theorem 8 to derive an explicit formula for the Lyapunov exponent attached to a stationary solution x_t^0 of (22). As explained in Section 3.1. we might reduce this case to the case of a fixed point at 0 by means of a moving coordinate frame but we will see that in the simple situation here it is no problem to work directly with the stationary solution.

Linearization of (22) at x_t^0 and transformation into Ito form yields:

$$dv_t = \left[\frac{\partial f_\alpha}{\partial x}(x_t^0) + \frac{\sigma^2}{2}\frac{\partial g}{\partial x}(x_t^0) \times \frac{\partial^2 g}{\partial x^2}(x_t^0)\right] v_t dt + \sigma \frac{\partial g}{\partial x}(x_t^0) v_t \, dW(t)$$

and as we have seen in Section 2.2 this implies the following theorem:

Theorem 14. *Assume that (22) has a unique stationary solution. Then its density is given by*

$$p(x) = \frac{C}{g(x)} \exp\left(\frac{2}{\sigma^2}\int \frac{f_\alpha(u)}{g^2(u)} du\right); \quad C \text{ a normalising constant}$$

and for the Lyapunoy exponent of this solution we obtain:

$$\lambda(\alpha) = \int \left[(f_\alpha)'(x) + \frac{\sigma^2}{2} g'(x) g''(x)\right] p(x) dx$$

Proof. The first part is a classical result because in the one-dimensional case the Fokker-Planck equation can be solved explicitly (if we assume natural boundaries); it may be found for example in Stratonovich [28]. If we combine this with the considerations above we will get the second statement.

If a one-dimensional deterministic equation is perturbed by noise this may be modelled in two different ways: either the noise is independent of the state of the system *(additive noise)* or we have so-called *parametric (= multiplicative)* noise. Their bifurcation behaviour is completely different, as we will show now. For examples the reader may consult, e. g., the paper of Wedig [30] in this volume.

a) Additive noise: We are interested in an equation of the following type:

$$dx_t = f_\alpha(x_t)dt + \sigma dW(t), \quad x_0 = x \in \mathbb{R} \tag{23}$$

which is supposed to have a stationary solution with density p. Then Theorem 14 enables us to deduce:

Theorem 15. *For all values of a the Lyapunov exponent of the stationary solution of (23) is negative and hence no stochastic bifurcation can occur.*

Proof. Because of Theorem 14 we obtain by means of an integration by parts:

$$\lambda(\alpha) = \int (f_\alpha)'(x)p(x)dx = -\frac{2C}{\sigma^2}\int ((f_\alpha(x))^2 p(x)ds < 0$$

This means that from the point of view of bifurcation theory one-dimensional systems perturbed by *additive* noise are not very exciting objects.

b) Parametric noise: We obtain an equation perturbed by multiplicative noise by requiring g in (22) to be a non-constant function. In this case we will always have $x_t^0 \equiv 0$ as a stationary solution. In contrast to the additive noise case, stochastic bifurcation is possible now.

Let us consider an example which is the so-called noisy Verhulst equation, which is used e. g. in biology:

$$dx_t = (\alpha x_t - x_t^2)dt + \sigma x_t \circ dW(t), \quad x_0 = x \in \mathbb{R}^+ \tag{24}$$

If we calculate the Lyapunov exponent of the zero solution (either directly by solving the linearized equation or by means of Theorem 14) we see that $\lambda(0) = 0$. Thus, at $\alpha_0 = 0$ a stochastic bifurcation may occur. In fact, for $\alpha > 0$ we will obtain a non-trivial stationary solution whose density is given by Theorem 14. If we determine the Lyapunov exponent attached to this bifurcating solution we see that it will be negative and hence the new solution is stable.

If we switch off the noise in equation (24) (i. e. $\sigma = 0$) we see that the deterministic bifurcation will also happen at $\alpha_0 = 0$. The next theorem will show, however, that this does not always hold in the case of parametric noise. Denote the deterministic bifurcation point by α_d and the stochastic one by α_s. Then we will immediately obtain

Theorem 16. *Let $x_t^0 := x_0$ be a stationary (equilibrium) solution of (22). Then we obtain for its Lyapunov exponent evaluated at the deterministic bifurcation point $\lambda(\alpha_d) = (\sigma^2/2)g'(x_0)g''(x_0)$, and hence the deterministic and the stochastic bifurcation point will only coincide if $g'(x_0) = 0$ or $g''(x_0) = 0$.*

Stochastic Bifurcation in Dimension $\geq$ 2. As pointed out earlier, Lyapunov exponents of higher dimensional systems may in general be only determined by means of asymptotic expansions or numerical methods. But once the biggest Lyapunov exponent is known in dependence on the bifurcation parameter we can also ask whether there may be a stochastic bifurcation.

We illustrate this by the example of the noisy *Van der Pol-Duffing oscillator:*

$$\ddot{x}_t = (\alpha + \sigma\xi_t(\omega))x_t + \beta\dot{x}_t - x_t^2\dot{x}_t - x_t^3, \ \beta < 0, \ \xi_t = \text{white noise}$$

which we have written in this form which should be more familiar to people who know this equation for $\sigma = 0$. A numerical calculation of the biggest Lyapunov exponent of the zero solution of this two-dimensional system yields the diagram Fig. 7.

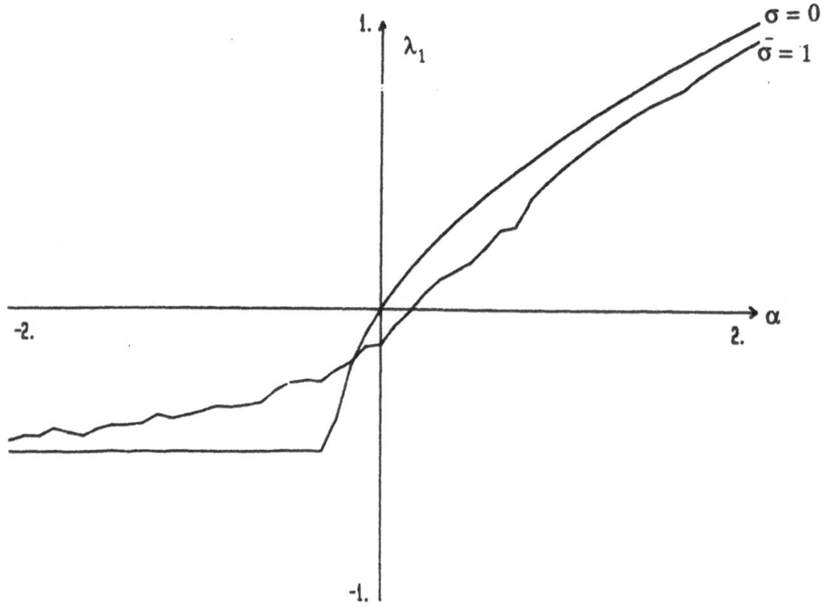

Fig. 7. Van der Pol-Duffing oscillator

This means that the stochastic bifurcation is retarded compared to the deterministic situation.

We see, however, that it would be preferable to deal mainly with low dimensional systems. For this reason it will be explained in Boxler [9] in this volume how a higher dimensional system may be reduced to a lower dimensional one.

4.3 Other Definitions of Stochastic Bifurcation

In the last section we have used the Lyapunov exponents as a tool to decide whether a stochastic bifurcation may occur. But especially in the physical literature another approach is suggested. For a stochastic bifurcation a "qualitative change" of the probability density is required there. Two problems connected with this approach should, however, not be neglected.

First, it is not at all obvious how a "qualitative change" is to be defined. Most people require a change in the *maximum* of the probability distribution (see, e. g., Horsthemke and Lefever [17] or Knobloch and Wiesenfeld [19]), others argue that one should consider the *moments* of the distribution (see Graham and Schenzle [14]). Thus, the situation is not very satisfactory.

Second, the probability distribution can change although the original solution does not lose its stability. An example is equation (20) perturbed by additive white noise, i. e.

$$dx_t = (\alpha x_t - X_t^3)dt + \sigma dW(t)$$

From Theorem 15 we know that the stationary solution which we obtain for $\alpha < 0$ does not lose its stability and hence there is no stochastic bifurcation in the sense of our definition. The probability density, which only describes the behaviour of *one* point whereas the Lyapunov exponents capture the behaviour of a vector, i. e. of *two* points, however, looks as in Fig. 8.

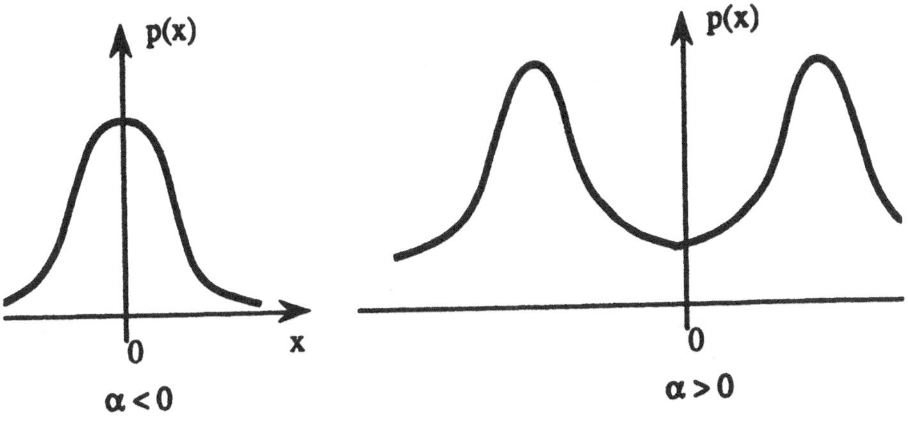

Fig. 8. Probability density

Thus, the position of the maxima changes and therefore Horsthemke and Lefever [17] call this a bifurcation. But does it make sense to speak of a bifurcation if there is no loss of stability of the original solution ?

References

1. Arnold, L.: Stochastic differential equations: theory and applications. Wiley, New York 1974.

2. Arnold, L., Kliemann, W.: Large deviations of linear stochastic differential equations. In: Engelbert (ed.): Proceedings of the Fifth IFIP Working Conference Eisenach, GDR 1986. Lecture Notes in Control and Information Sciences. Springer, Berlin, Heidelberg, New York, Tokyo 1987.
3. Arnold, L., Kliemann, W.: Qualitative theory of stochastic systems. In: A. T. Bharucha-Reid (ed.): Probabilistic analysis and related topics, vol. 3. Academic Press, New York 1981.
4. Arnold, L., Kliemann, W., Oeljeklaus, E.: Lyapunov exponents of linear stochastic systems. Proceedings of a workshop Bremen 1984. Lecture Notes in Mathematics vol. 1186. Springer, Berlin, Heidelberg, New York, Tokyo 1986.
5. Arnold, L., Kloeden, P.: Lyapunov exponents and rotation number of two- dimensional systems with telegraphic noise. Report No. 189, Institut für Dynamische Systeme, Universität Bremen 1988.
6. Arnold, L., Oel jeklaus, E., Pardoux, E.: Almost sure and moment stability for linear Ito equations. Proceedings of a workshop Bremen 1984. Lecture Notes in Mathematics vol. 1186. Springer, Berlin, Heidelberg, New York, Tokyo 1986.
7. Arnold, L., Papanicolaou, G., Wihstutz, V.: Asymptotic analysis of the Lyapunov exponent and rotation number of the random oscillator. SIAM J. Appl. Math. 46 (1986), 427 - 450.
8. Arnold, L., San Martin, L.: A multiplicative ergodic theorem for rotation numbers. Dynamics and Differential Equations 1(1988).
9. Boxler, P.: Stochastic center manifolds as a tool in a stochastic bifurcation theory. This volume.
10. Boxler, P.: Stochastische Zentrumsmannigfaltigkeiten. Ph. D. thesis, Institut für Dynamische Systeme, Universität Bremen 1988. (Condensed English version of [10]: Boxler, P.: A stochastic version of the center manifold theorem. Report No. 174, Institut für Dynamische Systeme, Universität Bremen 1988.)
11. Carverhill, A.: Flows of stochastic dynamical systems: ergodic theory. Stochastics 14 (1985), 273 - 317.
12. Coddington, E. A., Levinson, N.: Theory of ordinary differential equations. Tata McGraw-Hill, New Delhi, 8th ed. 1985.
13. Crauel, H.: Ergodentheorie linearer stochastischer Systeme. Report No. 59, Institut für Dynamische Systeme, Universität Bremen 1981.
14. Graham, R., Schenzle, A.: Stabilization by multiplicative noise. Phys. Rev. A 26,1676 (1982).
15. Guckenheimer, J., Holmes, Ph.: Nonlinear oscillations, dynamical systems and bifurcations of vector fields. Springer, Berlin, Heidelberg, New York, Tokyo, 2nd ed. 1986.
16. Has'minskĭı, R. Z.: Stochastic stability of differential equations. Sijthoff and Noordhoff, Al phen 1980.
17. Horsthemke, W., Lefever, R.: Noise-induced transitions. Springer, Berlin, Heidelberg, New York, Tokyo 1984.
18. Knobloch, H. W., Kappel, F.: Gewohnliche Differentialgleichungen. Teubner, Stuttgart 1974.
19. Knobloch, E., Wiesenfeld, K. A.: Bifurcations in fluctuating systems: The center-manifold approach. J. Stat. Phys., Vol. 33, no. 3 (1983), 611 - 637.
20. Kozin, F.: Stability of linear stochastic systems. In: R. Curtain (ed.): Stability of stochastic dynamical systems. Lecture Notes in Mathematics vol. 294. Springer, Berlin, Heidelberg, New York 1972.

21. Kunita, H.: Stochastic differential equations and stochastic flows of diffeomorphisms. Ecole d'été de Probabilités de Saint-Flour XII- 1982, 143 - 303. Lecture Notes in Mathematics vol. 1097. Springer, Berlin, Heidelberg, New York, Tokyo 1984.
22. Lyapunov, A. M.: Problème générale de la stabilité du mouvement. Comm. Soc. Math. Kharkov 2 (1892), 3 (1893), 265 - 272. Ann. Fac. Sci. Toulouse 9 (1907), 204 - 474. Reprint: Ann. of Math Stu dies 17, Princton University Press, Princton 1949.
23. Mañé, R.: Ergodic theory and differentiable dynamics. Springer, Berlin, Heidelberg, New York, London, Paris, Tokyo 1987.
24. Oseledec, V. I.: A multiplicative ergodic theorem. Lyapunov characteristic numbers for dynamical systems. Trans. Moscow Math. Soc. 19 (1968), 197 - 23 1.
25. Pardoux, E., Wihstutz, V.: Lyapunov exponent and rotation number of two dimensional linear stochastic systems with small diffusion. SIAM J. Appl. Math. 48, No. 2 (1988), 4 42 - 457.
26. Pinsky, M., Wihstutz, V.: Lyapunov exponents of nilpotent Ito-systems. Report No. 177, Institut für Dynamische Systeme, Universität Bremen 1987.
27. Ruelle, D.: Ergodic theory of differentiable dynamical systems. Publ. Math. I.H.E.S. 50 (1979), 27 - 58.
28. Stratonovich, R. L.: Topics in the theory of random noise, vol. 1. Gordon and Breach, New York 1963.
29. Talay, D.: Simulation and numerical analysis of stochastic differential equations. This volume.
30. Wedig, W.: Stochastic bifurcation of nonlinear parametric systems. This volume.

Pitchfork and Hopf Bifurcations in Stochastic Systems - Effective Methods to Calculate Lyapunov Exponents

Walter Wedig

Abstract

Wind turbulences or rough surfaces generate parameter fluctuations in dynamic systems which destabilize the system's equilibrium or a time-in-variant velocity distribution with increasing fluctuation intensity. The paper gives two typical examples of mechanical problems. To calculate associated Lyapunov exponents we distinguish Hopf or Pitchfork bifurcation systems and introduce polar or hyperbolic coordinates, respectively.

By this technique we obtain a separation of stationary angle or ratio processes the invariant measures of which are determined by singular diffusion equations. Their solutions give the Lyapunov exponents according to Oseledec's multiplicative ergodic theorem. To find best solution methods we discuss orthogonal Fourier and Hermite expansions, analytic solutions extended by singular perturbation methods and numerical integration procedures by backward differences.

1 Introduction to Stochastic Bifurcations

In technical mechanics it is quite common to model dynamic systems by time-invariant equations of motion and to investigate them applying external excitations by harmonic functions. In many practical situations, excitations are more general and generated by gusty winds, rough surfaces or turbulent layers such that they contain a broad band of excitation frequencies. These time fluctuations are called ergodic processes if they are stationary in the sense that linear and higher-order time averages exist and are finite. For a stochastic modelling of ergodic excitations it is convenient to make use of gaussian white noise $\dot{W}_t$ or the Wiener process W_t with zero mean and normed correlation functions as follows:

$$E(\dot{W}_t) = 0, \qquad E(\dot{W}_t \dot{W}_s) = \delta(t-s), \tag{1}$$

$$\dot{Z}_t + \omega_g Z_t = \sigma \dot{W}_t, \qquad \omega_g > 0,\ Z_t \in \mathbb{R}. \tag{2}$$

Obviously, the delta-correlation (1) leads to a power spectrum which is uniformly distributed in the entire frequency range. Hence, from this we obtain a more realistic band-limited process Zt by applying a first-order shaping filter (2) where ω_g is the limiting frequency and σ denotes the intensity parameter of white noise. Subscipt t denotes the time-dependency of stochastic processes. Dots are abbreviations for differentiations with respect to time.

Moreover, a serious modelling of many technical systems shows that ergodic perturbations go into the equations of motion in additive terms as well as in multiplicative forms. The latter produces parameter fluctuations in the sense that a deterministic system parameter is superimposed by corresponding time fluctuations. The paper gives two typical examples of structural or fluid-dynamic problems described by second-order differential equations of the following type:

$$\ddot{X}_t + 2D\omega_1(1+\delta\dot{X}_t^2/\omega_1^2)\dot{X}_t + \omega_1^2(1+\gamma X_t^2 + \sigma\dot{W}_t/\sqrt{\omega_1})X_t = 0 \tag{3}$$

It is a nonlinear oscillator parametrically excited by white noise. It has the natural frequency ω_1, the dimensionless damping measure D, and the parameters γ and δ for the cubic nonlinearities in the restoring and in the damping term, respectively.

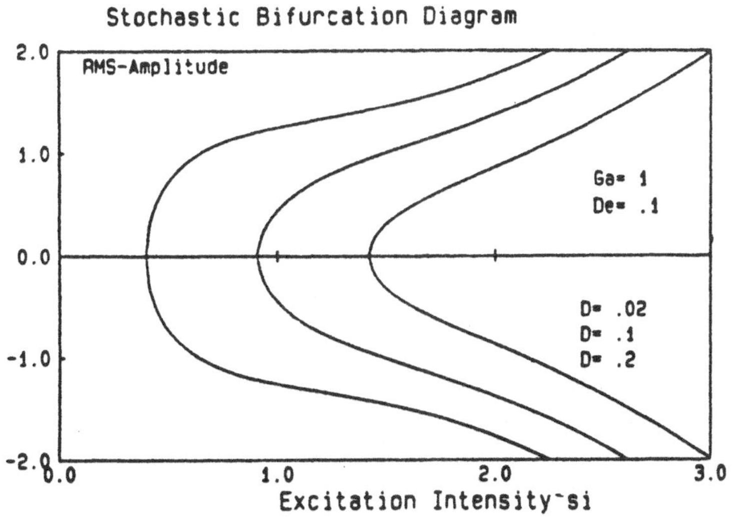

Fig. 1. Stochastic bifurcation with respect to noise

Equation (3) contains the deterministic solution $X_t \equiv 0$. It is the equilibrium position of the system. It is asymptotically stable for sufficiently small noise intensities σ and positive parameters D, ω_1, δ and $\gamma > 0$. However, for an increaing parameter fluctuation we reach a bifurcation point σ_{crit} at which the trivial solution becomes unstable with an exponential growth up to the nonlinear state space where the increasing behaviour is bounded by the cubic restoring and damping terms in (3). In Fig. 1 we find the corresponding bifurcation diagram calculated in a first approximation by Wedig [11]. From this it follows that equation (3) represents a mathematical model which possesses only two different states: a deterministic equilibrium solution and a stationary stochastic motion. The latter can also be called a chaotic motion in the sense that digital simulations are performed by applying a deterministic noise generator with a high sensitivity to initial conditions.

For practical applications it is important to calculate the bifurcation point σ_{crit} in dependence on given parameters D and ω_1 of the system. For this purpose it is sufficient to linearize the equation (3).

$$\ddot{X}_t + 2D\omega_1\dot{X}_t + \omega_1(\omega_1 + \sigma\sqrt{\omega_1}\dot{W}_t)X_t = 0, \qquad D, \omega_1 > 0, \tag{4}$$

$$\lambda = \lim_{t\to\infty} \frac{1}{t} \log \|\mathbf{X}_t\| \text{ a.s.}, \qquad \mathbf{X}_t = (X_t, \dot{X}_t)^T. \tag{5}$$

Introducing a suitable norm $\|.\|$ of the state vector $\mathbf{X}_t$ we investigate its growth behaviour by means of the Lyapunov exponent λ via Oseledec's multiplicative ergodic theorem [7]. For more details see Boxler [3] in this volume. Further results are given by Arnold and Kliemann [2], Kozin [5], Pardoux and Wihstutz [8] and Nishioka [6]. A comprehensive report on simulation and numerical analysis of higher-order systems is given by Talay [10] in this volume as well. It should be noted that analytic investigations of Lyapunov exponents are up to now restricted to two-dimensional problems where we obtain decoupled equations introducing polar coordinates. Then, the angular part of the solution rotates with a constant mean velocity called rotation number. The rotation possesses an invariant measure defined by the solution of singular Fokker-Planck equations. It determines the growth behaviour of the radial solution part and therewith the associated Lyapunov exponent as well.

Accordingly, in this paper we are most interested to calculate invariant measures of angle processes. This will be performed by different analytical and numerical approaches: orthogonal Fourier and Hermite expansions of the Fokker-Planck equations, analytlc solutions extended by singular perturbation methods and direct numerical integration procedures by backward differences. All solutions are compared with the results of Monte-Carlo simulations of the associated stochastic differential equations. Flnally, these investigations are extended to the three-dimensional case of parameter fluctuations by coloured noise (2). Moreover, there is a completely new aspect which may be important for future research. The transformation by polar coordinate is only appropriate in case of Hopf bifurcation systems like (4). In case of Pitchfork bifurcations, however, the rotation is stopped. Therefore, we replace the angular process by a ratio or direction process via hyperbolic coordinates to decouple it from the radial part of the solution. The ratio process has a well-defined invariant measure from which we calculate two different Lyapunov exponents of the entire solution. The almost sure (a.s.) stability boundary is evaluated in a simple closed-form expression in terms of the destabilizing noise intensity parameter and the stabilizing damping measure.

2 Mechanical Models with Parameter Fluctuations

In the following we give two typical examples of mechanical problems which lead to second-order systems with parameter fluctuations. They are classified by different eigenvalue distributions according to the deterministic theory of Hopf- and Pitchfork-bifurcations.

2.1 The Euler-Bernoulli Beam under Axial Loading

As a first example of stochastic bifurcation problems let us consider an Euler-Bernoulli beam under the axial loading P(t). Its bending and axial stiffness parameters are given by EI and EA, respectively. The beam has the mass μ per unit length 1 and an external damping denoted by β. Following Weidenhammer [14]. the transverse vibration w(x,t) of the range $0 \leq x \leq 1$ is described by the nonlinear partial integro-differential equation:

$$EIw_{xxxx} + \beta w_t + \mu w_{tt} - \frac{EA}{l}\left[u(t) + \int_0^1 \frac{1}{2} w_x^2 dx\right] w_{xx} = 0. \tag{6}$$

Here, subscipts t and x denote partlal derivatives with respect to time and space, respectively. Equation (6) holds under Kirchhoff's assumption (cf.[14]) that the axial extension of the beam is independent of x and only dependent upon time. Thus, the axial force P(t) produces an end displacement u(t) without any influence of longitudinal waves in the beam.

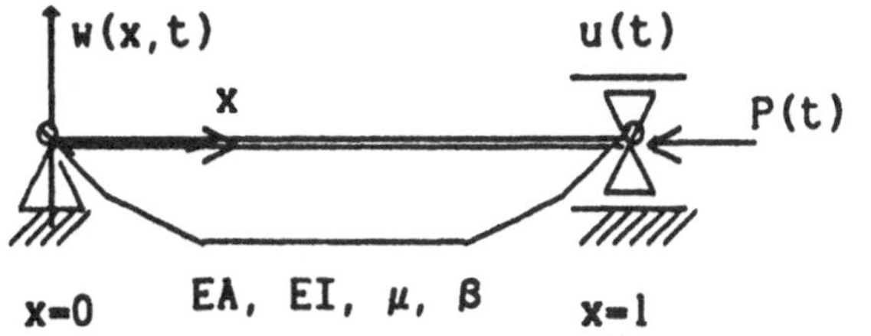

Fig. 2. Euler-Bernoulli beam under axial loading

As shown in Fig. 2, the beam is simply supported at both ends. Hence, the field equation (6) possesses the boundary condition

$$w(0,t) = w_{xx}(0,t) = w(l,t) = w_{xx}(l,t) = 0. \tag{7}$$

$$w(x,t) = T(t)\sin(\frac{\pi x}{l}), \qquad \omega_1^2 = \pi^4 \frac{EI}{\mu l^4}, \tag{8}$$

which is satisfied by the first mode $\sin(nx/l)$. It reduces equation (6) to the following nonlinear ordinary differential equation for T(t):

$$\ddot{T} + 2D\omega_1\dot{T} + \omega_1^2[1 + \varepsilon u(t)]Y + \gamma T^3 = 0. \tag{9}$$

In equation (9), ω_1 denotes the first natural frequency of the beam. It is given in (8). The two other parameters ε and γ can easily be computed. It is obvious that equation (9) is a first technical application of (3) in the sense mentioned before that the restoring parameter of the beam is superimposed by an ergodic time fluctuation u(t) coming from an external source through turbulent axial wind forces P(t). Hence, the straight equilibrium position $w(x,t) \equiv 0$ of the beam will be destabilized with increasing turbulence intensities ε of wind forces.

2.2 The Canal Flow on Rough Surfaces

As a next example, we consider the fluid-dynamic problem of a canal flow on rough surfaces to explain the transition of laminar and turbulent flows as a stochastic bifurcation phenomenon. Following Wedig [12], we reduce the fluid equation to a simple macroscopic model consisting of two masses m and two dashpots. Both masses are driven by constant forces $F/2$ to maintain stationary velocities v_1 and v_2. The viscosity of the fluid is modelled by the horizontal damper b. The second vertical one simulates the influence of the boundary layer and of the contact to the rough surface through the frictionless normal force N. Given a surface contour $u(s)$ and its derivatlve $u'(s) - \tan\alpha$ with respect to the way coordinate s we set up the following dynamic equations:

$$m\dot{v}_2 + b(v_2 - v_1) = \frac{F}{2}, \qquad N\cos\alpha = \beta b\dot{u}(s(t)),$$

$$m\dot{v}_1 + b(v_1 - v_2) + N\sin\alpha = \frac{F}{2}, \qquad \dot{u}(s(t)) = v_1 u'(s) \tag{10}$$

For the special case that the surface contour u(s) has the saw-tooth distribution shown in Fig. 3, its squared derivation is constant and not dependent upon s. Hence, the equations (10) yield the stationary solutions

$$[u'(s)]^2 = \alpha_0^2 \;\rightarrow\; V_{10} = \frac{F}{2b\beta\alpha_0^2}, \; V_{20} = (1 + \beta\alpha_0^2)v_{10}. \tag{11}$$

Results (11) represent two discrete values of a parabolic velocity distribution in the canal or in a pipe with the diameter D. The symmetric form sketched in Fig. 3 is obviously obtained if the applied model is en- larged by additional masses and supplemented by a symmetric boundary layer at the upper side of the mass chain. Then, the maximal value of this parabolic distribution is determined by the resistance coefficient $\gamma = \beta\alpha_0^2$ of the boundary layer.

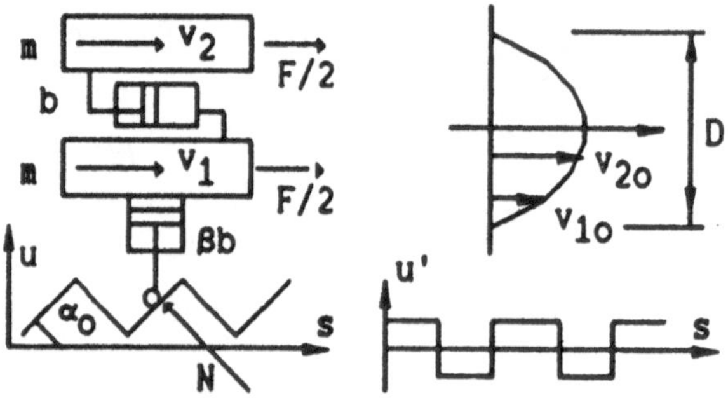

Fig. 3. Canal flow on rough surfaces

In practical environments, the regular boundary contour in Fig. 3 has to be generalized to irregular forms of natural surfaces. In a first step, these may be modelled by superimposing the saw-tooth distribution by spatial white noise W'_s derived from $\dot{W}_t$ by $s = v_{10}dt$ and $E(dW_s^2) = v_{10}dt$.

$$[u'(s)]^2 = \alpha_0^2 + \sigma W'_s, \qquad W'_s = \sqrt{v_{10}}\dot{W}_t, \tag{12}$$

$$v_1(t) = v_{10} + x_1(t), \qquad v_2(t) = v_{20} + x_2(t). \tag{13}$$

As a consequence, the stationary velocities, calculated in (11), can be destabilized. To check this we introduce the perturbations (13) and derive the following stability equations:.

$$\dot{X}_{2t} + \delta(X_{2t} - X_{1t}) = 0, \qquad \dot{X}_{1t} + \delta(X_{1t} - X_{2t}) + \delta\gamma X_{1t} + \sqrt{\delta\rho}\dot{W}_t X_{1t} = 0. \tag{14}$$

By elimination of X_{1t}, the first order system (14) can be written as

$$\ddot{X}_{2t} + 2\delta\dot{X}_{2t} + (\delta\gamma + \sqrt{\delta\rho}\dot{W}_t)(\dot{X}_{2t} + \delta X_{2t}) = 0, \tag{15}$$

$$\delta = b/m, \quad \gamma = \beta\alpha_0^2, \quad \rho = v_{10}(\beta\sigma)^2 b/m \tag{16}$$

where δ, γ and ρ are damping, resistance and noise intensity parameters noted in (16). Hence, with the equation (15) we finally obtained a second example where deterministic system parameters are perturbed by ergodic time fluctuations. Again, they are coming from an external noise source, here from an irregular surface contour. The noise is transported into the equation of motion through the boundary layer and the frictionless contact on the rough surfaces.

2.3 Eigenvalues of Pitchfork and Hopf Bifurcations

As already mentioned, the deterministic velocities (11) are destabilized with increasing noise parameter ρ and stabilized by the resistance coefficient γ of the boundary layer. In the limiting case that the system reaches the bifurcation point, both opposite influences are balanced. Therefore, we can characterize this critical situation by

$$\ddot{x}_2 + 2\delta\dot{x}_2 = 0, \qquad \lambda_1 = 0, \ \lambda_1 = -2\delta, \tag{17}$$

what is simply obtained from equation (15) by omitting both parameters and ρ. The reduced system has the eigenvalues λ_1 and λ_2 which are well-known from the situation of a deterministic pitchfork bifurcation. Since they are real-valued, the introduction of hyperbolic coordinates

$$a = \sqrt{x_1 x_2} \ (0 \le a < \infty), \qquad r = \frac{x_2}{x_1}, \ (0 < r < \infty) \tag{18}$$

is convenient to investigate corresponding stochastic eigenvalues for non-vanishing parameter fluctuations $\rho > 0$ and effected response fluctuations around the trivial solution $x_1 = x_2$.

An analoguous discussion can be performed for the bifurcation problem defined by equation (4). If we omit the destabilizing noise term σ and the stabilizing damping coefficient D we obtain the harmonic oscillator

$$\ddot{x} + \omega_1^2 x = 0, \qquad \lambda_1 = i\omega_1, \ \lambda_2 = -i\omega_1, \tag{19}$$

$$x_1 = a\cos\phi,\ x_2 = a\sin\phi,\quad (0 \leq a < \infty,\ -\pi \leq \phi \leq \pi) \tag{20}$$

with purely imaginary eigenvalues. Obviously, this is the well-known situation of a deterministic Hopf bifurcation. Due to Khasminskii [4] , the polar coordinates (20) are introduced to investigate the corresponding stochastic eigenvalues for non-vanishing parameter fluctuations $\sigma > 0$. A summarizing representation of both eigenvalue distributions and coordinate transformations is given in Fig. 4. More details on backgrounds and justifications are given in the following.

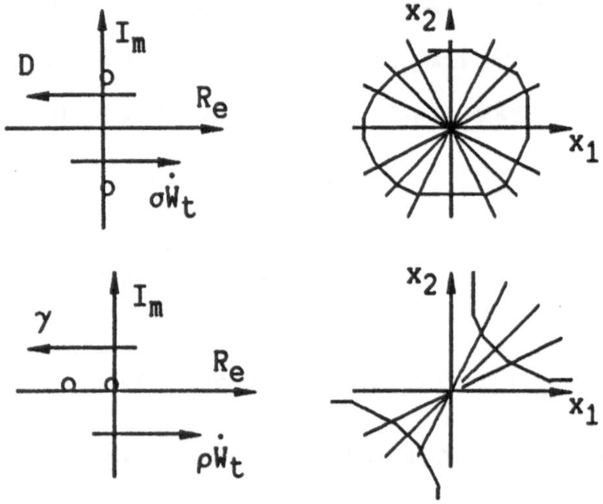

Fig. 4. Coordinate transformations and eigenvalue distributions of pitchfork and Hopf systems

3 Stability Analysis of the Pitchfork System

For the pitchfork system we introduce hyperbolic coordinates. Therewith, we derive an invariant measure for the stationary ratio process. It determines two different Lyapunov exponents and the almost sure stability boundary of the system.

3.1 Transformation via Hyperbolic Coordinates

Including Wong-Zakai [15] or Stratonovich [9] correction terms we rewrite the ordinary differential equations (14) in the form of stochastic Itô equations.

$$dX_{2t} = -\delta(X_{2t} - X_{1t})dt, \qquad E[(dW_t)^2] = dt,$$

$$dX_{1t} = -\delta(X_{1t} - X_{2t})dt - \delta(\gamma - \frac{\rho}{2})X_{1t}dt - \sqrt{\delta\rho}X_{1t}dW_t. \tag{21}$$

Subsequently, we introduce the ratio or direction process R_t and the radial part or norm process A_t according to the hyperbolic coordinates (18). Applying Itõ's calculus (cf. Arnold [1]) we derive the following stochastic differential equations for both processes:

$$dR_t = -\delta\Big[R_t^2 - 1 - (\gamma + \frac{\rho}{2})r_t\Big]dt + \sqrt{\rho\delta}R_t dW_t, \tag{22}$$

$$dA_t = -\frac{1}{2}\left[2 + \gamma - \frac{\rho}{2} - R_t - \frac{1}{R_t}\right]A_t dt - \frac{1}{2}\sqrt{\rho\delta}a_t dW_t. \tag{23}$$

Obviously, we obtained a separation of the ratio process R_t. It lives on a hyperbolic coordinate line without any influence of the radial process A_t.

For later evaluations, it is important to consider the limiting case of vanishing parameter fluctuations $\rho = 0$. Then, the trivial solutions X_{1t}, $X_{2t} \equiv 0$ are asymptotically stable for $\gamma > 0$ since the linear system (21) possesses the deterministic eigenvalues

$$\lambda_I = -\delta\left[1 + \frac{\gamma}{2} - \sqrt{1 + \left(\frac{\gamma}{2}\right)^2}\right], \; \lambda_{II} = -\delta\left[1 + \frac{\gamma}{2} + \sqrt{1 + \left(\frac{\gamma}{2}\right)^2}\right] \tag{24}$$

To derive now corresponding results from the transformed equations (22) and (23) we calculate their stationary solutions for $\rho = 0$

$$A_t \equiv 0, \; R_t \equiv r_0, \qquad r_0 = \frac{\gamma}{2} + \sqrt{1 + (\frac{\gamma}{2})^2} \tag{25}$$

Again, these solutions are asymptotically stable since the associated stability equations yield the eigenvalues

$$\lambda_a = -\frac{1}{2}\delta\Big(2 + \gamma - r_0 - \frac{1}{r_0}\Big) = -\delta\Big(1 + \gamma - r_0\Big), \quad \lambda_r = -\delta(2r_0 - \gamma). \tag{26}$$

From this it follows that $\lambda_I = \lambda_a$ and $\lambda_{II} = \lambda_r + \lambda_a$

3.2 Invariant Measure of the Ratio Process

To derive corresponding results for the stochastic case ($\rho > 0$) we set up the following Fokker-Planck equation for the ratio process R_t:

$$\frac{\partial p}{\partial t} + \frac{\partial}{\partial r}\left\{-\delta\Big[r^2 - 1 - (\gamma + \frac{\rho}{2})r\Big]p(r)\right\} - \frac{1}{2}\frac{\partial^2}{\partial r^2}\Big[\rho\delta r^2 p(r)\Big] = 0. \tag{27}$$

In the stationary case $\partial p/\partial t = 0$, equation (27) has the solution

$$p(r) = Cr^{-1+2\frac{\gamma}{\rho}} \exp\left[-\frac{2}{\rho}(r + \frac{1}{r})\right], \quad 0 < r < \infty \tag{28}$$

where C is an integration constant determined by the normalization condition. The density function (28) is only valid in the positive distribution range $r > 0$. Fig. 5 shows evaluations of p(r) for the resistance parameter $\gamma = 1$ and different noise intensities ρ. Hereby, we applied the linear scaling $z = r$ for low distributions $r \leq 1$ and a contraction by $z = 2 - 1/r$ in the upper range $r \geq 1$. In Fig. 5 we recognize that the process R_t possesses a well-defined mean value taking values in the positive range $0 < r < \infty$, only. The extreme values $(0, \infty)$ are excluded with probability one. Therefore, if the process R_t is started in the positive range, it remains in it with probability one. Starting conditions $r < 0$ are excluded in this stationary situation. If they occur nevertheless, R_t is unstable, i.e. there is a transient rotation in the state space up to a re-entering into the positive range $0 < r < \infty$.

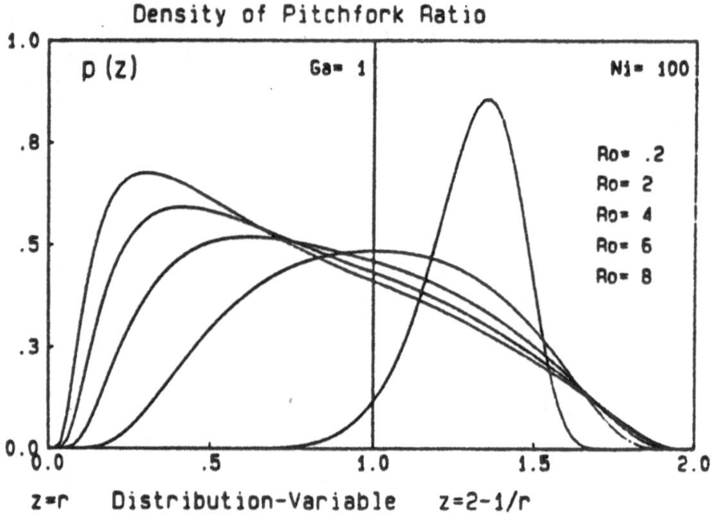

Fig. 5. Stationary density of the ratio process

To calculate the normalization constant in (28) and the mean value mentioned above we need evaluations of the density integrals

$$J_\nu = \int_0^\infty r^{\nu-1} \exp\left[-\frac{2}{\rho}\left(r + \frac{1}{r}\right)\right] dr = 2K_\nu\left(\frac{4}{\rho}\right), \tag{29}$$

$$J_\nu \int_0^1 \left(r^{\nu-1} + \frac{1}{r^{\nu+1}}\right) \exp\left[-\frac{2}{\rho}\left(r + \frac{1}{r}\right)\right] dr. \tag{30}$$

The integrations can be performed by means of Bessel functions K_ν or directly by numerical procedures according to formula (30). In the special cases $\nu = 1/2$ or $\nu = 3/2$ the Bessel functions are reducible to simple exponentials of the form

$$J_{1/2} = \sqrt{\frac{\pi\rho}{2}} \exp\left(-\frac{4}{\rho}\right), \qquad J_{3/2} = \sqrt{\frac{\pi\rho}{2}} \left(1 + \frac{\rho}{4}\right) \exp\left(-\frac{4}{\rho}\right). \tag{31}$$

Knowing the mean value $E(R_t)$, higher-order moments can be calculated by means of the recurrence formula

$$E(R_t^{n+1}) = \left(\gamma + \frac{\rho}{2}\right) E(R_t^n) + \left[1 + \rho\frac{(n-1)}{2}\right] E(R_t^{n-1}). \qquad (32)$$

It follows from equation (22) by applylng Itõ's calculus to the increment dR_t^n and holds for all natural numbers n except n = 0. For this we have to derive the increment of the natural logarithm of R_t.

$$d(\log R_t) = \delta\left(\frac{1}{R_t} - R_t + \gamma\right) dt + \sqrt{\rho\delta} dW_t, \rightarrow \ E\left(\frac{1}{R_t}\right) = E(R_t) - \gamma. \quad (33)$$

In the stationary case, this gives the expected value of the inverse process in dependence of its mean value and the resistance parameter γ.

3.3 Lyapunov Exponents and Almost Sure Stability

Provided that the separated ratio process R_t is positive and stationary, the growth behaviour of the radial norm process A_t can easily be determined with probability one. According to Oseledec's multiplicative ergodic theorem (5) we derive from equation (23) the increment of the natural logarithm of A_t and take the expected values on both sides of the equation (cf. [8]).

$$d(\log A_t) = -\frac{1}{2}\delta\left(2 + \gamma - R_t - \frac{1}{R_t}\right) dt - \frac{1}{2}\sqrt{\rho\delta} dW_t, \text{ for } t \rightarrow \infty, \qquad (34)$$

$$\dot{E}(\log A_t) = -\frac{1}{2}\delta\left[2 + \gamma - E(R_t) - E(\frac{1}{R_t})\right] dt \ \rightarrow \ E(\log A_t) = \lambda_I t. \qquad (35)$$

For $t \rightarrow \infty$, the ratio moments $E(R_t)$ and $E(1/R_t)$ are constant. Hence, the resulting $\log A_t$-moment is simply proportional to the time t. Its growth behaviour is determined by the Lyapunov exponent λ_I.

$$\lambda_I = -\frac{1}{2}\delta\left[2 + \gamma - E(R_t) - E(\frac{1}{R_t})\right] = -\delta\left[1 + \gamma - E(R_t)\right]. \qquad (36)$$

For $\rho = 0$, λ_I coincides with the deterministic result noted in (24).

In a next step, the second Lyapunov exponent is calculated. Similar as shown previously, we investigate the almost sure stability condition of the statlonary ratio process by $R_t = R_t + \Delta R_t$. Inserted into equation (22) we derive the linearized perturbation equation (37) and take the increment of the log-process to determine the growth behaviour of ΔR_t.

$$d(\Delta R_t) = \delta(\gamma - 2R_t + \frac{\rho}{2})\Delta R_t dt + \sqrt{\rho\delta}\Delta R_t dW_t, \text{ for } t \rightarrow \infty, \qquad (37)$$

$$d(\log \Delta R_t) = \delta(\gamma - 2R_t)dt + \sqrt{\rho\delta} dW_t, \rightarrow \ \lambda_2 = -\delta[2E(R_t) - \gamma]. \qquad (38)$$

Since λ_2 is always negative the ratio process R_t is stationary with probability one. Moreover, the second Lyapunov exponent is given by

$$\lambda_{II} = \lambda_I + \lambda_2, \rightarrow \ \lambda_{II} = -\delta[1 + E(R_t)]. \qquad (39)$$

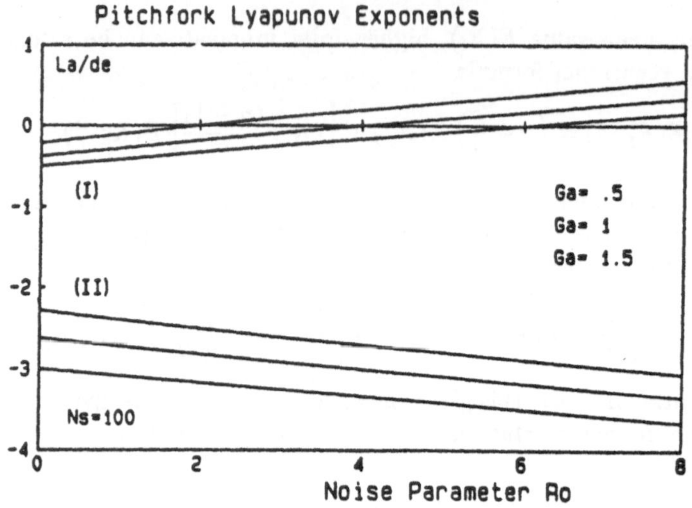

Fig. 6. Growth behaviour of the Lyapunov exponents

Replacing $E(R_t)$ by the deterministic value (25), λ_{II} coincides with the corresponding result for $\rho = 0$ which has been derived in equation (24).

In fig. 6, we find some evaluations of both Lyapunov exponents (36) and (39) in dependence of the noise parameter ρ for three different values of the resistance coefficient γ. The step size, applied in the numerical integration of the density integral (30), is given by $N_s = 100$. We recognize that the second Lyapunov exponent is decreasing with increasing noise parameter. The first Lyapunov exponent λ_I is also negative for small noise intensities. However, for increasing ρ, it reaches a bifurcation point ρ_{crit} at which there is a change of sign in λ_I. Consequently, the stationary velocity distribution v_{10} and v_{20} of the original pitchfork system (10) becomes unstable beyond this bifurcation point. The associated stochastic solutions have an exponential growth up to the nonlinear state range. In particular, Fig. 6 shows clearly that the zero crossings of λ_I are proportional to γ, i.e. the almost sure stability boundary is simply

$$\rho_{\text{crit}} = 4\gamma. \tag{40}$$

This can be verified by applying the special integrals (31) and inserting them into the top Lyapunov exponent λ_I.

To finish the stability analysis it is worthwile to introduce physical parameters like Reynold's number R_e, loss of pressure coefficient λ_{lam} of the laminar flow and a roughness parameter k_s which describes the influence of the stochastic boundary contour. Analoguous to a continuously distributed fluid of the diameter D they are defined as follows:

$$R_e = \frac{v_{10}}{Db/m}, \quad \lambda_{\text{lam}} = \frac{F/(mD)}{2(b/m)^2}, \quad k_s = D\left(\frac{\sigma\beta b}{m}\right)^2, \tag{41}$$

$$\gamma = \frac{\lambda_{\text{lam}}}{R_e} \quad \rho = k_s R_e \;\rightarrow\; k_s = 4\frac{\lambda_{\text{lam}}}{R_e^2}. \tag{42}$$

Therewith, we can express the parameters ρ and γ. Inserted into the stability condition (40) we arrive at the relation (42), i.e. the critical roughness k_s is inversely proportional to the squared Reynold's number R_e. The stability result (42) corresponds to main effects of laminar-turbulent transitions. According to well-known experiments we know that laminar flows are realizable up to high Reynold's numbers for sufficently smooth test canals. This property is shown in Fig. 7, where the roughness parameter k_s is plotted against the Reynold's number for different loss of pressure coefficients. However, there is still a lacking of the second important effect that there is always a lower critical Reynold's number under which the laminar flow can not be destabilized even in the limit that k_s tends to infinity. We expect this second effect when the applied roughness contour is more realistically modelled by filtered processes.

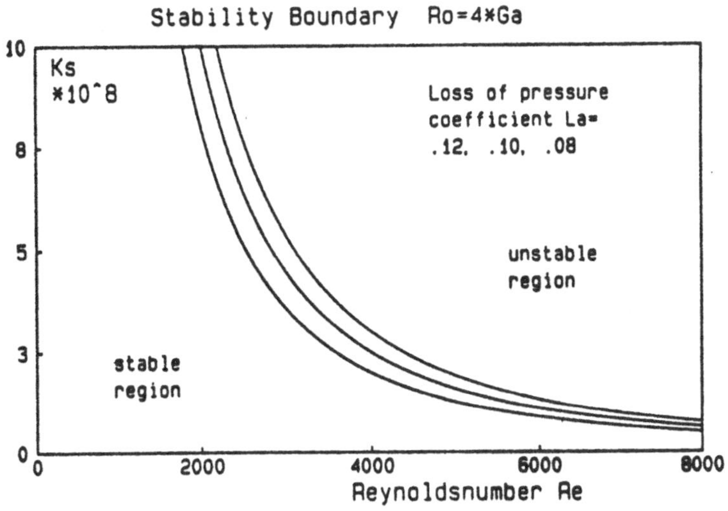

Fig. 7. Stability boundaries of the laminar flow

4 Eigenvalues of the Stochastic Hopf Oscillator

Due to the eigenvalue distribution of the Hopf oscillator there is a rotation in the state space. It is described by an invariant double-periodic measure of the angular process which determines the associated rotation number and the Lyapunov exponent of the system.

4.1 Transformation by Polar Coordinates

Introducing the dimensionless state processes X_{1t} and X_{2t}, the ordinary differential equation (4) of the Hopf oscillator goes over to a first order system of stochastic Itõ equations.

$$\left.\begin{aligned} dX_{1t} &= \omega_1 X_{2t} dt, \quad X_{1t} = X_t, X_{2t} = \dot{X}_t/\omega_1, \\ dX_{2t} &= -2D\omega_1 X_{2t} dt - \omega_1 X_{1t} dt - \sigma\sqrt{\omega_1} X_{1t} dW_t. \end{aligned}\right\} \tag{43}$$

According to (20) we apply the polar coordinate processes

$$A_t = \sqrt{X_{1t}^2 + X_{2t}^2}, \qquad \Psi_t = \mathrm{arctg}\frac{X_{2t}}{X_{1t}} \tag{44}$$

Here, Ψ_t is an angular process which rotates clockwise in the state space. The amplitude represents the euklidian norm of the state processes and gives a measure for their growth behaviour. According to Oseledec's multiplicative ergodic theorem we set up the increments of the natural logarithm of A_t and of the phase process Ψ_t.

$$d\log A_t = -\omega_1[2D\sin^2\Psi_t - \frac{1}{2}\sigma^2\cos 2\Psi_t\cos^2\Psi_t]dt - \frac{\sigma}{2}\sqrt{\omega_1}\sin 2\Psi_t dW_t, \tag{45}$$

$$d\Psi_t = -\omega_1[1 + 2D\sin\Psi_t\cos\Psi_t + \sigma^2\sin\Psi_t\cos^3\Psi_t]dt - \sigma\sqrt{\omega_1}\cos 2\Psi_t dW_t \tag{46}$$

Obviously, the angular process Ψ_t is decoupled from the growth behaviour of the amplitude process A_t. Provided there is a stationary rotation in the state space, the phase process Ψ_t possesses an invariant measure which determines the rotation number α and the Lyapunov exponent λ by

$$\lambda = \lim_{t\to\infty}\frac{1}{t}\log A_t, \text{ (a.s.)}, \qquad \alpha = \lim_{t\to\infty}\frac{1}{t}\log\Psi_t, \text{ (a.s.)} \tag{47}$$

$$\lambda = \lim_{t\to\infty}\frac{1}{t}E(\log A_t), \qquad \alpha = \lim_{t\to\infty}\frac{1}{t}E(\psi_t). \tag{48}$$

Moreover, the formulas (47) remain valid if we apply the expectation operator on both sides and interchange it with time averages. The expected values in (48) can be calculated from the Itõ equations (45) and (46).

4.2 Fourier Expansion of the Phase Distribution

We start setting up the Fokker-Planck equation of the phase (46).

$$\begin{aligned} &\frac{\partial p}{\partial t} + \frac{\partial}{\partial\psi}\Big[-\omega_1(1 + 2D\sin\psi_t\cos\psi_t + \sigma^2\sin\psi_t\cos^3\psi_t)p\Big] \\ &-\frac{1}{2}\frac{\partial^2}{\partial\psi^2}(\sigma^2\cos^4\psi p) = 0. \end{aligned} \tag{49}$$

For the stationary case, $\partial p/\partial t = 0$, we apply the Fourier expansion [13]

$$p(\psi) = \frac{1}{\pi}\left[\frac{1}{2} + \sum_{n=1}^{\infty}(c_n \cos 2n\psi + s_n \sin 2n\psi)\right], \quad 0 \le \psi \le 2\pi \tag{50}$$

to calculate a periodic solution of the phase density distribution for the angular range noted in (50). Obviously, the Fourier series in (51) satisfies the normalization condition for this range. Inserting the expansion into the stationary diffusion equation (49) and comparing all coefficients of the trigonometric functions we obtain the following infinite set of algebraic equations for the determination of c_n and s_n:

$$\begin{aligned} c_n &+ \frac{3}{8}\sigma^2 n s_n + \frac{1}{16}\sigma^2\Big[(n+1)s_{n+2} + (n-1)s_{n-2}\Big] \\ &+ \frac{1}{2}\Big[(2n+1)\frac{1}{4}\sigma^2 + D\Big]s_{n+1} + \frac{1}{2}\Big[(2n-1)\frac{1}{4}\sigma^4 - D\Big]s_{n-1} = 0 \end{aligned} \tag{51}$$

$$\begin{aligned} s_n &+ \frac{3}{8}\sigma^2 n c_n - \frac{1}{16}\sigma^2\Big[(n+1)c_{n+2} + (n-1)c_{n-2}\Big] \\ &- \frac{1}{2}\Big[(2n+1)\frac{1}{4}\sigma^2 + D\Big]c_{n+1} - \frac{1}{2}\Big[(2n-1)\frac{1}{4}\sigma^4 - D\Big]c_{n-1} = 0 \end{aligned} \tag{52}$$

Because of $c_0 = 1$ and $s_0 = 0$, equations (51) and (52) are inhomogeneous. According to [11], they are solved by truncating them at a certain order n with the closure conditions $c_n = s_n = 0$.

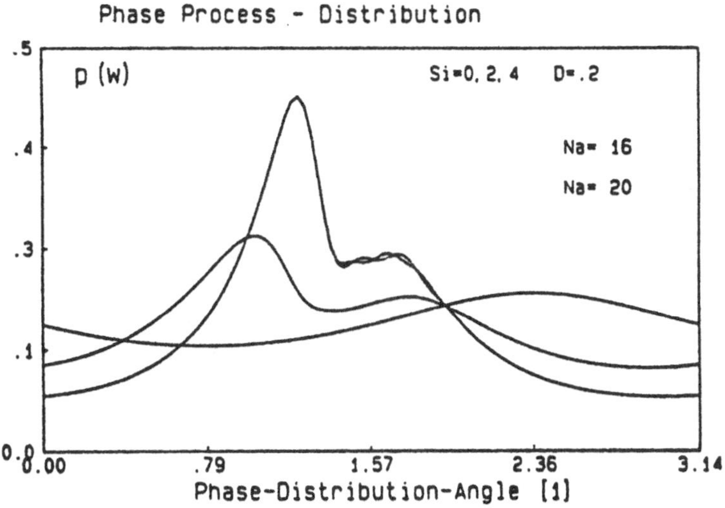

Fig. 8. Fourier expansion of the phase distribution.

Fig. 8 shows corresponding evaluation results for the expansion orders $N_a = 16$ and $N_a = 20$ performed by neglecting higer-order Fourier terms. Inserting the results, obtained from the equations (51) and (52), into the expansion (47) we get the phase process distribution in Fig. 8. This calculation is performed for the damping parameter $D = 0.2$ and for the three different intensities $\sigma = 0, 2$ and 4. The comparison of the two expansion orders $N_a = 16$ and $N_a = 20$ shows clearly that the Fourier series of the phase distribution has convergent properties. Only for the highest noise parameter $\sigma = 4$ we observe small deviations of the applied expansion orders in the range around $\psi = \pi/2$.

4.3 Lyapunov Exponent and Rotation Number

To apply the Fourier solution (50) of the periodic phase distribution we rewrite the equations (45) and (46) into the forms

$$d(\log A_t) = -\omega_1 \left[D - \frac{\sigma^2}{8} - (D + \frac{\sigma^2}{4}) \cos 2\Psi_t - \frac{\sigma^2}{8} \cos 4\Psi_t \right] dt \\ -\frac{\sigma}{2}\sqrt{\omega_1} \sin 2\Psi_t dW_t, \tag{53}$$

$$d\Psi_t = -\omega_1 \left[1 + (D + \frac{\sigma^2}{4}) \sin 2\Psi_t + \frac{\sigma^2}{8} \sin 4\Psi_t \right] dt \\ -\frac{\sigma}{2}\sqrt{\omega_1}(1 + \cos 2\Psi_t) dW_t \tag{54}$$

and take the expectations in both stochastic Itõ equations. Therewith, the diffusion terms are vanishing. Only, the drift terms are remaining. They lead to deterministic differential moment equations uhich can easily be integrated. For $t \to \infty$, their solutions are simply proportional to the time t. Consequently, the application of the formulas (48) leads to

$$\lambda = -\omega_1 \left[D - \frac{\sigma^2}{8} - (D + \frac{\sigma^2}{4}) E(\cos 2\Psi_t) - \frac{\sigma^2}{8} E(\cos 4\Psi_t) \right], \tag{55}$$

$$\alpha = -\omega_1 \left[1 + (D + \frac{\sigma^2}{4}) E(\sin 2\Psi_t) + \frac{\sigma^2}{8} E(\sin 4\Psi_t) \right] \tag{56}$$

This is a close-form result for the Lyapunov exponent λ and the rotation number a of the Hopf oscillator (4). Both are evaluable when the Fourier coefficients c_n and s_n of the phase distribution $p(\psi)$ are known.

According to the orthogonality condition of Fourier expansions they are calculable by

$$c_n = \int_0^{2\pi} \cos(2n\psi) p(\psi) d\psi = E(\cos 2n\psi_t), \tag{57}$$

$$s_n = \int_0^{2\pi} \sin(2n\psi) p(\psi) d\psi = E(\sin 2n\psi_t). \tag{58}$$

Hence, the coefficients c_n and s_n are the Fourier moments of the phase process which are also calculable from the algebraic equations (51) and (52), directly without any knowledge of $p(\psi)$. Moreover, we need only the first four Fourier moments to determine the Lyapunov exponent (55) and the rotation number (56). Both stochastic eigenvalues are evaluated in Fig. 9 for the damping coefficients $D = 0$, 0.4 and 0.8 and for the expansion orders $N_a = 8$ and $N_a = 10$ of corresponding Fourier terms. Differences of the two evaluation numbers are only observable in the range of higher intensities $\sigma > 4$. Furthermore, we get the well-known deterministic eigenvalues $\lambda = -D\omega_1$ and $\alpha = \omega_1$ in the limiting case of vanishing parameter fluctuation.

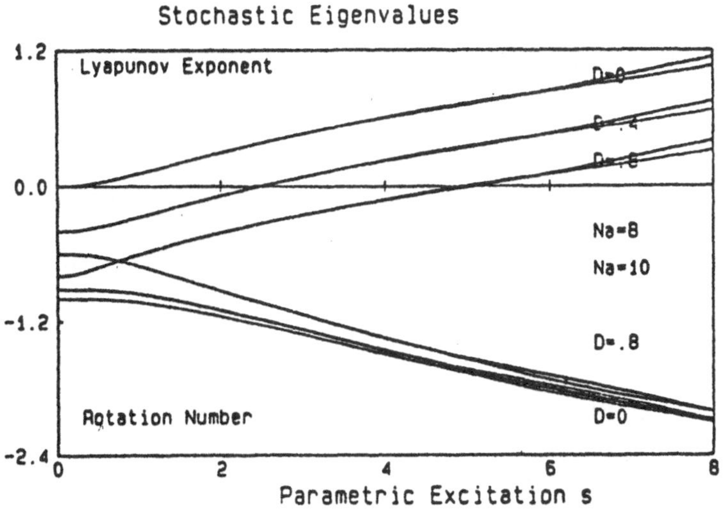

Fig. 9. Lyapunov exponent and rotation number in dependence of the fluctuation intensity

5 Singular Diffusion Equation of the Phase

In the following we verify the physical existence of the invariant phase distribution and the convergence of the applied Fourier expansion. For this purpose, further analytical and numerical approaches are applied.

5.1 Analytic Solution and Singular Perturbation Method

The stationary condition $\partial p/\partial t = 0$ of the phase distribution leads to a parabolic degeneration with the consequence that the Fokker-Planck equation (49) possesses two singularities in $\psi_0 = \pm\pi/2$. This becomes obvious when we perform a first integration in the differential equation (49), which is linear, ordinary and of second order for $\partial p/\partial t = 0$.

$$\frac{\sigma^2}{2}\cos^4\psi p'(\psi) + \Big[1 + (2D - \sigma^2\cos^2\psi)\sin\psi\cos\psi\Big]p(\psi) = C \tag{59}$$

$$p'(\psi) = \frac{dp}{d\psi}, \quad -\frac{\pi}{2} \leq \psi \leq +\frac{\pi}{2}, \quad C = p(\pm\frac{\pi}{2}. \tag{60}$$

Here, C is a constant of integration which has to be regarded to remove the singularities mentioned above. To determine C we insert the angles $\psi_0 = \pm\pi/2$ into the singular equation (59). Provided that $p'(\pi/2)$ is finite, the integration constant C is determined by $p(\pm\pi/2)$ which is finally determined by the normalization condition of the periodic density $p(\psi)$.

Outside the singularity range of a certain angle ψ_s a second integration of the equation (59) can be performed by means of classical rules of ordinary differential equations. Accordingly, we obtain an homogeneous solution of the form

$$p_h(\psi) = \frac{1}{\cos^2\psi}\exp\left[-\frac{2}{\sigma^2}(\tan\psi + Dtg^2\psi + \frac{1}{3}\tan^3\psi)\right] \tag{61}$$

and complete it by the associated inhomogeneous part. This result in

$$p(\psi) = p(\psi_s) + \frac{2}{\sigma^2}p\left(-\frac{\pi}{2}\right)\int_{\psi_s}^{\psi}\frac{1}{\cos^2\phi\cos^2\psi}\exp\left[\frac{2}{\sigma^2}\Big[(\tan\phi - \tan\psi)\right.$$
$$\left. + D(\tan^2\phi - \tan^2\psi) + \frac{(\tan^3\phi - \tan^3\psi)}{3}\Big]\right]d\phi,$$
$$\psi_s - \frac{\pi}{2} \leq \psi \leq \frac{\pi}{2} - \psi_s. \tag{62}$$

As already mentioned, the analytical solution (62) is valid in the non-singular range, noted in (62).

To calculate now the solution inside the singular range we apply the singular perturbation method by means of a Taylor series. This expansion is exactly started in the singular point $\psi_0 = -\pi/2$ or in $\psi_0 = +\pi/2$.

$$p(\psi) = \sum_{n=0}^{\infty}\frac{1}{n!}p^{(n)}(\frac{\pi}{2})(\psi - \frac{\pi}{2})^n, \quad \frac{\pi}{2} - \psi_s \leq \psi \leq \frac{\pi}{2} + \psi_s \tag{63}$$

We determine the Taylor coefficients $p^{(n)}(\pi/2)$ by differentiating the linear equation (59) with respect to the phase angle ψ.

$$\left[\frac{\sigma^2}{2}\cos^4\psi p'(\psi) + [1 + (2D - \sigma^2\cos^2\psi)\sin\psi\cos\psi]p(\psi)\right]^{(n)} = 0 \tag{64}$$

$$p^{(1)}(\frac{\pi}{2}) = 2Dp(\frac{\pi}{2}),$$
$$p^{(2)}(\frac{\pi}{2}) = 8D^2p(\frac{\pi}{2}), \tag{65}$$

$$p^{(3)}(\frac{\pi}{2}) = \left[8D(6D^2-1) - \frac{3}{2}\sigma^2\right]p(\frac{\pi}{2}), \ldots \qquad (66)$$

which gives us the derivatives of the phase distribution in dependence of the normalization constant $p(\pi/2)$. The continuation of this calculation leads to the following recurrence formula by which the derivatives in the singular points $\pm\pi/2$ are determined for all n = 0,1,2,....

$$p^{(n)}(\frac{\pi}{2} = 2\sum_{m=0}^{(n-1)/2}(-4)^m(2^n_{m+1})\left[D + \frac{n-m}{m+1}(4^m-1)\frac{\sigma^2}{4}\right]p^{n-2m-1}(\frac{\pi}{2}) \qquad (67)$$

In Fig. 10 we show an evaluation of both solutions (62) and (63) in comparison with the Fourier expansion (50) for the damping measure D=0.2 and the two parameters $\sigma = 2$ and 4. We recognize that both representations of the diffusion solution coincide graphically. The applied Fourier terms are $N_f = 30$. $N_s = 500$ gives the step size of integration in (62) and $N_f = 30$ the Tayor terms in (63). The complete analytic solution is started with the Taylor representation in $-\pi/2$ up to the end value $p(\psi_s)$ at the first interrupted line at ψ_s. Subsequently, it is continued by the integration form (62) with the initial value $p(\psi_s)$.

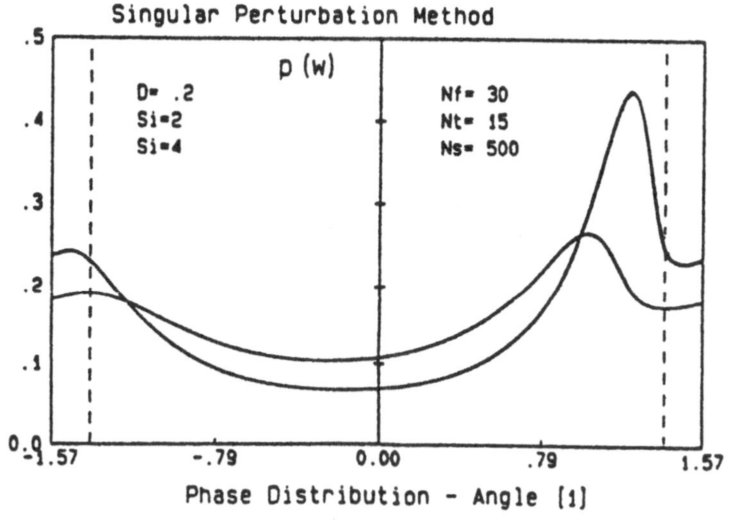

Fig. 10. Analytic Fokker-Planck solutions in comparison with the Fourier representation

5.2 Monte-Carlo simulations of the phase distribution

The physical existence of periodic phase densities and associated invariant measures can be proofed by calculating the Lyapunov exponent of the stochastic

Ito equation (46), similar as already done in section 3.3. Therefore, it is more interesting to verify this existence simply by means of Monte-Carlo simulations of the stochastic phase equation. For this purpose we apply an explicit Euler scheme via forward differences. The increments of the Wiener process are discretized by random number Z_n which are normally distributed and statistically independent with zero mean and the normalized square mean $E(Z_n^2) = 1$.

$$\psi_{n+1} = \psi_n - \omega_1(1 + 2D\sin\psi_n\cos\psi_n + \sigma^2\sin\psi_n\cos^3\psi_n)\Delta t$$
$$-\sigma\sqrt{\omega_1}\cos^2\psi_n Z_n\sqrt{\Delta t} \qquad (68)$$

The numbers Z_n are produced, as usually, via the noise generators of uniformly distributed numbers (69).

$$U_{n+1} = \text{frac}(aU_n + b), \qquad V_{n+1} = \text{frac}(aV_n + b), \qquad (69)$$

$$Z_n = \sqrt{-2\log U_n}\cos(2\pi V_n), \qquad n = 0, 1, 2, \ldots \qquad (70)$$

Both are started with $U_0 = 0.3$, $V_0 = 0.7$ and the parameters $a = 9821$ and $b = 0.211322$. Inserted into (70) they generate the normally distributed numbers Z_n.

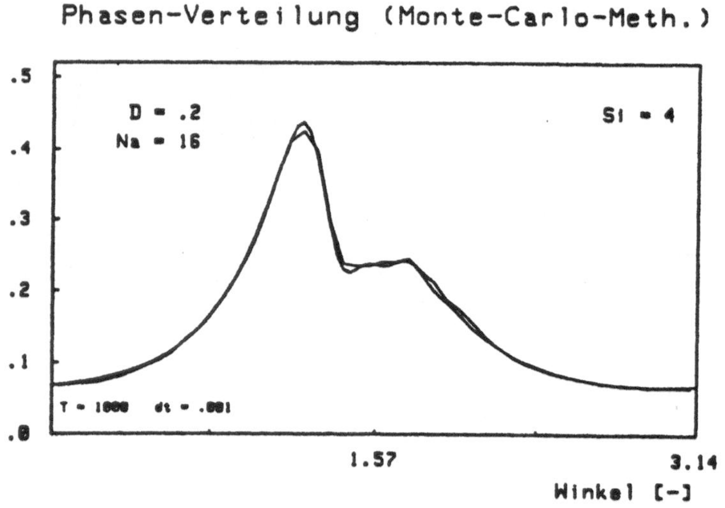

Fig. 11. Monte-Carlo simulation and estimation of $p(\psi)$

In Fig. 11 we show typical numerical results of the Monte-Carlo simulation (68) and compare them with the analytical solution of the Fokker-Planck equation. We recognize only small deviations in the singularity range at $\psi_0 = +\pi/2$ and around the peak of the phase distribution. The results are valid for the damping coefficient $D = 0.2$ and the noise parameter $\sigma = 4$. The step size, applied in (68), was $\Delta t = 0.001$. To estimate the density distribution $p(\psi)$ we produced alltogether 1,000,000 simulation points what takes about 40 min. at the personal computer HP 320. Naturally, the computing time for calculating and drawing the analytic solution is much less, about 2 min. at the same computer.

5.3 Numerical Integration of Singular Diffusion Equations

There is now a fourth method to integrate the singular diffusion equation (59). This can be performed in a purely numerical way applying simple backward differences of the form

$$p(\psi) = p_n, \quad p'(\psi) = \frac{p_n - p_{n-1}}{\Delta\psi}, \quad n = 1, 2, \ldots N_s. \tag{71}$$

Inserted into (59), they give a recurrence formula of the form

$$p_n = \frac{p_0 + \frac{\sigma^2}{2\Delta\psi}\cos^4\psi_n p_{n-1}}{1 + (2D - \sigma^2\cos^2\psi_n)\sin\psi_n\cos\psi_n + \frac{\sigma^2}{2\Delta\psi}\cos^4\psi_n}, \quad \Delta\psi = \frac{\pi}{N_s} \tag{72}$$

for the step size $\Delta\psi$. This Euler scheme is started at the left side with the normalization constant $p_0 = p(-\pi/2)$ and ends at the right side of the periodic interval $-\pi/2 \leq \psi \leq +\pi/2$. The normalization of p_n however, is related to the entire angular range 2π.

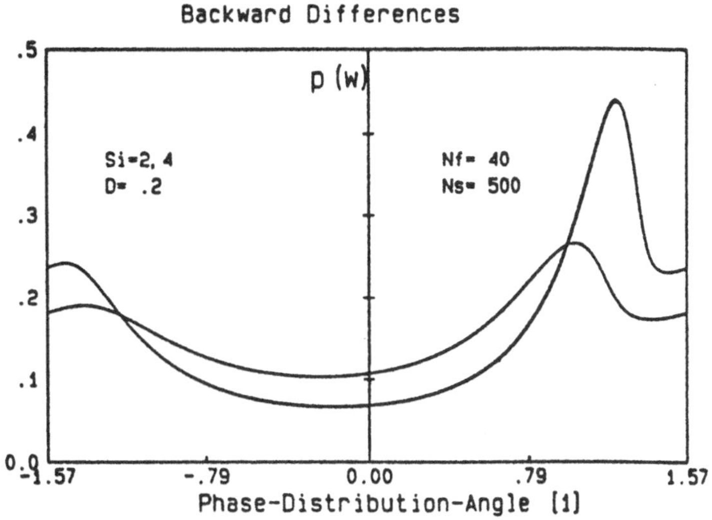

Fig. 12. Integration by backward differences

Fig. 12 shows a comparison of such numerical integration results with solution representation by Fourier series. The applied parameters are $D = 0.2$, $\sigma = 2$ and

$\sigma = 4$. Here, the Fourier series is restricted to $N_f = 40$ terms. The step size of the backward Euler scheme is $N_s = 500$. Clearly, we recognize that there is graphically no difference between both drawings. Moreover, it is worthwhile to mention that the integration routine (72) is numerically stable also in the limiting case of highest possible step sizes like $\Delta\psi = \pi/2$. Naturally, this excellent stability property is lost if we apply forward instead of backward differences. In this case, the numerical integration becomes unstable near the singularities of equation (59) through the coefficient $\cos^4\psi$ which vanishes at $\psi_0 = \pm\pi/2$.

6 Stability Analysis of the Stochastic Hopf Oscillator

Numerical integration by backward differences are most effective methods to calculate Lyapunov exponents for given damping and noise parameters. In the stability analysis, there is an additional speeding up by eliminating damping terms. This is performed by the exponential function

$$X_t = Y_t \exp(-D\omega_1 t), \quad \ddot{Y}_t + \omega_1^2(1-D^2)y_t + \sigma\omega_1\sqrt{\omega_1}\dot{W}_t Y_t = 0 \tag{73}$$

which reduces the Hopf oscillator (4) to the form, above. If λ_r is the Lyapunov exponent of the new process Y_t, then $\lambda = -D\omega_1 + \lambda_r$ is holding, i.e. the Lyapunov exponent of the original process X_t is diminished by the damping term $D\omega_1$.

6.1 Density and Stability for Low Damping $D < 1$

As already explained in section 4.1, we introduce the dimensionless state processes $Y_{1t} = Y_t$ and $Y_{2t} = \dot{Y}_t/\nu_1$ and go over to first-order stochastic equation, as follows.

$$\begin{aligned} dY_{1t} &= \nu_1 Y_{2t} dt, \\ \nu_1 &= \omega_1\sqrt{1-D^2}, \\ dY_{2t} &= -\nu_1 Y_{1t} dt, -\rho\sqrt{\nu_1} Y_{1t} dW_t, \\ \rho &= \sigma/\sqrt{(1-D^2)\sqrt{1-D^2}} \end{aligned} \tag{74}$$

Provided that $D < 1$, the parameter ν_1 denotes the damped natural frequency of the Hopf oscillator and ρ is a modified noise intensity parameter of the reduced system (73). Subsequently, we introduce polar coordinates for the state processes Y_{1t} and Y_{2t} leading to the following Itõ equations:

$$d\Psi_t = -\nu_1(1 + \rho^2 \sin\psi_t \cos^3\psi_t)dt - \rho\sqrt{\nu_1}\cos^2\psi_t dW_t, \tag{75}$$

$$d(\log A_t) = \frac{1}{2}\left[\rho^2\nu_1 \cos 2\psi_t \cos^2\psi_t dt - \rho\sqrt{\nu_1}\sin^2\psi_t dW_t\right] \tag{76}$$

Taking the expectation in the $\log A_t$-equation (76) we obtain the exponents λ_r and also λ of the systems (73) and (4), respectively.

$$\lambda = -D\omega_1 + \rho^2\nu_1 E/2, \qquad E = E(\cos 2\psi_t \cos^2\psi_t), \tag{77}$$

$$\lambda = -D\omega_1 + \frac{1}{8}\frac{\sigma^2\omega_1}{1-D^2}\left[1 - \frac{2/3}{1+(1-D^2)^3(8/3\sigma)^4}\right]. \tag{78}$$

In a first evaluation, the expected value E, noted in (77), can analytically be calculated by means of a Fourier expansion like (51) and (52). If it is truncated with the first two moments we obtain the result (78) which is asymptotically valid for small parameters D, $\sigma << 1$.

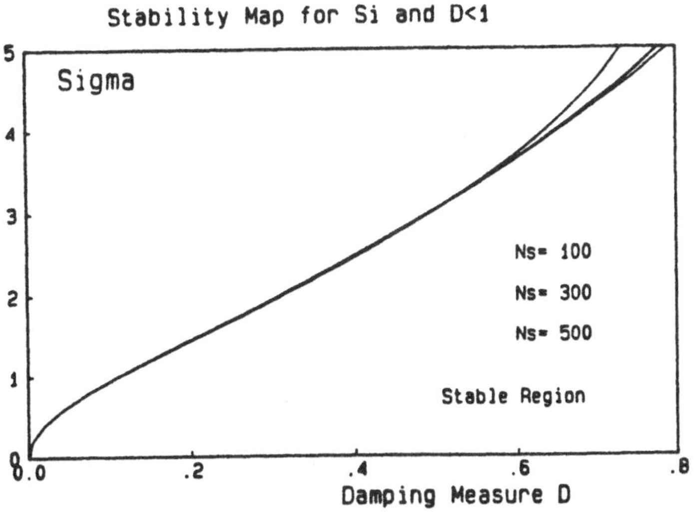

Fig. 13. Stability map for low damping $D < 1$

Exact evaluations of the Lyapunov exponent (77) have to be performed numerically. In particular, we are interested in those values of the noise and damping parameters which result in vanishing Lyapunov exponents $\lambda = 0$. For this stability analysis, the critical values D_{crit} and σ_{crit} of the original Hopf oscillator (4) are determined by

$$D_{\text{crit}} = \frac{\rho^2 E/2}{\sqrt{1+(\rho^2 E/2)^2}}, \qquad \sigma_{\text{crit}} = \rho(1 - D_{\text{crit}}^2)^{3/4} \tag{79}$$

in dependence of a given parameter ρ. Consequently, we can take any value for ρ and calculate the associated Fourier moment E which is now independent upon D. Insertion of E gives finally the critical parameters (79) yielding a stability map plotted in Fig. 13. As already mentioned, the evaluation of the Fourler moment E is performed via the Fokker-Planck equation of the stationary phase process (75).

$$\frac{1}{2}\rho^2\cos^4\psi p'(\psi) + (1 - \rho^2\sin\psi\cos^3\psi)p(\psi) = p(-\frac{\pi}{2}). \tag{80}$$

Applying again backward differences in the equation (80) we obtain a one-step recurrence formula of the form:

$$p_n = \frac{p_0 + (\frac{\rho^2}{2\Delta\psi})\cos^4\psi_n p_{n-1}}{1-\rho^2\sin\psi_n\cos^3\psi_n + (\frac{\rho^2}{2\Delta\psi})\cos^4\psi_n}, \tag{81}$$

$$n = 1,2,3,\ldots, \qquad p_0 = p(-\frac{\pi}{2}).$$

The recursion (81) is started in the left end of the periodic interval $-\pi/2 \leq \psi \leq +\pi/2$. The step sizes $\Delta\psi = \pi/N_s$, applied in Fig. 13, are given by $N_s = 100$, 300 and 500.

6.2 Density and stability for high damping $D > 1$

For high values $D > 1$, the damped natural frequency ν_1 is replaced by the frequency parameter μ_1 and correspondingly, ρ by the noise intensity τ. Accordingly, there is a change of sign in equations (74). The new ones read as follows:

$$\begin{aligned} dY_{1t} &= \mu_1 Y_{2t}dt, \\ \mu_1 &= \omega_1\sqrt{D^2-1}, \\ dY_{2t} &= -\mu_1 Y_{1t}dt, -\tau\sqrt{\mu_1}Y_{1t}dW_t, \\ \tau &= \frac{\sigma}{\sqrt{(D^2-1)\sqrt{D^2-1}}} \end{aligned} \tag{82}$$

Applying polar coordinates to the system (82) we obtain

$$d\Psi_t = (\cos 2\Psi_t - \tau^2 \sin\Psi_t \cos 3\Psi_t)dt - \tau\sqrt{\mu_1}\cos^2\psi_t dW_t, \tag{83}$$

$$d(\log A_t) = \mu_1(\sin 2\Psi_t + \frac{1}{2}\tau^2\cos 2\psi_t\cos^2\psi_t)dt - \frac{1}{2}\tau\sqrt{\mu_1}\sin 2\psi_t dWt. \tag{84}$$

With the transformed equations (83) we are ready to investigate density and stability for the case $D > 1$.

For this purpose we set up the stationary Fokker-Planck equation which belongs to the phase process (83).

$$\frac{1}{2}\tau\cos^4\psi p'(\psi) + (\sin^2\psi - \cos^2\psi - \tau^2\sin\psi\cos^3\psi)p(\psi) = p(-\frac{\pi}{2}). \tag{85}$$

It is valid in the periodic interval $-\pi/2 \leq \psi \leq +\pi/2$ and discretized by backward differences of the step size $\Delta\psi = \pi/N_s$.

$$p_n = \frac{p_0 + (\frac{\tau^2}{2\Delta\psi})\cos^4\psi_n p_{n-1}}{\sin^2\psi_n - \cos^2\psi_n - \tau^2\sin\psi_n\cos^3\psi_n + (\frac{\tau^2}{2\Delta\psi})\cos^4\psi_n}, \tag{86}$$

$$n = 1,2,3,\ldots$$

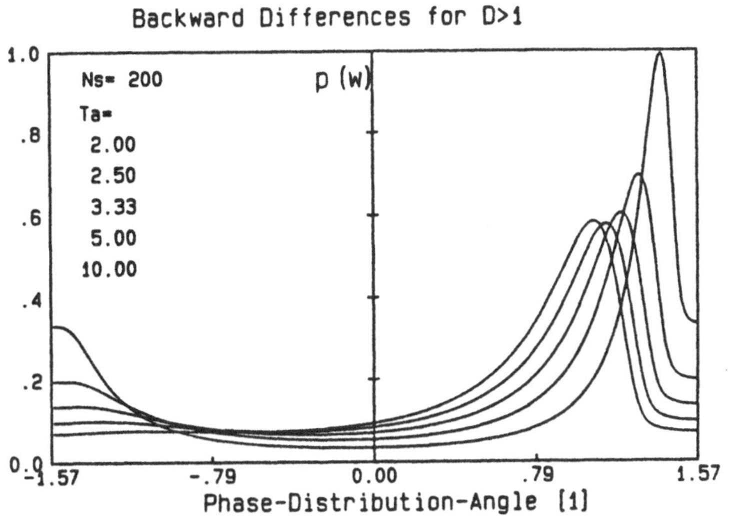

Fig. 14. Phase density distribution for $D > 1$

The recurrence formula (86) is started with the initial value $p_0 = p(-n/2)$ and applied for all $n = 1, 2, \ldots N_s$. Calculating p_n-values we sum them up according the Simpson rule for simultaneously obtaining of the normalization constant p_0. Fig. 14 shows results for $N_s = 200$ and different noise parameters τ.

In a second step we apply these results to calculate the Fourier moment G of the Lyapunov exponent assciated to the $\log A_t$ equation (84).

$$\lambda = -D\omega_1 + \mu_1 G, \quad G = E(\sin 2\Psi_t) + \frac{1}{2}\tau^2 E(\cos 2\Psi_t \cos^2 \psi_t) \tag{87}$$

$$D_{\text{crit}} = \frac{G}{\sqrt{G^2 - 1}}, \quad \sigma_{\text{crit}} = \tau(D_{\text{crit}}^2 - 1)^{3/4}. \tag{88}$$

Finally, from the vanishing Lyapunov exponent (87) we get the critical damping and noise intensity (88), as already explained in section 6.1. In Fig. 15 we find the corresponding stability boundaries calculated for $N_s = 100$, 300 and 500.

7 Parameter Fluctuations by Coloured Noise

Following Wedig [13] we extend the stability analysis to the case that parameter fluctuations are modelled by filtered noise. In this case, there exist a two-dimensional invariant measure which can be calculated by Fourier and Hermite expansions.

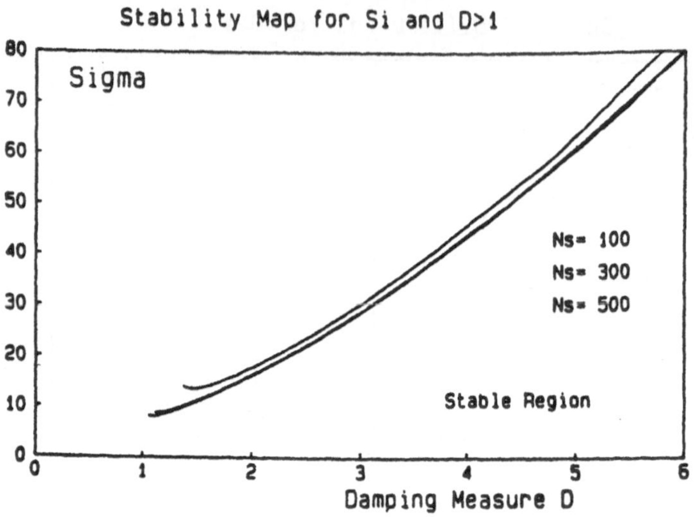

Fig. 15. Stability map for high damping $D > 1$

7.1 Lyapunov Exponent and Rotation Number

We consider a Hopf oscillator like (4) with parameter fluctuations by coloured noise according to equation (2).

$$\ddot{X}_t + 2D\omega_1 \dot{X}_t + \omega_1^2(1 + Z_t)X_t = 0,$$

$$\dot{Z}_t + \omega_g Z_t = \sigma \dot{W}_t \tag{89}$$

$$U_t = \frac{\sqrt{\omega_g} Z_t}{\sigma},$$

$$dU_t = -\omega_g U_t dt + \sqrt{\omega_g} dW_t. \tag{90}$$

We normalize the perturbation process Z_t by U_t so that the stationary distribution of the new process U_t is given by $p(u) = C\exp(-u^2)$. Obviously, the normal density p(u) represents the weighting function of the Hermite orthogonality condition. Subsequently, we go over to a first order system in correspondence to (43) and introduce the polar coordinates (44). This results into the following Itõ equations for the phase process Ψ_t and the $\log A_t$ process of the oscillator (89).

$$d\Psi_t = -\omega_1[1 + 2D\sin\Psi_t + \beta U_t(1 + \cos 2\Psi_t)]dt \tag{91}$$

$$d\log A_t = -\omega_1[D(1 - \cos 2\Psi_t) + \beta U_t \sin 2\Psi_t]dt \tag{92}$$

$$\beta = \frac{\sigma}{2\sqrt{\omega_g}}$$

Here, the parameter β denotes the fluctuation intensity of the parameter perturbation (90).

According to Oseledec's multiplicative ergodic theorem we apply the expectation operator in both equations (91) and (92) in arriving at

$$\dot{E}(\Psi_t) = \alpha, \quad \alpha = -\omega_1[1 + DE(\sin 2\Psi_t) + \beta E(U_t \cos 2\Psi_t)] \tag{93}$$

$$\dot{E}(\log A_t) = \lambda, \quad -\omega_1[D(D - DE(\cos 2\Psi_t) + \beta E(U_t \sin 2\Psi_t)]. \tag{94}$$

Hence, both Lyapunov moments, the expected values of the phase and of the $\log A_t$ process, have time solutions simply proportional to α and to λ, respectively. Furthermore, both, the Lyapunov exponent λ and the rotation number α, are completely determined by equations (93) and (94) if the expected values of the angular process Ψ_t and of the excitation process U_t are known. As already mentioned, they are calculated by means of Fourier series and Hermite polynomials with the expansion order N_a and K, respectively. In Fig. 16, there are typical results of λ and α in dependence of the fluctuation intensity σ for the system parameters $D = 0.1$, $\omega_1 = 1$ and $\omega_g = 5$.

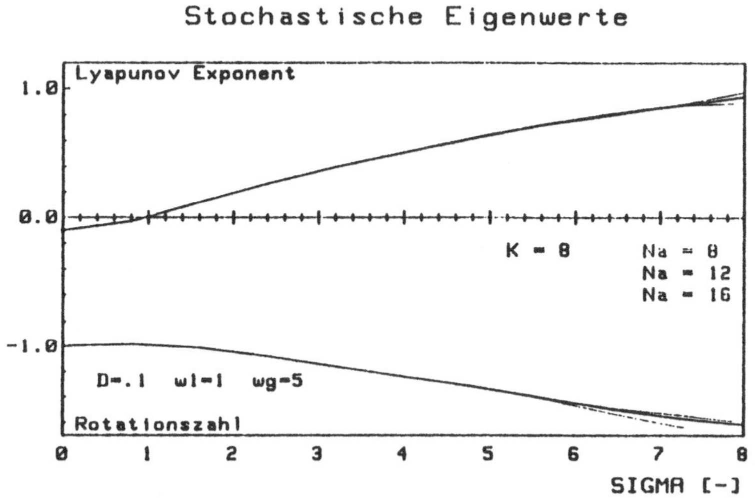

Fig. 16. Stochastic eigenvalues for coloured noise

7.2 Joint Density by Fourier and Hermite Expansion

The joint density of the periodic phase process Ψ_t and the stationary excitation process U_t is determined by their two-dimensional Fokker-Planck equation, which is given by the corresponding Ito equations (90) and (91).

$$\frac{\partial}{\partial\psi}\Big(-\omega_1[1+D\sin 2\psi+\beta u(1+\cos 2\psi)]p(\psi,u)\Big)$$
$$+\frac{\partial}{\partial p}\Big[-\omega_g u p(\psi,u)\Big]-\frac{1}{2}\frac{\partial^2}{\partial u^2}\Big[\omega_g^2 p(\psi,u)\Big]=0. \tag{95}$$

One possibility to solve the partial differential equation (95), is given by Fourier and Hermite expansions, as follows:

$$p(\psi,u)=\frac{1}{\pi}\exp(-u^2)\Big[\frac{1}{2}+\sum_{k=1}^{\infty}\frac{1}{\sqrt{\pi}2^k k!}h_k(u)$$
$$\sum_{n=1}^{\infty}(c_{k,n}\cos 2n\psi+s_{k,n}\sin 2n\psi)\Big] \tag{96}$$
$$(0\le\psi\le 2\pi)\qquad(-\infty<u<+\infty)$$

Applying the orthogonality relations of Fourier series and Hermite polynomials to the two-dimensional density (96) we obtain

$$c_{c,k}=E\Big(H_k(U_t)\cos 2n\Psi_t\Big)$$
$$=\int_{-\infty}^{+\infty}\int_0^{2\pi}H_k(u)\cos(2n\psi)p(\psi,u)d\psi du,$$
$$s_{k,n}=E\Big(H_k(U_t)\sin 2n\Psi_t\Big)$$
$$=\int_{-\infty}^{+\infty}\int_0^{2\pi}H_k(u)\sin(2n\psi)p(\psi,u)d\psi du, \tag{97}$$

Hence, the coefficients $c_{n,k}$ and $s_{n,k}$ of the expansion (96) represent the joint moments of the Hermite processes $H_k(U_t)$ and of the Fourier processes $\cos(2n\Psi_t)$ and $\sin(2n\Psi_t)$, respectively. The application of Fourier expansion is motivated by periodic properties in the range $-\pi\le\psi\le\pi$. The second expansion in the range $-\infty<u<+\infty$ is justified by the fact that Hermite polynomials are eigenfunctions of the parabolic u-operator in the Fokker-Plank equation (95).

Inserting the expansion (96) into the Fokker Plank equation (95) and comparing all Hermite and Fourier terms, we finally arrive at the following algebraic equations for the determination of the coefficients $c_{k,n}$ and $s_{k,n}$:

$$\left.\begin{aligned}
&k\omega_g c_{k,n}-2n\omega_1\Big[\frac{1}{2}k\beta(s_{k-1,n-1}+s_{k-1,n+1})+\beta(\frac{1}{2}s_{k+1,n}+ks_{k-1,n})\\
&\quad+s_{k,n}+\frac{1}{2}D(c_{k,n-1}-c_{k,n+1})+\frac{1}{4}\beta(s_{k+1,n-1}+s_{k+1,n+1})\Big]=0\\
&k\omega_g s_{k,n}-2n\omega_1\Big[\frac{1}{2}k\beta(c_{k-1,n-1}+c_{k-1,n+1})+\beta(\frac{1}{2}c_{k+1,n}+kc_{k-1,n})\\
&\quad+c_{k,n}+\frac{1}{2}D(-s_{k,n-1}+s_{k,n+1})+\frac{1}{4}\beta(c_{k+1,n-1}+c_{k+1,n+1})\Big]=0
\end{aligned}\right\} \tag{98}$$

Because of $c_{0,0}=1$ and $s_{0,0}=0$, the equations (98) are inhomogeneous. They are solved by truncating at N_a Fourier coefficients and at K Hermite polynomials. The solutions deliver the two-dimensional distribution density (96) which is shown in Fig. 17.

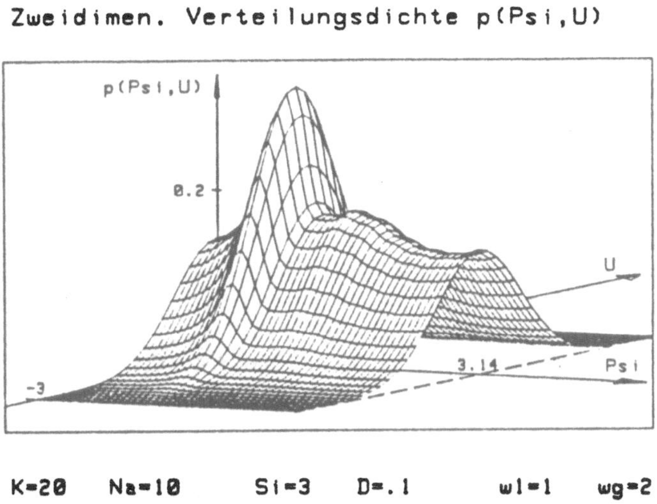

Fig. 17. Two-dimensional distribution density

8 Conclusions

Parameter fluctuations in dynamic structures are physically explained as external noise sources by wind turbulences or rough surfaces. With increasing intensities they destabilize the equilibrium position of the system or a time-invariant velocity distribution. In two-dimensional systems, there are two different situations classified by typical eigenvalue distributions of Hopf or pitchfork bifurcations. Accordingly, polar or hyperbolic coordinates are appropriate to separate a stationary solution part from the exponential growth behaviour of the state processes.

In a pitchfork system, the two eigenvalues are real-valued and lead to trivial solutions on the diagonal line of the state space. The stationary part remains in the positive section of the state space. It possesses an invariant measure from which we derive two Lyapunov exponents in dependence of the noise intensity of parameter fluctuations.

In a Hopf system, the two eigenvalues are purely imaginary. Hence, there exists a trivial solution on a circle in correspondence to polar coordinates. The angular part rotates clockwise in the state space described by double-periodic density distributions. They determine both, the rotation number and the Lyapunov exponent.

The invariant measures of stationary solution parts follow from singular diffusion equations. Best ways to solve them are direct integrations by means of backward differences. They are numerically stable and computerized by simple recurrence formulas. The goal is now to combine all techniques, presented here, and to extend them to higer-dimensional systems of interest.

References

1. L. Arnold, Stochastic Differential Equations: Theory and Application. Wiley, New York, 1974.
2. L. Arnold and W. Kliemann, Qualitative theory of stochastic systems. In: A.T. Bharucha-Reid, ed., Probabilistic Analysis and Related Topics, Vol. 3, Academic Press, New York, 1981.
3. P. Boxler, Lyapunov exponents indicate stability and detect stochastic bifurcations, this volume.
4. R.Z. Khasminskii, Necessary and sufficient conditions for asymptotic stability of linear stochastic systems. Theor. Prob. and Appls., 12 (1967) 144-147.
5. F. Kozin, Stability of linear stochastic systems. In: R.Curtain, ed., Stability of Stochastic Dynamical Systems. Lecture Notes in Mathematics, Vol. 294, Springer, New York, 1972, 186-229.
6. K. Nishioka, On the stability of two-dimensional linear stochastic systems. J. Kodai Math. Sem. Rep. 27 (1976) 211-230.
7. V.I. Oseledec, A multiplicative ergodic theorem, Lyapunov characteristic numbers for dynamical systems, Trans. Moscow Math. Soc. 19 (1968) 197-231.
8. E. Pardoux and V. Wihstutz, Lyapunov exponent and rotation number of two dimensional linear stochastic systems with small diffusion. SIAM J. Appl. Math. 48, No.2 (1988) 441-457.
9. R.L. Stratonovich, Topics in the Theory of Random Noise, Gordon and Breach, 1963.
10. D. Talay, Simulation and numerical analysis of stochastic differential equations, this volume.
11. W. Wedig, Stochastic bifurcations of nonlinear parametric systems, In: P.D. Spanos, ed., Proc. ASCE Specialty Conference on Probabilistic Methods in Civil Engineering, ASCE, New York, 1988, 297-300.
12. W.Wedig, Bifurcations in stochastic systems-models, analysis and simulation. J. Math. and Computers in Simulation 31 (1989) 1-15.
13. W. Wedig, Lyapunov Exponent und Rotationszahl der stochastischen Eigenwerttheorie, GAMM-Tagung Wien, 1988, ZAMM 69 (1989) T542.
14. F. Weidenhammer, Biegeschuingungen des Stabes unter axial pulsierender Zufallslast, VDI-Berichte Nr. 135 (1969) 101-107.
15. W. Wong and M. Zakai, On the relation between ordinary and stochastic equations. Int. J. Eng. Sci., 2 (1965) 213 -229.

Stochastic Center as a Tool in a Stochastic Bifurcation Theory

Petra Boxler

1 Introduction: Why do we need Stochastic Center Manifolds ?

In Section 4 of Boxler [2] in this volume we have seen that the Lyapunov exponents attached to a nonlinear cocyle that depends on a parameter α are able to detect possible bifurcation points: a Lyapunov exponent vanishing at $\alpha = \alpha_0$ is a necessary condition for a stochastic bifurcation at α_0. But we have also explained there that for the purpose of bifurcation theory it is not enough to consider only the linearized cocycle because it is just the nonlinear part of the system that is responsible for the existence of stable bifurcating solutions.

What do we know up to now about the behaviour of the entire (nonlinear) system ? We have to distinguish between two situations:

1. Before having reached the stochastic bifurcation point:
All the Lyapunov exponents are negative and hence the linearized system is asymptotically stable. Furthermore, Theorem 11 in [2] tells us that this behaviour carries over to the nonlinear system which is therefore asymptotically stable, too.
2. At the stochasnc bifurcation point:
The biggest Lyapunov exponent vanishes but all the other Lyapunov exponents remain negative. But then it is no longer possible to apply Corollary 7 in [2] and, even worse, we do not know anything about the relationship between the linearized and the original nonlinear cocycle.

The idea to deal with case 2 is now the following:

Since systems with negative Lyapunov exponents are well understood (because of their asymptotic stability) we ask whether at the bifurcation point the part of the cocycle corresponding to the negative Lyapunov exponents can be isolated. Its dynamical behaviour being clear it should then be possible to restrict the further analysis to the remaining part of the system where the Lyapunov exponent vanishes.

In fact, this heuristic idea can be made mathematically rigorous, and the device to do so is the concept of a stochastic center manifold. The idea of decomposing the cocycle with the help of stochastic center manifolds will also yield a considerable reduction of the dimension of the problem we will have to study if we are interested in bifurcation phenomena; very often it will be sufficient to deal with a one- or two-dimensional system although the original equation has been rather high dimensional.

2 Definition of Stochastic Center Manifolds

Let us assume the situation described in Section 3 of Boxler [2], i.e. our basic object is a nonlinear cocycle $\phi(t,\omega,x)$. In bifurcation theory it will depend on a parameter but for notational simplicity we will drop this dependence for the moment. Recall that we are allowed to assume without loss of generality that the cocycle has got a fixed point at 0. Furthermore, the multiplicative ergodic theorem applied to the linearized cocyle $\Psi(t,\omega)$ ensures the existence of $r \le d$ real numbers $\lambda_1 > \ldots > \lambda_r$, the Lyapunov exponents, and of r random linear subspaces $E_i(\omega) \subset \mathbb{R}^d$, the Oseledec spaces, which replace the real parts of the eigenvalues and the eigenspaces used in the deterministic case.

Being mainly interested in bifurcation theoretical applications we will assume that one Lyapunov exponent is zero (possibly with multiplicity > 1) and that the others are negative. For the general case, where positive exponents are considered as well, see Boxler [3].

In a first step we collect the Oseledec spaces corresponding to zero resp. negative Lyapunov exponents in order to obtain a new random coordinate system. Hence we define:

$$E_c(\omega) \stackrel{\text{def}}{=} \bigoplus_{\lambda_i=0} E_i(\omega) \qquad E_s(\omega) \stackrel{\text{def}}{=} \bigoplus_{\lambda_i<0} E_i(\omega) \tag{1}$$

For all $t \in \mathbb{R}$ Oseledec's theorem will then yield the following decomposition:

$$\mathbb{R}^d = E_c(v_t\omega) \oplus E_s(v_t\omega) \quad \text{(recall that: } v_0\omega = \omega) \tag{2}$$

Thus, for any t we may decompose the cocycle into its components with respect to this coordinate system. The visualization of this projection at time 0 and at time t might look as in Fig. 1.

Next we may decompose ϕ_c and ϕ_s into its linear and its nonlinear part; this yields the system which will be the starting point for our further investigations:

$$\phi_c(t,\omega,x_c,x_s) = \Psi_c(t,\omega)x_c + \Phi_c(t,\omega,x_c,x_s) \tag{3}$$
$$\phi_s(t,\omega,x_c,x_s) = \Psi_s(t,\omega)x_c + \Phi_s(t,\omega,x_c,x_s) \tag{4}$$

If these equations were decoupled it would be sufficient to examine (3) because the stability behaviour of (4) would be well-known thanks to Theorem 11 in [2]. The stochastic center manifold will enable us to carry out such a decoupling, i.e. to eliminate x_s from equation (3) by replacing it by a function of x_c. How should such stochastic center manifolds be defined ?

Forget about the nonlinear part of (3) and (4) for a moment. Then the equations are decoupled and we only have to study what happens in the directions of $E_c(\omega), E_s(\omega)$ resp. Thus, if there is a nonlinear part which is not too big (or, equivalently, if we restrict ourselves to a neighborhood of the origin) we will expect that locally, i.e. near the origin, the linearized system will be a good approximation and that we obtain the picture 2.

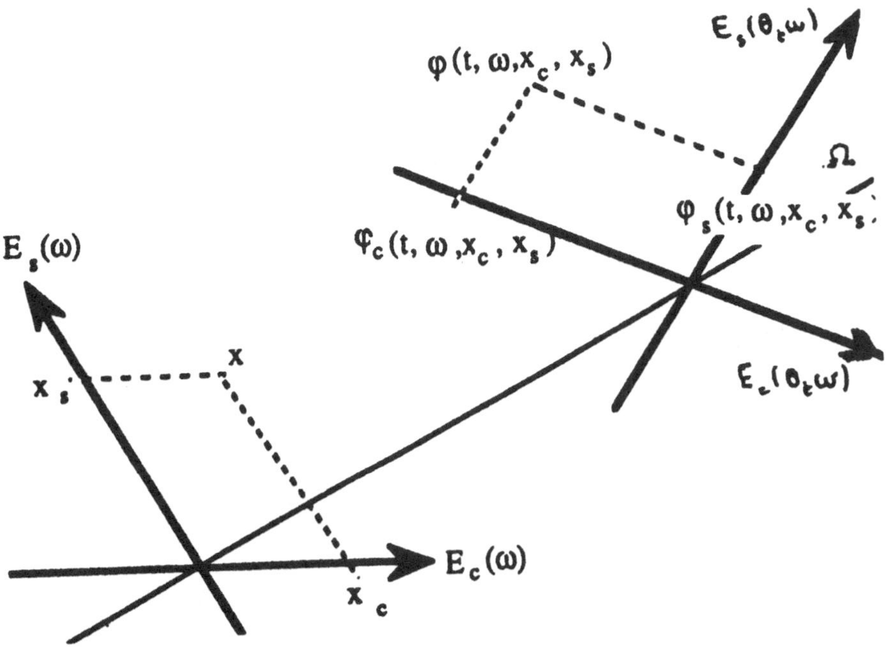

Fig. 1. Projections

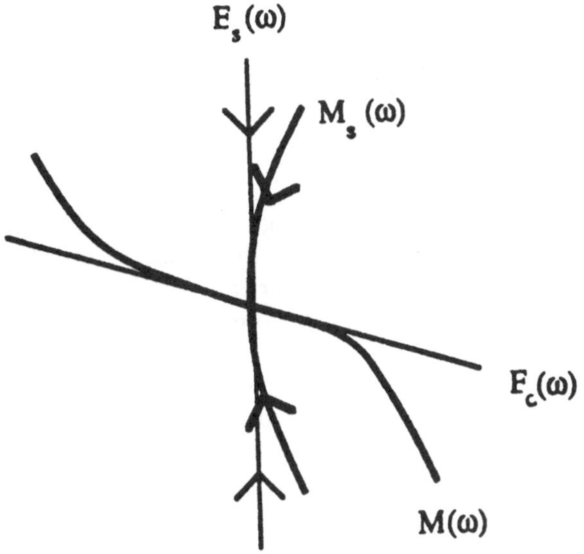

Fig. 2. Linearized system

In mathematical terms this means that we will require the stochastic center manifols $M(\omega)$ to be tangent to $E_c(\omega)$. Furthermore we will try to express it as a graph on $E_c(\omega)$ an finally it will have to be invariant with respect to the cocyle, i.e. if we start on the stochastic center manifold we wish to stay there for ever. This property is necessary to obtain a correct decoupling not only at time 0 but for any time t.

The multiplicative ergodic theorem tells us that the Oseledec spaces are not fixed spaces but move according to the property of stochastic invariance. Hence, if the stochastic center manifold is to be tangent to $E_c(\omega)$ its required invariance will have to be adapted to the stochastic invariance of $E_c(\omega)$. For this reason $M(\omega)$ will have to satisfy:

$$\phi(t,\omega,.)M(\omega) \subset M(v_t\omega) \quad \text{for all } t \in \mathbb{R} \text{ a.s.} \tag{5}$$

As a consequence we define:

Definition 1. A (global) stochastic center manifold *for the cocycle* ϕ is a *graph*

$$M(\omega) = \{(x, h(\omega, x))|\ x \in E_c(\omega)\} \tag{6}$$

where h *a suitable map, such that* $M(\omega)$ *is tangent to* $E_c(\omega)$ *and invariant (in the sense of (5)).*

Of course this definition raises the natural question whether it is possible to find a map h living on a suitable space such that the object just defined, which we have called a stochastic center manifold, does exist. Thus, let us start by defining an appropriate space to work in:

Let

$$E = \{(\omega, x) \in \Omega \times \mathbb{R}^d|\ x \in E_c(\omega)\} \tag{7}$$

$$\mathcal{X} = \{h : E \to \mathbb{R}^d \text{ meas.} \tag{8}$$

$$|\ h(\omega,.) : E_c(\omega] \to E_s(\omega) \text{ a. s. cont., bounded}\} \tag{9}$$

This means that the map h does not assign values to any pair (ω, x); once you have chosen an ω you will only have the choice among the elements of $E_c(\omega)$ (mathematicians call this a *bundle*, and for each $\omega \in \Omega$ $E_c(\omega)$ is called a *fibre* of the bundle E). The situation might therefore look as in Fig. 3.

In the definition of the space $\mathcal{X}$ we required $h(\omega,.)$ to be bounded. This boundedness is to be understood with respect to the norm $|.|_{\infty,\omega}$ where

$$|h(\omega,.)|_{\infty,\omega} \stackrel{\text{def}}{=} \sup_{x \in E_c(\omega)} |h(\omega, x)|_\omega^s \tag{10}$$

and where $|.|_\omega^s$ denotes a norm on space $E_s(\omega)$ which we will define below. Hence, on each fibre $E_s(\omega)$ above $\omega \in \Omega$ we introduce a norm that depends on ω. Why can we not use the usual Euclidian norm ?

$M_s(\omega)$ (see Fig. 3) and the stochatic center manifold, the latter being described by a map $h \in \mathcal{X}$, are supposed to capture the dynamical behaviour of

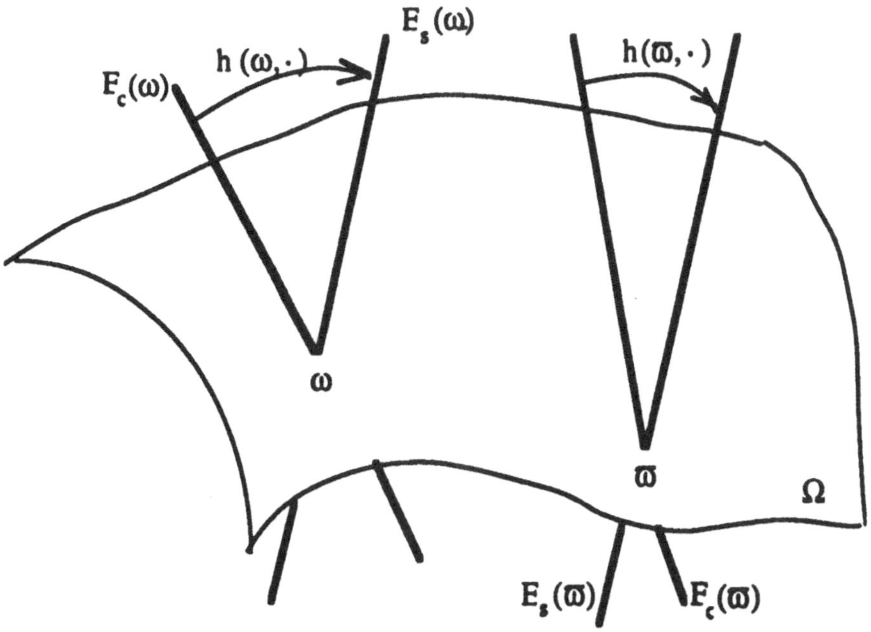

Fig. 3. Bundle

the *nonlinear* system, as do the spaces $E_s(\omega)$ and $E_c(\omega)$ in the case of the *linearized* cocycle. But the only thing we know about the dynamics of the system is the description of the behaviour for t tending to $\pm\infty$ given by Osedelc's theorem:

$$\lim_{t\to\pm\infty} \frac{1}{t} \log \|\Psi(t,\omega)x\| = \lambda_i \quad \text{if an only if} \quad x \in E_i(\omega) \tag{11}$$

This means, however, that for each $\omega \in \Omega$ we will only be able to tell what happens for $|t|$ bigger than a certain $t_0(\omega)$ whereas we do not know anything about the behaviour of the system for the smaller values of t for which all kinds of transients are possible.

If we started at time 0 and took the Euclidean norm then we would measure all these transient influences which do not give us the information about the asymptotic behaviour of the system we are looking for. On the other hand, the idea of simply restricting our investigation of the cocycle to sufficiently big values of t will not besuccessful because for each ω we would have to take possibly different values $t_0(\omega)$. For this reason the only chance we have is to try to use our information about the long-term behaviour already at time 0. This is precisely what is realized by the norm $|.|^s_\omega$ (together with a norm $|.|^c_\omega$ on $E_c(\omega)$). For its definition we denote $\lambda_s := \max_{\lambda_i<0} \lambda_i$

In the sequel we will always choose $\beta > 0$ such that $\lambda_s + 4\beta < 0$.

Lemma 2. *Let*

$$|x|^s_\omega = \int_0^\infty e^{-(\lambda_s+2\beta)\tau} \|\Psi_s(\tau,\omega)x\| d\tau \text{ for all } x \in E_s(\omega)$$

$$|x|_\omega^c = \int_{-\infty}^{\infty} e^{-2\beta\tau}\|\Psi_c(\tau,\omega)x\|d\tau \text{ for all } x \in E_c(\omega)$$

Then $|.|_\omega^s$ *and* $|.|_\omega^c$ *define norms on* $E_s(\omega)$, $E_c(\omega)$, *resp. and the following estimates hold:*

$$|\Psi_s(t,\omega)x|_{v_t\omega}^s \leq e^{(\lambda_s+2\beta)t}|x|_\omega^s \quad \text{for all } x \in E_s(\omega) \text{ and } t \geq 0 \text{ a.s.} \tag{12}$$

$$|\Psi_c(-t,v_t\omega)x|_\omega^c \leq e^{(2\beta)|t|}|x|_{v_t\omega}^c \quad \text{for all } x \in E_c(v_t\omega) \text{ and } t \in \mathbb{R} \text{ a.s.} \tag{13}$$

Proof. From equation (11) we can derive an estimate for the Euclidean norm which is similar to (12), (13) but with a random variable $C(\omega)$ multiplying the right-hand side. $C(\omega)$ copes with the values of the linearized cocycle taken before time $t_0(\omega)$. Using these estimates one can directly verify the statements of the lemma. For details see Boxler [3], Lemma 4.2.

Since we are working in $\mathbb{R}^d$ the Euclidian norm and the norm $|.|_\omega$ are of course equivalent, i.e. $C_1(\omega)\|x\| \leq |x|_\omega \leq C_2(\omega)\|x\|$, and C_1 and C_2 are explicitly determined.

You might wonder whether Lemma 2 is not just a technicality. It will, however, turn out in the subsequent sections that the estimates provided there are one of the keys not only for the proof of existence of stochastic center manifolds described by maps h belonging to the space $\mathcal{X}$ but also for the proofs of the various properties of these objects which we will get to know later on.

3 Existence of Stochastic Center Manifolds

3.1 Which Assumptions Do We Need and Why ?

Up to now we have specified a space $\mathcal{X}$ in which we want to look for a map h characterizing the desired stochastic center manifold. Before stating the existence theorem we will single out a subset of $\mathcal{X}$ reflecting certain properties the map h will be required to have.

For any constants $k \geq 1$ and $L > 0$ let us consider the following properties a map $e \in \mathcal{X}$ might have:

$$D^j h(\omega,.) \text{ exists for all } j = 0,\ldots,k \tag{14}$$

$$h(\omega,0) = D^1 h(\omega,0) = 0 \tag{15}$$

$$|D^j h(\omega,.)|_{\infty,\omega} \leq L \text{ for all } j = 0,\ldots,k \tag{16}$$

$$|D^k h(\omega,x) - D^k h(\omega,\tilde{x}) - |_\omega^s \leq L|x-\tilde{x}|_\omega^c \text{ for all } x,\tilde{x} \in E_c(\omega) \tag{17}$$

The second property expresses the fact that the stochastic center manifold is required to be tangent to $E_c(\omega)$, property 16 means (uniform) boundedness of all the derivatives, and 17 is a global Lipschitz condition for the k-th order derivative. Hence the constant k measures the smoothness of h. The set we will be interested in is

$$A_k(L) := \{h \in \mathcal{X} | \; h \text{ satisfies } (14) - (17) \text{ a.s.} \tag{18}$$

Our task will be accomplished once we will have found a map h in this set which produces a graph $M(\omega)$ that is invariant in the sense of 5. Which assumptions does the cocycle ϕ have to satisfy in order to ensure the existence of the desired h?

As far as the linear part Ψ of the cocycle is concenred the estimates given in Lemma 2 will do. Thus we will have to control the nonlinear part Φ. For this reason we determine the class of nonlinearities we can allow; it is clear that if the influence of Φ becomes too big then the nice behaviour of the linear part will be distroyed and will no longer carry over to the entire cocycle ϕ. Furthermore, the smoothness of ϕ will obviously decide about the smoothness we can hope to obtain for the stochastic center manifold.

We are interested in those $C^{k,1}$-cocycles ϕ (i.e. ϕ a C^k-map such that the k-th order derivative satisfies a global Lipschitz condition) whose nonlinear parts have the following properties:

For any $x, y \in \mathbb{R}^d, t \in \mathbb{R}$ with $-1 < t < 1$ we consider:

$$(NL1)(\varepsilon) \qquad \|D^j\Phi(t,\omega,x)\| \leq \varepsilon(v_t\omega) \text{ for all } j = 0,\dots,k$$

$$(NL2)(\varepsilon) \qquad \|D^k\Phi(t,\omega,x) - D^k\Phi(t,\omega,y)\| \leq C^{-1}(\omega)\varepsilon(v_t\omega)\|x-y\|$$

Here $\varepsilon(.)$ is a random varaible of the form $\varepsilon(\omega) := \varepsilon/(C(\omega)\|p(\omega)\|)$, $\varepsilon > 0$ a constant.

C has already been mentioned in the proof of Lemma 2, i.e. it copes with those values of the linearized cocycle for which the dynamical characterization expressed in Oseledec's theorem does not apply yet. If Ψ may take big values for $|t| < t_0(\omega)$ or if $t_0(\omega)$ is big for a certain ω then it will be intuitively clear that we will not have much freedom for the nonlinear part, i.e. $\varepsilon(\omega)$ will have to be small for this value of ω.

$\|p(\omega)\|$ describes the maximal value the projection from $\mathbb{R}^d$ to $E_c(\omega), E_s(\omega)$ resp. can take, this value not being constant as we have seen in FIG. 1. If for a certain ω the spaces $E_c(\omega)$ and $E_s(\omega)$ come close together we will have to take this into account by choosing the corresponding $\varepsilon(\omega)$ small.

Thus, it is nothing mysterious that will force the nonlinear part of the cocycle to be sufficiently small. In the course of the proof we will only have to determine the size of the constant ε; then we will be able to decide for a given cocycle whether it falls into the class just described or not.

Definition 3. A $C^{k,1}$-cocycle ϕ whose nonlinear part satisfies the conditions (NL 1, 2)(ε) is called a cocycle of class (NL 1, 2)(ε).

In general the conditions we will have to impose on the nonlinearities to stay within the appropriate class are rather restrictive. But if we restrict ourselves to a suitable neighborhood of the origin (and that is completely sufficient for the purpose of (local) bifurcation theory) then the necessary assumptions will always be satisfied. This means that we have two possibilities: either the nonlinear part of the cocycle has to be small; in this case we are able to show the existence of *global* stochastic center manifolds. Or we work on a neighborhood of the origin and do not impose conditions on Φ; then we can prove the existence of *local* stochastic center manifolds. We will come back to this point later on.

3.2 An Existence Theorem for Global Stochastic Center Manifolds

After these rather lengthy preparations we are now ready to prove the following theorem:

Theorem 4. *For each $L > 0$ sufficiently small there is a constant $\varepsilon(L) > 0$ such that for any cocycle ϕ belonging to the class (NL 1, 2)($\varepsilon(L)$) there exists a global stochastic $C^{k,1}$-center manifold $M(\omega)$ which may be written in the form*

$$M(\omega) = \{(x, h(\omega, x)) | \; x \in E_c(\omega)\} \tag{19}$$

with $h \in A_k(L)$.

Proof. Since the complete proof is quite long and technical we will only give the main ideas; all the details may be found in Boxler [3], Chapter 5.

We have already pointed out in Section 3.1 that we will have to find a map h in the set $A_k(L)$ such that the set $M(\omega)$ constructed with this map will be invariant. Thus, it is reasonable to start writing this invariance (5) in terms of h. (5) requires in particular:

$$\phi_s(t, v_{-t}\omega, x_c, h(v_{-t}\omega, x_c)) = h(\omega, \phi_c(t, v_{-t}\omega, x_c))) \text{ for all } t \in \mathbb{R} \tag{20}$$

This is illustrated by Fig. 4.

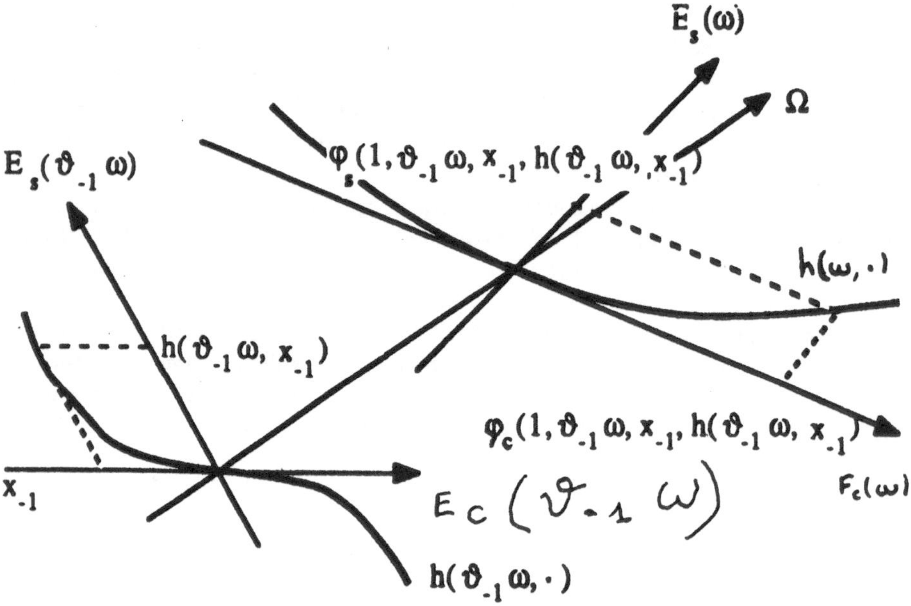

Fig. 4. Invariance

How can we find a map $h \in A_k(L)$ such that (20) is satisfied ? The idea consists in constructing h as a fixed point of a suitable operator acting on the set $A_k(L)$. For this let us consider an operator

$$T: \; A_k(L) \to \mathcal{X}$$
$$h \to Th$$

Th being a map from E to $\mathbb{R}^d$ we will have to define $(Th)(\omega, y)$ for any $y \in E_c(\omega)$.

To do this we put

$$(Th)(\omega, y) := \phi_s(1, v_{-1}\omega, x_1, h(v_{-1}\omega, x_1)) \tag{21}$$

where $x_{-1} \in E_c(\omega)$ is chosen such that

$$y = \phi_c(1, v_{-1}\omega, x_1, h(v_{-1}\omega, x_1)). \tag{22}$$

From this definition it is clear that a fixed point of T (recall that this means that $Th = h$) satisfies (20) for $t = 1$. It will then have to be shown that such an h will have the required invariance property not only for this value of t but for all $t \in \mathbb{R}$.

What will we have to do to ensure the existence of a fixed point of the operator T ? We proceed in several steps:

(1) We want to apply *Banach's fixed point theorem* (= contraction mapping theorem, see e.g. Dieudonné [4] or Kantorowitsch and Akilow [7]). But for this we have to work in a complete metric space whereas up to now our operator T is defined on $\mathcal{X}$ which does not have a metric structure yet. How can we endow it with such a structure ? We put for any $h, \tilde{h} \in \mathcal{X}$:

$$d(h, \tilde{h}) := E \frac{|h(\omega, .) - \tilde{h}(\omega, .)_{\infty,\omega}|}{1 + |h(\omega, .) - h(\omega, .)_{\infty,\omega}|} = \int_\Omega \frac{|h(\omega, .) - \tilde{h}(\omega, .)_{\infty,\omega}|}{1 + |h(\omega, .) - \tilde{h}(\omega, .)_{\infty,\omega}|} P(d\omega) \tag{23}$$

It is not difficult to check that this defines a pseudometric on $\mathcal{X}$, and if we denote by X the space of equivalence classes of $\mathcal{X}$ with respect to almost sure equality (i.e. we identify those elements of $\mathcal{X}$ which are a.s. equal) then we can prove:

Lemma *(X, d) is a complete metric space, and d is called the metric of convergence in probability.*

Proof. This is a straightforward adaptation of a proof in Federer [5], page 79. The difference is due to the fact that for different values of ω we use (possibly) different norms on $E_s(\omega)$. For details see Boxler [3], Lemma 4.3.

(2) Looking at (21) we see that in order to evaluate Th at (ω, y) (for a given it h) we will have to be able to determine an initial value x_{-1} such that (22) holds. In mathematical terms this means that we have to prove that for almost all ω the map $x \to \phi_c(1, v_{-1}\omega, x, h(v_{-1}\omega, x))$ is a bijection if we choose the $\tilde{\varepsilon} > 0$ in a assumption (NL 1, 2) $(\tilde{\varepsilon})$ sufficiently small.
This is in fact possible because $\tilde{\varepsilon}$ small implies thanks to (NL 1)$(\tilde{\varepsilon})$ that the nonlinear part Φ_c of ϕ_c may be considered as a small perturbation of the linear isomorphism Ψ_c. The inverse function theorem even enables us to deduce that the map above is a $C^{k,1}$-diffeomorphism (which is important to show the required smoothness of h).

(3) In Banach's fixed point theorem the fixed point is obtained by means of an approximating sequence $(h_n)_{n\in\mathbb{N}}$ where $h_n \in A_k(L)$ and which converges with respect to the towards a map h which will be the desired fixed point. In order to have a chance that h will have the required properties we will have to make sure that the operator T maps the set $A_k(L)$ onto itself because in this set we have collected properties we would like to preserve. More precisely we will show:
For any given constant $L > 0$ there is an $0 < \varepsilon(L) < \tilde{\varepsilon}$ (step 2) such that for a cocycle of class (NL 1, 2)($\varepsilon(L)$) we have: $T(A_k(L)) \subset A_k(L)$.
This is rather lengthy but not very difficult and hence we refer to Boxler [3] for details. One has to make use of the estimate provided by Lemma 2 (for the linear part of the cocycle) and of the boundedness and Lipschitz conditions which hold for the nonlinearpart of the cocycle because of assumptions (NL 1, 2) and for $h \in A_k(L)$ by definition of this set.

(4) The contraction mapping theorem tells us that the existence of a fixed point of the operator T (i.e. the convergence of the approximating sequence $(h_n)_{n\in\mathbb{N}}$ to a certain h) will be established as soon as we will have shown that T is a contraction on $A_k(L)$, i.e. that there is a constant c, $0 < c < 1$, such that
$$d(Th, T\tilde{h}) \leq c\, d(h, \tilde{h}) \text{ for any } h, \tilde{h} \in a_k(L) \tag{24}$$
For this we will prove that if we choose $L > 0$ sufficiently small then the assumptions (NL 1, 2)(ε(L)), where ε(L) has to be determined by step 3) for this value of L, enable us to verify (24).
This proof is carried out by first estimating
$$|Th(\omega,.) - T\tilde{h}(\omega,.)|_{\infty,\omega} = \sup_{y\in E_c(\omega)} |Th(\omega,y) - T\tilde{h}(\omega,y)|^s_\omega \tag{25}$$
by methods similar to those used in step 3) and then deducing from this an estimate for $d(Th, Th)$ by means of Hölder's inequality and the invariance of the measure with respect to the shift.

(5) We do not know yet whether the properties of the approximating sequence (h_n) carry over to its limit h. In fact we can show that the properties (14), (15) and (17) are preserved for $n \to \infty$ (and these are the properties we need for a stochastic center manifold whereas (16) is included in the set $A_k(L)$ for technical reasons). To do so we make use of the fact that (h_n) converges to h with respect to the metric d and hence in probability. For this reason there is an almost surely convergent subsequence and for this subsequence the statement follows thanks to Ascoli's theorem and a lemma concerning the derivative of a limit function (for details and references see Boxler [3], Lemma 4.4).
For $k = 0$ the property just derived simply means that $A_k(L)$ is a closed subset of X but this does not hold true for $k > 0$ because the derivatives are not included in the defmition of the norms $|.|_{\infty,\omega}$.

The existence of a (unique) fixed point h of the operator T is now established and as explained earlier it remains to be shown that h satisfies the invariance property (20) *for any* $t \in \mathbb{R}$.

Assume that at time 0 we start on the stochastic center manifold, i.e. at $(x, h(\omega, x))$. Then we know up to now that

$$\phi_s(1, v_{-1}\omega, x_c, h(v_{-1}\omega, x_c)) = h(\omega, \phi_c(1, v_{-1}\omega, x_c))) \tag{26}$$

But will we still be on the center manifold if we go ***backwards*** in time, i.e. at $t = -1$? The answer is yes and this is a consequence of the cocycle property together with (26). The intuitive reason is that if one starts on the stochastic center manifold and is not there at $t = -1$ then one would also not be there at time $t = 1$, in contrast to (26).

For $t \in \mathbb{Z}$ the invariance property will follow from this by an induction argument combined with the cocycle property.

Intuitively this carries over to any $t \in \mathbb{R}$ because it is impossible to be on the stochastic center manifold at all integer values of time and to leave the center manifold in between. Mathematically this can be shown by first using the cocycle property together with the uniqueness of the fixed point h to extend the invariance property to any $t \in \mathbb{Q}$. An arbitrary $t \in \mathbb{R}$ is then approximated by rational values and the assertion follows from a continuity argument.

Theorem 4 is now completely proven because if we construct the graph $M(\omega, .)$ of the map it h, which we have obtained as a fixed point of the operator T, then $M(\omega, .)$ will be invariant by construction of h and it will be tangent to $E_c(\omega)$ because this is a property of all the elements of the set $A_k(L)$. Finally the desired smoothness of the stochastic center manifold follows from properties (14) and (17) of $A_k(L)$.

3.3 Existence of Local Stochastic Center Manifolds

After having seen the assumptions of Theorem 4 you will certainly be convinced that they are rather restrictive. But as pointed out earlier the bifurcation theory we are interested in is a local one which is restricted to a neighborhood of the origin (by the way, this is the same point of view as in the deterministic theory; compare for example Guckenheimer and Holmes [6] or Vanderbauwhede [9]).

For such a local analysis, however, our existence theorem is completely satisfactory: since our cocycle has a fixed point at the origin and is continuous its nonlinear part will always be as small as we want if we restrict ourselves to a sufficiently small neighborhood of the origin. We have seen in the global existence theorem that theallowed size of the nonlinear part depends on ω. This implies that in the local case the neighborhood of the origin on which we work will be random, and we have explained in Section 3.1. which influences will determine its size for the different values of ω. We can thus prove the following theorem:

Theorem 5. *Assume the situation described in the beginning of Section 2. If ϕ is a cocycle of C^k-diffeomorphisms ($k \geq 2$) then there exists a local stochastic C^k-center manifold for ϕ.*

Remarks:

1. Here "local" means that there is a (random) neighborhood of the origin such that all the points in the neighborhood satisfy the invariance property as long as they stay in this neighborhood.
2. There is not really a loss of smoothness because the local stochastic center manifold is described by a map h such that $D^{k-1}h(\omega,.)$ satisfies a local Lipschitz condition by construction.

Proof. We have already pointed out that inside a suitable neighborhood of the origin the nonlinear part of the cocycle satisfies the desired global Lipschitz conditions. Hence the idea of the proof mainly consists in a cut-off procedure such that the nonlinearities are zero outside an appropriate neighborhood. Then one might try to repeat the global proof for the altered system. But let us be a bit more precise !

In a first step one has to make sure that the object which we obtain by means of the cut-off procedure will be a so-called *local cocycle* such that the essential properties of ϕ are preserved.

The problem is, however,that the cocycle property is lost as soon as we leave the neighborhood where the cocycle has not been changed. But if we examine the proof of Theorem 4 then we see that this property is not needed to construct the stochastic center manifold but only to show its invariance at any instant of time. Therefore we may repeat steps 1) to 5) of the proof and it is then immediately seen that the object we obtain is invariant until our neighborhood of the origin will be left for the first time. Compare this to Remark 1. to see that this is exactly what we had claimed.

More details concerning the rigorous formulation of this localization procedure may be found in the English version of Boxler [3], Section 6.

4 Stochastic Center Manifolds Applied to Bifurcation Theory

4.1 Attractivity of stochastic center manifolds

We have already seen in the last section that our restriction to *global* stochastic center manifolds does not restrict the class of systems we will be able to examine because we can always switch to *local* stochastic center manifolds. For notational simplicity we will thus only deal with the global case.

Among the properties of global stochastic center manifolds there is one which is particularly important for practical (i.e. numerical) purposes, and this is their attractivity. Up to now we know that if we start *on* the stochastic center manifold we will stay there. But what happens if we can not manage to start there ? This question is settled by the following theorem:

Theorem 6 Attractivity of global stochastic center manifolds. *For any $t \geq 0$ the following estimate holds a.s.:*

$$\|\phi_s(t,\omega,x_c,x_s) - h(v_t\omega,\phi_c(t,\omega,x_c,x_s))\| \leq c(t,\omega)\|x_s - h(\omega,x_c\| \tag{27}$$

Furthermore, for $t \to \infty$ $c(t,\omega)$ converges to 0 exponentially fast with respect to the metric of convergence in probability.

Proof. After having written the desired estimate by means of the norm $|.|^s_\omega$ we start by estimating the left-hand side for $t = 1$. This can be done by methods similar to those used in the proof of Theorem 4. The cocycle property enables us to iterate the procedure and to deduce the result. Details may be found in Boxler [3], Theorem 7.1.

Hence we learn from this theorem that if we start away from the stochastic center manifold we will be attracted by it exponentially fast. But this also means that the stochastic center manifold is precisely that part of the system that will survive in the long run.

This reflects the fact that the stochastic center manifold is the nonlinear counterpart to the space $E_c(\omega)$ the latter corresponds to the biggest Lyapunov exponent, and for this reason almost all initial values of the linearized system will be attracted by the direction of this space.

4.2 The Reduction Principle

Recall that we have claimed in the beginning of this paper that stochastic center manifolds are a tool to reduce the dimension of that bifurcation problem which will finally have to be studied in order to obtain infornation about the original system. Let us explain this in detail now.

Consider once again equations (3) and (4) but this time we replace x_S by $h(\omega, x_c)$ which means that we are interested in the restriction of these equations to the stochastic center manifold. Hence we obtain:

$$\phi_c(t,\omega,x_c,h(\omega,x_c)) = \Psi_c(t,\omega)x_c + \Phi_c(t,\omega,x_c,h(\omega,x_c)) \tag{28}$$

$$\begin{aligned}\phi_s(t,\omega,x_c,h(\omega,x_c)) &= \Psi_s(t,\omega)h(\omega,x_c) + \Phi_s(t,\omega,x_c,h(\omega,x_c)) \\ &= h(v_t\omega, \phi_c(t,\omega,x_c,h(\omega,x_c)))\end{aligned} \tag{29}$$

Of course the last equality is nothing but the invariance property for stochastic center manifolds. Then these equations are decoupled, and this has been one of the reasons why we have introduced stochastic center manifolds because the heuristic idea was that it should be sufficient to examine (28) because the stability behaviour of the second equation is well-known. This is in fact true, as we see from the following theorem:

Theorem 7. *The zero solution of system (3) and (4) is asymptotically stable (resp. unstable) if and only if the zero solution of equation (28) has this property.*

Proof. Obviously the stability behaviour of the entire system (3) and (4) will carry over to the reduced system described by equation (28).

On the other hand, if the zero solution of (28) is asymptotically stable (resp. unstable) this property will also hold for the entire system because of the attractivity and the invariance of the stochastic center manifold.

Remark: It can also be shown that the error that is made by considering system (28) and (29) instead of (3) and (4) tends to zero exponentially fast. We will not make explicit use of this result here and so we omit its proof, which is rather lengthy. The interested reader may consult Boxler [3], Section 7.

Because of Theorem 7 it will thus be sufficient to study a system that has the same dimension as the space $E_c(\omega)$. What does this mean in the context of bifurcation theory ?

Let us assume a parameter depending nonlinear system (either white or real noise) whose solution generates a (nonlinear) cocycle ϕ_α which will also depend on the bifurcation parameter α. Before we reach the bifurcation point the stationary solution (which may be reduced to the case of a fixed point at 0 as explained earlier) of the system is supposed to be asymptotically stable and hence all the Lyapunov exponents will be negative. This means that $E_{c,\alpha}(\omega) = \{0\}$ for these values of α.

At the bifurcation point α_0 the biggest Lyapunov exponent will become zero (recall that this was just a necessary condition for a stochastic bifurcation) whereas the other Lyapunov exponents will remain negative. It is intuitively clear that in the usual cases this procedure will generate a space $E_{c,\alpha_0}(\omega)$ of dimension 1 or 2 because for situations which are not too degenerate it is improbable that for a certain parameter value all the Lyapunov exponents or at least most of them will vanish at the same time.

Hence in bifurcation theory equation (28) will in general be of considerably lower dimension than the original system and we have seen that the stability behaviour of the entire system can be determined from the stability behaviour of this equation.

Furthermore, as pointed out in Section 4.1, the stochastic center manifold is the only object that survives in the long run because of its attractivity. But this means that the bifurcating solutions will have to live on the center manifold and this implies that if we know the stochastic center manifold we also know where we might have a chance to find bifurcating solutions. This, however, raises the question of how to determine the stochastic center manifold. Is it possible to calculate it ?

Before answering this question in the next section another consequence of the theory of stochastic center manifolds is to be mentioned:

If we are able to cope with one- and two-dimensional systems, i.e. if we can determine their bifurcation behaviour, then we will already have solved a much larger class of problems because, as we have explained, all the "non-degenerate" systems will yield an equation on the stochastic center manifold of type (28) which will be one- or two-dimensional and we have seen that it is sufficient to study this equation and its bifurcation behaviour. Thus, thanks to stochastic center manifold theory it is justified that we concentrate our efforts in stochastic bifurcation theory mainly on one- and two- dimensional systems !

5 Approximation of Stochastic Center Manifols

5.1 An Explicit Second Order Approximation

As in the deterministic situation it is usually difficult to calculate stochastic center manifolds explicitly. It is, however, possible to approximate them and there is a theorem (see Boxler [3], Theorem 9.1) which says that, in principle, an approximation with any desired degree of accuracy may be obtained.

This theorem not being constructive we will omit the details here. Instead, we will derive an explicit approximation by means of a Taylor series up to second order. The same procedure would also work for higher orders but for notational simplicity we restrict ourselves to the second order case.

By construction, we have: $h(\omega,0) = D^1h(\omega,0) = 0$. This enables us to write the following Taylor series expansion for h:

$$h(\omega,y) = \frac{1}{2}D^2h(\omega,0)(y,y) + O(\|y\|^3),\ y \in E_c(\omega) \tag{30}$$

Here the second order derivative is understood as a bilinear form and $f(\omega,x) = O(\|x\|)$ means that $f(\omega,x) < C(\omega)\|x\|$ a. s for a positive-valued random variable C and sufficiently small x.

By construction of the stochastic center manifold we know that

$$h(\omega,y) = \Psi_s(1,v_{-1}\omega)h(v_{-1}\omega,x) + \Phi_s(1,v_{-1}\omega,x,h(v_{-1}\omega,x)) \tag{31}$$

where

$$x = \Psi_c^{-1}(1,v_{-1}\omega)y - \Psi_c^{-1}(1,v_{-1}\omega)\Phi_c(1,v_{-1}\omega,x,h(v_{-1}\omega,x)) \tag{32}$$

Now we have to calculate the second order derivative of this expression and this can be done by using the following consequence of the chain rule (see e.g. Abraham, Marsden and Ratiu [1], page 92):

$$\begin{aligned} D^2(g\circ f)(0)(v,w) = D^1g(f(0))(D^2f(0)v) + \\ D^2g(f(0))(D^1f(0)v, D^1f(0)w) \end{aligned} \tag{33}$$

We take into account that

$$h(\omega,0) = D^1h(\omega,0) = 0 \tag{34}$$

$$\Phi_{c,s}(t,\omega,0) = D^1\Phi_{c,s}(t,\omega,0) = 0 \tag{35}$$

and obtain by an iteration of this procedure because of the cocycle property:

$$\begin{aligned} D^2h(\omega,0)(v,v) = {} & \Psi_s(n,v_{-n}\omega)D^2h(v_{-n}\omega,0)(\Psi_c(-n,\omega)v,\Psi_c(-n,\omega)v) + \\ & \sum_{k=0}^{n-1}\Psi_s(k,v_{-k}\omega)D^2_{x_c}\Phi_s(1,v_{-k-1}\omega,0,0) \\ & (\Psi_c(-k-1,\omega)v,\Psi_c(-k-1),\omega)v) \end{aligned} \tag{36}$$

At first glance this does not seem to be helpful because the right-hand side still contains the unknown second-order derivative of the map h. But what happens if we consider $n \to \infty$? For that we make use of the estimates derived in Lemma 2. Together with assumption (NL 1)(ε) they yield:

$$|D^2 h(\omega,0)(v,v)|^s_\omega \leq L e^{(\lambda_s+2\beta)n} + \sum_{k=0}^{n-1} \varepsilon e^{(\lambda_s+2\beta)k} \tag{37}$$

Theorem 8. *The map,* h *describing the stochastic center manifold may be written as follows:*

$$\begin{aligned} h(\omega,y) = \frac{1}{2}\sum_{n=0}^{\infty} & \Psi_s(n, v_{-n}\omega) D^2_{x_c}\Phi_s(1, v_{-n-1}\omega, 0, 0) \\ & +(\Psi_c(-n-1,\omega)y, \Psi_c(-n-1),\omega)y) \\ & +O(\|y\|^3),\ y \in E_c(\omega) \end{aligned} \tag{38}$$

5.2 Stochastic Center Manifolds = Deterministic Center Manifolds + noise ?

Perhaps you have already wondered whether the stochastic center manifold theory presented up to now could not be replaced by a theory derived by simply adding noise to the deterministic center manifold, the latter being obtained by switching off the noise. This approach has been suggested in the physical literature by Knobloch and Wiesenfeld [8] and at first glance it seems to be appealing.

In this section we will, however, describe a simple example which shows that such an approach would not yield very satisfactory results, in particular in case the noise is not small. In our example we will calculate the second order approximation of the stochastic center manifold. If the stochastic center manifold could really be obtained as the deterministic center manifold plus fluctuations then at least the expectation of the stochastic center manifold and the deterministic center manifold should coincide. We will see that this is not the case.

Thus, let us consider the following system:

$$dx_1 = \alpha x_1 dt + \sigma_1 x_1 \circ dW_1(t) \tag{39}$$

$$dx_2 = (bx_2 + x_1^2)dt + \sigma_2 x_2 \circ dW_2(t) \tag{40}$$

where $\alpha \in \mathbb{R}$ is the bifurcation parameter, $\sigma_1, \sigma_2 \geq 0$ describe the noise intensides, and b is supposed to be < 0.

If we linearize this equation at its fixed point $x = 0$ we obtain:

$$\Psi_\alpha(t,\omega)\begin{pmatrix} x_1 \\ x_2 \end{pmatrix} = \begin{pmatrix} e^{(\alpha t+\sigma_1 W_1(t,\omega))}x_1 \\ e^{(bt+\sigma_2 W_2(t,\omega))}x_2 \end{pmatrix} \tag{41}$$

This immediately yields the Lyapunov exponents because of $W_i(t,\omega)/t \to 0$ for $t \to \infty$. We determine $\lambda_1(\alpha) = \alpha, \lambda_2(\alpha) = b$ and this means that the stochastic bifurcation point is reached at $\alpha_0 = 0$. From now on we will restrict ourselves to this point and drop the subscript α for notational simplicity.

Here the spaces E_c, E_s resp. coincide with the x_1, x_2 axis, resp. and they are thus independent of ω. In the sequel we will write $x_c := x_1, x_s := x_2$. Then the solution of equation (40) may be written as

$$\begin{aligned}\phi_s(t,\omega,x) &= \Psi_s(t,\omega)x_s + \int_0^t \Psi_s(t-s,\omega)e^{2\sigma_1 W_1(s,\omega)}x_c^2 ds \\ &= \Psi_s(t,\omega)x_s + \Phi_s(t,\omega,x_c,x_s) \end{aligned} \tag{42}$$

We assume conditions on the size of σ_l, σ_2 and b relative to each other (we will omit the details) such that the existence of a stochastic center manifold is ensured by

Theorem 9. *In the next step we will seek an approximation for it of the form*

$$h(\omega, x_c) = \frac{1}{2}h''(\omega,0)x_c^2 + O(\|x_c\|^3) \tag{43}$$

Because of Theorem 8 we know that

$$h\prime\prime(\omega,0) = \sum_{n=0}^{\infty} c_n(\omega) = \Psi_s(n, v_{-n}\omega)\frac{\partial^2 \Phi_s}{\partial x_c^2}(1, v_{-n-1}\omega, 0, 0)\Psi_c^2(-n-1,\omega) \tag{44}$$

But on the other hand we have

$$\Psi_s(n, v_{-n}\omega) = e^{bn+\sigma_2 W_2(n, v_{-n}\omega)} = e^{bn-\sigma_2 W_2(-n,\omega)} \tag{45}$$

$$\Psi_c^2(-n-1,\omega) = e^{2\sigma_1 W_1(-n-1,\omega)} \tag{46}$$

$$\begin{aligned}&\frac{\partial^2 \Phi_s}{\partial x_c^2}(1, v_{-n-1}\omega, 0, 0) \\ &= 2\int_{s=0}^1 \Psi_s(1-s, v_{-n-1}\omega)e^{s\sigma_1 W_1(s, v_{-n-1}\omega)}ds \\ &= 2\int_{s=0}^1 e^{b(1-s)+\sigma_2(W_2(-s-n)-W_2(-n-1))+2\sigma_1(W_1(s-n-1)-W_1(-n-1))}ds\end{aligned} \tag{47}$$

Consequently this inplies that

$$c_n(\omega) = 2e^{bn}e^{-\sigma_2 W_2(-n)}\int_{s=0}^1 e^{b(1-s)}e^{\sigma_2(W_2(-s-n)-W_2(-n-1))}e^{2\sigma_1 W_1(s-n-1)} \tag{48}$$

Switching off noise, i.e. putting $\sigma_1 = \sigma_2 = 0$, we obtain

$$h_{\text{det}}(x_c) = -\frac{1}{b}x_c^2 + O(\|x_c\|^3) \tag{49}$$

and this is just the expression for the deterministic center manifold. This value does, however, not coincide with the expectation of the stochastic center manifold which we are going to calculate now.

Since $W_1(s-n-1), W_2(-n)$ and $W_2(-s-n) - W_2(-n-1)$ are independent and due to the well-known fact that $Ee^{\sigma(W(t+s)-W(s))} = e^{(\sigma^2/2)|t|}$ we may write:

$$Ec_n = 2e^{bn}Ee^{-\sigma_2 W_2(-n)}$$
$$\int_{s=0}^{1} e^{b(1-s)}Ee^{\sigma_2(W_2(-s-n)-W_2(-n-1))}Ee^{2\sigma_1 W_1(s-n-1)}ds$$
$$= 2e^{bn}e^{\frac{\sigma_2^2}{2}n}\int_{s=0}^{1} e^{b(1-s)}e^{\frac{\sigma_2^2}{2}(1-s)}e^{2\sigma_1^2|s-n-1|}ds = 2\frac{1-e^{-a}}{a}e^{-an} \quad (50)$$

where we have put $a := -(b + \sigma_2^2/2 + 2\sigma_1^2)$.

Finally this means that for $a > 0$ we obtain the epextation of the stochastic cneter manifold:

$$Eh(\omega, x_c) = \left[\frac{1-e^{-a}}{a}\sum_{n=0}^{\infty} e^{-an}\right] x_c^2 + O(\|x_c\|^3)$$
$$= \frac{1}{a}x_c^2 + O(\|x_c\|^3)$$
$$= \frac{-1}{\left(b + \frac{\sigma_2^2}{2} + 2\sigma_1^2\right)}x_c^2 + O(\|x_c\|^3) \quad (51)$$

Hence we see that it is *not* possible to explain the influence of the noise simply as a fluctuation around a deterministic situation. New phenomena appear, and they can only be understood with the tools provided by stochastic analysis.

References

1. Abraham, R., Marsden, J. E., Ratiu, T.: Manifolds, tensor analysis and applications. Addison-Wesley Reading, Massachusetts 1983.
2. Boxler, P.: Lyapunov exponents indicate stability and detect stochastic bifurcations. This volume.
3. Boxler, P.: Stochastische Zentrumsmannigfaltigkeiten. Ph. D. thesis, Institut für Dynamische Systeme, Universität Bremen 1988. (Condensed English version of [3]: Boxler, P.: A stochastic version of center manifold theory. Probab. Th. Rel. Fields (to appear).)
4. Dieudonné: Foundations of modern analysis. Academic Press, New York, London 1969.
5. Federer, H.: Geometric measure theory. Springer, Berlin, Heidelberg, New York 1969.
6. Guckenheimer, J., Holmes, Ph.: Nonlinearoscillations, dynamical systems and bifurcations of vector fields. Springer, Berlin, Heidelberg, New York, Tokyo, 2nd ed. 1986.
7. Kantorowitsch, L. W., Akilow, G. P.: Funktionalanalysis in normierten Räumen. Harri Deutsch, Thun, Frankfurt 1964.
8. nobloch, E., Wiesenfeld, K. A.: Bifurcations in fluctuating systems: The center manifold approach. J. Stat. Phys., Vol. 33, no. 3 (1983), 611 - 637.
9. Vanderbauwhede, A.: Center manifolds, normal forms and elementary bifurcations. In: U. Kirchgraber, H. 0. Walther (eds.): Dynamics Reported, Vol. 2, Teubner and Wiley 1989, 89 - 169.

Lyapunov Exponents for a Class of Hyperbolic Random Equations

Ina Lindemann

Abstract

In this paper we are going to study the Lyapunov exponents of infinite dimensional systems $\dot{x} = A(t,\omega)x$, where $A(t,\omega) = \mathrm{diag}(A_1(t,\omega), A_2(t,\omega), \ldots)$ has a certain infinite matrix modal representation. It will be shown that the finite dimensional systems $\dot{x} = A_n(t,\omega)x$ determine the Lyapunov exponents of the original system.

1 Introduction

We consider a class of hyperbolic random equations

$$\frac{\partial^2}{\partial t^2} w(t,x) + (A + A(f_t))w(t,x) = 0$$

on the real Hilbert space $L^2([0,1])$, where (f_t) is a stationary stochastic process. On the assumption that A and $A(f_t)$ are selfadjoint unbounded operators and meet some further requirements, the problem can be transferred onto the Hilbert space l^2 using spectral theory. Thus, we arrive at a system of the following form:

$$\dot{x}(t) = A(t,\omega)x(t) = \begin{pmatrix} A_1(t,\omega) & & 0 \\ & A_2(t,\omega) & \\ 0 & & \ddots \end{pmatrix} x(t) \tag{1}$$

i.e. with a particularly convenient infinite matrix representation of $A(t,\omega)$. This class of problems models e.g. large flexible space structures [3]. The aim is to determine the Lyapunov exponents

$$\lambda_\infty(x_0) = \overline{\lim_{t\to\infty}} \frac{1}{t} \log \|x(t;x_0)\|$$

of solutions to the problem (1).

We pursue the following idea: the stability of the system (1) should be determined by the behaviour of the finite dimensional block systems

$$\dot{x}_n(t) = A_n(t,\omega)x_n(t) \tag{2}$$

But in general the infinite dimensional system might create new phenomena. We will show that in our special situation

$$\lambda_\infty(x_0) = \sup_{n\geq 1} \lambda_n(x_n^0) \tag{3}$$

holds, where

$$\lambda_n(x_n^0) = \overline{\lim_{t\to\infty}} \frac{1}{t} \log \|x_n(t; x_n^0)\|$$

is the Lyapunov exponent for solutions of the problem (3) and x_n^0 is the restriction of x_0 to the nth block.

In section 2 we will formulate the framework in detail. In section 3 we will go ahead with the transformation into an l^2-problem and we will see that a unique random evolution operator exists. Section 4 contains the proof of equation (3). In section 5 we will consider the beam equation

$$\frac{\partial^2 w}{\partial t^2} + 2\beta \frac{\partial w}{\partial t} + (f_0 + f(\xi_t)) \frac{\partial^2 w}{\partial x^2} + \frac{\partial^4 w}{\partial x^4} = 0$$

with certain boundary conditions. Sufficient conditions for stability of this system were given by Kozin [6]. We will investigate the Lyapunov exponents λ_n for large n. In that special case we will calculate the asymptotic behaviour of λ_n when n tends to infinity while we follow methods given in [2].

2 Formulation of the problem

A class of hyperbolic random equations

$$\left.\begin{aligned} \frac{\partial^2}{\partial t^2} w(t,x) + (A + A(t_t))w(t,x) &= 0 \\ w(0,x) &= w_0 \in D(A) \\ \frac{\partial w}{\partial t}(0,x) &= w_0' \end{aligned}\right\} \tag{4}$$

with $x \in [0,1]$ and $t \in [0,\infty)$ will be investigated.

Let A be a linear, positive definite, seladjoint, unbounded operator on $H = L^2([0,1])$ with pure point spectrum. The complete orthonormal system of eigenfunctions of A and the corresponding set of eigenvalues will be denoted by $\{\phi_n\}_{n\in\mathbb{N}}$ and $\{\mu_n\}_{n\in\mathbb{N}}$, respectively. The operator $A(f_t)$ statifies the followind conditions:

(A) For every $t \geq 0$, $A(f_t)$ is selfadjoint and has the same eigenfunctions as A; i.e. for every $n \in \mathbb{N}$ there exist $\{\kappa_n(f_t)\} \subseteq \mathbb{R}$ such that $A(f_t)\phi_n = \kappa_n(f_t)\phi_n$. Furthermore, we suppose $D(A) \subseteq D(A(f_t))$ and we set $D(A+A(f_t)) = D(A)$ for all $t \geq 0$.

(B) For

$$a_n(f_t) := \frac{\kappa_n(f_t)}{\sqrt{\mu_n}}$$

there exist constants K_1 and K_2 such that

$$\sup_{t\geq 0, n\in\mathbb{N}} |a_n(f_t)| \leq K_1 \leq \infty$$

and for all $0 \leq s,t \leq \infty$ and all $n \in \mathbb{N}$ we have

$$|a_n(f_t) - a_n(f_s)| \leq K_2 |f_f - f_s|.$$

Furthermore, let $(f_t)_{t\geq 0}$ be a stationary ergodic stochastic process with f_0 integrable, whose paths are bounded, continuous functions with probability one. Additionally, (f_t) satifies the condition

(C) there exist constants $\gamma \geq 1, C, \alpha > 1$ such that $E|f_t - f_s|^\gamma \leq C|t-s|^\alpha$ for all $s, t \geq 0$, where $E(.)$ denotes expectation with respect to P on a complete probability space $(\Omega, \mathcal{F}, P)$.

In the usual way the problem (4) can be transferred into a first order problem. With the notations

$$z := \begin{pmatrix} w \\ \frac{\partial w}{\partial t} \end{pmatrix}, \mathcal{A} := \begin{pmatrix} 0 & Id \\ -A & 0 \end{pmatrix} \text{ and } \mathcal{A}(t) := \begin{pmatrix} 0 & 0 \\ -A(f_t) & 0 \end{pmatrix}$$

we obtain the problem

$$\frac{\partial z}{\partial t} = (\mathcal{A} + \mathcal{A}(t)z \quad z(0,x) = z_0 \in D(\mathcal{A}) \tag{5}$$

where $D(\mathcal{A}) = D(A) \times D(A^{1/2})$.

3 Description on the Hilbert space l^2

Using spectral properties of A and $A(f_t)$, the problem (5) will be formulated on l^2. The energetic space of A, cf. Triebel [9] p.275, is

$$D(A^{1/2}) := \{w \in L^2([0,1]); \sum_{j=1}^{\infty} \mu_j |\langle w, \phi_j\rangle|^2 < \infty\}$$

with the inner product

$$\langle x, y\rangle_{D(A^{1/2})} = \langle A^{1/2}x, A^{1/2}y\rangle.$$

Then $\{\mu_j^{1/2}\phi_j\}$ is a complete orthonormal system in $D(A^{1/2})$. Note that μ_n tends to infinity as $n \to \infty$, since A is unbounded.

Introducing the Hilbert space

$$Z := (D(A^{1/2}) \times L^2([0,1]); \langle .,.\rangle_{D(A^{1/2})} + \langle .,.\rangle_{L^2})$$

with the complete orthonormal system

$$\{\psi_{2n-1}, \psi_{2n}\}_{n\in\mathbb{N}} := \left\{ \begin{pmatrix} \mu_n^{1/2}\phi_n \\ 0 \end{pmatrix}, \begin{pmatrix} 0 \\ \phi_n \end{pmatrix} \right\},$$

we obtain that every $z \in Z$ has the representation

$$z = \begin{pmatrix} z_1 \\ z_2 \end{pmatrix} = \sum_{n=1}^{\infty} \langle z, \psi_n\rangle_Z \psi_n = \sum_{n=1}^{\infty} \sqrt{\mu_n}\langle z_1, \phi_n\rangle_{L^2}\psi_{2n-1} + \langle z_2, \phi_n\rangle_{L^2}\psi_{2n}$$

and for $z \in D(\mathcal{A})$ we have

$$\mathcal{A}_z = \begin{pmatrix} z_2 \\ -Az_1 \end{pmatrix} = \sum_{n=1}^{\infty} \sqrt{\mu_n} \langle z_2, \phi_n \rangle_{L^2} \psi_{2n-1} - \mu_n \langle z_1, \phi_n \rangle_{L^2} \psi_{2n}.$$

Thus, $\mathcal{A}$ corresponds to an operator (denoted by $\mathcal{A}$ as well) defined on

$$h := \{x \in l^2 : \|\mathcal{A}x\|_{l_2} < \infty\},$$

a dense subspace of l^2, of the following form

$$\mathcal{A} = \begin{pmatrix} \ddots & & 0 \\ & \boxed{\begin{matrix} 0 & \sqrt{\mu_n} \\ -\sqrt{\mu_n} & 0 \end{matrix}} & \\ 0 & & \ddots \end{pmatrix}$$

Remark. Since A is closed, $\overline{h} = l^2$ and since for all $\lambda < 0$ it holds that

$$\|(\lambda Id - \mathcal{A})^{-1}\| \le 1/\lambda,$$

it follows from the Hille-Yosida theorem that $\mathcal{A}$ generates a strongly continuous semigroup of contractions on l^2 (cf Pazy [8]) which is given by

$$T_t = \begin{pmatrix} \ddots & & 0 \\ & \boxed{\begin{matrix} \cos\sqrt{\mu_n}t & \sin\sqrt{\mu_n}t \\ -\sin\sqrt{\mu_n}t & \cos\sqrt{\mu_n}t \end{matrix}} & \\ 0 & & \ddots \end{pmatrix}$$

Since

$$\mathcal{A}(t)z = \begin{pmatrix} 0 \\ A(f_t)z_1 \end{pmatrix} = \sum_{n=1}^{\infty} -a_n(f_t)\sqrt{\mu_n} \langle z_1, \phi_n \rangle_{L^2} \psi_{2n}$$

$\mathcal{A}(t)$ has the representation

$$\mathcal{A}(t) = \begin{pmatrix} \ddots & & 0 \\ & \boxed{\begin{matrix} 0 & 0 \\ -a_n(f_t) & 0 \end{matrix}} & \\ 0 & & \ddots \end{pmatrix}$$

on l^2. Thus, we want to consider the system

$$\dot{x}(t) = \mathrm{diag}(A_n(t))x(t) = \begin{pmatrix} \ddots & & 0 \\ & \boxed{\begin{matrix} 0 & \sqrt{\mu_n} \\ -(\sqrt{\mu_n} + a_n(f_t)) & 0 \end{matrix}} & \\ 0 & & \ddots \end{pmatrix} x(t), \qquad (6)$$

$x(0) = x_0 \in h$ which is equivalent to (4).

Remark. Since $\mathcal{A}(t), 0 \le t < \infty$, are bounded linear operators on l^2 with $\|\mathcal{A}(t)\| \le K_1$ for every $0 \le t < \infty$, $\{\mathcal{A} + \mathcal{A}(t)\}_{t \ge 0}$ is a stable family of infinitesimal generators. The domains $D(\mathcal{A} + \mathcal{A}(t)) = h$ are independent of t and since (f_t) is continuous almost surely it follows that there exists a unique evolution system $U(t, s), 0 \le s, t < \infty$ on l^2. For $x_0 \in h$ the unique h-valued solution of (6) with initial condition $x(s) = x_0$ is $U(t, s)x_0$ (cf. Pazy [8]).

4 Stability properties

Problem (6) consists of decoupled two dimensional systems. Thus, the evolution operaotr $U(t, s)$ has the structure

$$U(t,s) = \begin{pmatrix} \ddots & & 0 \\ & U_n(t,s) & \\ 0 & & \ddots \end{pmatrix}$$

as well, where $U_n(t, s)$ is the solution for $\dot{x}_n(t) = A_n(t)x_n(t)$. This induces the question whether the stability of (6) is already determined by the behaviour of the two dimensional block systems. In general this is not clear. In order to find an answer to this we consider the Lyapunov exponent

$$\lambda_\infty(x_0) := \overline{\lim_{t \to \infty}} \frac{1}{t} \log \|U(t, 0)x_0\|.$$

Lemma 1. *If $\mathcal{L}(l^2)$ and $\mathcal{L}(\mathbb{R}^2)$ are the spaces of linear operators on l^2 and $\mathbb{R}^2$, respectively, and if $U = \operatorname{diag}(U_n)$ is an operator in two dimensional block structure, then*

$$\|U\|_{\mathcal{L}(l^2)} = \sup_{n \ge 0} \|U_n\|_{\mathcal{L}(\mathbb{R}^2)}.$$

For an arbitrary $x = (x_1, x_2, \ldots) \in l^2$ we denote $(x_{2n-1}, x_{2n}) \in \mathbb{R}^2$ by $\tilde{x}_n$ and $(0, \ldots, 0, x_{2n-1}, x_{2n}, 0, \ldots) \in l^2$ by $\overline{x_n}$.

Proof. On the one hand we have

$$\|U\|_{\mathcal{L}(l^2)} = \sup_{\|x\|_{l^2}=1} \|Ux\|_{l^2} \ge \sup_{\|\overline{x}_n\|_{l^2}=1} \|U\overline{x}_n\|_{l^2} = \|U_n\|_{\mathcal{L}(\mathbb{R}^2)}$$

for all $n \in \mathbb{N}$, on the other hand for every $x \in l^2$

$$\begin{aligned} \|Ux\|_{l^2} = \left(\sum_{n=1}^{\infty} \|U_n\tilde{x}_n\|^2_{\mathbb{R}^2}\right)^{1/2} &\le \left(\sum_{n=1}^{\infty} \|U_n\|^2_{\mathcal{L}(\mathbb{R}^2)} \|\tilde{x}_n\|^2_{\mathbb{R}^2}\right)^{1/2} \\ &\le \sup_{n \ge 0} \|U_n\|_{\mathcal{L}(\mathbb{R}^2)} \|x\|_{l^2}, \end{aligned}$$

which proves the lemma. □

If for the initial value x_0 the number of nonzero entries is finite the theory of finite dimensionnal systems can be applied. Let $\lambda_n(x_0)$ be the Lyapunov exponent of the nth block system i.e.

$$\lambda_n = \lambda_n(x_0) := \overline{\lim_{t\to\infty}} \frac{1}{t} \log \|U_n(t)x_n^0\|$$

The special structure of $A_n(t)$ allows us to state that the only invariant control set of the associated control system on S^1 for large n is S^1 itself. Since all solutions with initial values in that invariant control set have one and the same Lyapunov exponent and the trace of $A_n(t)$ is zero, every block system realizes one nunnegative Lyapunov exponent. A detailed treatment of the 2×2 system may be found in Arnold and Kliemann [1].

An important tool for investigating the Lyapunov exponents of finite dimensional systems was provided by Hasminskii [5].

We project a nontrivial solution $x_n(t)$ of the two dimensional block system $\dot{x}(t) = A_n(t)x_n(t)$ onto the unit sphere S^1 by setting

$$s_n(t) = \frac{x_n(t)}{\|x_n(t)\|}.$$

The projected process s_n is a solution of the nonlinear differential equation

$$\dot{s}_n = h(s_n, A_n(t)) = (A_n(t) - q(s_n, A_n(t))Id)s_n,$$

$$s_n^0 = \frac{x_n^0}{\|x_n^0\|},$$

where

$$q(s, A) = s'As, \; s \in S^1$$

(prime denoting the transpose).

We are now interested in initial values of the following form:

$$x_0 = (0, \ldots, 0, x_{2N-1}, x_{2N}, \ldots)$$

for large $N \in \mathbb{N}$ and we would also like to find out whether $\lambda_\infty(x_0)$ will get small in N for those x_0.

Proposition 2. *There exist constants $C, \beta, 0 \leq \beta \leq 1$ such that for $N \in \mathbb{N}$ and $x_0 = (0, \ldots, 0, x_{2N-1}, x_{2N}, x_{2N+1}, \ldots)$*

$$0 \leq \lambda_\infty(x_0) \leq \frac{C}{(\sqrt{\mu_N})^\beta}$$

holds.

Proof. By definition we have $0 \leq \lambda_n(x_0) \leq \lambda_\infty(x_0)$ for all $n \in \mathbb{N}$. Furthermore, the following holds:

$$\begin{aligned}
\lambda_\infty(x_0) &= \overline{\lim_{t\to\infty}} \frac{1}{2t} \log \|U(t)x_0\|_{l^2}^2 \\
&= \overline{\lim_{t\to\infty}} \frac{1}{2t} \log \sum_{n=N}^{\infty} \|U_n(t)x_n^0\|_{\mathbb{R}^2}^2 \\
&= \overline{\lim_{t\to\infty}} \frac{1}{2t} \log \sum_{n=N}^{\infty} \|x_n^0\|^2 \exp\left(2\int_0^t q(s_n(\tau), A_n(\tau))d\tau\right) \\
&\leq \overline{\lim_{t\to\infty}} \frac{1}{2t} \log(\sup_{n\geq N} \exp\left(2\int_0^t q(s_n(\tau), A_n(\tau))d\tau\right) \|x_0\|_{l^2}^2 \\
&= \overline{\lim_{t\to\infty}} \sup_{\dot{N}\geq N} \frac{1}{t}\int_0^t q(s_n(\tau), A_n(\tau))d\tau.
\end{aligned}$$

In what follows let $n \geq N$. Setting $s_n(\tau) = (\cos\phi_\tau^{(n)}, sin\phi_\tau^{(n)})$ we have

$$q(s_n(\tau), A_n(\tau)) = \frac{1}{2} a_n(f_\tau) \sin 2\phi_\tau^{(n)},$$

where the angular process $(\phi_\tau^{(n)})$ is a solution of the differential equation

$$\dot{\phi}_\tau^{(n)} = -\sqrt{\mu_n} + a_n(f_\tau)\cos^2\phi_\tau^{(n)},\ \phi_0^{(n)} = \arccos s_n^0$$

(cf. Wihstutz [11]). We estimate $\lambda_\infty(x_0)$ using the fact that compared with $a_n(f_\tau)$ the process $(\phi_\tau^{(n)})$ is for large n "very fast".

Let $\Psi_t^{(n)} := \sqrt{\mu_n}t + \phi_t^{(n)}$. Then the process $(\Psi_t^{(n)})$ statifies

$$\Psi_t^{(n)} = \Psi_0^{(n)} + \int_0^t a_n(f_\tau)\cos^2(\Psi_\tau^{(n)} - \sqrt{\mu_n}\tau)d\tau.$$

At this point an important inequality is not to be ignored.Since

$$|\Psi_t^{(n)} - \Psi_s^{(n)}| \leq K_1|t-s|$$

we have

$$\begin{aligned}
&|a_n(f_t)\sin 2\Psi_t^{(n)} - a_n(f_s)\sin 2\Psi_s^{(n)}| \\
&= |a_n(f_t)(\sin 2\Psi_t^{(n)} - \sin 2\Psi_s^{(n)}) + (a_n(f_t) - a_n(f_s))\sin 2\Psi_s^{(n)}| \qquad (7) \\
&\leq 2K_1^2|t-s| + K_2|f_t - f_s|.
\end{aligned}$$

Now

$$\begin{aligned}
B &:= \frac{1}{t}\int_0^t a_n(f_\tau)\sin 2\phi_\tau^{(n)} d\tau \\
&= \underbrace{\frac{1}{t}\int_0^t a_n(f_\tau)\sin 2\Psi_\tau^{(n)}\cos 2\sqrt{\mu_n}\tau\, d\tau}_{:=B_1} - \frac{1}{t}\int_0^t a_n(f_\tau)\cos 2\Psi_\tau^{(n)}\sin 2\sqrt{\mu_n}\tau\, d\tau.
\end{aligned}$$

The estimation for both terms on the right-hand side is similar. Thus, we only consider B_1:

$$B_1 = \frac{\pi}{\sqrt{\mu_n}t}\int_0^{\frac{\sqrt{\mu_n}t}{\pi}} a_n(f_{\frac{r\pi}{\sqrt{\mu_n}}})\sin 2\Psi_{\frac{r\pi}{\sqrt{\mu_n}}}\cos 2\pi r dr.$$

With the notations $\nu_n := \frac{\sqrt{\mu_n}}{\pi}$ and $[\nu_n]$ the integer part of ν_n it holds that

$$B_1 = \underbrace{\frac{1}{\nu_n t}\int_0^{[\nu_n t]} a_n(f_{\frac{r}{\nu_n}})\sin 2\Psi_{\frac{r}{\nu_n}}\cos 2\pi r\, dr}_{:=B_2} + \frac{1}{\nu_n t}\int_{[\nu_n t]}^{\nu_n t} a_n(f_{\frac{r}{\nu_n}})\sin 2\Psi_{\frac{r}{\nu_n}}\cos 2\pi r\, dr.$$

It is easy to deduce that the second term is less than $K_1/\nu_n t$. Turning to B_2 we have

$$\begin{aligned} B_2 &= \frac{1}{\nu_n t}\sum_{k=1}^{[\nu_n t]}\int_0^1 a_n(f_{\frac{(k-1)+\tau}{\nu_n}})\sin 2\Psi_{\frac{(k-1)+\tau}{\nu_n}}\cos 2\pi\tau\, d\tau \\ &= \frac{1}{\nu_n t}\sum_{k=1}^{[\nu_n t]}\int_0^1 [a_n(f_{\frac{(k-1)+\tau}{\nu_n}})\sin 2\Psi_{\frac{(k-1)+\tau}{\nu_n}} - a_n(f_{\frac{k}{\nu_n}})\sin 2\Psi_{\frac{k}{\nu_n}}]\cos 2\pi\tau\, d\tau \end{aligned}$$

Using inequality (7) we obtain:

$$|B_2| \le \frac{2K_1^2}{\nu_n} + K_2\frac{1}{\nu_n t}\sum_{k=1}^{[\nu_n t]}\int_0^1 |f_{\frac{(k-1)+\tau}{\nu_n}} - f_{\frac{k}{\nu_n}}|\, d\tau$$

Thus, we have shown so far

$$|\frac{1}{t}\int_0^t q(s_n(\tau), A_n(\tau))d\tau| \le \frac{K_1}{\nu_n t} + \frac{2K_1^2}{\nu_n} + K_2\underbrace{\frac{1}{\nu_n t}\sum_{k=1}^{[\nu_n t]}\int_{k-1}^k |f_{\frac{k}{\nu_n}} - f_{\frac{r}{\nu_n}}|\, dr}_{:=B_3}.$$

In order to estimate B_3 the following property of the process (f_t) —caused by the condition (C)— should be noticed. If β is an arbitrary positive number less than $(\alpha - 1)/\gamma$ then for any interval $[a, b)$, there exists a positive random variable $C(\omega)$ with $E(C^\gamma) \le \infty$ such that

$$|f_t(\omega) - f_s(\omega)| \le C(\omega)|t - s|^\beta$$

holds for all $s, t \in [a, b)$ and for almost all ω (cf. Kunita [7]).

Now we fix $\beta \le (\alpha - 1)/\gamma$ and set

$$C_n(r,\omega) := \sup_{u,v\in[r,r+\frac{1}{n})} \frac{|f_u(\omega) - f_v(\omega)|}{|u - v|^\beta}.$$

Then $C_n(r, \omega)$ has the following properties:

- $C_n(r,\omega)$ is decreasing in n (for r fixed),
- $(C_n(r,\omega))_{r\in[0,\infty)}$ is a stationary ergodic process for every $n \in \mathbb{N}$.

Therefore

$$\begin{aligned}
B_3 &\leq \frac{1}{\nu_n t}\sum_{k=1}^{[\nu_n t]}\int_{k-1}^{k} C_n\left(\frac{r}{\nu_n},\omega\right)\left(\frac{k-r}{\nu_n}\right)^{\beta} dr \\
&\leq \left(\frac{1}{\nu_n}\right)^{\beta}\frac{1}{\nu_n t}\int_0^{\nu_n t} C_n\left(\frac{r}{\nu_n},\omega\right) dr \\
&= \left(\frac{1}{\nu_n}\right)^{\beta}\frac{1}{t}\int_0^{t} C_n(\tau,\omega)\, d\tau \\
&\leq \left(\frac{1}{\nu_N}\right)^{\beta}\frac{1}{t}\int_0^{t} C_N(\tau,\omega)\, d\tau.
\end{aligned}$$

Since $(C_N(\tau,\omega))$ is a stationary ergodic process, we have almost surely

$$\lim_{t\to\infty}\frac{1}{t}\int_0^t C_N(\tau,\omega)\, d\tau = E(C_N(0,\omega)) < \infty.$$

Finally, we get

$$\lambda_\infty(x_0) \leq \frac{2K_1^2}{\nu_N} + \frac{K_1}{\nu_N t} + K_2 E(C_N(0))\left(\frac{1}{\nu_N}\right)^{\beta}$$

which complete the proof. □

Theorem 3. *For all initial values $x_0 \in l^2$ we have*

$$\lambda_\infty(x_0) = \sup_{n\geq 1}\lambda_\nu.$$

Proof. By definition of the Lyapunov exponent it follows that

$$\lambda_\infty(x_0) \geq \sup_{n\geq 1}\lambda_n.$$

By propostion (2) if hold that

$$\begin{aligned}
\lambda_\infty(x_0) &\leq \overline{\lim_{t\to\infty}}\sup_{n\geq 1}\frac{1}{t}\int_0^t q(s_n(\tau),A_n(\tau))\, d\tau \\
&= \overline{\lim_{t\to\infty}}\max\left\{\max_{1\leq n\leq N}\frac{1}{t}\int_0^t q(s_n(\tau),A_n(\tau))\, d\tau;\right. \\
&\qquad \left.\sup_{n\geq N}\frac{1}{t}\int_0^t q(s_n(\tau),A_n(\tau))\, d\tau\right\} \\
&\leq \max\left\{\max_{1\leq n\leq N}\lambda_n\, ;\, \frac{C}{(\sqrt{\mu_N})^{\beta}}\right\}
\end{aligned}$$

for all $N \in \mathbb{N}$. Since $(\max_{1\leq n\leq N}\lambda_n)$ is an increasing sequence in N, and $(C/\sqrt{\mu_N}^{\beta})$ tends to zero if $N\to\infty$, we have $\lambda_\infty(x_0) \leq \sup_{n\geq 1}\lambda_n$. □

5 An example

Let $x \in [0,1], t \in [0,\infty)$ and consider the beam equation

$$\left.\begin{aligned} \frac{\partial^2 w}{\partial t^2} + 2\beta\frac{\partial w}{\partial t} + (f_0 + f_t)\frac{\partial^2 w}{\partial x^2} + \frac{\partial^4 w}{\partial x^4} = 0 \\ w(0,t) = w(1,t) = \frac{\partial^2 w}{\partial x^2}(0,t) = \frac{\partial^2 w}{\partial x^2}(1,t) = 0 \end{aligned}\right\} \tag{8}$$

The coefficient β may represent damping, f_0 a constant axial load and f_t random excitation of the axial load.

Let f_t fulfil the same conditions as before. Sufficient conditions for stability were studied by Kozin [6].

In order to apply the above we consider $r(t,x) := \exp(\beta t)w(t,x)$ which is a solution of the transformed system

$$\frac{\partial^2 r}{\partial t^2} + (f_0 + f_t)\frac{\partial^2 r}{\partial x^2} + \frac{\partial^4 r}{\partial x^4} - \beta^2 r = 0$$

with the same boundary conditions. Notice that the Lyapunov exponents therefore change as follows:

$$\lambda_r^\infty(x_0) = \overline{\lim_{t\to\infty}}\frac{1}{t}\log\|r(t,x;x_0)\| = \lambda_w^\infty(x_0) + \beta.$$

The same procedure as in section (3) with

$$A = \frac{\partial^4}{\partial x^4} \text{ and } A(f_t) = (f_0 + f_t)\frac{\partial^2}{\partial x^2} - \beta^2 Id$$

and a suitable domain yields the following l^2- system:

$$\dot{x}(t) = \begin{pmatrix} \ddots & & 0 \\ & \boxed{\begin{matrix} 0 & (n\pi)^2 \\ -(n\pi)^2 + (f_0 + f_t) + \frac{\beta^2}{(n\pi)^2} & 0 \end{matrix}} & \\ 0 & & \ddots \end{pmatrix} x(t),$$

$x(0) = x_0$. Because of theorem 3 it is sufficient to concentrate on the stability of the block systems

$$\dot{x}_n(t) = \begin{pmatrix} 0 & (n\pi)^2 \\ -(n\pi)^2 + (f_0 + f_t) + \frac{\beta^2}{(n\pi)^2} & 0 \end{pmatrix} x_n(t), \tag{9}$$

$x_n(0) = x_n^0$. We will analyze the random oscillator equation (9) depending on the parameter $\sigma := (n\pi)^2$. On the assumption of certain conditions pertaining to the random process (f_t) an asymptotic expansion for the Lyapunov exponent $\lambda_\sigma := \lambda_n$ will be developed. With regard to the methods used we are indebted to Arnold, Papanicolaou, Wihstutz (cf. [2]).

Let (f_t) be of the form $f_t = f(\xi_t)$ where $\xi(.)$ is assumed to be an ergodic Markov process on a smooth connected Riemannian manifold M (with or without boundary) with invariant probability $\rho(\xi)d\xi$, and $f : M \to \mathbb{R}$ is a smooth bounded nonconstant function. Let G denote the infenitesimal generator of $\xi(t)$. We will assume the following condition:

(D) M is a compact manifold. G is a selfadjoint elliptic diffusion operator on M with zero an isolated, simple eigenvalue.

Here, the angular process introduced in section 4 in the proof of proposition (2) is a solution of the differential equation

$$\dot{\phi}_t = h(\xi_t, \phi_t) = -\sigma + (f_0 + f(\xi_t)) \cos^2 \phi_t + \frac{\beta^2}{\sigma} \cos^2 \phi_t.$$

The pair (ϕ_t, ξ_t) is a diffusion process on $S^1 \times M$ with generator

$$L = G + h(\phi, \xi)\frac{\partial}{\partial \phi}. \tag{10}$$

On the assumption (D), L has a unique (up to π-periodicity), smooth invariant probability $p(\phi, \xi)\, d\phi\, d\xi$ on $S^1 \times M$ such that its marginal on M is $\rho(\xi)$, and its support is $S^1 \times M$. Under the above conditions the Lyapunov exponent λ_σ exists and is given by

$$\begin{aligned} \lambda_\sigma = (q, p) &= \int_{S^1 \times M} q(\phi, \xi) p(\phi, \xi)\, d\phi\, d\xi \\ &= \int_{S^1 \times M} \frac{1}{2}(f_0 + f_\xi + \frac{\beta^2}{\sigma}) \sin 2\phi\, p(\phi, \xi)\, d\phi\, d\xi. \end{aligned} \tag{11}$$

The invariant density p statisfies the Fokker-Planck equation

$$L^* p = 0.$$

The operator L defined by (10) has the form

$$L_\sigma = \sigma L_0 + L_1 + \frac{1}{\sigma} L_2.$$

We now want to construct a formal expansion of the density p:

$$p_\sigma = p_0 + \frac{1}{\sigma} p_1 + \ldots + \frac{1}{\sigma^N} p_N + \ldots,$$

such that

$$\begin{aligned} L_0^* p_0 &= 0 \\ L_0^* p_1 + L_1^* p_0 &= 0 \\ L_0^* p_2 + L_1^* p_1 + L_2^* p_0 &= 0 \\ &\cdots \\ L_0^* p_N + L_1^* p_{N-1} + L_2^* p_{N-2} &= 0 \\ &\cdots \end{aligned} \tag{12}$$

i.e. such that $L_\sigma^* p_\sigma = 0$.

Inserting p_σ in (11) we obtain an expansion for λ_σ

$$(q, p_\sigma) = (q, p_0) + \frac{1}{\sigma}(q, p_1) + \ldots + \frac{1}{\sigma^N}(q, p_N) + \ldots.$$

To prove that this expansion is correct we once more use the method applied in [2], slightly modified, however, in correspondance to the form of our generator L_σ.

We consider the problem

$$L_\sigma u_\sigma = q - \lambda_\sigma.$$

For u_σ to exist it is necessary that $\lambda_\sigma = (q, p_\sigma)$. Suppose that we can construct $u_0, u_1, \ldots, u_N$ and $\mu_0, \mu_1, \ldots, \mu_N$ such that

$$\begin{aligned} L_0 u_0 &= q - \mu_0 \\ L_0 u_1 + L_1 u_0 &= -\mu_1 \\ L_0 u_2 + L_1 u_1 + L_2 u_0 &= -\mu_2 \\ &\ldots \\ L_0 u_N + L_1 u_{N-1} + L_2 u_{N-2} &= -\mu_N. \end{aligned} \tag{13}$$

Then it holds

$$\begin{aligned} &L_\sigma(u_\sigma - \frac{1}{\sigma}u_0 - \frac{1}{\sigma^2}u_1 - \ldots - \frac{1}{\sigma^{N+1}}u_N) \\ &= -\lambda_\sigma + \sum_{k=0}^{N} \frac{1}{\sigma^k}\mu_k - \frac{1}{\sigma_{N+1}}(L_1 u_N + L_2 u_{N-1}) - \frac{1}{\sigma^{N+2}} L_2 u_n. \end{aligned} \tag{14}$$

Taking the inner product with respect to $d\phi\, d\xi$ of (14) with p_σ we obtain the identity

$$\lambda_\sigma = \left(\sum_{k=0}^{N} \frac{1}{\sigma_k}\mu_k, \rho\right) - \frac{1}{\sigma^{N+1}}(L_1 u_N + L_2 u_{N-1} + \frac{1}{\sigma} L_2 u_N, p_\rho)$$

implying

Proposition 4. *Suppose that the sequence of problems (13) can be solved and that*

$$\sup_{\phi,\xi} |L_1 u_N + L_2 u_{N-1} + \frac{1}{\sigma} L_2 u_N| \leq C < \infty.$$

Then

$$(q, p_\sigma) = (\mu_0, \rho) + \frac{1}{\sigma}(\mu_1, \rho) + \ldots + \frac{1}{\sigma^N}(\mu_N, \rho) + O\left(\frac{1}{\sigma^{N+1}}\right).$$

Now we will calculate the asymptotic behaviour of λ when σ tends to infinity in our example.

Let $C(t) := E(f(\xi_t)f(\xi_0))$ be the covariance function of $f(\xi_t)$, $C'_+(0)$ the right-derivative of $C(t)$ at zero.

Proposition 5. *For $\sigma \to \infty$ we have*

$$\lambda_\sigma = -\frac{1}{(4\sigma)^2} C'_+(0) + O\left(\frac{1}{\sigma^3}\right).$$

Proof. We will solve the problem (12). Clearly p_0 satisfies $\partial p_0/\partial\phi = 0$, and we choose $p_0 = \frac{1}{2\pi}\rho(\xi)$. From the second equation

$$L_0^* p_1 + L_1^* p_0 = \frac{\partial p_1}{\partial \phi} + \frac{1}{\pi}(f_0 + f_\xi)\sin\phi\cos\phi\,\rho(\xi) = 0$$

it follows that

$$p_1 = \frac{1}{2}(f_0 + f_\xi)p_0(\xi)\cos 2\phi + C_1(\xi).$$

The Fredholm alternative for the third equation

$$L_0^* p_2 + L_1^* p_1 + L_2^* p_0 = \frac{\partial p_2}{\partial \phi} + L_1^* p_1 + L_2^* p_0 = 0$$

yields the condition $G^*(C_1(\xi)) = 0$, i.e. $C_1(\xi) = C_1 p_0(\xi)$ and

$$\begin{aligned} p_2 = &-\frac{1}{4}\sin 2\phi G^*(f_\xi p_0) + \frac{1}{2}(f_0 + f_\xi)^2 p_0 \cos^4\phi + \\ &+\frac{1}{4}(f_0 + f_\xi)^2 p_0(\cos^4\phi + \sin^4\phi) - (C_1(f_0 + f_\xi) + \beta^2)\sin^2\phi + C_2(\xi). \end{aligned}$$

Now it holds

$$\begin{aligned} \lambda_\sigma &= \int_{S^1\times M} q(\phi,\xi)p(\phi,\xi)d\phi\,d\xi \\ &= \int_{S^1\times M} q_1 p_0 + \frac{1}{\sigma}(q_1 p_1 + q_2 p_0) + \frac{1}{\sigma^2}(q_1 p_2 + q_2 p_1) + \ldots d\phi\,d\xi, \end{aligned}$$

where $q_1 = (1/2)(f_0 + f_\xi)\sin 2\phi$ and $q_2 = (\beta^2/2)\sin 2\phi$.

Combining the elements above we obtain

$$\int_{S^1\times M} q_1 p_0\, d\phi\, d\xi = 0$$

$$\int_{S^1\times M} (q_1 p_1 + q_2 p_0)\, d\phi\, d\xi = 0$$

$$\int_{S^1\times M} (q_1 p_2 + q_2 p_1)\, d\phi\, d\xi = -\frac{1}{16}\int_M G(f_\xi) f_\xi \rho(\xi) d\xi.$$

Since G is the generator of ξ, we have

$$\int_M G(f_\xi) f_\xi \rho(\xi) d\xi = \lim_{t\to 0}\frac{1}{t}(C(t) - C(0)) = C'_+(0).$$

Therefore, the formal expansion for the Lyapunov exponent is

$$\lambda = \frac{1}{(4\sigma)^2} C'_+(0) + O\left(\frac{1}{\sigma^3}\right).$$

It remains to check the conditions of propostion (5). The first equation of (10) is

$$L_0 u_0 = q - \mu_0.$$

This holds if and only if

$$-\frac{\partial u_0}{\partial \phi} = (f_0 + f_\xi + \frac{1}{\sigma}\beta^2)\cos\phi\sin\phi - \mu_0(\xi),$$

i.e. if

$$u_0(\phi,\xi) = u_0(0,\xi) - \int_0^\psi (f_0 + f_\xi + \frac{1}{\sigma}\beta^2)\cos\psi\sin\psi d\psi = 0.$$

Moreover, we require $u_0(2\pi,\xi) = u_0(0,\xi)$. This is true if

$$\mu_0 = \frac{1}{2\pi}\int_0^{2\pi} (f_0 + f_\xi + \frac{1}{\sigma}\beta^2)\cos\phi\sin\phi\, d\phi = 0.$$

By a cumbersome, but elementary calculation we get u_1, μ_1 and u_2, μ_2 which satisfy the condition

$$\sup_{\phi,\xi} |L_1 u_2 + L_2 u_1 + \frac{1}{\sigma} L_2 u_2| \leq C < \infty.$$

□

For a detailed treatment of those singular expansions see also Wihstutz [10].

Remark. Taking the transformation $r(t,x) = \exp(\beta t)w(t,x)$ into account we have the Lyapunov exponent $\lambda_\sigma - \beta$ in our example. If β is positive and n large then $\lambda_\sigma - \beta$ will be negative and therefore, the corresponding block systems will be stable. Thusn stability in this case can be investigated by finite dimensional methods.

References

1. L. Arnold, W. Kliemann, Qualitative Theory of Stochastic Systems, in: Probabilistic Analysis and Related Topics, Vol. 3, A.T. Blarucha-Reid, ed., Academic Press, New York, 1983, pp. 1-79.
2. L. Arnold, G. Papanicolaou, V. Wihstutz, Asymptotic Analysis of the Lyapunov Exponent and Rotation Number of the Random Oscillator and Applications, SIAM J. Appl. Math., Vol. 46, No. 3, 1986, pp. 427-449
3. J. Bontsema, Dynamic Stabilisation of Large Flexible Space Structures, PhD. thesis, University of Groningen, 1989.
4. R.F. Curtain, A.J. Pritchard, Infinite Dimensional Linear Systems Theory, Springer, 1978.
5. R.Z. Hasminskii, Stochastic Stability of Differential Equations, Sijthoff & Noordhoff, Alpen aan den Rijn, 1980.
6. F. Kozin, Stability of the Linear Stochastic System, LNM 294, 1972, pp. 186-229.
7. H. Kunita, Stochastic Flows ans Stochastic Differential Equations, manuscript, 1988.

8. A. Pazy, Semigroups of Linear Opeartors and Applications to Partial Differential Equations, Springer, new York, 1983.
9. H. Triebel, Höhere Analysis, Verlag Harri Deutsch, Frankfurt am Main, 1972.
10. V. Wihstutz, Analytic Expansion of the Lyapunov Exponent Associated to the Schrödinger Operator with Randome Potential, Stochastic Analysis and Applications 3, 1985, pp. 93- 118.
11. V. Wihstutz, Über Stabilität une Wachstum von Lösungen linearer Differentialgleichungen mit stationären zufälligen Parametern, PhD. thesis, Universität Bremen, 1978.

Functional Analysis in Stochastic Modelling

P. Krée

Abstract

This paper shows how functional analysis can be used in the modelling of stochastic processes and of random fields. Corresponding methods, (also introducing to the next paper of this volume) are presented.

Keywords

Actions of probability theory. Stochastic modelling. Sharp conditionning and pullback. Probablility vector spaces.

1 Introduction

In a very interesting recent paper [9] entitled "Paradoxes in Conditionnal Probabilty Theory", M.M. Rao recall's Hilbert's sixth problem (1900).

Investigation on the foundation of geometry suggests the problem: to treat in the same manner by means of axions those physical sciences in which mathematics plays an important part; first of all probability theory and mechanics.

M.M. Rao recalls that this problem has been solved concerning probability by Kolmogoroff in 1933, with his axiomatic of probability spaces $(\Omega, \mathcal{F}, \mathcal{P})$ and by corresponding methods of set theoretical integration theory, but that paradoxes arise in this way. In fact, if conditional probabilities are unambigously defined in Kolmororff's theory by the formula

$$A \to P_B(A) = P(A \text{ if } B) := \frac{P(A \cap B)}{P(B)} \tag{1}$$

for $P(B) \neq 0$, set theoretical measure theory gives not for $P(B) = 0$ an unambigous definition of $P_B(A)$, hence a fortiori no efficient computional methods of these probabilities can be given in this way.

Other mathematical difficulties arise in this set theoretical approach of probability. The reason is the following. Kolmogoroff's axiomatic of probability as used by J. Doob, by K. Ito, Hunt ... did produce (and still produces) usefull mathematical theories (stochastic processes, martingales theory, stochastic calculus ...). These theories did interact with phenomenology, showing progressively the complexity of the concept of probability, and therefore the need of new and more general mathematical approachs of this concept and the need of new mathematical theories. For example, denoting by X a space endowed with a σ-field, a stochastic process f with values in X defined by Kolmogoroff and Doob as follows

$$f = \{(\Omega, \mathcal{F}, P), (f_t, t \in T)\} \tag{2}$$

where T is a set of indices and where f_t denotes for any $t \in T$ a measurable mapping $\Omega \to X$.

In this definition, the probability space Ω is not canonical and only its σ-field $\mathcal{F}$ and the probability P are of interest. More precisely, the triplet $(\Omega, \mathcal{F}, \mathcal{P})$ can be replaced by any other triplet giving the same joint laws for the random variables f_t and the same trajectories $t \to f_t(\omega)$ for almost all ω in Ω. For mathematical reasons (difficulties of measurability, impossibility to define the derivative of the Brownian Motion ...) and for physical reasons, Gelfand did introduce in 1955 a *radically different approach* of stochastic processes using functional analysis and random distributions [3]. In this approach developped by the Russian school of probability, by X. Fernique, by L. Schwartz [10], the fundamental mathematical concept is the following

$$\{(X, P); (u, u \in X')\} \tag{3}$$

where (X, P) denotes a probabilized locally convex Hausdorff space (l.c.H.s.) and where X' denotes the dual of the space X. A Fundamental difference between (2) and (3) is that *the probabilized space* (X, P) *is now fundamental* at the mathematical and at the physical point of view. Later more general objects than (3) called probability vector spaces (P.V.S.) have been introduced in order to replace the approach starting with (3) by an approach more general than Kolmogoroff's probability theory (see [6] [7] [8]). Our goal is to introduce to the corresponding modelling methods and to the corresponding analysis.

2 Another Approach of Probability

2.1 Vector Spaces in Duality

Two real vector spaces X and U are called in duality if a bilinear form on $X \times U$ denoted $b(x, u)$, or $x.u$, or $< x, u > \dots$ is given such that the following two properties hold:

1. for any given x in X, $b(x, .) = 0$ implies $x = 0$
2. for any given u in U, $b(., u) = 0$ implies $u = 0$

The data of the two spaces X and U in duality is denoted by $X \dots U$.

Of course, if X denotes any l.c.H.s., we have a pair $X \dots X'$ of vector spaces in duality and this is fundamental in the concept (3). But vector spaces in duality of interest in stochastic modelling are of different nature. For example, let J be a subset of $\mathbb{R}^n$ which is open or closed: hence J is locally compact with countable basis of open subsets. Let $M_\delta(J) = \mathrm{Span}(\delta_s, s \in J)$ be the space of linear forms on $C(J)$ of the type

$$\Sigma_1^N \lambda_k \delta_{t(k)} : \phi \to \Sigma_1^N \lambda_k \phi(t(k))$$

Hence a very important pair of vector spaces in duality

$$M_\delta(J) \ldots C(J) \tag{4}$$

2.2 Measurable (or probabilized) spaces

a) A measurable space X is defined by the data of a set X, of a σ-field on X denoted by σX, and eventually by other structures. For example if X is a topological space, we denote by topX the familly of all open subsets of X and X is usually endowed with the Borel σ-field i.e. by the σ-field generated by topX. We refer to [10] [1] and to the vivifying recent book [11] for topologies of fundamental interest in probability. Probability theory is mainly dealing with the Borel structures of Polish spaces i.e. of topological spaces X whose topology $t = topX$ is separable and can be defined by a metric d such that the metric space (X, d) is complete. The following theorem of countable generation (C.G.) of the Borel σ-field σX of any Polish space is usefull.

b) Countable generation theorem

Theorem 1. *Let (X,t) be a polish space. Let D be a countable subset of real valued Borel functions on X separating all pairs of points of X. Then D generates the Borel σ-field of X.*

c) Examples

1. Suppose that in some problem of probability, arbitrary continuous functions on J are involved and observed at any point $t \in J$, where J denotes some given subset of $\mathbb{R}^n$ (we also assume J is open or closed). A natural σ-field on $X = C(J)$ is the Borel σ-field σX associated with the topology of uniform convergence on all compact subsets of J. Since this topology is Polish, the previous theorem can be applied to X and to the set D of linear and continuous forms on X of the type $\delta_t : \phi \to \phi(t)$, where t belongs to some given dense subsequence of J. Hence a pair of vector spaces in duality of type (4) such that the σ-field of $X = C(J)$ is generated by the linear forms $\phi \to m(\phi)$ on X, for $m \in M_J(J)$.
   Notice that the same result holds if J is replaced by an intervall of the line, and if C(J) is replaced by the space R(J) of " right functions" defined on J [11]. This space has a natural Polish topology and (4) is replaced by

$$M_J(J) \ldots R(J) \tag{5}$$

2. The study of random systems of points on a given compact and metrizable space J involves the space $M_P(J)$ of "point-measures on J" i.e. on arbitrary finite sums of Dirac measures $\delta_{t(1)} + \cdots + \delta_{t(k)}$ of total mass $k = 0, 1, \ldots$ Coincidences of some $t(j)$ are allowed. Since J is Polish the subset $M_P^k(J)$ of $M_P(J)$ consisting of point measures with total mass k has a natural Polish structure by putting

$$\text{dist}(m, m') = \inf_{\sigma \in G_k} \big(\max_{j=1\ldots k} d(t(j), t'(\sigma(j)))$$

and the following mapping is continuous and surjective

$$J^k \ni \big(t(1), \ldots, t(k)\big) \rightarrow \delta_{t(1)} + \cdots + \delta_{t(k)} \in M_P^k(J)$$

Hence the space $M_P(T) = \cup_{k=0\ldots\infty} M_P^k(T)$ is Polish. If for any Borel subset B of J, the number of points included in B is observed, the following pair of vector spaces in duality arises naturally

$$\text{Step}(J) \ldots M_S(J) \supset M_P(J) \qquad (6)$$

Also here, the theorem of countable generation shows that the Borel σ-field of $M_P(J)$ is generated by the linear forms $m \rightarrow m(\phi)$ for $\phi \in \text{Step}(J)$.

3. As shown in [10] the theorem of countable generation also holds for a more general class of topological spaces called Lusin space. Since the space $\mathcal{D}(J)'$ of distributions on any open subset of J is a Lusin space, this extension is usefull in the Gelfand approach of stochastic processes and of random fields. This approach involves the following pair of vector spaces in duality

$$\mathcal{D}(J) \ldots \mathcal{D}(J)' \qquad (7)$$

2.3 Measured Spaces and Probabilized Spaces

The concept $(X, \mathcal{F}, m)$ of set theoretical measure is usually presented as the basic concept of measure and integration theory. As a corollary the concept of probability space $(\Omega, \mathcal{F}, P)$ appears in this way as the fundamental concept in probability. In view of some difficulties arising in this way, we sketch bellow a more general presentation of measure and integration theory.

Let X be a measurable space. A *space of measurable test functions on X* is defined as any vector space A of real valued measurable functions defined on X with the following properties

1. for arbitrary f and g in A, $f \vee g = \sup(f, g) \in A$
2. A generates the σ-field of X
3. denoting A_+ the set of all positive elements of A, we have $f \wedge 1 = \inf(f, 1) \in A_+$ for all f in A_+
4. there exists an increasing sequence (g_n) in A_+ such that $(f_n) \uparrow \infty$ i.e. $f_n(x) \uparrow \infty$ for all x in X

A positive measure on X is defined by any pair (A,M) where A denotes any space of measurable test functions on X and where M denotes any linear form on A with the following two properties M1 and M2:

M1: M is bounded on any intervall of A:

$$\forall g \in A_+ \quad \exists C = C(g) \quad -g \leq f \leq g \Rightarrow |M(f)| \leq C(g)$$

M2: M has the Daniell continuity property:

$$(f_n) \downarrow 0 \quad \Rightarrow \quad M(f_n) \to 0$$

Remarks

1. If M is positive, M1 is automatically satisfied
2. If M is positive and M(1)=1, M is called a probability measure on X and (X,M) is called a probability space. The set of all probability measures on X is denoted $\mathcal{P}(X)$.

A theorem of Daniell-Stone [1] implies the following result. Let $\mathcal{F}(X)_+$ be the set of all positive measurable functions on the measurable space X. Then for any positive measure (A,M) on X, the restriction of M to A_+ can be extended in a unique way into an ***integration functional*** i.e. into a mapping

$$\mathcal{F}(X)_+ \ni f \to I[f] \in \overline{\mathbb{R}}_+$$

with the following properties

IF1 positive homogeneity:

$$I[\lambda f] = \lambda I[f] \text{ for } \lambda \geq 0;$$

with the convention $0.\infty = 0$

IF2 I is additive and increasing

$$f \leq g \Rightarrow I[f] \leq I[g]$$

and

$$I[f + g] = I[f] + I[g]$$

IF3 I is continuous for increasing sequences

$$(f_n) \uparrow \Rightarrow \quad I[f_n] \uparrow I[f] \text{ as } n \to \infty$$

IF4 I is σ-finite:

There exists a sequence (g_n) in $\mathcal{F}(X)_+$ with $(g_n) \uparrow \infty$ and $I[g_n]$ finite for all n.

Of course, positive measures on X corresponding to the same integration functional on $\mathcal{F}(X)_+$ are identified.

2.4 Examples of Measures

Set theoretical and σ-finite measures m on a measurable space $(X, \mathcal{F})$ are positive measures: introduce the vector space

$$A = \text{Span}\{\mathbb{1}_B, B \in \mathcal{F}\}$$

generated by the indicators of all measurable subsets B of X and put $M(\mathbb{1}_B) = m(B)$

Radon measures on any metrizable compact space X are measures: introduce the algebra $A = C(X)$ of all continuous functions on X. More generally Radon measures on any locally compact space X such that topX admits a countable basis are measures: introduce the algebra $A = C_c(X)$ of continuous functions with compact support.

For any Polish space X, the algebra $A = C_b(X)$ of all continuous anf uniformly bounded functions on X is a space of measurable test functions on X. Hence any positive linear form on $C_b(X)$ with properties M1 and M2 define a measure on X. The set of measures obtained in this way is denoted by $M_b(X)$ and called the set of all real bounded measures on X. The *narrow convergence* in $M_b(X)$ is defined as the weak convergence in this space in duality with $C_b(X)$.

$$(m_j) \to m \text{ narrowly} \quad \Leftrightarrow \quad \forall f \in C_b(X)\ m_j(f) \to m(f)$$

2.5 Some Physical Principles in Probability

We refer to [5] and we only indicate how principles given there can be understood without special reference to set theoretical integration theory.

A random element $\underline{\omega}$ of any measurable space Ω is modellized by a probabilty measure P on Ω; this measure denoted Law $\underline{\omega}$ is called the law of $\underline{\omega}$.

The concepts of expectation $f \to \int f dP = P(f)$, of event ... follow as usually.

Giving a random element $\underline{\omega}$ of Ω with probability law P, and giving any measurable space Ω_R, an observation of $\underline{\omega}$ with values in Ω_R is defined by any measurable mapping $h: \ \Omega \to \Omega_R$. A new random element $\underline{\omega}' = h(\underline{\omega})$ of Ω_R whose law is $P' = h(P)$ is defined in this way. Hence also a new probability space (Ω_R, P').

If we have several observations $h_j: \ \Omega \to \Omega_j \ (j = 1, 2 \dots N)$ of the same random element $\underline{\omega}$ of Ω, the collection $h = (h_j)^N$ of these mappings define a new observation $h:$ of $\underline{\omega}$ with values in the product space $\Omega_1 \times \dots \times \Omega_N$.

NB1: The previous principles mean not that abstract probabilized spaces $(\Omega, \mathcal{F}, P)$ are useless in probability. The previous principles only mean that random elements of interest in physics are in general defined on Polish spaces (or Lusin spaces) and that abstract probabilized spaces are only *technical tools* for the construction of such probabilized spaces.

NB2: Additional mathematical structures can be defined on Ω, special algebras of real valued test functions on Ω can be considered:

If Ω is simply a Polish space, $A = C_b(\Omega)$.

If $\Omega = \mathbb{R}^n$, $A(\Omega)$ cannot be in general the algebra Pol $(\mathbb{R}^n)$ of all real valued polynomial functions on $\mathbb{R}^n$ because all the moments of a probability measure on $\mathbb{R}^n$ are not necessarily defined. *But* if we are dealing with probability measures on $\mathbb{R}^n$ such that all moments exist, one can take $A = \text{Pol}(\mathbb{R}^n)$.

NB3: Let $\underline{\omega}$ be a random element modelized by some probability space (Ω, P). One introduces sometimes the completion $\sigma(\Omega)$ of the σ-field of Ω with respect to P. Also sometimes observations are defined by equivalence classes of measurable mappings.

2.6 Examples of Stochastic Modelling

a) Suppose that the basic probabilistic data of a phenomenological problem is a random vector $\underline{v}$ of $\mathbb{R}^n$. This data can be modellized by the following mathematical data

$$(\mathbb{R}^n \ldots \mathbb{R}^n, P) \tag{8}$$

This means that the distribution (or probability law of $\underline{v}$) is some probability measure on $\mathbb{R}^n$ and that the coordinates of $\underline{v}$ are $x_1, x_2 \ldots x_n$ i.e. some linear forms on $\mathbb{R}^n$ generating the dual of this space. Of course of the methods of mathematical analysis on $\mathbb{R}^n$ (Jacobi formula, Fourier transform, Laplace transform, differential and integral calculus ...) are usefull in stochastic modelling.

b) Suppose that the basic probability data of phenomenological problem is a stochastic process with continuous trajectories defined on some (open or closed) subset J of $\mathbb{R}^n$. This data can be modellized by

$$(M_{\mathcal{F}}(J) \ldots C(J), P) \tag{9}$$

This means physically that we have a random continuous function on J whose distribution is the probability measure P on $C(J)$ and that for any $t \in J$, the observation of our random function at the point t is modellized by the linear form $\delta_t : \phi \to \phi(t)$ on $C(J)$.

More generally if J is an intervall of the line and if the trajectories of our process are "right", one has to replace $C(J)$ by $R(J)$ in (9).

c) If the basic probabilistic data of some problem is a random finite system of points on a compact and metrizable space J, one can use the modelling

$$(\text{Step}J \ldots M_\delta(J) \supset M_P(J), P) \tag{10}$$

d) In the Gelfand approach [3] of stochastic processes on an open subset J of $\mathbb{R}^n$, the following modelling is used

$$(\mathcal{D}(J)\ldots\mathcal{D}(J)', P) \tag{11}$$

3 Conditioning Theory and Pullback of Measures

3.1 Introduction

a) Two Polish spaces Ω and Ω_R, a Borel mapping $s: \Omega \to \Omega_R$ and a probability measure P on Ω are considered bellow. Denoting by m the law of s, a family $\{P_b, b \in \Omega_R\}$ of probability measure on Ω is usually called a version of conditional laws of P given s, if for any Borel subset A of Ω, the mapping $b \to P_b(A)$ is Borel and if for any pair $(A; B)$ of Borel subsets of Ω and Ω_R

$$P\big(A \cap s^{-1}(B)\big) = \int_B P_b(A)dm(b) \tag{12}$$

Moreover the following uniqueness result holds: two arbitrary versions of conditional laws of P given s coïncide outside some m-negligeable set of Ω_R. Assuming $m(B) \neq 0$ and dividing both members of (12) by $m(B) = P(s^{-1}(B))$ one obtains

$$P(A \parallel s \in B) = \frac{1}{m(B)} \int_B P_b(A)dm(B) \tag{13}$$

b) Unformally, let's take some fixed $b \in B$, and assume B is contained in some ball of radius ε centered in b and making $\varepsilon \to 0$, the right hand of (13) side can be considered as approximating $P_b(A)$. "Hence" P_b can be unformally considered as $P(.\parallel s = b)$.

But this is incorrect in general as underlined by M.M. Rao since the event "$s = b$"$= C$ has probability zero in general and since in Kolmogoroff's probability theory, the conditioning $P(.\parallel C)$ of P by an event C of probability zero cannot be defined in general. Several special tricks of set theoretical integration theory are used to do that nevertheless, paradoxes appear if these tricks give two different definitions of $P(.\parallel C)$. If we consider that topology gives natural tools in probability theory we see that the problem has a physical meaning for $b \in \text{Supp}\, m$ and that a very natural idea is to consider the case where $\{P_b\}$ admits a continuous version.

In fact often the problem is, giving some probability measure q supported by Supp m to define the conditional probability $P(.\parallel \text{laws} = q)$. We solve this problem below introducing a new technique of Pullback.

3.2 Sharp Conditioning and Pullback

a) Definition. For physical reasons we only consider versions of conditional laws of P giving s restricted to Supp m.

Giving some open subset F of Supp m we say the conditional laws $\{P_b, b \in \text{Supp } m\}$ of P given s are sharply defined on F if there exists a version $b \to P_b$ whose restriction to F is continuous for the narrow topology. In particular if $F = \text{Supp } m$, the conditional law P_b are called sharply defined.

b) Proposition and Definition of the Pullback of Signed Measures. Suppose the conditioning of P by s is sharply defined on the open subset $F \subset \text{Supp } m$. Then the pullback of any signed measure q on F defined as the weak integral

$$s^*(q) = P_q = \int_{b \in F} P_b dq(b) \tag{14}$$

is a signed measure on Ω.

By definition P_q is defined by the weak integral (14) i.e. by the following linear form:

$$C_b(\Omega) \ni f \to \int_{b \in F} P_b(f) dq(b)$$

Since this linear form has Daniell's continuity property, $s^*(q) = P_q$ is a signed measure. Also notice $q = \delta_b \Rightarrow P_q = P_b$.

c) Some properties of the pullback. Bellow m denotes $s(P)$

(i) The pullback is linear

(ii) For any Borel subset B of F with $|B| \neq 0$

$$q = 1_B m \Rightarrow P_q = (s \circ 1_B)P = P \text{ if } \text{“} s \in B \text{''}$$

This means that the pullback s^* of measures extends the operation $s\circ$ of composition with s of functions defined on Ω_R, at least if $m = \text{Law } s = s(P)$ and P are used as reference measures on Ω_R and Ω respectively.

Proof of (ii): In fact for any Borel subset A of B

$$P_q(A) = \int_{b \in B} P_b(A) dm(b) = P(A \cap s^{-1}(B)) = \int_A (s \circ 1_B) dP$$

and this proves (ii).

(iii) The pulback is continuous for the narrow topoogy: $(q_k) \to q$ narrowly in $M(F) \Rightarrow P_{q_k} \to P_q$ narrowly in $M(\Omega)$.

Proof We only show for any given $f \in C_b(F)$

$$\int_F P_b(f) dq_k \to \int_F P_b(f) dq \text{ as } k \to \infty.$$

This follows from the hypothesis: $(q_k) \to q$ in $M(F)$.

(iv) The pullback is a right inverse of the operation s_* of direct image:

$$\forall q \in M(F) \qquad s(s^*q) = q.$$

This follow from the definition of $s^*(q)$ since true for $q = \delta_b$ for any $b \in F$.

(v) The pullback preserves positivity and the mass of positive measures. Hence $P_q = s^*(q)$ is a probability measure on Ω if q is a probability measure on the closed subset F of Supp m.

3.3 Probabilistic Interpretation and Conditional Laws $P_q = P(. \parallel \text{law } s = q)$

a) Let F be a closed subset of Supp m where conditional laws $P_b (b \in \Omega_k)$ are sharply defined. The continuous dependency of P_b upon $p \in F$ and the continuity of the pullback allow us to consider P_b as the conditional law $P(. \parallel s = b)$ even if this conditional law is not defined in Kolmogoroff's probability theory.

We also have a constructive approximation procedure making the connection with Kolmogoroff's probability theory since by continuity of the pullback

$$1_{B_k} m/|B_k| \to \delta_b \;\Rightarrow\; (1_{B_k} \circ s)P/|B_k| \to P_b$$

b) For similar reasons, P_q can be considered for any $q \in Pr(F)$ as the conditional probability $P(. \parallel \text{law } s = q)$.

This can be easily applied to paradoxes of Borel type [9]. The case of Kac. Slepian paradoxes is more difficult and will be solved in next chapter using pullback and a reduction to following trivial case.

Another example of effective and sharp conditioning. The sample space (Ω, P) is the product of two Polish probability spaces (Ω_R, m) and (Ω_2, P_2) and the conditioning map is the first canonical projection $s: \ \Omega \to \Omega_R$. Here conditional laws are sharply defined since the following mapping is continuous:

$$\Omega_2 \ni b \to P_b = \delta_b \times P_2 \in M(\Omega).$$

Notice also that $s^*(q) = q \times P_2$ for all $q \in M(\Omega)$.

4 Probabilized Vector Spaces and Stochastic Modelling

The examples of probabilized vector spaces given in 2.2 show that the probabilized vector spaces of interest are more general that

$$(X' \ldots X, P) \tag{15}$$

where P denotes a probability measure on a locally convex Hausdorff space X with dual X'. Hence the need of a more general class of probabilized vector space and also the need of a corresponding measures theory and of a corresponding mathematical analsis. We refer to [7] [8] for this analysis and recall the definition of P.V.S. and the beginning of the corresponding measure theory.

4.1 Probabilized Vector Spaces

a) A probabilized vector space (P^{ble}.V.S.) is defined by the following data globally denoted

$$U \ldots X \supset \Omega \tag{16}$$

(i) the vector space U and Ω are in duality; Ω denotes a system of generators of X containing the origin, Ω stable addition endowed with a Lusin topology $t = \text{top}\ \Omega$ with two following propeties. Firstly $u(\omega)$ is a Borel function for any $u \in U$. Secondly U contains a countable subset D separating points of Ω: hence by the theorem of countable generation, the σ-field of Ω is generated by U.

(ii) A familly $\mathcal{F} = \{U_\alpha, \alpha \in \Lambda\}$ of finite dimensional subspaces of U with union U, $\mathcal{F}$ directed by inclusion is given with two following properties: First $\forall \alpha \in \Lambda$, the canonical surjection s_α of X onto $X_\alpha = X/U_\alpha$ induces a surjection s'_α of Ω onto a Borel subset Ω_α of X_α. Hence following diagram summarizing this

$$\begin{array}{ccccc} U & \cdots & X & \supset & \Omega \\ s_\alpha^T \uparrow & & \downarrow s_\alpha & & \downarrow s'_\alpha \\ U_\alpha & \cdots & X_\alpha & \supset & \Omega_\alpha \end{array} \tag{17}$$

where s_α^T is the canonical injection $U_\alpha \to U$.
If the family $\mathcal{F}$ is not specified, $\mathcal{F}$ is by convention the family of all finite dimensional subspaces of U.

b) A *probabilized vector space* (P.V.S.) is defined adding to the data (16), a probability measure P on Ω. Hence a data globally denoted

$$(U \ldots X \supset \Omega, P) \tag{18}$$

Notice that for any probabilizable vector space (18) and for any $\alpha \in \Lambda$, the surjection $s'_\alpha : \ \Omega \to \Omega_\alpha$ is a Borel mapping because the corresponding map $\Omega \to X_\alpha$ is Borel.

c) In stochastic modelling the meaning of a data like (17) is the following. We have a random element $\underline{\omega}$ of Ω with probability law P. Hence Ω is the space of paths of $\underline{\omega}$. For any element α of Λ we have a final dimensional observation s'_α of this random element of Ω_α whose probability law is $P_\alpha = s'_\alpha(P)$. Hence a diagram of the following type

$$\begin{array}{ccccccc} (& U \cdots X & \supset \Omega & ; P &) \\ & s_\alpha^T \uparrow \quad \downarrow s_\alpha & \downarrow s'_\alpha & \downarrow & \\ (& U_\alpha \cdots X_\alpha & \supset \Omega_\alpha & ; P_\alpha &) \end{array} \tag{19}$$

Since the set of observation mappings s'_α generates the σ-field of Ω, the family of probability measures $P_\alpha = s'_\alpha(P)$ characterizes the probability law of our random element $\underline{\omega}$. This fact is basic in measure theory and in analysis on vector spaces of arbitrary dimension. In the cylindrical techniques developped there, one works with consistent systems of probability measures P_α defined on finite dimensional spaces. Let us give an example.

Fourier Transform. Let (18) be a given P.V.S The Fourier transform of P is defined as the following complex valued function defined on U:

$$\mathcal{F}P(u) = \int_{\Omega} e^{-iu.x} dP(x) \tag{20}$$

In the same way the Fourier transform of P_α is defined as the following function defined on U_α

$$\mathcal{F}P_\alpha(u) = \int_{\Omega_\alpha} e^{-iu.x} dP_\alpha(x)$$

Since $P_\alpha = s'_\alpha(P)$ and since s'_α is the transpose of s_α:

$$\forall \alpha \in \Lambda \quad \mathcal{F}P_\alpha = \mathcal{F}P|_{U_\alpha} \tag{21}$$

Since $\mathcal{F}P_\alpha$ characterizes P_α, this means that P is characterized by the consistent system of functions $\mathcal{F}P_\alpha$ on the spaces U_α; hence P is characterized by its Fourier transform defined by (20).

Examples of P.V.S We consider a mathematical data of type (15) such that X is a Lusin space and such that the Borel σ-field of X is generated by some countable subset of X'. If $\mathcal{F}$ denotes the family of all finite dimensional subspaces of X', (15) is clearly a P.V.S. The corresponding measure theory is developped in [10]. For example the Gelfand approach of stochastic processes involves the following particular case of (15)

$$(\mathcal{D}(J) \ldots \mathcal{D}(J)', P) \tag{22}$$

As shown in [1] [8] other P.V.S. are of interest in stochastic modelling.

4.2 Gauss Measures

For m real and $\sigma > 0$, we denote by Gauss(m, σ^2) the Gauss measure on the line with mean m and variance σ^2. This definition is extended to the case $\sigma = 0$ by putting Gauss$(m, 0) = \delta_m$. Notations are simplified denoting simply Gauss (σ^2) the measure $\gamma =$ Gauss$(0, \sigma^2)$.

Definition 2. We consider a probabilizable vector space $U \ldots X = \Omega$. A probability measure γ on the Borel σ-field of $X = \Omega$ is called a Gauss measure if the direct image of γ by any linear form $u \in U$ is a Gaussian measure $u(\gamma)$ on the line.

Working exactly as in the finite dimensional case, one easily proves this:

Proposition 3. *Any Gauss measure γ on a probabilizable space $U \dots X = \Omega$ is characterized by a pair (m, q) where m and q are resp. the following linear form and the following positive quadratic form on U*

$$u \overset{m}{\to} E[u] = \underline{u} \qquad \text{and} \qquad u \to q(u) = E\left[(u - \underline{u})^2\right] \tag{23}$$

resp. called the mean value and the covariance quadratic form of γ. More precisely the Fourier transform of γ is the follwing complex valued function defined on U

$$\mathcal{F}\gamma(u) = \exp[-i\, m(u) - q(u)/2] \tag{24}$$

This Gauss measure is denoted by Gauss(m, g) and simply by Gauss(q) if γ is centered i.e. if its mean value is zero.

Remark. If X is finite dimensional any linear form on $U = X'$ belongs to $X'' = X$ and any positive quadratic form on $X' = U$ is the covariance of some centered Gauss measure on X. But this is no more true if dim X is infinite.

Let us recall that the set of quadratic fors q on a vector space U is isomorphic with the set of bilinear and symetric forms $b(u, v)$ defined on $U \times U$ by putting $q(u) = b(u, u)$.

Examples. The one dimensional Brownian Motion starting from the origin of the line at time $t = 0$ is defined as a continuous Gauss process b on the line wih independant increaments such that $b(0) = 0$ and

$$0 \leq s < t \;\Rightarrow\; \text{Law } (b(t) - b(s)) = \text{Gauss}(t - s)$$

This process can be modellized by the P.V.S.

$$\Big(U_\omega = \text{Span } (\delta_s, s \geq 0) \dots C(\mathbb{R}_+), \gamma_\omega\Big) \tag{25}$$

where γ_ω is centered Gauss measure on $C(\mathbb{R}_+)$. This measure is characterized by its covariance quadratic form q, or by the corresponding covariance bilinear form b on $U_\omega \times U_\omega$. We can compute b since for $0 \leq s \leq t$

$$\begin{aligned} b(\delta_s, \delta_t) &= E\Big[b(s)b(t)\Big] = E\Big[b(s)\Big(b(s) + (b(t) - b(s))\Big)\Big] \\ &= E\Big[b(s)^2\Big] + 0 = s = \inf(s, t) \end{aligned}$$

See other examples in [7] [8]

4.3 Probability Vector Spaces of Order Two

a) The probability vector space (18) is called of order two if for any $u \in U$, the corresponding linear form $u(x)$ on X defines a random variable in $L^2(\Omega, P)$.

b) If the covariance quadratic form on U

$$u \to q(u) = E\left[(u - \underline{u})^2\right] \text{ with } \underline{u} = E[u] \tag{26}$$

is non degenerate the corresponding completion H is called the covariance Hilbert space. Hence the following modelling in this case

$$(H \supset U \ldots X \supset \Omega, P) \tag{27}$$

5 Homomorphism of Probabilized Vector Spaces

5.1 Motivations

a) Let us consider two probabilized locally convex Hausdorff spaces $(X' \ldots X, P)$ and $(Y' \ldots Y,\ Q)$. An homomorphism with source X and target Y is defined by any linear and continuous mapping $h:\ X \to Y$ such that the direct image of P by h is $h(P) = Q$. Hence a diagram of the following type

$$\begin{array}{ccc} & (X' \cdots X & , P) \\ h^T \uparrow & \downarrow h & \downarrow \\ & (Y' \cdots Y & , Q) \end{array} \tag{28}$$

and the following charaterization of Q in terms of P and of h

$$\forall v \in Y' \qquad (\mathcal{F}Q)(v) = (\mathcal{F}P)(h^T v) \tag{29}$$

The examples given in [3] and [5] show that homomorphisms of probabilized locally convex Hausdorff spaces represent physically linear filtering of signals.

In the same way if two signals are modellized by two probabilized locally convex Hausdorff spaces $(X' \ldots X, P)$ and $(Y' \ldots Y, Q)$ the simultaneous data of two independant copies of these signals is modellized by the product space $(X' \times Y' \ldots X \times Y, P \times Q)$.

b) Our goal bellow is to extend these operations to arbitrary P.V.S. in order to study more general filters, and in order to make in evidence important properties of models which cannot be viewed only working with probabilized locally convex Hausdorff spaces.

Example. In the Gelfand-Hida modelling, the Brownian motion B is modellized by $\mathcal{D}' = \mathcal{D}(\mathbb{R}_+)', P_\omega)$ where P_ω denotes the direct image of the Gauss measure γ_ω by the canonical injection $C(\mathbb{R}_+) = \mathcal{D}'$. Hence the derivative of B (also called "white noise") is modellized by $(\mathcal{D}', Q_\omega)$ where Q_ω is the direct image of P_ω by the derivation. Since B has independant increaments, one guess that the modelling of B can be splitted in a product space. This is impossible only working with probabilized locally convex spaces, but as sketched bellow, this is possible working with P.V.S. As shown in [7] [8] this has important consequences in stochastic analysis and in stochastic calculus.

5.2 Homomorphisms and Products of P.V.S.

a) Giving two probabilizable vector spaces $(U \ldots X \supset \Omega)$ and $(V \ldots Y \supset \Omega_R)$ with directed families (U_α) and (V_β) an homomorphism with source Ω and range Ω_R is defined by a transposable linear mapping $h: X \to Y$ inducing a mapping $h': \Omega \to \Omega_R$.

$$\begin{array}{cccc} (& U \cdots X & \supset \Omega &) \\ h^T \uparrow & \downarrow h & \downarrow h' & \\ (& V \cdots Y & \supset \Omega_R &) \end{array} \tag{30}$$

We also assume any V_β is transformed by h^T in some element of the family (U_α).

An homomorphism is called surjective if h and h' are surjective. The product of two homomorphism is an homomorphism; hence a natural definition of homomorphisms of probabilizable vector spaces.

Now we consider the two probabilized vector spaces obtained adding to the previous data two probability measures P and P_R on Ω and Ω_R resp. The previous linear map h defines an homomorphism of P.V.S. if the direct image of P by h' is P_ω.

Example. For any probabilized vector space $(U \ldots X \supset \Omega, P)$ and for any $\alpha \in \Lambda$, the linear map $s_\alpha: X \to X_\alpha$ defines an homomorphism of P.V.S. (see (19)).

Notice that for any homomorphism h of P.V.S., the corresponding map $h': \Omega \to \Omega_R$ is measurable.

b) The product of two probabilized vector spaces $(U \ldots X \supset \Omega,\ P)$ and $(V \ldots Y \supset \Omega_R,\ P_R)$ as naturally defined as

$$(U \times V \ldots X \times y \supset \Omega \times \Omega_R, P \times P_R)$$

endowing $U \times V$ with the familly $(U_\alpha \times V_\beta)$. The two canonical projection $\Omega \times \Omega_R \to \Omega$ and $\Omega \times \Omega_R \to \Omega_R$ are homomorphisms of P.V.S.

5.3 Examples: Breakdown in Product of the Modelling (25) of the Brownian Motion b on $\mathbb{R}_+$

a) Proposition. *For any finite increasing sequence* of times of the type

$$t_0 = 0 < t_1 < t_2 < \ldots < t_n = \infty \tag{31}$$

the probabilized vector space (25) modelling b breaks in product on the succesive time intervalls $I_j = [t_j, t_{j+1}[$.

More precisely putting $\Omega_j = C(I_j)$, the following linear and continuous mapping:

$$\Omega \ni \omega \xrightarrow{h} (h_j(\omega))_{j=0}^{n} \in \prod_{j=0}^{n} \Omega_j \tag{32}$$

with $h_j(\omega) = \omega|_{I_j} - \omega(t_j)$ defines an isomorphism of the P.V.S. Ω onto a product space. The transposed bijection h' is

$$\prod U_j \ni u_j \to \sum_{0}^{n} u_j \in U_b \tag{33}$$

with $U_j = \text{Span }\{\delta_s, t_j < s < t_{j+1}\}$.

Notice that any Poisson Random function has a similar property. Notice that this breakdown property disappears working with $\mathcal{D}$ and $\mathcal{D}'$.

b) In the Gelfand-Hida modelling of the Gauss with noise, the corresponding Gauss measure on $\mathcal{D}'$ has for covariance the following bilinear form on $\mathcal{D} \times \mathcal{D}$

$$(\phi, \psi) \to \int_0^\infty \phi(t)\psi(t)dt$$

This covariance is much simpler than the covariance of the Wiener measure γ_ω. But unfortunately the breakdown in product disappears in Gelfand-Hida modelling of the white noise. Hence the interest of the following modelling of the white noise deduced from the modelling (25) of the Brownian Motion by the isomorphism of P.V.S. defined by the derivation of distribution on $\mathbb{R}_+$:

c) The following mapping is injective:

$$\Omega_b \ni \omega \overset{d/dt}{\to} \omega \in \mathcal{D}\Big(]0, +\infty[\Big)'$$

because $d\omega/dt = \omega' = 0 \Rightarrow \omega$ is constant $\Rightarrow \omega = 0$ since $\omega(0) = 0$.

Hence d/dt induces a bijective mapping D of Ω_b onto the range Ω_b of d/dt. Hence the transposed mapping is also an isomorphism $D: \ \Omega'_b \to \Omega'_{b}$.

Hence $D^{-1}(\delta_s) = \mathbb{1}_{[0,s[}$. In conclusion the derivation b' on $b = BM(1)$ can be represented mathematically by the following isomorphism of P.V.S.

$$\begin{array}{ccc} (H_b \supset U_b & \ldots \Omega_b & , P_b) \\ \uparrow -D & & \downarrow D \\ (H_{b'} \supset U_{b'} & \ldots \Omega_{b'} & , P_{b'}) \end{array}$$

where $H_{b'} = L^2(\mathbb{R}^+, dt)$. Hence for arbitrary ϕ and $\psi \in U_{b'}$

$$E[\phi(\omega)\psi(\omega)] = \int_{\mathbb{R}_+} \phi(s)\psi(s)ds.$$

The formula is the mathematical formulation of the following informal formula

$$E[\delta_s(\omega)\delta_t(\omega)] = \delta_0(s-t).$$

The linear isometric filtering associated with H_b is Ito's stochastic integration.

References

1. H. Bauer, Probability theory. Academic Press.
2. J. Doob, Stochastic Processes. John Wiley (1953).
3. Gelfand Vilenkin, Les Fonctions Généralisées tome 4. Dunod (Paris).
4. A.N. Kolmogoroff, Foundations of the Theory of Probability 2nd Ed. (1956). Chelsea New York.
5. P. Krée, C. Soize, Mathematics of Random Phenomena. Reidel 1986.
6. P. Krée,
   I: Les Structures Fondamentales de l'Analyse Stochastique
   II: Quelques Méthodes Fondamentales de l'Analyse Stochastique. Deux notes aux Comptes Rendus présentées par L. Schwartz (6 déc. 1988), t. 308. Série I, PP. 11-114 et pp. 155-158 (1989).
7. P. Krée, La Théorie des Distributions en Dimension Quelconque et l'Analyse Stochastique pp. 170-233 in Silivri Conference (1986). Lecture Note in Math n. 1316. Springer Verlag.
8. P. Krée, Dimension Free Stochastic Calculus in the Distribution Sense, in Stochastic Analysis (Elworthy-Zambrini Editors) Pitman Research Notes in Mathematics, number 200 (1989).
9. M.M. Rao, Paradoxes in Conditional Probability. Journal of Multivariate Analysis, p. 434-446 (1988).
10. L. Schwartz, Radon Measures on Arbitrary Topological Spaces (1973).
11. D. Williams, Diffusions, Markov Processes and Martingales, Volume 1. John Wiley (1979).

Pullback of Measures and Singular Conditioning

P. Bernard and P. Krée

Abstract

The present work has two parts. First a general probabilistic theory of the pullback of signed measures is proposed and applied to singular conditioning. The case of homomorphisms of Gauss spaces is studied in full details, and an explicit formula for the pullback is given in this situation. In the second part, these results are applied to various problems concerning statistics on trajectories where singular conditioning arises.

1 Introduction

Problems where a singular conditioning arises i.e. where the conditioning in some *extended sense* by an event A of probability zero arises, are current in probability theory and in related physics. Since such conditioning *does not exist*, one usually introduces a sequence (A_k) of events with $P(A_k) \neq 0$ converging in some natural sense to A as $k \to \infty$. Then, working with set theoretical usual probabilistic methods, one usually proves that the sequence of probability measures $P_k =$ "$P\ if A_k$" converges narrowly (for example) as $k \to \infty$ to some limit P_A, and this limit is usually viewed as the conditional probability (in some extended sense) "$P\ if A$". Working in this way, the limit P_A has no direct interpretation in terms of set theoretical integration theory. And in fact, in numerous natural examples, different approximation procedures of the same event A with $P(A) = 0$ may produce different limits $P'_A, P''_A \cdots$ or no limit at all [28]. Also, since the situations where singular conditionings arise are infinite, the general way of presentation of the solutions (giving one special treatment for each particular case, using methods relatively far from the current physical intuition) is perhaps not fully effective; therefore, as noticed recently by M.M. Rao [28] more general results and more general methods of treatment are needed.

The goal of the present work is to show that the pullback of measures provides methods of this type.

More specifically, assuming that (Ω, P) is a probabilized Polish space, and introducing another Polish space Ω_R, some situations of the following two types are considered.

Non-parametric Case. Given an equivalence class f of P-measurable mappings $\Omega \to \Omega_R$, and denoting by $m = Law(f)$ the events A_k are of the type "$f \in B_k$" where (B_k) denotes any sequence of Borel subsets of Ω_R such that $|B_k| = m(B_k) \neq 0$, shrinking to some point b of Ω_R.

Parametric Case. Given a sequence of equivalence classes of P- measurable mappings $f_k : \Omega \to \Omega_R$, with $m_k = Law(f_k)$, the events A_k are of the type $(f_k \in B_k)$ where $B_1, B_2, \cdots$ are Borel subsets of Ω_R.

These problems will be studied introducing the general notion of pullback of measures and reducing the problem to a problem of convergence of pullbacked measures. Several results are established in this direction, and then applied to some problems of singular conditioning.

2 Pullback of Measures

The general notations of this part are the following : (Ω, P) denotes a probabilized Polish space, f an equivalence class of P-measurable mappings $\Omega \to \Omega_R$ taking values in some Polish space Ω_R and $m = f(P) =$ Law (f) denotes the probability law of f. As well known, there exists essentially one regular version of the conditional probability of P given f i.e. a family $\{P_b, b \in \Omega_R\}$ of probability measures on Ω such that for any A in the Borel σ – field $\sigma(\Omega)$ of Ω the mapping $b \to P_b(A)$ is Borel and

$$\forall B \in \sigma(\Omega_R) \quad P(A \cap f^{-1}(B)) = \int_B P_b(A) dm(b). \tag{1}$$

Note that any regular version $b \to P_b$ of the conditional probability of P given f is essentially characterized by its restriction to the support $Supp(m)$ of m i.e. the intersection of all closed subsets F of Ω_R such that $m(A \cap F) = m(A)$ for all A in the Borel σ-field of Ω_R. Also P_b vanishes outside $f^{-1}(\{b\})$ for m-almost all b in Supp(m).

The following basic asumption, refered as asumption $H(W)$, will be assumed: *[H(W)] There exists an open subset W of the topological subspace $Supp(m)$ of Ω_R such that a regular version of the conditional probability of P given f is defined and narrowly continuous on W.*

Let us recall some facts concerning narrow convergence and Daniell's integral. In general, we denote by $F \cdots G$ the data of two real vector spaces in duality. This means that a bilinear form $b(f, g) = f.g = \langle f, g \rangle$ is defined on $F \times G$ such that the two following linear mappings are injective:

$$F \ni f \mapsto b(f, .) \in G^* \quad and \quad G \ni g \mapsto b(., g) \in F^*.$$

For example, let $F = C_b(\Omega)$ be the space of all continuous and bounded real valued functions on the Polish space Ω. By the theory of Daniell integral, the space $M(\Omega)$ of all signed measures on Ω is isomorphic with the space of all linear forms q on $C_b(\Omega)$ uniformly bounded on the unit ball of this space and having Daniell's continuity property: $(f_k) \downarrow 0 \Rightarrow q(f) \to 0$. Hence $C_b(\Omega) \cdots M(\Omega)$. The corresponding weak topology on $M(\Omega)$ is called the narrow topology. This gives an imbedding of the set $Pr(\Omega)$ of probability measures on Ω in the space $M(\Omega)$.

2.1 Definition of the Pullback of Measures

For any signed measure q defined on W, the following linear form

$$C_b(\Omega) \ni \varphi \mapsto P_q(\varphi) = \int_W P_b(\varphi) dq(b). \tag{2}$$

is uniformly bounded on the unit ball of $C_b(\Omega)$ and has the Daniell continuity property i.e. $\varphi_n \downarrow 0$ implies $P_q(\varphi_n) \to 0$. Therefore P_q defines a unique signed measure on Ω also denoted P_q. Hence a linear mapping called the pullback by f of measures defined on $W \subset \Omega_R$

$$f^* : M(W) \ni q \mapsto P_q \in M(\Omega).$$

Remark. By the theorem of monotone classes, the mapping $b \mapsto P_b(f)$ is Borel for any bounded Borel function f defined on Ω and moreover

$$P_q(f) = \int_W P_b(f) dq(f). \tag{3}$$

Taking for example $f = 1_A$ with $A \in \Omega$, this gives a direct set theoretical definition of P_q.

2.2 Some properties of the pullback

a) The pullback is narrowly continous since by (2)

$$(q_j) \to q \text{ narrowly in } M(W) \Longrightarrow \forall\varphi \in C_b(\Omega), \;\; P_{q_j}(\varphi) \to P_q(\varphi) \tag{4}$$

b) The pullback preserves positivity since by (2)

$$\varphi \geq 0 \text{ and } q \geq 0 \Longrightarrow < f^*(q), \varphi >= P_q(\varphi) \geq 0 \tag{5}$$

Therefore P_q is a probability measure if q is a probability measure since by (2)

$$P_q(1) = \int_W P_b(1) dq(b) = 1$$

c) The pullback f^* extends the composition of functions defined on W with f, in the following sense : for any Borel subset B of W

$$q = 1_B.m \Longrightarrow f^*(q) = (1_B \circ f)P \tag{6}$$

since for any Borel subset A of Ω, (1) gives:

$$\int_A f^*(q) = \int_B P_b(A) dm(b) = P(A \cap f^{-1}(B)) = \int_A (1_B \circ f) dP \tag{7}$$

d) Assuming f continuous, the pullback is a right inverse of the operation $f_* = f$ of direct image of measures i.e.

$$\forall q \in M(W) \qquad f f^*(q) = q$$

In fact, by c), this relation holds for any measure on W of type $1_B.m$. By linearity and continuity, the relation $ff^*(q) = q$ holds for all q in the closed subspace C of $M(W)$ generated by the measures $1_B.m$ with $B \in \sigma(W)$. Finally, $C = M(W)$. e) For any Borel subset B of W with $|B| = m(B) \neq 0$, f^* induces the conditioning of P by the event $(f \in B)$ in the following sense

$$q = (1_B m)/|B| \Longrightarrow f^*(q) = (P \; if \; (f \in B)) \tag{8}$$

This last probability is denoted P_B. Note that e) follows from c).

Now, for any fixed element $b \in W$, we have for any $\varphi \in C_b(\Omega)$:

$$P_B(\varphi) - P_b(\varphi) = \frac{1}{|B|}\int_B P_c(\varphi)dm(c) - P_b(\varphi) = \frac{1}{|B|}\int_B (P_c(\varphi) - P_b(\varphi))dm(c).$$

This relation and the continuity of $c \mapsto P_c$ imply the following approximation property: f) Let $B_1, B_2 \cdots$ be any sequence of Borel subsets of W with $|B_j| \neq 0$ and shrinking to some fixed element b of W i.e. such that for any neighborhood V of b with $V \subset W$, $B_k \subset V$ for k large enough. Then the conditioning of P by the event $A_k = (f \in B_k)$ produces a sequence of probability measures P_k converging narrowly to P_b as $k \to \infty$. The limit is the same for all approximation procedures satisfying these conditions. This approximation property shows that P_b is unambiguously defined for any $b \in W$.

2.3 First Examples and a Counterexample

a) Numerous examples can be found in elementary probability. Notice the following general class of examples where $H(W)$ is satisfied with $W = Supp(m)$. Suppose that (Ω, P) is the product $(\Omega_R, m) \times (Y, n)$ of two probabilized Polish spaces, and that f is the first canonical projection π_1 of $\Omega = \Omega_R \times Y$ onto Ω_R. In this case, the following mapping

$$W = Supp(m) \ni b \mapsto \delta_b \otimes n \in M(\Omega) \tag{9}$$

is continuous since for any $\varphi \in C_b(\Omega)$, Lebesgue's theorem shows that the following mapping is continuous:

$$W \ni b \mapsto < \delta_b \otimes n, \varphi > = \int_Y \varphi(b, y)dn(y). \tag{10}$$

Since $\delta_b \otimes n$ is a regular version of the conditional probability of P given f, this shows that the corresponding pullback f^* of measures is defined. Moreover, $P_q = q \otimes n$.

b) **Federer Co-aera Formula [7].** Let us consider two Euclidean spaces Ω and Y, and a C^1-mapping with surjective derivative $\nabla f(\omega) : \Omega \to Y$ for all $\omega \in \Omega$. Let P be a probability measure on Ω with continuous density such that $m = f(P)$ has a continuous density $m(y)$ for all $y \in Y$. Then the pullback is defined on $W = \{b \in Y, m(b) \neq 0\}$ and Federer's coaera formula gives an explicit expression of $\tilde{\delta}_b = m(b)P_b$ for all $b \in Y$. Notice that the following argument of functional analysis gives directly the existence of the pullback: δ_b is the image of δ_b by the transposed mapping of the following linear and continuous mapping:

$$f_* : C_k(\Omega) \ni g \mapsto f_*(gP) \in C_k(Y)dy. \tag{11}$$

c) The following counterexample shows that the approximation property is not necessarily true for $b \in Supp(m)$: f is the first canonical projection of $\Omega = [0,1]^2$ onto $\Omega_R = [0,1]$, $b = 1/2$, and the probability P on Ω is the joint law of two random variables X and Y such that X is uniformly distributed on $[0,1]$, $Y = 0$ for $X < 1/2$ and $Y = 1$ for $X \geq 1/2$. If the event $A = (f = 1/2)$ is approximated by the events $A_k = (f \in B_k)$ where the intervalls B_k tend to b remaining in $[0,1/2[$ (resp. remaining in $[1/2,1]$), one obtains the limit $P_b' = \delta_{1/2} \otimes \delta_0$ (resp. $P_b'' = \delta_{1/2} \otimes \delta_1$). Switching two sequences of intervalls of the previous type, one obtains a sequence of conditional probabilities without limit. See [28] for other examples of this type.
Let us now establish the relation with singular conditioning. In order to avoid paradoxes and confusion with the usual conditioning, a new terminology is introduced concerning the statistical interpretation of pullbacked probabilities.

2.4 Statistical Interpretation

a) Let us consider a probabilized Polish space (Ω, P), a Polish space Ω_R, and an equivalence class f of P-measurable mappings: $\Omega \to \Omega_R$ satisfying $H(W)$ for some open subset W of the support of $m = f(P)$. For any probability q on W, $f^*(q) = P_q$ can be called *P with the constraint "Law(f) = q"* and denoted *P with* $(Law(f) = q)$. For any $b \in W$ and taking $q = \delta_b$, P_q is denoted by P_b.
b) **Relation with Singular Conditioning in the Parametric Case** With the notations of the introduction, we will prove in some cases the existence of an equivalence class $f : \Omega \to \Omega_R$ such that the sequence (P_{A_k}) converges narrowly to $f^*(q)$. This means that *P with* $(Law(f) = q)$ is the limit of the sequence of usual conditional probabilities *P if* $(f_k \in B_k)$.
c) **Supplement in the particular case 2.3.a)**. Let us now assume that $\Omega_R \times Y$ is endowed with a structure of topological vector space, such that f is continuous. Then, identifying the product with a sum, for any $q \in M(W)$, the random vector $\underline{\omega}$ of Ω represented by P_q is the sum of two independant random vectors of Ω_R and Y:

$$\underline{\omega} = \underline{\omega}_1 + \underline{y}$$

with $Law(\underline{\omega}_1) = q$ and $Law(\underline{y}) = n$.

Let us now check some general results concerning the pullback (section 3) and the parametric pullback (section 4), and show on specific examples that the present approach is effective (sections 5 and 6).

3 Some Results Concerning the Pullback

In this section, we present a general class of cases reductible to the product case 2.3.a), and also the relation with S.Watanabe's pullback. [33]

3.1 Probabilizable Vector Spaces

Definition. A *probabilizable vector space* $U \ldots \Omega$ is defined by a Frechet separable locally convex space Ω, and a subspace U of the dual space Ω', with a directed family (U_α) of finite dimensional subspaces of U with union U, such that U contains some countable subset D separating the points of Ω. Given two probabilizable vector spaces $U \ldots \Omega$ and $V \ldots \Omega_R$, a continuous linear mapping $\ell : \Omega \to \Omega_R$ is called an homomorphism of probabilizable vector spaces if there exists a linear mapping $\ell^T : V \to U$ transposing ℓ, and such that ℓ^T maps any element V_β of the directed family on V onto some element U_α of the directed family on U.

Some Facts. a) A priori, if $\sigma_D(\Omega)$ denotes the σ-field on Ω generated by the elements of D and $\sigma_U(\Omega)$ the σ-field on Ω generated by the elements of U, we have

$$\sigma_D(\Omega) \subset \sigma_U(\Omega) \subset \sigma(\Omega).$$

But, in view of a theorem due to X. Fernique [8], these three σ-fields coïncide.
b) Giving the probabilizable vector space $U \ldots \Omega$ and any element U_α of the directed family of subspaces of U, the canonical surjection s_α of Ω onto $\Omega_\alpha = \Omega / U_\alpha^\perp$ is a surjective homomorphism. We have the diagram

$$\begin{array}{ccc} U & \ldots & \Omega \\ \uparrow i_\alpha & & \downarrow s_\alpha \\ U_\alpha & \ldots & \Omega_\alpha \end{array}$$

where i_α denotes the canonical injection. Moreover, by a), the Borel σ-field $\sigma(\Omega)$ on Ω is generated by the family (s_α).
c) Since (s_α) generates $\sigma(\Omega)$, any probability measure q on $\sigma(\Omega)$ is characterized by a consistent system of probability measures (q_α) on the quotient spaces (Ω_α), i.e. by a cylindrical probability. For example, given $m \in \Omega$ and a linear symmetric mapping $U \to U^*$, the Gaussian cylindrical probability $P = Gauss(m, C)$ on Ω with mean m and covariance operator C is defined by its Laplace transform:

$$U \ni u \to \int e^{u.\omega} dP(\omega) = exp(< m, u > +1/2 < Cu, u >) \qquad (1)$$

3.2 Probability Vector Spaces (P.V.S.)

Definitions. A *probability vector space* $(H \supset U \cdots \Omega, P)$ is defined by a probabilizable vector space, by a centered probability P on Ω having moments of all order and such that the covariance bilinear form defined on $U \times U$ by :

$$b(u, v) = E(uv) = \int_\Omega u(\omega) v(\omega) P(d\omega)$$

is a scalar product on U, i.e. $b(u, u) = 0$ implies $u = 0$; and by the completion H of U for this scalar product is the *covariance Hilbert space of P*.

In practice, Ω is a space of paths of a centered random field, P is the law of the field, and U is a distinguished set of linear random variables $u(\omega) = u.\omega$. A linear process arises in this way:

$$U \ni u \mapsto u(\omega) \in \mathcal{L}^o(\Omega)$$

Since the following linear mapping is isometric:

$$H \supset U \ni u \mapsto u(\omega) \in L^2(\Omega, P)$$

elements of H can be identified with centered second order random variables. The data $(H \supset U \cdots \Omega, P)$ is sometimes simply denoted Ω.

The *covariance operator* is defined as the linear injective mapping with values in the algebraic dual of U:

$$C : U \ni u \mapsto b(u, .) \in U^*.$$

Carrying over structures by C, the covariance scalar product defined on U gives the following reproducing scalar product defined on the range $Im\ C$ of C

$$[Cu, Cv] =< u, v >_H =< Cu, v > .$$

As well known [21], if $Im\ C \subset \Omega$, the completion H' of $Im\ C$ for this scalar product is an Hilbertian subspace of Ω i.e. the natural mapping $H' \to \Omega$ is injective and continuous: H' is called the *reproducing Hilbert space of* P.

Given $(H \supset U \cdots \Omega,\ P)$, a corresponding excentered P.V.S. can be defined simply substituing P by $\tau_m P$ where τ_m denotes the translation by the vector m in Ω.

Homomorphisms of P.V.S. - Regular Homomorphisms. Given two P.V.S. Ω and $(V \ldots \Omega_R, P_R)$, a linear mapping $h : \Omega \to \Omega_R$ is an *homomorphism of P.V.S.* if h is an homomorphism of the corresponding probabilizable vector spaces and if $h(P) = P_R$. Homomorphisms of P.V.S. can be composed, hence a natural definition of isomorphisms of P.V.S.
The homomorphism $h : \Omega \to \Omega_R$ is called *regular* if the following conditions are satisfied:
i) $h : \Omega \to \Omega_R$ is surjective (hence h^T is injective)
ii) Identifying V with the subspace $\mathrm{Im}(h^T)$ of U, the orthogonal projector of H onto the closed subspace K of H induces a mapping $\ell : U \to V$
iii) Moreover, the transposed mapping of ℓ induces a continuous mapping ℓ^T : $\Omega_R \to \Omega$.

The data of a regular homomorphism $h : \Omega \to \Omega_R$ can be summarized by the following diagram:

$$\begin{array}{c} (H \supset\ U \ldots \Omega\ ,\ P) \\ \ell \downarrow\uparrow h^T \qquad h \downarrow\uparrow \ell^T \\ (K \supset\ V \ldots \Omega_R\ ,\ P_R) \end{array} \tag{2}$$

Notice that ℓ is a left inverse of h^T, and that ℓ^T is a right inverse of h. Notice also that for a Gauss space (Ω, P) and for any regular homomorphism of finite rank, *i)* implies *ii)* and *iii)* since $V = K$ is closed and $H'_R \simeq \Omega_R$, hence ℓ^T induces a continuous linear mapping $K' \to H'$.

3.3 Decomposition in Product of Ω Associated with any Regular Homomorphism $h : \Omega \to \Omega_R$ of Gauss spaces

Theorem 1. (Decomposition in product for a P.V.S.)
Let h be a regular homomorphism of P.V.S. (2). Then:
a) The space U is the algebraic sum of $U_1 = Im(h^T)$ and of $U_2 = ker(\ell)$. Moreover, $U_1 \perp U_2$;
b) The space Ω is the topological sum of $\Omega_1 = U_2^\perp \simeq Im(\ell^T)$ and of $\Omega_2 = U_1^\perp \simeq ker(h)$; and h induces an isomorphism of Frechet spaces $\Omega_1 \to \Omega_R$ whose inverse is ℓ^T;
c) In particular, if the source space Ω is a Gauss space, then it is the product of two Gauss spaces

$$(H_i \supset U_i \cdots \Omega_i, P_i) \quad i = 1, 2$$

where H_i is the closure of U_i in H and

$$< z, z >_{H_2} = < z, z >_H - \|\ell z\|_K^2.$$

This theorem is an extension of a theorem of L. Gross [10] available in the particular case where the source space is a Wiener space and where Ω_R is finite dimensional.
Proof. Let $U_1 = h^T(V)$, and p_{U_1} be the orthogonal projector onto U_1. Then, $U = Im(p_{U_1}) \overset{\perp}{\oplus} ker(p_{U_1})$, and $ker(p_{U_1}) = ker(\ell)$. Now, for any u in U, $u_1 = h^t \ell u$ belongs to U_1 and $u_2 = u - u_1$ belongs to $U_2 = ker \ell$ since

$$\ell(u_2) = \ell u - \ell(h^T u) = \ell u - \ell u = 0.$$

This proves a). Denoting by H_i the closure of U_i in H for $i = 1, 2$, we have a decomposition in Hilbertian sum $H = H_1 \oplus H_2$. Since $U = U_1 + U_2$ is in duality with Ω, the intersection of $\Omega_1 = U_2^\perp$ and of $\Omega_2 = U_1^\perp$ is $\{0\}$. For any element ω in Ω, $\omega_1 = \ell^T h(\omega)$ belongs to Ω_1 since

$$\ell u = 0 \Rightarrow < \omega_1, u > = < \ell^T h(u), u > = < h(\omega), l(u) > = 0.$$

and $\omega_2 = \omega - \omega_1$ belongs to Ω_2 since for any $v \in V$

$$< h^T v, \omega_2 > = < h^T v, \omega > - < h^T v, \omega_1 > = < h^T v, \omega > - < h^T v, \ell^T h(\omega) >$$

$$= < h^T v, \omega > - < h^T \ell h^T v, \omega > = 0.$$

Hence $\Omega = \Omega_1 \oplus \Omega_2$ algebraically. This also holds topologically since the first canonical projection $\ell^T h$ is continuous. Therefore the second canonical projection

$Id_\Omega - \ell^T h$ is also continuous. Since $\Omega_2 = ker\ h$, the canonical factorization of the linear and continuous mapping h is $h = bos$ where

$$\Omega \xrightarrow{s} \Omega_1 = \Omega/ker(h) \xrightarrow{b} \Omega_2$$

where the bijective continuous mapping b is bicontinuous since open (by the Banach theorem).

Now, combining theorem 1 with 2.3 a), we obtain the following result.

Application to the Pullback by Regular Homomorphisms of Gauss Spaces.

Lemma 2. *Let $h : \Omega \to \Omega_R$ be a regular homomorphism of Gauss spaces. Then, for any probability q on Ω_R, the random vector $\underline{\omega}$ of Ω represented by $P_q = h^*(q)$ is the sum of two independant random vectors $\underline{\omega}_1$ and $\underline{\omega}_2$ whose laws are explicitely known:*

$$Law(\underline{\omega}_1) = \ell^T(q) \text{ and } Law(\underline{\omega}_2) = g_h \text{ where } g_h = Gauss(0, \|z\|^2 - \|h^T \ell z\|^2).$$

Hence,

$$P_q = \ell^T(q) * g_h. \tag{3}$$

Note incidentally that, if Ω is endowed with an excentered probability $P =$ Gauss (μ, C) where $\mu \in \Omega$ and $< Cz, z >= \|z\|_H^2$, we get a decomposition of (Ω, P) into a product of two excentered Gauss spaces (Ω_i, P_i) where $P_i =$ Gauss (μ_i, C_i) and $\mu_1 = \ell^T h(\mu)$. Note also that, as a first application, for any $b \in \Omega_R$, $P_b = Gauss(\ell^T b; \|z\|^2 - \|h^T \ell z\|^2)$.

3.4 Note Concerning Some Recent Non-linear Results

Let us for example consider a Gauss space $(H \supset U \cdots \Omega, P)$ and an equivalence class of non linear mappings $f : \Omega \to Y$ taking values in some Euclidean space Y. Let us sketch briefly how the existence of the pullback and even a counterpart of Federer's co-aera formula [7] for $\tilde{\delta}_b$ have been recently established, just in the line of the results sketched in 2.3 c). Since the usual differential calculus is no more available, one has to use the calculus in Gauss Sobolev spaces $W^{k,p}(\Omega)$ first developped in [15] [16] [17] [23], and later applied to stochastic analysis as explained in [32]. Note that the pullback was missing in this theory and has been first defined and studied by S.Watanabe [33] (see also [22] and the note [18] giving an argument of functional analysis similar to the argument sketched in 2.3 c)). In order to write some co-aera formula [5], one needs the theory of capacities on Gauss Sobolev spaces first developped by P.Paclet [27]. Following [18] we briefly hint how the hypothesis $H(W)$ can be checked using mainly results of [33]. We introduce the Watanabe space $W(\Omega) = \cap_{1<p<\infty,\ k\in Z} W^{k,p}(\Omega)$. Given an Euclidean space $Y = \Omega_R$, let f be an equivalence class of mappings $\Omega \to Y$ belonging to $W(\Omega, Y)$, such that $\nabla f(\omega)$ is almost surely surjective, and such that the inverse of the Gram matrix of f belongs to $\cap_{1<p<\infty} L^p(\Omega)$. Since the following mapping is continuous:

$$f_* : W(\Omega) \ni g \mapsto f_*(gP) \in \mathcal{S}(Y)dy,$$

this first show, taking $g \equiv 1$, that $m = Law(f)$ has a density $m(y)$ which belongs not only to $C^\infty(Y)$ but also to $\mathcal{S}(Y)$, the Schwartz space of test functions. Transposing f_*, we obtain the Watanabe mapping constructed directly in [33]:

$$f_*^T : \mathcal{S}(Y)' \ni T \mapsto \tilde{T} \in W(\Omega)'$$

hence, in particular, for any $b \in \mathbb{R}^d$, a positive measure $\tilde{\delta_b}$, such that

$$\forall g \in W(\Omega) \quad \tilde{\delta_b}(g) = (f_*(gP))(b). \tag{4}$$

Therefore the mass of $\tilde{\delta_b}$ is $f_*(1P)(b) = m(b)$. For any open subset W of $Supp(m)$ where $m \neq 0$ and for any $b \in \Omega$, $P_b = m(b)^{-1}\tilde{\delta_b}$ is a probability measure on Ω such that for any Borel subset B of W

$$\int_B P_b(g)m(b)db = \int_B f_*(gP)(b)db = \int_\Omega g\, 1_B o f dP. \tag{5}$$

Therefore there exist a regular version of the conditional probability P given$f$, which is continuous on W.

Note that the pullback by an homomorphism of Gauss spaces as defined by lemma 2 is slightly different from the Watanabe pullback.

4 Properties of the Parametric Pullback

4.1 Convergence of Regular Homomorphisms

Definition. Let us consider a fixed Gauss space $(H \supset U \cdots \Omega, P)$ and a probabilizable vector space $V \cdots \Omega_R$. We say that a sequence of regular homomorphisms $h_\varepsilon : \Omega \to \Omega_R$ converges to the regular homomorphism $h : \Omega \to \Omega_R$ if the following conditions are satisfied:

1. For any z in U, $h_\varepsilon^T l_\varepsilon z \to h^T l z$ in H as $\varepsilon \to 0$.
2. For any compact subset $K \subset \Omega_R$, $l_\varepsilon^T(\omega) \to l^T(\omega)$ in Ω, uniformly on K.

Theorem 3. Continuity of the pullback.
With the notations of the previous definition, let us assume that $h_\varepsilon \to h$ as $\varepsilon \to 0$. Then, for any sequence q_ε of signed measures on Ω_R converging narrowly to some limit q, the sequence $h_\varepsilon^*(q_\varepsilon)$ narrowly converges to $h^*(q)$ as $\varepsilon \to 0$.

In view of lemma 2, we only prove $l_\varepsilon^T(q_\varepsilon) * g_{h_\varepsilon} \to l^T(q) * g_h$ narrowly as $\varepsilon \to 0$, where $g_{h_\varepsilon} = Gauss(0, \|z\|^2 - \|h_\varepsilon^T \ell_\varepsilon z\|^2)$ and $g_h = Gauss(0, \|z\|^2 - \|h^T \ell z\|^2)$. This will be done using the two following lemmas.

Lemma 4. A compactness lemma for Gauss measures.
Let $\gamma = Gauss(\ell, Q)$ be a Gaussian Radon measure on the probabilizable vector space $U \ldots \Omega$. Then, the set $G(Q)$ of all centered Gaussian measures on $U \ldots \Omega$ with covariance quadratic form dominated by Q is a set of Radon measures and is narrowly compact.

Proof. Slepian's Lemma ([2]) implies that any $\mu \in G(Q)$ is a Radon measure. Prokhorov's theorem combined with Slepian's Lemma shows that $G(Q)$ is narrowly relatively compact. In order to prove that $G(Q)$ is narrowly closed, let us consider a sequence (γ_j), $\gamma_j = Gauss(Q_j)$ in $G(Q)$ converging narrowly to μ in $M_b(\Omega)$.
Hence $exp(-iQ_j(u)/2) \to \mathcal{F}\mu(u)$ for any u in U. By P. Levy's theorem, μ is a cylindrical Gauss measure. Since $Q_j(u) \leq Q(u)$ for all j and all u in U, the covariance quadratic form Q_0 of μ satisfies $Q_0(u) \leq Q(u)$ for all u in U. Hence μ is in $G(Q)$, and $G(Q)$ is closed, hence narrowly compact.

Lemma 5. Equivalence of Several Topologies on $G(Q)$
The set $G(Q)$ is defined as previously. For arbitrary elements m and m_j of $G(Q)$ whose covariance are denoted by Q and Q_j, the following are equivalent :

i) $m_j \to m$ *narrowly .*
ii) $m_j \to m$ *in the sense of convergence of moments, i.e.*

$$\forall u \in U \ \ \forall k \in \mathbb{N} \ \ \int_\Omega (u \cdot \omega)^k dm_j(\omega) \to \int (u \cdot \omega)^k dm(\omega)$$

iii) $Q_j(u) \to Q(u)$ *for any u in U.*

Proof. The Laplace transform Lm of m is

$$Lm\ (u) = e^{Q(u)/2} = \int_\Omega e^{u \cdot \omega} dm(\omega). \tag{1}$$

Hence, expanding the exponentials :

$$\int_\Omega (u \cdot \omega)^k dm(\omega) = \begin{cases} 0 & \text{if } k \text{ is odd} \\ \dfrac{Q^\ell(u)(2\ell)\,!}{2^\ell \ell\,!} & \text{if } k = 2\ell. \end{cases} \tag{2}$$

This shows that ii) is equivalent to iii), and, combined with (1), that the topology of convergence of moments is Hausdorff. Since the topology of narrow convergence restricted to $G(Q)$ is compact, we only have to show that i) implies iii). This is clear since *i)* implies $\mathcal{F}m_j \to \mathcal{F}m$ for all $u \in U$. ($\mathcal{F}$ denotes Fourier transform).

Now, theorem 3 can be proved easily.

Lemmas 4 and 5 show that the mapping $h_\varepsilon \mapsto g_{h_\varepsilon}$ is narrowly continuous. For any given f in $C_b(\Omega)$,

$$< \ell_\varepsilon^T q_\varepsilon, f > = < q_\varepsilon, f \circ \ell_\varepsilon^T > .$$

Since $(q_\varepsilon) \to q$, for any given $\eta > 0$, there exists a compact subset K of Ω_R such that, for all ε

$$\int_{\Omega_R \setminus K} \|q_\varepsilon\| < \eta/2.$$

Hence, using condition 2 of 4.1, an argument in $\eta/2$ gives $< \ell_\varepsilon^T q_\varepsilon, f > \to < \ell^T q, f >$ as $\varepsilon \to 0$. Hence the mapping $\varepsilon \mapsto \ell_\varepsilon^T q_\varepsilon$ is narrowly continuous.
Now theorem 3 follows since P_ε is the convolution product of two bounded measures depending continuously on ε, and since the convolution of two probability measures is sequentially narrowly continuous.

4.2 Integrals of Pullbacked Measures

Lemma 6. Integrals of Pullback of Measures
Let M be a metric and compact space endowed with a bounded positive measure μ. Let Ω be a Gauss space, and $(h_t)_{t\in M}$ be a family of regular homomorphisms of Gauss spaces $\Omega \to \Omega_R$ depending continuously upon t :

$$t \in M \text{ and } t_j \to t \quad \Longrightarrow \quad h_{t_j} \to h_t.$$

Then, for any sequence q_n of bounded measures on Ω_R converging narrowly to q in $M(\Omega_R)$, the integrals

$$I_n = \int_M h_t^*(q_n)d\mu(t) \quad \in M(\Omega_R)$$

converge narrowly, as $n \to \infty$, to the integral

$$I = \int_M h_t^*(q)d\mu(t).$$

In the last two expressions, I_n and I are weak integrals. More precisely the signed measure I_n on Ω is represented (in view of the theory of Daniell integral) by the linear form

$$C_b(\Omega) \ni f \to I_n(f) = \int_M h_t^*(q_n)fd\mu(t).$$

Proof. We only check that for any given $f \in C_b(\Omega)$:

$$\int_M h_t^*(q_n)fd\mu(t) \to \int_M h_t^*(q)fd\mu(t) \text{ as } n \to \infty.$$

By theorem 3 we know that for all n, $t \mapsto h_t^*(q_n)f$ is continuous; and also that for all $t \in M$ $h_t^*(q_n)f \to h_t^*(q)f$ as $n \to \infty$. By Lebesgue's theorem, we only have to check $sup_{t,n}|h_t^*(q_n)f| < \infty$.
Since the subspace $N = \{q, q_1, q_2, \cdots\}$ of $M(\Omega_R)$ is compact, we only have to check that the following mapping is continuous on $M \times N$:

$$(t, q_n) \mapsto h_t^*(q_n)f$$

i.e. for any converging sequence (t_k) of M with some limit t,

$$h_{t_k}(q_n)f \to h_t(q)f \text{ as } k \text{ and } n \to \infty.$$

This also follows from theorem 3.

5 Application to the Vertical Windowing

Let us consider a real continuous Gaussian field $t \to \xi_t$ indexed by some open subset J of $\mathbb{R}^d$. Hence the probability measure P representing ξ lives on $\Omega = C(J)$. For any closed subset F of J, we have a surjective and continuous mapping :

$$\Omega \ni \omega \xrightarrow{h} \omega' = \omega/_F \in \Omega_R := C(F). \tag{1}$$

Hence, if h is a regular homomorphism of Gauss spaces, for any probability measure q on Ω_R, the theorem 1 gives explicitely the decomposition in sum of two independant vectors of the random vector of Ω represented by $P_q = P\, with\, (Law(f) = q)$. In particular, if $q = \delta_b$ for any $b \in \Omega_R$, theorem 3 applied to $h_\varepsilon \equiv h$ with $q_\varepsilon = 1_{B_\varepsilon} m/|B_\varepsilon|$ can be interpreted as a result concerning "Vertical Windowing" (V.W.) for Gauss fields, in the sense of [12].

Let us give a classical example concerning the Brownian motion $BM(d)$.

5.1 Example

The space of trajectories is the Frechet space Ω_W of all $\mathbb{R}^d$-valued continuous functions $\omega(t)$ defined on $[0,\infty[$ such that $\omega(0) = 0$. Denoting e_j the jth element of the canonical basis of $\mathbb{R}^d$, Ω_W is in duality with the space U_W generated by the linear forms

$$\delta_t \otimes e_j^* : \quad \Omega_W \ni \omega = (\omega_j)_{j=1}^d \to \omega_j(t) \in \mathbb{R}$$

with $t > 0$ and $j = 1 \cdots d$. The covariance Hilbert space H_W of $X =$ BM (d) is the completion of U_W for the covariance scalar product $(\delta_s \otimes e_j^*) \cdot (\delta_t \otimes e_k^*) = s \wedge t\, \delta_j^k$. Here U_W is the union of the finite dimensional subspaces U_α whose indices $\alpha = \{t_1 < \cdots < t_d\}$ are all finite subsets of $\mathbb{R}_+$ and

$$U_\alpha = \text{span}\,\{\delta_s \otimes e_j^*\ ,\ s \in \alpha, j = 1, \cdots, d\}.$$

For any $A > 0$, the following mapping is a regular homomorphism of Gauss spaces

$$h_A : \Omega_W \ni \omega \mapsto \omega(A) \in \mathbb{R}^d = \Omega_R \tag{2}$$

as soon as $\Omega_R \simeq \Omega_R'$ is endowed with $P_R = h_A(P) =$ Gauss $(0, \sqrt{A}\, I_{\Omega_R})$.
As can easily be verified:

$$\ell_A : \ \delta_s \otimes e_j^* \mapsto \frac{s \wedge A}{A} \delta_A \otimes e_j^*$$

and

$$\ell_A^T : \ \omega_1 \mapsto \frac{\cdot \wedge A}{A} \omega_1.$$

Using the theorem of decomposition in product 1 in this situation, the following result is obtained:

the Brownian motion $BM(d)$ can be decomposed into the sum of two independant terms:

$$b = \frac{\cdot \wedge A}{A} b(A) + \tilde{b}$$

where $\tilde{b}$ is a process extending the Brownian Bridge between 0 and A. This is true since $ker(\ell_A^T) = \{\omega \in \Omega : \omega(A) = 0\}$. The Gaussian probability g_{h_A} appears here as the law of this extended Brownian bridge.
Some more applications to Markov fields will be considered elsewhere.

6 Applications of the parametric case

6.1 Oblique Windowing (O.W.) Problems and Slepian's Model

Oblique Windowing problems have various forms [24] [12]. We only consider one typical example. Let J be an open intervall of $\mathbb{R}$. Let X be a zero mean Gauss process defined on J with C^1 trajectories and covariance function $C(s,t) = E[X(s)X(t)]$. Given a level a, a slope γ, and $\varepsilon > 0$ such that $I_\varepsilon = [t, t+\varepsilon] \subset J$, the oblique window at level a with slope γ on I_ε is defined as the graph of the affine real valued function ψ on I_ε such that $\psi(t) = a$ and $\psi(t+\varepsilon) = a + \gamma\varepsilon$. The event of upcrossing by X of this window is defined by:

$$A_\varepsilon = \{\omega \in \Omega \;:\; \omega(t) < a \text{ and } \omega(t+\varepsilon) > a + \gamma\varepsilon\}. \tag{1}$$

The corresponding O.W. problem has practically two parts. One first has to prove that the conditional probabilities $P_\varepsilon = (P \; if \; A_\varepsilon)$ converge narrowly to some limit P' as $\varepsilon \to 0$. One also has to identify the limit P' and the corresponding so called "Slepian model" describing explicitely the process associated to P'. These problems are generally treated assuming X stationary and $\gamma = 0$, in order to apply a technique of Palm measures; each special case needs a particular approach and the limit P' has no probabilistic interpretation. We show below how the results described before give a general method of resolution of these problems and give also a probabilistic interpretation of the limit P'.

6.2 The Method of Resolution

Let $\Omega = C^1(J)$, the Gauss process X being represented by its law P on Ω. Let ε go to 0 in (1),then heuristically, $\omega(t) = a$ and $\dot{\omega}(t) \geq \gamma$. Hence we have to introduce a family of regular homomorphisms h_ε with source (Ω, P) converging as $\varepsilon \to 0$ to the regular homomorphism

$$\Omega \ni \omega \xrightarrow{h} (\omega(t), \dot{\omega}(t)) \in \mathbb{R}^2 = \Omega_R. \tag{2}$$

Natural candidates for that are the following homomorphisms h_ε:

$$h_\varepsilon : \Omega \ni \omega \to \left(\omega(t), \frac{\omega(t+\varepsilon) - \omega(t)}{\varepsilon}\right) \in \Omega_R = \mathbb{R}^2. \tag{3}$$

Since a derivative at time t is involved, the U space is choosen as follows:

$$U = span\{(\delta_s, \dot{\delta}_t),\ s \in J\}$$

where $\dot{\delta}_t$ is the opposite of the distribution derivative of δ_t with respect to t. This means that the following duality is used between Ω and U:

$$< \omega, (\delta_s, \dot{\delta}_t) >= \omega(s) + \dot{\omega}\,(t) \tag{4}$$

The following notations will be used:

$$\partial_1 C(s,u) = \partial C(s,u)/\partial s;\ \partial_2 C(s,u) = \partial C(s,u)/\partial u;\ \partial_{1,2} C(s,u) = \frac{\partial^2}{\partial s \partial u} C(s,u).$$

Assuming that the covariance bilinear form on U is non degenerated, a Gauss space $(H \supset U \cdots \Omega, P)$ is defined. Identifying as usually $\mathbb{R}^2$ with its dual space, the diagram concerning h_ε is the following:

$$\begin{array}{ccccc} H \supset & U & \ldots & \Omega\ ; & P \\ & \ell_\varepsilon \downarrow\uparrow h_\varepsilon^T & & h_\varepsilon \downarrow\uparrow \ell_\varepsilon^T\ ; & \downarrow \\ V = & (\mathbb{R}^2)^* & \ldots & \Omega_R = \mathbb{R}^2\ ; & \gamma_\varepsilon = h_\varepsilon(P) \end{array}$$

A similar diagram holds for the homomorphism h. In view of the remark following the definition of regular homomorphisms in 3.2.4, the homomorphisms h_ε and h are regular. We have $P_\varepsilon = P\ if\ (h_\varepsilon \in B_\varepsilon)$, where B_ε denotes the angular domain of the (x, y)-plane where $x < a$ and $y > \gamma - \frac{x-a}{\varepsilon}$. In view of the property 2.2.e), $P_\varepsilon = h_\varepsilon^*(q_\varepsilon)$, where $q_\varepsilon = 1_{B_\varepsilon}\gamma_\varepsilon/\gamma_\varepsilon(B_\varepsilon)$.

In view of the theorem 3 of continuity of the pullback, the narrow convergence of P_ε as $\varepsilon \to 0$ is proven and the limit P' is computed in four steps.

Step 1: Computation of the Limit of (q_ε). We first compute the limit of $q_\varepsilon(\varphi)$ for any $\varphi \in \mathcal{D}(\mathbb{R}^{\in})$ with compact support K.

$$q_\varepsilon(\varphi) = \int\int_{B_\varepsilon} \varphi(x,y) p_\varepsilon(x,y) dx dy \Big/ \int\int_{B_\varepsilon} p_\varepsilon(x,y) dx dy \tag{5}$$

where p_ε denotes the density of γ_ε. As $\varepsilon \to 0$, this density converges uniformly on K to the density $p(x,y)$ of $\gamma = h(P)$. The covariance of γ is:

$$Cov(\gamma) = \begin{bmatrix} C(t,t) & \partial_1 C(t,t) \\ \partial_1 C(t,t) & \partial_{1,2} C(t,t) \end{bmatrix} \tag{6}$$

As $\varepsilon \to 0$, the angular sector B_ε is shrinking to the half line $x = a$ and $y > \gamma$. Hence, performing first the integrals with respect to x in the two double integrals (5), we obtain:

$$q_\varepsilon(\varphi) \sim (\varepsilon \int_\gamma^\infty (y-\gamma) p(a,y) \varphi(a,y) dy)/(\varepsilon \int_\gamma^\infty (y-\gamma) p(a,y) dy). \tag{7}$$

The positive measure

$$r(dx, dy) = \delta_a(dx) \otimes ((y-\gamma)_+ p(a,y)dy) \tag{8}$$

has total mass

$$\|r\| = \int_\gamma^\infty p(a,y)dy.$$

The relation (7) means that $q_\varepsilon(\varphi) \to r(\varphi)/\|r\|$ as $\varepsilon \to 0$. Since $r/\|r\|$ is a probability measure, this means that $q_\varepsilon \to r/\|r\|$ narrowly.

Step 2: Proof of the Convergence $h_\varepsilon^T \ell_\varepsilon z \to h^T \ell z$. Denoting by $(\boldsymbol{i}, \boldsymbol{j})$ the canonical basis of $\mathbb{R}^2$:

$$\begin{aligned} h_\varepsilon^T(x\boldsymbol{i} + y\boldsymbol{j}) &= x\delta_t + y(\delta_{t+\varepsilon} - \delta_t)/\varepsilon \\ h^T(x\boldsymbol{i} + y\boldsymbol{j}) &= x\delta_t + y\dot{\delta}_t \end{aligned} \tag{9}$$

This means that for any $\varepsilon > 0$,

$$h_\varepsilon^T(\boldsymbol{i}) = \delta_t \ , \ \ h_\varepsilon^T(\boldsymbol{j}) = (\delta_{t+\varepsilon} - \delta_t)/\varepsilon \tag{10}$$

and, at the limit $\varepsilon \to 0$,

$$h^T(\boldsymbol{i}) = \delta_t \ , \ \ h^T(\boldsymbol{j}) = \dot{\delta}_t. \tag{11}$$

In order to compute $h_\varepsilon^T \ell_\varepsilon z$, the Gram-Schmidt orthogonalization technique is applied to the set $(\delta_t, \delta_{t+\varepsilon})$ of generators of the subspace $h_\varepsilon^T(V)$ of H. This produces an orthonomal basis $(\boldsymbol{e}, \boldsymbol{f}_\varepsilon)$ of $h_\varepsilon^T(V)$:

$$\begin{aligned} \boldsymbol{e} &= \delta_t / \sqrt{C(t,t)} \\ \boldsymbol{f}_\varepsilon &= \frac{\varepsilon}{\alpha(\varepsilon)}\left(\frac{\delta_{t+\varepsilon} - \delta_t}{\varepsilon} - \frac{C(t,t+\varepsilon) - C(t,t)}{\varepsilon C(t,t)}\delta_t\right) \end{aligned} \tag{12}$$

where

$$\begin{aligned} \alpha(\varepsilon) &= \left(C(t+\varepsilon, t+\varepsilon) - \frac{C^2(t,t+\varepsilon)}{C(t,t)}\right)^{1/2}. \\ \beta &= (\partial_{1,2}C(t,t) - (\partial_1 C(t,t))^2 C^{-1}(t,t))^{1/2}. \end{aligned} \tag{13}$$

Since the covariance bilinear form on U is non degenerated, we have $C(t,t) \neq 0$ and $\beta \neq 0$. Straightforward computations give

$$\alpha(\varepsilon) = \varepsilon\beta + o(\varepsilon) \ \text{ as } \varepsilon \to 0$$

and

$$\boldsymbol{f}_\varepsilon \to \boldsymbol{f} = \beta^{-1}\left(\dot{\delta}_t - \frac{\partial_1 C(t,t)}{C(t,t)}\delta_t\right) \ \text{ in } H.$$

Since the scalar product is continuous in $H \times H$, the Gram-Schmidt algorithm applied to $(\delta_t, \dot{\delta}_t)$ produces the orthonormal basis $(\boldsymbol{e}, \boldsymbol{f})$ of $span(\delta_t, \dot{\delta}_t)$. Since

$h_\varepsilon^T \ell_\varepsilon$ and $h^T \ell$ are respectively the orthogonal projectors in H onto $span(\delta_t, \delta_{t+\varepsilon})$ and onto $span(\delta_t, \dot{\delta}_t)$, we obtain that $h_\varepsilon^T \ell_\varepsilon z$ converges, as $\varepsilon \to 0$, to $h^T \ell z$ where:

$$\begin{aligned} h^T \ell z &= (z.e)e + (z.f)f \\ &= C(t,t)^{-1}(z.\delta_t)\delta_t + (z.(\dot{\delta}_t - \frac{\partial_1 C(t,t)}{C(t,t)}\delta_t)\beta^{-2}[(\dot{\delta}_t - \frac{\partial_1 C(t,t)}{C(t,t)}\delta_t]. \quad (14) \end{aligned}$$

In the particular case where $z = \delta_s$, we obtain:

$$h^T \ell \delta_s = \frac{C(s,t)}{C(t,t)}\delta_t + \frac{1}{\beta^2}(\partial_2 C(s,t) - \frac{\partial_1 C(t,t)}{C(t,t)}C(s,t))[(\dot{\delta}_t - \frac{\partial_1 C(t,t)}{C(t,t)}\delta_t]. \quad (15)$$

Step 3: Proof of the Convergence $\ell_\varepsilon^T i$ to $\ell^T i$ and of $\ell_\varepsilon^T j$ to $\ell^T j$. Using (11),

$$\ell \delta_s = \frac{C(s,t)}{C(t,t)} i + \frac{1}{\beta^2}(\partial_2 C(s,t) - \frac{\partial_1 C(t,t)}{C(t,t)}C(s,t))(j - \frac{\partial_1 C(t,t)}{C(t,t)} i). \quad (16)$$

Hence:

$$\begin{aligned} \ell^T i(s) &= < \ell^T i, \delta_s > = < i, \ell(\delta_s) > \\ &= \frac{C(s,t)}{C(t,t)} - \frac{1}{\beta^2}(\partial_2 C(s,t) - \frac{\partial_1 C(t,t)}{C(t,t)}C(s,t))\frac{\partial_1 C(t,t)}{C(t,t)}. \\ \ell^T j(s) &= \frac{1}{\beta^2}(\partial_2 C(s,t) - \frac{\partial_1 C(t,t)}{C(t,t)}C(s,t)). \quad (17) \end{aligned}$$

The components of $\ell_\varepsilon \delta_s$ on i and j are computed precisely as the corresponding components of $\ell_\varepsilon^T \delta$ in step 2. More precisely, and using (10):

$$\begin{aligned} \ell_\varepsilon \delta_s = \frac{C(s,t)}{C(t,t)} i &+ \frac{\varepsilon^2}{\alpha(\varepsilon)^2}[\frac{C(s,t+\varepsilon) - C(s,t)}{\varepsilon} - C(s,t)\frac{C(t,t+\varepsilon) - C(t,t)}{\varepsilon C(t,t)}] \\ &\times [j - \frac{C(t,t+\varepsilon) - C(t,t)}{\varepsilon C(t,t)} i]. \quad (18) \end{aligned}$$

From this expression, one can see that $\ell_\varepsilon^T i \to \ell^T i$ and $\ell_\varepsilon^T j \to \ell^T j$.
The theorem 3 shows that the probabilities $h_\varepsilon^*(q_\varepsilon) = P_\varepsilon$ narrowly converge to $P_q = h^*(q)$ as $\varepsilon \to 0$.

Step 4: Description of the Random Vector of Ω Characterized by P_q. By theorem 1, the process with C^1-trajectories characterized by P_q is the sum of two independant processes whose laws are respectively $\ell^T q$ and $g_h = Gauss(0, \|z\|^2 - \|h^T \ell z\|^2)$. Since ℓ^T is explicitely expressed by (17), we only have to characterize the Gauss process ζ with law g_h. The bilinear form $b(z,z')$ on U associated with the quadratic form $\|z\|^2 - \|h^T \ell z\|^2$ on U is:

$$b(z,z') = z \cdot z' - h^T \ell z \cdot h^T \ell z'. \quad (19)$$

Hence the covariance of ζ is the following function defined on $U \times U$:

$$K(u,v) = b(\delta_u, \delta_v)$$
$$= C(u,v) - \frac{1}{\beta^2}\partial_2 C(u,t)\partial_2 C(v,t) - \{\frac{1}{C(t,t)} + \frac{1}{\beta^2}[\frac{\partial_1 C(t,t)}{C(t,t)}]^2\}C(u,t)C(v,t)$$
$$+ \frac{\partial_1 C(t,t)}{\beta^2 C(t,t)}[\partial_2 C(u,t)C(v,t) + C(u,t)\partial_2 C(v,t)]. \tag{20}$$

These results are summarized by the folllowing theorem:

Theorem 7. O.W. Problem and Related Slepian Model *Let us consider the O.W. problem described in 6.1.*

a) A_ε *denotes the event of upcrossing by* X *of the oblique window with level* a *and slope* γ *on the intervall* $(t, t+\varepsilon)$. *As* $\varepsilon \to 0$, *the conditional probability* $P_\varepsilon = P \, if \, A_\varepsilon$ *converges narrowly to* $P_q = h^*(q) = P$ *with* $(law(h) = q)$, *where* h *is defined by (2) , and where* $q = r/\|r\|$, r *defined by (8).*
b) Denoting by $C(s,t)$ *the covariance of* X, *the process* $\eta(s)$ *with* C^1*-trajectories represented by* P_q *can be written:*

$$\eta(s) = a\{\frac{C(s,t)}{C(t,t)} - \frac{1}{\beta^2}\frac{\partial_1 C(t,t)}{C(t,t)}[\partial_2 C(s,t) - \frac{\partial_1 C(t,t)}{C(t,t)}C(s,t)]\}$$
$$+ Y\frac{1}{\beta^2}[\partial_2 C(s,t) - \frac{\partial_1 C(t,t)}{C(t,t)}C(s,t)] + \zeta(s), \tag{21}$$

where Y *denotes a random variable with law* $(y-\gamma)_+ p(a,y)dy$, *and* ζ *denotes a centered Gaussian process, independant of* Y, *with covariance* $K(u,v)$ *defined by (20).*

6.3 Example, and an Application

Example. In the stationary case, with $J = \mathbb{R}$, $t = 0$, $\gamma = 0$, $C(u,v) = R(u-v)$, since $E(X(t)^2 = \lambda_0$ and $\beta^2 = \lambda_2$, the second spectral moment, the previous formula gives the classical Slepian model:

$$K(u,v) = R(u-v) - \frac{1}{\lambda_2}\dot{R}(u)\dot{R}(v) - \frac{1}{\lambda_0}R(u)R(v) \tag{22}$$

$$\eta(s) = a\frac{R(s)}{\lambda_0} - Y\frac{\dot{R}(s)}{\lambda_2} + \zeta(s) \tag{23}$$

where Y has a Rayleigh law with parameter λ_2 and $\zeta(s)$ is a centered nonstationary Gaussian process with covariance defined by (22).

The Random Decrement Method. As an illustration, we show below how an harmonic oscillator can be identified observing its stationary motion when driven by a white noise. The dynamical equation is:

$$\ddot{X}_t + 2\alpha\omega_0 \dot{X}_t + \omega_0^2 X_t = \sigma^2 \dot{B}_t. \tag{24}$$

where B_t is a Brownian motion ($\dot{B}_t$ is a white noise).

The problem is to identify the characteristic eigenvalue ω_0, and the damping ratio α.

In this case, the stationary solution (ξ_t) is a stationary ergodic process with correlation function $R(t) = h * \check{h}$ where h is the impulse response of the oscillator, and $\check{h}(t) = h(-t)$:

$$h(t) = \frac{1}{\omega_0\sqrt{1-\alpha^2}} e^{-\alpha\omega_0 t} sin(\omega_0\sqrt{1-\alpha^2}t) 1_{(t>0)}. \tag{25}$$

The Slepian model associated with upcrossings of level 0 is

$$\eta_t = -\frac{Y}{\lambda_2}\dot{R}(t) + \zeta(t) \tag{26}$$

where Y is a random variable with Rayleigh distribution $Rayl(\lambda_2)$ and where the centered Gaussian process ζ is independant of Y. Hence:

$$E(\eta_t) = -\dot{R}(t).$$

An easy computation shows that this mean value is, up to a multiplicative constant, the impulse function $h(t)$ of the oscillator. In fact, by ergodicity, only one trajectory has to be observed in order to estimate $E(\eta_t)$. This is the principle of an important method for the identification of huge structures in civil engineering, aeronautics, $\cdots$ This is the so called random decrement method [3].

7 Terminal Remarks

7.1 Local Times

Let X be a continuous Gauss process defined on some compact intervall T of the line with values in some Euclidean space Y with orthonormal basis $e_1, \cdots, e_n$; X is represented by a Gauss space $(H \supset U \cdots \Omega, P)$ where $\Omega = C(T, Y)$ is in duality with

$$U = span\{\delta_s \otimes e_j; s \in T, j = 1, \cdots, n\}.$$

Denoting by $|B(a,\varepsilon)|$ the volume of the ball $B(a,\varepsilon)$ of Y with center a and radius ε, the following random variable is well defined for $\varepsilon > 0$

$$L_{a,\varepsilon}(X) = (B(a,\varepsilon))^{-1} \int_T 1_{B(a,\varepsilon)}(X(t))dt. \tag{1}$$

Let us denote by $\gamma_t = \gamma_t(y)dy$ the law of $X(t)$ and by $\delta_a(X(t))$ the pullback of $\gamma_t(a)\delta_a$ by the regular homomorphism h_t

$$h_t : \Omega \ni \omega \to \omega(t) \in Y. \tag{2}$$

Since these homomorphisms depend continuously on t, the lemma 6 shows that the following positive measure on Ω

$$L_a(X) = \int_T \delta_a(X(t))dt \tag{3}$$

is the narrow limit as $\varepsilon \to 0$ of the measures $L_{a,\varepsilon}(X)P$. Therefore (3) defines the local time in a weak sense. Further arguments would be necessary in order to study the regularity of $L_a(X)$. For example, an argument of Ito stochastic calculus shows that for $X = BM(1)$, $L_a(X)$ has a density.

More generally, the examples given below concerning upcrossing rates show that generalized random variables defined by formulas like (3) give directly a "weak" definition to difficult probabilistic concepts, and new approaches of some questions concerning statistics on trajectories.

7.2 Rice Formula

Let us consider a Gaussian stationary process X defined on the line, with C^1 trajectories and non degenerated covariance. The usual starting point for the study of upcrossings by X of the level a during the time interval [0,1] is the following formula first written by Rice:

$$(N_+^a)(\omega) = \lim_{\varepsilon \downarrow 0} \frac{1}{2\varepsilon} \int_0^1 1_{[a-\varepsilon, a+\varepsilon]}(\omega(t))(\dot{\omega}(t))_+ dt. \tag{4}$$

(See [11], [21]).
Let us give a simple proof of (4) as an equality of positive measures. Modelling X by a Gauss space $\Omega = C^1(T)$ and considering the regular homomorphisms h_t

$$\Omega \ni \omega \to (\omega(t), \dot{\omega}(t)) \in \mathbb{R}^2 \tag{5}$$

the right hand side of (4) can be written

$$N_T^{a,\varepsilon}(\omega) = \int_0^1 h_t^*(q_\varepsilon)dt \tag{6}$$

where $q_\varepsilon = (2\varepsilon)^{-1} 1_{(a-\varepsilon,a+\varepsilon)}(x) \otimes y_+ dp(x,y)$ where $dp(x,y) = p(x,y)dxdy$ is the joint law of $\omega(t)$ and $\dot{\omega}(t)$. As $\varepsilon \to 0$, q_ε converges narrowly to

$$q = \delta_a(x) \otimes y_+ p(a,y)dy \tag{7}$$

Hence, lemma 3 shows that the measures $(N_T^{a,\varepsilon}P)$ narrowly converge to the positive measure

$$N_T^a = \int_0^1 h_t^*(q)dt. \tag{8}$$

The formula can be written in the following heuristic way:

$$N_T^a = \int_0^1 \delta(\omega(t))(\dot{\omega}(t))_+ dt. \tag{9}$$

The total mass of this measure can be interpreted as the mean upcrossing rate of level a and computed as follows:

$$\begin{aligned} \|N_T^a\| = \|h_t^*(q)\| = \|q\| &= \int_0^\infty yp(a,y)dy \\ &= \frac{1}{2\pi}\sqrt{\frac{\lambda_2}{\lambda_0}} exp(-\frac{a^2}{2\lambda_0}) \end{aligned} \tag{10}$$

Extension to Fields. Let us first describe the Gauss space $(H \supset U \ldots \Omega\, , P)$ modeling a stationary and centered scalar Gaussian field $\xi = (\xi(t), t \in \mathbb{R}^d)$ with C^1 trajectories.
A structure of separable Frechet space is defined on $\Omega = C^1(\mathbb{R}^d)$ by the topology of uniform convergence of ω and $\nabla\omega$ on all compact subsets of $\mathbb{R}^d$.
The duality between Ω and $U = span\{(\delta_s, \partial_j \delta_t)\ ;\ j = 1, \ldots, d\ ;\ t, s \in \mathbb{R}\}$ is defined in the following way

$$< \omega, (\delta_s, \partial_j \delta_t) >= \omega(s) + \partial_j \omega(t) \tag{11}$$

$(\partial_j \omega(t) = \frac{\partial}{\partial t_j}\omega(t))$.
The law of ξ is a Gauss measure γ on Ω invariant by all shifts $\omega \to \tau_h \omega = \omega(\cdot + h)$.
This measure is characterized by the covariance $R(s-t) = E[\xi_s \xi_t]$, or equivalently by the spectral measure m :

$$R(t) = \int e^{it\cdot u} dm(u). \tag{12}$$

Let us introduce the standard notations $\lambda_0 = \int dm(u) = R(0)$ and the $d \times d$ matrix $\Lambda = (\Lambda_{j,k})$ with coefficients

$$\Lambda_{j,k} = \int u_j u_k m(du) = -\partial_{jk}^2 R(0). \tag{13}$$

The covariance scalar product on U satisfies the following relations :

$$\begin{aligned} &< \delta_s, \delta_t > &&= R(s-t) \\ &< \partial_j \delta_s, \delta_t > &&= E[\partial_j \xi_s \cdot \xi_t] \quad = \frac{\partial}{\partial s_j} R[s-t] \\ &< \partial_j \delta_s, \partial_k \delta_t > &&= E[\delta_j \xi_s \cdot \partial_k \xi_t] = -\partial_{jk}^2 R[s-t]. \end{aligned} \tag{14}$$

Let us now have a glance to problems of level crossing for fields .

Let T be a bounded open set in $\mathbb{R}^d$ and, for any a in $\mathbb{R}$, consider the level manifold

$$A_T^a(\omega) = \{t \in T : \omega(t) = a\} \quad \omega \in \Omega.$$

Let $H_T = H(A_T^a)$ be the $(d-1)$-dimensionnal Hausdorff measure of these sets. When $d = 1$, A_T^a is just a set of isolated points, and Hausdorff measure is then counting measure. By geometrical arguments that can be found in ([1]) and ([34]), we can write the following :

$$H_T = \int_T \int_{\mathbb{R}^d} \delta_a[\xi(t)]\|\nabla\xi(t)\|dt \tag{15}$$

where $\| \ \|$ is the natural Euclidian norm in $\mathbb{R}^d$. Therefore, in a weak sense, we first consider the following measure on $\mathbb{R}^{d+1}$: $q = r/\|r\|$, with

$$r(dx, dy) = \delta_a(dx) \otimes \|y\|p(a, y)dy. \tag{16}$$

We associate to q the measure $\mu_t^+ = h_t^*(q)$, where h_t is defined as in the preceding section, and then define

$$H_T = \int_T h_t^*(q)dt. \tag{17}$$

Corresponding to upcrossing problems of the case $d = 1$, we can consider conditioning by the existence of a crossing of level a at index t with $\nabla\omega(t)$ is some half-space K of $\mathbb{R}^d$. Then, we shall use :

$$r_t^K = \delta_a(dx) \otimes \|y\|1_K(y)p(a, y)dy. \tag{18}$$

References

1. R. J. Adler: The Geometry of Random Fields. John Wiley and Sons, (1981)
2. A. Badrikian, S. Chevet: Measures cylindriques, Espaces de Wiener et Fonctions Aléatoires Gaussiennes. Lectures Notes in Mathematics 379. Springer,(1974)
3. P. Bernard: Identification de Grandes Structures : une remarque sur la méthode du décrément aléatoire. Jal. de Méca. Théo. et Appl. **7** (3) (1988) 1-12
4. J.L. Besson: Sur les Processus Ponctuels des Passages par un Niveau d'une fonction Aléatoire à Trajectoires Absolument Continues. Thèse de Doctorat. Université Claude Bernard (Lyon) (1984)
5. M.De La Pradelle - D.Feyel: Sur la formule des co-aires de Federer. Exposé au séminaire de théorie du potentiel. Université de Paris VI, (1990)
6. M.D. Donsker, J. L. Lions: Fréchet Volterra Variational Equations, Boundary Value Problems and Function Space Integrals. Acta Mathematica. **107** (1962) 147-228
7. Federer: Geometric Measure Theory. Springer (1969)
8. X. Fernique: Processus linéaires. Processus généralisés. Annales de l'Institut Fourier. **XVII** (1) (1967) 1-92
9. X. Fernique: Régularité des trajectoires des fonctions aléatoires gaussiennes. Lectures Notes in Math 480. Springer (1975) - Ecole d'été de Probabilités de St-Flour (1974) 1-96

10. L. Gross: Potential Theory on Hilbert Space. Jal. of Functional Analysis. **1** (1967) 123-181
11. M. Kac: On the Average Number of Real Roots of a Random Algebraic Equation. Proc. Amer. Math. Soc. **49** 314-320
12. M. Kac, D. Slepian: Large Excursions of Gaussian Processes. Ann. Math. Stat. **30** (1959) 1215-1228
13. P. Kopff: Using Vibrational Responses to Wind for Modal Annalysis of Huge Structures such as Natural Convection Cooling Towers. Proceedings 3rd IMAC (International Modal Analysis Conference), Orlando, Fev. 1985. Union College Schenectody, NY12308
14. M. Krée: CRAS **279** Série A (1974) 157-160
15. M. Krée: Propriété de Trace en Dimension Infinie d'espaces du type Sobolev. Bull. Soc. Math. France. **105** (1977) 141-163
16. P. Krée: CRAS **278** Série A (1974) 753-755 and Lecture Notes in Math. **410** (1974) and **473** (1975)
17. P. Krée: Introduction à la Théorie des Distributions en Dimension Infinie. Bull. Soc. Math. France. Mémoire 46 (1976) 143-162
18. P. Krée: Régularité C^∞ des lois conditionnelles par rapport à certaines variables aléatoires. CRAS. **296** Série 1 (31 Janv. 83) 223-225
19. P. Krée: Dimension Free Stochastic Calculus in the Distribution Sense. Stochastic Analysis, Path Integration and Dynamics. Edited by D. Elworthy and J.C. Zambrini.Pitman Research Notes in Mathematics Series **200** (1989)
20. P. Krée: Les structures Fondamentales de l'Analyse Stochastique. Quelques méthodes fondamentales de l'Analyse Stochastique. C.R. Acad. Sci. Paris, 308 série I (1989) 111-114 and 155-158
21. P. Krée, C. Soize: Mathematics of Random Phenomena. D. Reidel Publishing Compagny (1986)
22. H. H. Kuo: Donsker's delta function as a generalized Brownian functional and its application. Theory and Applications of Random Fields. Ed. Kallianpur. Lecture Notes in Control and Information Sciences 49. Springer (1983) 167-178
23. B. Lascar: Propriétés locales d'espaces de Sobolev en dimension infinie. CRAS (1976) and Communications in Part. Diff. Eq. **I** Ch.6 (1976) 561-584 and Sem. E.D.P.∞. of Poincaré Institute (1974-1975)
24. B. Leadbetter, G. Lindgren, H. Rootzen: Extremes and Related Properties of Random Sequences and Processes Springer Verlag (1983)
25. J. Neveu: Processus Aléatoires Gaussiens. Les Presses de l'Université de Montréal (1968)
26. J. Neveu: Processus Ponctuels. Ecole d'été de Probabilité de Saint-Flour (1976) Lecture Notes in Math. 598. Ed. P.L. Hennequin (1977)
27. P.Paclet: Théorie de la capacité en dimension quelconque. Séminaire EDP en dimension infinie (1977-1978). Institut poincaré, et CRAS série A **288** 981 - 983 et **289** 337 - 340 (1979)
28. M.M. Rao: Paradoxes in Conditional Probability. Journal of Multivariate Analysis. **27** 434-446 (1988)
29. S. O. Rice: Mathematical Analysis of Random Noïse. Bell. System Techn. Jal. **24** (1945) 46-156
30. L. Schwartz: Radon Measures on Topological Vector Spaces. Tata Institute (1973)
31. D. Stroock: The Malliavin Calculus, a Functional Analytic Approach. Jal. Functal Anal. **44** (1981) 212-257

32. D. Stroock: Rewiew of Bell's book on Malliavin's Calculus. American scientist. **76** (4) (1988) 410-440 (1988)
33. S. Watanabe: Malliavin's Calculus in Terms of Generalized Wiener Functionals. Theory and Applications of Random Fields. Lecture Notes in Control and Information Sciences. Springer 49 (1983).
34. M. Wschebor: Surfaces Aléatoires. Lecture Notes in Math. **1147** Springer (1985)

Adaptive Sub-Optimal Parametric Control for Non-Linear Stochastic Systems. Application to Semi-Active Isolators

S. Bellizzi and R. Bouc

Abstract

The main objective of this paper is to study adaptive control for systems in which the control acts on some multiplicative parameters of the state vector. The example of a semi-active suspension with dry-friction is considered. The force in the damper is generated by modulating its orifice areas for fluid flow, which gives rise to variations in the damping coefficient. The dynamics of the controlled system is expected to be close to the behaviour of a "reference" (or "target") linear system, having some prescribed properties, such as those ensuring satisfactory dynamic comfort. The parameters in the feedback law are continuously adapted, so that the reference linear model will constitute the linearized system corresponding to the controlled non-linear one, as defined in the framework of the "true" stochastic linearization method, [5]. It should be noted that the exact form of the nonlinear terms arising in the mathematical description are not required for adaptation. In practical applications, only measured response processes will be used.

1 Introduction

The new concept of "semi-active" suspensions for road vehicles, or for shock and vibration isolation systems in general, gives rise to control problems in which the control acts on some multiplicative parameter of the system. The force in the damper is generated by modulating its orifice areas for fluid flow [1], which leads to variations in the damping coefficient. Since the isolator is essentially a passive device, virtually no external power is required. Semi-active suspensions are a compromise between active and passive isolation systems. It is shown in [2], [3], that a semi-active isolator can perform nearly as efficiently as a fully active isolator and far more efficiently than passive ones. The case of a stochastic excitation was considered in [4], where similar conclusions were reached.

To our knowledge, the stochastic "optimal" parametric control problem, in connexion with semi-active dampers, was first solved in [4], via the numerical resolution of the Hamilton-Jacobi-Bellman equation [5]. It can be easily seen that, even in case where the system is linear, the feedback control problem is non-linear and no exact solution exists. For a cost function associated with a mean criterion of dynamic comfort, it was shown that the optimal law acts mainly in a domain where the sign of the product of the relative displacement and the relative velocity is negative. The existence of a class of sub-optimal control laws was also suggested by this result. This class contains the control scheme studied in [2], [3]. In order to avoid the discontinuity of the force resulting from these laws, another class, depending on a finite number of parameters, was also derived. Furthermore, this class is easily implementable in practical applications.

The main objective of this paper is to study adaptive parametric control for non-linear stochastic systems. A control law must be able to provide adequate control, despite any local non-stationarity in the forcing stochastic input, any variations in, or uncertainty in knowledge of true parameters values, etc. To achieve this, the control problem will be set in an unusual way. The dynamics of the controlled system will be expected to be close to the behaviour of a "reference" (or "target") linear system, having some prescribed properties, such as those ensuring satisfactory dynamic comfort. The parameters in the feedback law will be continuously adapted, so that the the reference model will be the linearized system of the controlled non-linear one, as defined in the framework of the "true" stochastic linearization method, [6]. It is worth mentionning, that the exact form of the nonlinear terms in the mathematical description will not be required for adaptation, only measured response processes will be used.

These ideas are presented in a general context in section 2. An adaptive, semi-active suspension in the presence of dry friction is considered in section 3. A simulated numerical example is given in section 5.

2 General Background

2.1 Stochastic Linearization for Non-Linear Stochastic Systems

The stochastic linearization method was first introduced for non-linear stochastic control and non-linear random vibrations and further developed in a wide range of non-linear mechanical engineering problems. Here we shall briefly describe the method and the new results published recently in [6].

Let us consider the real non-linear vector stochastic system

$$\dot{y}_t = X(y_t) + n_t \tag{1}$$

where n(t) is a stationary vector process with zero mean. We assume that a stationary probability ergodic measure exists for (1) and, for simplicity, that $E(y(t)) = 0$. We seek an equivalent linear system having the same dimension (see remark below),

$$\dot{x}_t = A^* x_t + n_t \tag{2}$$

with the constant matrix A^* given by

$$A^* = \text{Argmin}_A d(A), \tag{3}$$

where d(A) is defined by

$$d(A) = \text{trace} E[(X(y) - Ay)(X(y) - Ay)^T] \tag{4}$$

d(A) can be viewed as the distance between the two systems. From (3) we obtain

$$A^* E(yy^T) = E(X(y)y^T) \tag{5}$$

As shown in [6], the linear system (2) and the non-linear system (1) have identical first and second moment solutions if n(t) is the standard white noise vector process and *if and only if the expectations in (4) (5) are calculated with the invariant measure of the system (1)*. Unfortunately, the invariant measure of (1) is generally unknown. Thus, the matrix A^* is usually approximated using gaussian law to calculate (5), as suggested by (2) for gaussian input. Nevertheless, the explicit form of the non-linear vector function X(y) must be known for computation.

In engineering estimation or control problems the exact form of X(y) is generally unknown whereas the joint input-output (n,y)-process is often observable. A very interesting result for applications is also established in [6]. The "true" linearized matrix A^* can be obtained from the trajectories of the observed input output processes as the almost sure limit of the following statistical estimators:

$$A_T = \text{Argmin}_A \left[\left(\frac{1}{T}\right) \int_0^T \|\dot{y}_t - n_t - Ay_t\|^2 dt \right] \tag{6}$$

$$A^* = rma.s. \lim_{T \to \infty} A_T \tag{7}$$

Remark: If the expectation of y(t) is not zero, we replace in (4) Ay by $Ay+a$ and we look for $(A^*, a^*) = \min d(A, a)$.

2.2 Stochastic Linearization and Non-Linear Parametric Control

Our interest is now focused on a parametric stochastic control problem of the form

$$\dot{y}_t = X(y_t) + [Y(y_t)]u_t + n_t \tag{8}$$

[Y(y)] is a rectangular matrix with possibly non-linear terms. n_t is a stationary stochastic input, ut is the control vector.

Generally the problem is the following: Find a feedback control $u_t = u(y_t)$ which minimize the average of a given cost function among a class U of admissible feedback laws taken values in U.

Here our objective will be a little different: A "reference" or "target" linear system of the form (2) being given (with A^*y possibly replaced by $A^*y + a^*$ according to Remark above), we look for a feedback control $u^* \in U$, such that system (2) is the "best" linearized system of (8) in the sense of the "true" stochastic linearization. More precisely : Assume that for each $u \in U$, a stationary ergodic probability measure exists for system (8). Let A^* *be given*, find $u^* = u^*(y)$ such that

$$u^* = \text{Argmin}_{u \in U} d(u, A^*) \tag{9}$$

$$d(u, A^*) = \text{trace} E[(X(y) + [Y(y)]u(y) - A^*y)$$
$$(X(y) + [Y(y)]u(y) - A^*y)^T] \tag{10}$$

under the constraints

$$E[(X(y) + [Y(y)]u(y) - A^*y)y^T] = 0 \tag{11}$$

$$y = y_u(.) \text{ satisfies (8).} \tag{12}$$

The expectations in (10) (11) must be calculated with the invariant measure generated by (8). However, as above, using trajectortories of the observed (controlled) processes (y_u, n), these expectations will be replaced by their temporal average in practical applications.

2.3 A Class of Sub-Optimal Controls

More simply, we can only ask that the constraint (11) (12) be satisfied. In this case the "target" linear system (2) will be the linearized system of (8) in the sense of the "true" stochastic linearization but not necessarily the "best". This sub-optimal control problem is easier to solve than the previous optimal one.

Let p be the dimension of y, m the dimension of u with $m \leq p$. Assume that U be a closed convex set in $\mathbb{R}^m$ and denote by P_U the projection operator of $\mathbb{R}^m$ onto U. Let $[Y(y)]^\#$ be the left-hand (pseudo)inverse of [Y(y)] assuming that $[Y]^T[Y]$ is non singular. It follows that

$$u(y) = P_U[Y(y)]^\#(A^*y - X(y)) \tag{13}$$

is a possible sub-optimal control law, see (10) (11) (12), because

$$v(y) = [Y(y)]^\#(A^*y - X(y)) \tag{14}$$

is the best v-solution of the instantaneous equation

$$X(y) + [Y(y)]v - A^*y = 0 \tag{15}$$

in a least-squares sense, and often the exact solution. It is worth noting that in the case where $U = \mathbb{R}^m$, and if (14) is the exact solution of (15), then (14) is also the unique solution of the optimal problem (8) (9) (10) (11) .

However, in general, it is clear that equation (11) is not satisfied with the control (13). Nevertheless an explicit dependence in y is proposed. Moreover it is always possible to consider that v(y) in (14) is a member of a (non unique) family $v(y, \theta)$ depending on a finite number of parameters, arrayed as the components of a real q-dimensional vector θ . For example, we can always include in θ some of the uncertain parameters of the pair ([Y(y)], X(y)). By choosing in (8) u(y) of the form,

$$u(y, \theta) = P_U[v(y, \theta)], \tag{16}$$

we then have an explicit feedback control law which depends on q parameters. Next, we look for θ in order to simultaneously solve (8) and (2.11). This procedure can be repeated with any special form $u(y, \theta)$ suggested by (14) depending on a finite number of parameters.

2.4 Adaptive Sub-Optimal Control

In many engineering control problems, a control law must be able to provide adequate control, despite any local non-stationarity in the forcing stochastic input, any variations in, or uncertainty in knowledge of true parameters values, etc. As we shall see later in detail in an example, it is possible to periodic ally update the θ-vector in (16) from the observed (n,y) process, using the modified recursive least squares algorithm in order to (11) be continuously "almost" satisfied.

The procedure may be as follows: From result (6), for fixed $\theta = \theta_m$, an on-line identification of the linearized system associated with (8) (16) in the sense of the true stochastic linearization method, is carried out according to the stochastic gradient algorithm [7],[8]

$$A_k(\theta_m) = A_{k-1}(\theta_m) + \rho_k[\dot{y}_k - n_k - A_{k-1} = (\theta_m)y_k]y_k^T \tag{17}$$

where ρ_k is a well-chosen sequence of scalar gain. At the same time, the vector θ_m is periodically recursively updated in such a way that the difference between the sequence of estimate $A_k(\theta_m)$ and the desired matrix-value A^* is minimized.

3 Application to Semi-active Suspension with Dry Friction

3.1 Mathematical Formulation

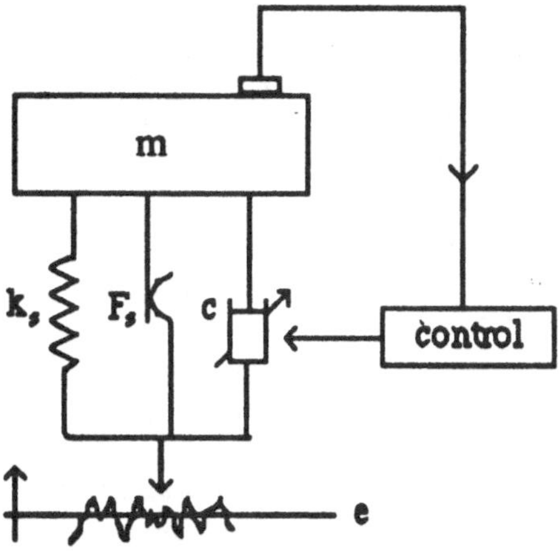

Fig. 1. Mass-sprin-damper system with dry friction

Let us consider the one-degree-of-freedom "quarter vehicle model", as shown in Figure 1. The suspension is controllable in the sense that the characteristics of the shock-absorber can be modified by modulating its orifice areas for fluid flow. The dynamic equation is

$$m\ddot{x}_t + c_t f(\dot{x}_t) + k_s g(x_t) + z_t = -m\ddot{e}_t \tag{18}$$

where x denotes the relative displacement of the total mass m, e denotes the stochastic input due to the road profile. $k_s g(x_t)$, $k_s > 0$, is a non-linear restoring force due to the spring and, possibly, to an elastic stop. It will be assumed that g(x) is a smooth, strictly increasing monotone function, such that g(0) = 0. The total dry friction force due to friction in the shock-absorber and in the spring is denoted by z. It can be modelled by the following differential hysteresis model, [9],[10],

$$\dot{z}_t + b[|\dot{x}_t|z_t + \dot{x}_t|z_t|] = (k - k_s)\dot{x}_t \tag{19}$$

where $b > 0$ and $k - ks > 0$. It is easy to see that

$$|z_t| < F_s = \frac{(k - k_s)}{2\beta} \tag{20}$$

and also that z_t evolves as a pure Coulomb friction force,

$$z_t = F_s \operatorname{sgn}\dot{x}_t \tag{21}$$

for "large" displacement x_t, or with β and $k - k_s$ large enough but of the same order of magnitude.

Basically, a shock-absorber consists of a piston moving in a closed cylinder. The two chambers of the cylinder are filled with a liquid (oil) and interconnected by a channel with diameter d . The dynamic force due to the acceleration of the piston is included in the total acceleration term $m\ddot{x}$. The damping force generated is denoted by $c_t f(\dot{x})$. Typically, see [11], a basic model leads to the form $f(\dot{x}) = \dot{x} + a\dot{x}|\dot{x}|$, $a > 0$. In fact, $f(\dot{x})$ may be non symmetrical. We shall assume that $f(\dot{x})$ is a smooth, strictly increasing monotone function, such that $f(0) = 0$. It will be assumed also that the diameter d can be modulated, thus causing variations in the damping coefficient $c_t = c(d)$. This coefficient will be our control term. A shock-absorber being essentially a passive device, we shall look for constrained feedback laws of the form:

$$0 < c_m \le c_t = c(x_t, \dot{x}_t) \le c_M \tag{22}$$

The latter inequality ensures the complete stability of the free oscillations of (18). For this purpose, setting $\ddot{e}t = 0$ and multiplying both sides of (18) with $\dot{x}$ yields

$$\begin{aligned}
&\frac{d}{dt}[m\dot{x}_t^2 + (k - ks)^{-1}z_t^2 + k_s G(x_t)] \\
&= -2c(x_t, \dot{x}_t)f(\dot{x}_t)\dot{x}_t - 2\beta[|\dot{x}_t|z_t z_t + \dot{x}_t z_t|z_t|] \\
&\le -2c_m f(\dot{x}_t)\dot{x}_t \le -2c_m\gamma\dot{x}_t^2, \quad (\gamma > 0)
\end{aligned} \tag{23}$$

This inequality follows from the fact that the second term in the above equality is always non-negative and, by hypothesis, $f(\dot{x})\dot{x} \ge \gamma\dot{x}^2$. $G(x)$ is defined by, $g(x) = d(G(x))/dx$.

In the case where $\ddot{e}_t$ is the gaussian white noise, $\ddot{e}_t = \sigma \dot{w}_t$, $s > 0$, where w_t is the standard brownian motion, we obtain

$$\begin{aligned} &\frac{d}{dt}E[m(\dot{x}_t^2) + (k - k_s)^{-1}(z_t^2) + k_s G(x_t)] \\ &\leq -2c_m E[f(\dot{x}_t)\dot{x}_t] + m\sigma^2 \end{aligned} \tag{24}$$

from which we can easily deduce, using (20) and Gronwall's Lemma, that

$$E(\dot{x}_t^2) + \alpha E(G(x_t)) \leq C, \ \forall t \tag{25}$$

for some positive constants α, C. Using the same argument as in [12, 5], it can be shown that there exists, for any $c_t = c(x_t, \dot{x}_t)$ satisfying (22), a unique invariant probability measure generated by (18).

3.2 Optimal Ergodic Control. [4],[5]

We look for a feedback control of the shock-absorber such that the square of the absolute acceleration of the mass is minimized in mean square. More precisely, we look for a control $c_t = c(x_t, \dot{x}_t) \geq c_m$ such as the ergodic cost

$$J(c) = \lim_{T \to \infty} E\left[\frac{1}{T}(\ddot{x}_t + \ddot{e}_t)^2 dt\right] \tag{26}$$

is minimized and (18) is satisfied.

This problem is solved in [4], via the numerical resolution of the Hamilton-Jacobi- Bellman equation, for the two dimensional simplified model : $f(\dot{x}) = \dot{x}$, $g(x) = x$, $z = F_s \operatorname{sgn}\dot{x}$, $\ddot{e}_t = \sigma \dot{w}_t$ (white noise), $c_m = 0$. We refer the reader to [5] (this symposium) for more details. *It can be observed that the optimal feedback is principally non zero in the domain where $x\dot{x} < 0$.*

3.3 The Reference Linear System Associated with (18)(19)

With the constant damping coefficient $c_t = c$, the linear system associated with (18) (19) in the sense of the true stochastic linearization is of the form (see [13], assuming symmetrical non-linearities and zero mean input),

$$\ddot{x} + x\dot{x} + \beta x + z = -\ddot{e} \tag{27}$$

$$\dot{z} + az = \nu \dot{x} \tag{28}$$

where ξ, β, α, ν, are positive constant parameters.

Systems (27) (28) and (18) (19) can be put into the form (2) and (8) respectively, by introducing the vector $y = (y_1, y_2, y_3)^T$, $y_1 = x$, $y_2 = \dot{x}$, $y_3 = z$, and

$$c_t = c_m + u_t \tag{29}$$

It follows that:

$$A^* = \begin{pmatrix} 0 & 1 & 0 \\ -\beta & -\xi & -1 \\ 0 & \nu & -\alpha \end{pmatrix} \tag{30}$$

$$X(y) = [y_2, -m^{-1}(c_m f(y_2) + k_s g(y_1) + y_3),$$
$$(k - k_s)y_2 - \beta(y_2|y_3| + |y_2|y_3)]$$

$$Y(y) = [0, -m^{-1} f(y_2), 0]^T \tag{31}$$

$$n = [0, -\ddot{e}, 0]^T \tag{32}$$

$$u \in U \leftrightarrow u_t = u(x_t, \dot{x}_t) \geq 0 \tag{33}$$

It is now proposed to develop the general ideas outlined in section 2.

The parameters $\xi = \xi^*$, $\beta = \beta^*$, $\alpha = \alpha^*$, $\nu = \nu^*$ are first chosen in such a way that corresponding linear system has some prescribed properties (performances), such as those ensuring satisfactory dynamic comfort. Note that both the stiffness and the damping are fixed. Secondly, an adaptive feedback law is sough such that the true non-linear system is continuously close to the reference linear system in the sense of the true stochastic linearization.

As we have already mentioned, recursive on-line algorithms will be carried out from the observed (easured) trajectories of the response processes. Unfortunately, the dry friction force is difficult to measure and models like (19) (21) are often uncertain. If our objective is to achieve a practical control (and not simply a numerical simulation of the dynamics), a procedure which does not require the observation of the z-process must be used.

A hysteresis force such as z in (18) contributes to the damping force as well as to the restoring force. The dry friction force will therefore be assumed in (18) to be a given functional of the relative displacement

$$z = z(x(.)) \tag{34}$$

the exact mathematical description of which has not been established. It follows that the corresponding reference linear system will be taken, in place of (27) (28), in the simplified form

$$\ddot{x} + \xi^* \dot{x} + \beta^* x = -\ddot{e} \tag{35}$$

where the total damping is governed by the parameter $\tau^* = \xi^*/2(\beta^*)^{1/2}$, whereas the square of the natural undamped frequency is governed by the parameter β^*.

3.4 Sub-Optimal Control of the Shock-Absorber

Systems (18) and (35) can be written as two dimensional functional-differential and differential systems (8) (2), respectively. Assume for a time that the simplified hypothesis of the paragraph 3.2 be satisfied. The corresponding equation (14) reduces to

$$\begin{aligned} v(y) &= v(x,\dot{x}) \\ &= -m^{-1}c_m + \xi^* - [(m^{-1}k_s - \beta^*)x + m^{-1}F_s \operatorname{sgn}\dot{x}]/\dot{x} \end{aligned} \tag{36}$$

which may be parametrized according to

$$v(x,\dot{x},\theta) = \theta_0 + (\theta_1 x \operatorname{sgn}\dot{x} + \theta_2)|\dot{x}|^{-1} \tag{37}$$

Assuming for simplicity that $c_M = \infty$ in (22), the sub-optimal control (16) is found to be

$$u = P_U v, \quad u(x,\dot{x},\theta) = \text{“positive part of” } v(x,\dot{x},\theta) \tag{38}$$

(38) may be written

$$u(x,\dot{x},\theta) = \big(v(x,\dot{x},\theta) + |v(x,\dot{x},\theta)|\big)/2 \tag{39}$$

showing the force term to have a discontinuous behaviour [2],[3]:

$$u(x,\dot{x},\theta)\dot{x} = \frac{1}{2}\big[\theta_0\dot{x} + \theta_1 x + \theta_2 \operatorname{sgn}\dot{x} \tag{40}$$

$$+|\theta_0|\dot{x}| + \theta_1 x \operatorname{sgn}\dot{x} + \theta_2| \operatorname{sgn}\dot{x}\big] \tag{41}$$

To avoid the discontinuity for $\dot{x} = 0$, (37) may be simplified as folllows

$$v(x,\dot{x},\theta) = \theta_0 + \theta_1 x \operatorname{sgn}\dot{x} \tag{42}$$

which leads to the following continuous expression for $\dot{x} = 0$:

$$u(x,\dot{x},\theta)\dot{x} = \frac{1}{2}\big(\theta_0\dot{x} + \theta_1 x|\dot{x}| + |\theta_0 + \theta_1 x \operatorname{sgn}\dot{x}|\dot{x}\big) \tag{43}$$

Comments: The force is non zero only when $(\theta_0 + \theta_1 x \operatorname{sgn}\dot{x}) > 0$. Hence, if the parameters are chosen such that $\theta_0 \leq 0$ and $\theta_1 < 0$, the control will act principally when $x\dot{x} < 0$ and never in the domain where $x\dot{x} > 0$, minimizing, as in paragraph 3.2, the absolute acceleration of the mass $\ddot{a} = \ddot{x} + \ddot{e}$. Furthermore, assuming for simplicity that $f(\dot{x}) = \dot{x}$, $g(x) = x$, $z = 0$, ${}_cm = 0$, (18) becomes with (43),

$$m\ddot{x}_t + \frac{1}{2}\big(\theta_0 + |\theta_0 + \theta_1 x_t \operatorname{sgn}\dot{x}_t|\big)\dot{x}_t \tag{44}$$

$$+ (k_s + \theta_1|\dot{x}_t|/2)x_t = -m\ddot{e}_t \tag{45}$$

showing that the choice $\theta_1 < 0$, mainly contributes to the decrease in the stiffness and thus in the resonant frequency, while θ_0 mainly affects the damping coefficient (see Figure 4 in section 5).

To end this paragraph, it can be observed that (42) is a particular case of the more general law

$$v(x,\dot{x},\theta) = \theta_0 + \theta_1 x \ \mathrm{sgn}\dot{x} + \theta_2 \ddot{a} \ \mathrm{sgn}\dot{x} \tag{46}$$

$$u(x,\dot{x},\theta) = \text{"positive part of" } v(x,\dot{x},\theta) \tag{47}$$

where all the easily measurable response processes are taken in account. The choice $\theta_2 > 0$, contributes also to the decrease in the resonant frequency (see Figure 5 in section 5).

4 Adaptive Control Based on the Tracking of the Reference Linear System

Let the structure of fig (1) be approximately described by equation (18), in the sense that the form of the non-linearities and the description of the friction term are not exactly known. Recall that $c_t = c_m + u_t$. Let the processes $\ddot{a}$, $\dot{x}$, x, be observable and consider a feedback control law of the form (46) (47). Let (35) be the reference linear system.

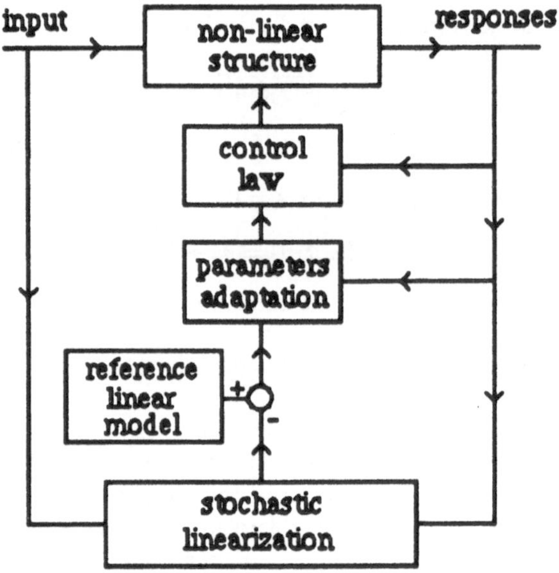

Fig. 2. Principle of the adaptive control

The adaptative procedure is summerazed in Figure 2. An on-line stochastic linearization of the structure, based on the measured responses, was carried out. The sequence of estimates $\hat{\xi}_n$, $\hat{\beta}_n$, was compared to the prescribed values ξ^*, β^*, recursively updating the vector θ in the feedback law, in order to minimize the difference.

The sequence of estimates $\hat{\xi}_n$, $\hat{\beta}_n$ is defined by

$$\hat{\xi}_n = \hat{\xi}_{n-1} - \mu_n\big[\dot{x}_n(\ddot{a}_n + \hat{\xi}_{n-1}\dot{x}_n + \hat{\beta}_{n-1}x_n)\big] \tag{48}$$

$$\hat{\beta}_n = \hat{\beta}_{n-1} - \mu_n\big[x_n(\ddot{a}_n + \hat{\xi}_{n-1}\dot{x}_n + \hat{\beta}_{n-1}x_n)\big] \tag{49}$$

where $\ddot{a}_n = \ddot{a}(t_0 + n\Delta t)$, $\dot{x}_n = \dot{x}(t_0 + n\Delta t)$, $x_n = x(t_0 + n\Delta t)$, $n = 1, 2, \ldots$, denote a suitable sampling of the response processes and μ_n a sequence of scalar gains. These recursions can be viewed as a stochastic gradient method, [7] [8], recursively solving the minimum problem

$$(\hat{\xi}, \hat{\beta}) = \lim_{T\to\infty}\big[\min_{(\xi,\beta)} \frac{1}{T}\int_0^T (\ddot{a}_t + \xi\dot{x}_t + \beta x_t)^2 dt \tag{50}$$

associated with the "true" stochastic linearization method (see (6) (7)). The usual choices of the sequence of gains are as follows, [7], [8]:

$$\mu_n = \mu = \text{ constant} \tag{51}$$

$$\mu_n = \mu/|\dot{x}_n|^2 \text{ and } \mu_n = \mu/|x_n|^2 \text{ (normed gain)}$$

$$\mu_n = \mu/r_n$$

$$\text{with } r_{n+1} = r_n + \alpha[|\dot{x}_n|^2 - r_n]$$

$$\text{and } r_{n+1} = r_n + \alpha[|xn|^2 - r_n] \tag{52}$$

(normed gain with exponential fading memory, $1 > \alpha > 0$).

The choice of the constant value μ results from a compromise between a high speed of convergence (μ large), and small residual fluctuations near equilibrium (μ small).

Let us now define the recursions on $\theta = (\theta_0, \theta_1)$, assuming for simplicity $\theta_2 = 0$.

Recall (see paragraph 3.4) that θ_1 acts mainly on the natural frequency (β) whereas θ_0 mainly affects the damping coefficient (ξ). It follows that the recursions, $k = 1, 2, \ldots$,

$$\hat{\theta}_1^k = \hat{\theta}_1^{k-1} + \rho_k\big[\partial_1\hat{\beta}_k(\beta^* - \hat{\beta}_k)\big] \tag{53}$$

$$\hat{\theta}_0^k = \hat{\theta}_0^{k-1} + \rho_k\big[\partial_0\hat{\xi}_k(\xi^* - \hat{\xi}_k)\big] \tag{54}$$

define a sequence of estimates, minimizing the expectations $E(\beta^* - \hat{\beta}_n)^2$ and $E(\xi^* - \hat{\xi}_n)^2$ separately. ρ_k is a sequence of scalar gains defined as μ_n in (51). $\partial_1\hat{\beta}_k$ and $\partial_0\hat{\xi}_k$ denote the gradient of $\hat{\beta}$ and $\hat{\xi}$ with respect to θ_1 and θ_0, respectively. Neglecting the derivative of x and $\dot{x}$ with respect to θ_1 and θ_0 and the product $x_n\dot{x}_n$ (which rapidly reaches zero due to stationarity), we can write, from (48) (49),

$$\partial_0\hat{\xi}_n = \partial_0\hat{\xi}_{n-1} - \mu_n[\dot{x}_n(\partial_0\ddot{a}_n + \dot{x}_n\partial_0\hat{\xi}_{n-1})] \tag{55}$$

$$\partial_1\hat{\beta}_n = \partial 1\hat{\beta}_{n-1} - \mu_n[x_n(\partial_1\ddot{a}_n + x_n\partial_1\hat{\beta}_{n-1})] \tag{56}$$

while the model (18) with (43) leads to

$$\partial_0 \ddot{a}_n = (2m)^{-1}[(1+ \ \mathrm{sgn}(\theta 0 + theta_1 x_n \ \mathrm{sgn}\dot{x}_n)]\dot{x} \tag{57}$$

$$\partial_1 \ddot{a}_n = (2m)^{-1}[(1+ \ \mathrm{sgn}(\theta 0 + theta_1 x_n \ \mathrm{sgn}\dot{x}_n)]|\dot{x}|x \tag{58}$$

Comments: It is worth noting that we have only called the mathematical model for providing analytical expressions for the above approximate derivatives. All the quantities appearing in the above recursions will be measured on the actual mechanical suspension.

The time step chosen on the θ-recursion will be greater than (but a multiple of) the time step on the (ξ, β)-recursions, so that the response process can be established before updating θ.

5 Simulated Illustrative Example

The model (18) (19) was numerically solved in the case where $f(\dot{x}) = \dot{x}$, $g(x) = x$. The input acceleration $\ddot{e}_t$ was a simulated low-passed white noise, with a high intensity level in the frequency range 0-10 Hz. The modulus of the acceleration transmissibility curve, for various sub-optimal controls (37) (46) (47), is plotted in Figures 3, 4, 5. The transmissibility curve was classically defined as the coherent transfer function between $\ddot{e}$ and the absolute acceleration of the mass $\ddot{a}$ (Figure 1), according to the formula

$$T(f) = \frac{S_{\ddot{e}\ddot{a}}(f)}{S_{\ddot{e}}(f)} \tag{59}$$

where $S_{\ddot{e}\ddot{a}}$ and $S_{\ddot{e}}$ denote the cross-spectral density function between $\ddot{e}$ and $\ddot{a}$ and the spectral density function of $\ddot{e}$, respectively. Figure 3 was obtained from (37) (38) with negative θ-values, Figure 4 from (46) (47) with $\theta_0 < 0$, $\theta_1 < 0$, $\theta_2 = 0$, whereas Figure 5 was obtained from (46) (47) with $\theta_0 < 0$, $\theta_1 = 0$ and $\theta_2 > 0$. On each of these figures, the transmissibility curve, corresponding to a *constant damper ensuring the same filtering beyond 4 Hz*, was also plotted (see curves entitled "without control").

It can be concluded from this example that a suitable feedback control of the shock-absorber is able to transform an "undesirable" passive suspension stage (the amplitude and frequency values of the resonance were too high), into an "acceptable" suspension stage (appreciable decrease in the levels of these two parameters). Furthermore, there exists no constant control producing the same performances simultaneously.

To illustrate the efficiency of the adaptive control algorithm (48) to (58), consider an excitation term of the form

$$\ddot{e}_t* = q(t)\ddot{e}_t \tag{60}$$

where q(t) is a deterministic jump-function: $q(t) = 1$ for $t \leq T$, $q(t) = 1.4$ for $t > T$. The results are plotted on the Figures 6 and 7. The control law was (46) (47) with $\theta_2 = 0$. The sequences of gains μ_n and ρ_n were chosen according to

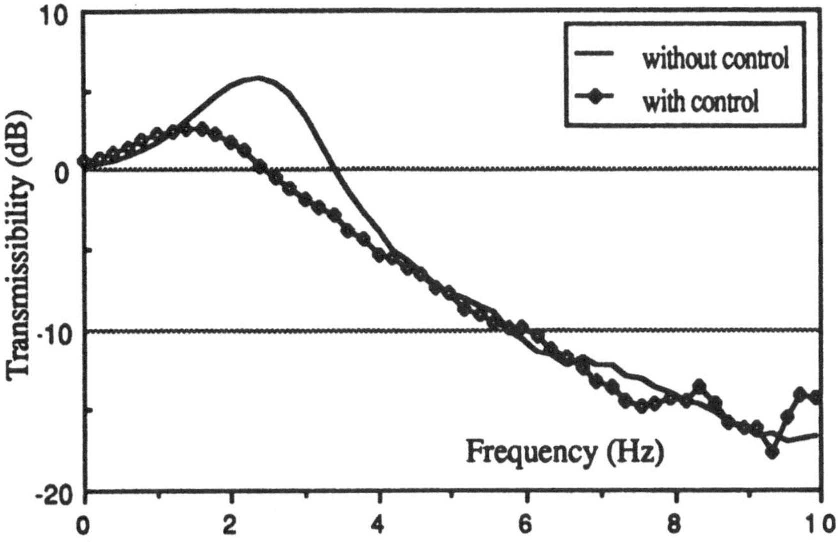

Fig. 3. Control with (37) (38)

the last formula in (51). The constant values μ and ρ were chosen to ensure that the residual fluctuation in the estimates $\hat{\xi}_n$, $\hat{\beta}_n$ would not exceed 10prescribed values ξ^*, β^*.

The curves entitled "without control" were obtained with the constant damper already used in figures 3-5. It was observed that the equivalent stiffness coefficient $\hat{\beta}$ is not affected by the jump in the input acceleration, while the *total* equivalent damping coefficient $\hat{\xi}$ decreases. Due to the high excitation level, the dry friction term (19) evolves as a pure Coulomb friction force (21), so that only the damping is modified. The decrease in $\hat{\xi}$ along with the increase in the input level, clearly indicates the presence of dry friction in the suspension. In the curves entitled "control without adaptation", the parameters θ_0 and θ_1 of the control law were fixed at values corresponding to those of the level $q(t) = 1$, which led to the desired values ξ^*, β^*. Since these parameter values were *not updated*, a decrease in the stiffness coefficient $\hat{\beta}$ and an increase in the damping coefficient $\hat{\xi}$ occurred (see equation (44) with the choice $\theta_1 < 0$ and $\theta_0 < 0$). Finally, it can be seen from the figures, that the jump in the input acceleration has little influence on the time trajectories of the estimates $\hat{\xi}_n$, $\hat{\beta}_n$ when the adaptive control is performed.

6 Concluding Remarks

The adaptive parametric control problem for non-linear stochastic systems is studied here by tracking a prescribed reference linear system. The parameters in the feedback law are continuously adapted so that the reference linear system constitutes the linearized system corresponding to the controlled one, as defined

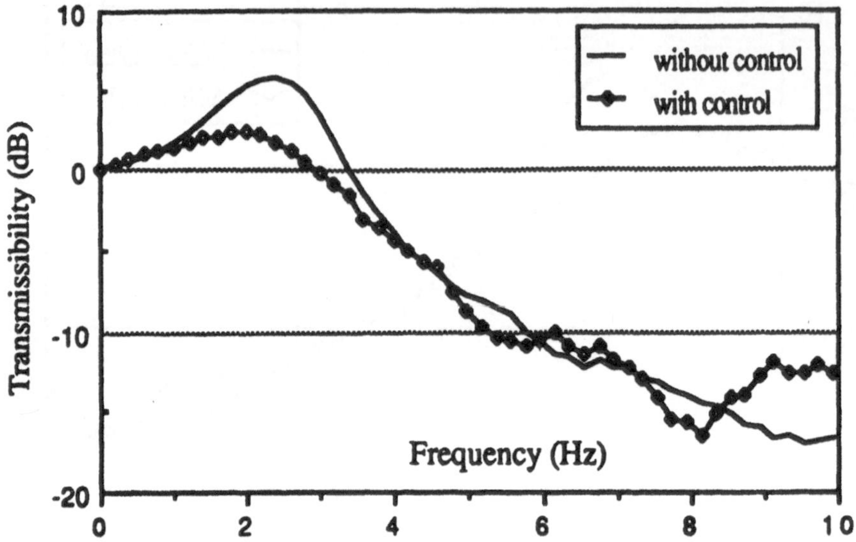

Fig. 4. Control with (46) (47; $\theta_2 = 0$)

in the framework of the stochastic linearization method. The exact mathematical formulation of the non-linearities does not require to be known for adaptation. In future practical applications, only measured responses need to be used. The example of a semi-active suspension with dry friction was studied along these lines.

References

1. M.J. Crosby and D.C. Karnopp "The active damper-a new concept for shock and vibration control" The shock and vibration bulletin, 43 (4), 119-133, 1973.
2. S. Rakheja and S. Sankar "Vibration and shock isolation performance of a semi-active on-off damper" J. of Vibr., Acoustics, Stress, and Reliability in Design, 107, 398-40 3, 1985.
3. J. Alanoly and S. Sankar "Semi-Active Force Generators for Shock Isolation". Journal of sound and vibration 126 (1), 145-156, 1988.
4. S. Bellizzi, R. Bouc, F. Campillo, E. Pardoux "Contrôle optimal Semi-Actif de suspension de véhicules". Lectures Notes in Control and Information Sciences, n. 111. Springer Verlag. 1988.
5. F. Campillo "Optimal ergodic control of non-linear stochastic systems" This Volume.
6. F. Kozin, "The method of statistical linearization for non-lin ear stochastic vibrations". Proceedings of the IUTAM symposium, Innsbruck, Juin 1987, " Non-linear stochastic dynamic engineering systems, Springer-Verlag, 1988.
7. L. Ljung, T. Söderström, "Theory and practice of recursive identification" MIT Press, 1983.
8. A. Benveniste, M. Métivier, P. Priouret "Algorithmes adaptatifs et approximations stochastiques" Masson, Paris, 1987.

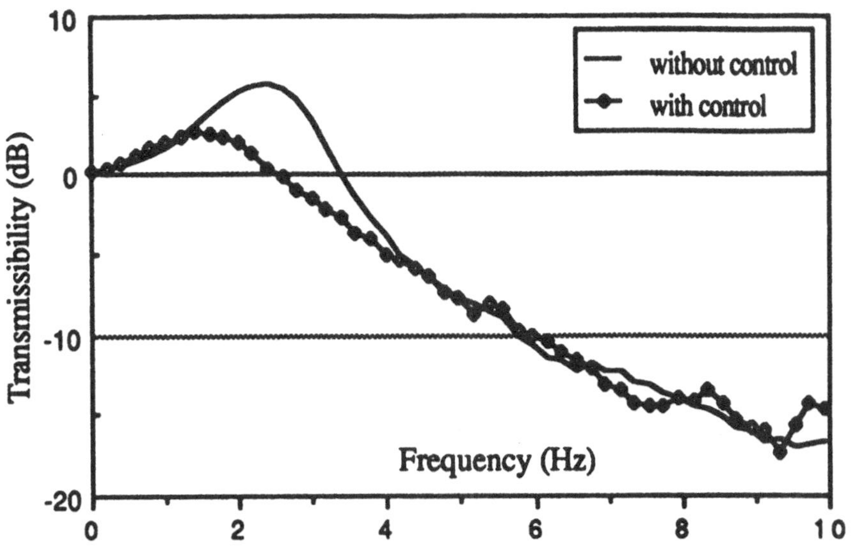

Fig. 5. Control with (46)(47; $\theta_1 = 0$)

9. R. Bouc "Modèle mathématique d'hystérésis". Acustica, Vol. 24, 16-25, 1971. See also "Forced vibration of a mechanical system with hysteresis". Proc. 4ème conf. ICNO, Prague, 1967, (Summary only).
10. S. Bellizzi et R. Bouc "Identification of the hysteresis parameters of a non-linear vehicle suspension under random excitation" Proceedings of the IUTAM symposium, Innsbruck, Juin 1987, "Non-linear stochastic dynamic engineering systems, Springer- Verlag, 1988.
11. P. Hagedorn and J. Wallaschek "On equivalent harmonic and stochastic linearization for non-linear shock-absorbers "Proceedings of the IU TAM symposium", Innsbruck, Juin 1987, "Non-linear stochastic dynamic engineering systems", Springer-Verlag, 1988.
12. F. Campillo, F. Le Gland, E. Pardoux, "Approximation d'un problème de contrôle dégénéré" Actes du colloque "Automatique non-linéaire", CNRS, Nantes, France, Juin 1988.
13. Y.K. Wen, "Equivalent linearization for hysteretic systems under random excitation", J. of Applied Mechanics, Trans. of ASME, 47, n.1, 150-154, (1980).

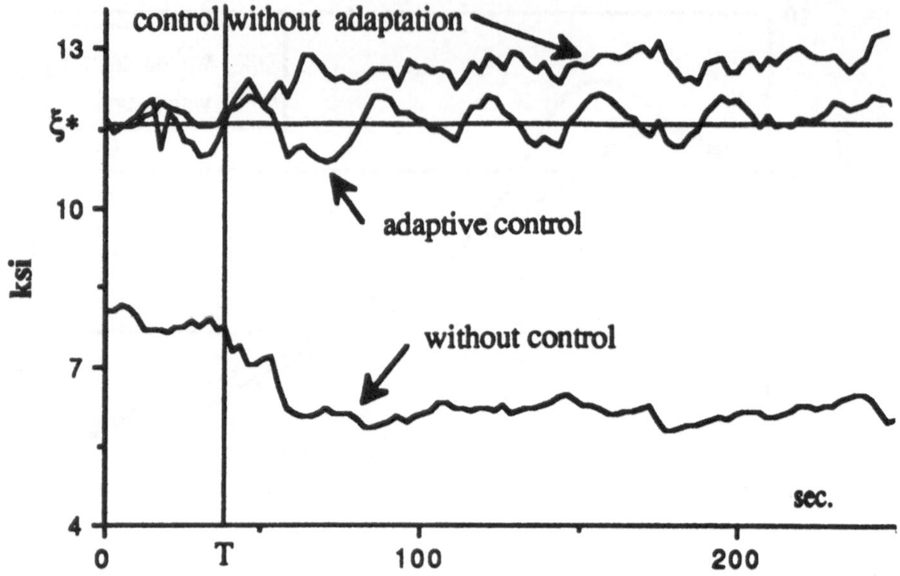

Fig. 6. Time evolution of the estimator $\hat{\xi}_n$

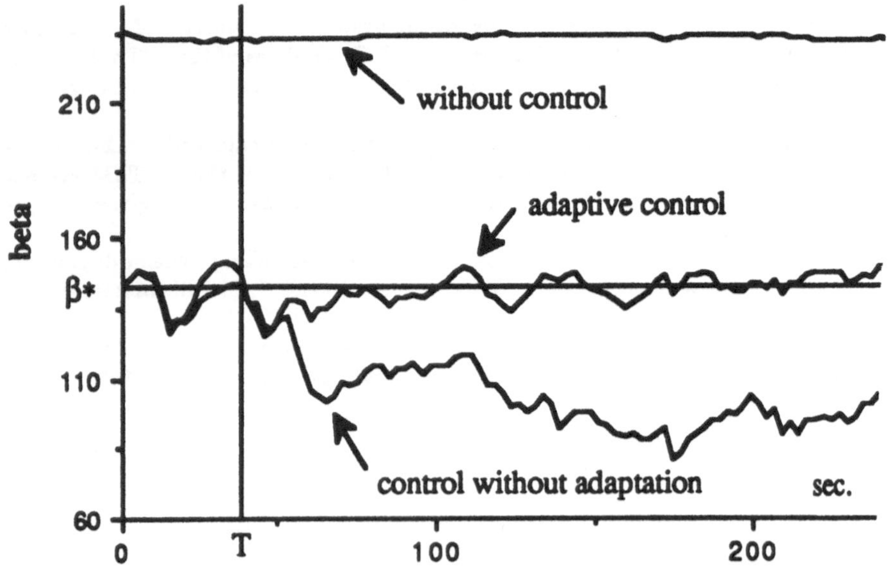

Fig. 7. Time evolution of the estimator $\hat{\beta}_n$

Optimal Ergodic Control of Nonlinear Stochastic Systems

Fabien Campillo

Abstract

We study a class of ergodic stochastic control problems for diffusion processes. We describe the basic ideas concerning the Hamilton–Jacobi–Bellman equation. For a given class of control problems we establish an existence and uniqueness property of the invariant measure. Then we present a numerical approximation to the optimal feedback control based on the discretization of the infinitesimal generator using finite difference schemes. Finally, we apply these techniques to the control of semi–active suspensions for road vehicle.

1 Introduction

This paper deals with a numerical procedure for optimal stochastic control problems and its application to a non trivial example. This procedure consists in approximating the non linear Hamilton–Jacobi–Bellman partial differential equation which is formally satisfied by the minimal cost function. We use finite difference techniques and with a suitable choice of the schemes, the resulting discrete equation can be viewed as the dynamic programming equation for the minimal cost function for the optimal control of a certain Markov process with finite state space [26].

In section 2, we present the different Hamilton–Jacobi–Bellman (HJB in short) equations which arise in the optimal control of diffusion processes in $\mathbb{R}^n$ (for different cases: finite horizon, infinite horizon with discounted or undiscounted cost functions). We stress the intuitive setup of the HJB equations rather than the mathematical aspects.

In section 3, we introduce a particular class — denoted by $\mathcal{C}$ — of ergodic control problems. Some characteristics of this problem are non classical (the diffusion is degenerate, the coefficients are non linear and discontinuous) and there is no available result concerning the HJB equation. Even the first step — giving a meaning to the cost function — is non trivial; this point is treated here (we prove the existence and the uniqueness of the invariant measure associated with the system, for any given admissible control). This class of problems derives from a particular application in control of suspension systems [4].

In section 4, the approximation procedure is detailed in a more general context than the class $\mathcal{C}$. For the special case of the class $\mathcal{C}$ we have already stated two types of results [4]: existence and uniqueness property for the discrete HJB equation (with convergence of the algorithm used for solving it) and a convergence property of the approximation as the discretization step tends to 0. Finally, we apply these techniques to the suspension problem [4, 3] and perform some numerical tests; related suboptimal and adaptive techniques may be found in [3].

2 The HJB Equation

In this section we give an *intuitive* presentation of the HJB equations. For mathematical treatments of this problem one can consult: for the deterministic control [7, 17, 30], for the stochastic control of diffusion processes on finite horizon [7, 5, 17], for the control of diffusion processes on infinite horizon [7, 31, 36], for the control of Markov chains [10], for the control of Markov processes on infinite horizon [15], for the probabilistic aspects [24] and for numerical aspects [1, 6, 14, 21, 25, 26, 28, 32, 34, 35]. For other aspects of the stochastic control theory of diffusion processes one can consult [9] for the impulse control problem, [8] for the optimal stopping time problem and [20] for the stochastic maximum principle.

2.1 Finite–horizon problem

The problem We consider a diffusion process on $\mathbb{R}^n$

$$dX_t = b(X_t, u(X_t,t))\,dt + \sigma(X_t)\,dW_t\,, \quad 0 \le t \le T\,, \quad X_0 = x_0 \in \mathbb{R}^n\,, \tag{1}$$

where W is a n–dimensional standard Wiener process. We suppose that $u \in \mathcal{U}$, $\mathcal{U}$ is a given class of admissible controls which take values in $U \subset \mathbb{R}^k$. Suppose that for any (x,t), $\{u(x,t); u \in \mathcal{U}\} = U$. We will give later an example of such a class.

We fix the instant $T > 0$ and we define the following cost functional

$$J(u) \triangleq E\left[\int_0^T f(X_t, u(X_t,t))\,dt + h(X_T)\right]\,. \tag{2}$$

The stochastic control problem is to find $\hat{u} \in \mathcal{U}$ which minimizes the cost functional J among all the admissible controls.

We introduce the infinitesimal generator associated with the system (1)

$$\mathcal{L}^u\phi(x) \triangleq \sum_{i=1}^n b_i(x,u)\,\frac{\partial\phi(x)}{\partial x_i} + \frac{1}{2}\sum_{i,j=1}^n a_{ij}(x)\,\frac{\partial^2\phi(x)}{\partial x_i\,\partial x_j}\,. \tag{3}$$

with $a(x) \triangleq \sigma(x)\,\sigma(x)^*$.

The HJB equation

Definition 1. The *value function* is defined by

$$v(x,t) \triangleq \inf_{u\in\mathcal{U}} J_{x,t}(u) , \tag{4}$$

with[4]

$$J_{x,t}(u) \triangleq E^u_{x,t}\left[\int_t^T f(X_s,u(X_s,s))\,ds + h(X_T)\right] . \tag{5}$$

Optimality Principle For any $h>0$,

$$v(x,t) = \inf_{u\in\mathcal{U}} E^u_{x,t}\left[\int_t^{t+h} f(X_s,u(X_s,s))\,ds + v(X_{t+h},t+h)\right] . \tag{6}$$

Proof. First we introduce (a version of) the *Dynamic Programming Principle*: suppose that $(x,s)\to\hat{u}^t(x,s)$ is the optimal feedback for the control problem over the time interval $[t,T]$ starting at point x, i.e.

$$J_{x,t}(\hat{u}^t) = v(x,t)$$

then, given $h>0$, $\hat{u}^t|_{[t+h,T]}$ is the optimal feedback for the control problem over the time interval $[t,T]$ starting at point X_{t+h}, i.e.

$$J_{X_{t+h},t+h}(\hat{u}^t) = v(X_{t+h},t+h) \quad \text{a.s.} . \tag{7}$$

Now we consider

$$\overline{u}(x,s) = \begin{cases} \tilde{u}(x,s) \ , & s\in[t,t+h] \ ,\\ \hat{u}^t(x,s) \ , & s\in]t+h,T] \ ,\end{cases}$$

for some control $\tilde{u}$. By definition of $v(x,t)$ and using (7), we have

$$\begin{aligned} v(x,t) &\le J_{x,t}(\overline{u})\\ &= E^u_{x,t}\left[\int_t^{t+h} f(X,\tilde{u})\,ds + \int_{t+h}^T f(X,\hat{u}^t)\,ds + h(X_T)\right]\\ &= E^u_{x,t}\left[\int_t^{t+h} f(X,\tilde{u})\,ds + E^{\hat{u}^t}_{X_{t+h},t+h}\left(\int_{t+h}^T f(X,\hat{u}^t)\,ds + h(X_T)\right)\right]\\ &= E^u_{x,t}\left[\int_t^{t+h} f(X,\tilde{u})\,ds + v(X_{t+h},t+h)\right] . \end{aligned}$$

[4] *Notations*: The term $E^u_{x,t}$ in expressions of the form

$$E^u_{x,t}\int_{t_1}^{t_2} \Psi(X_t,t)\,dt \text{ or } dW_t ,$$

means that we consider the diffusion process X_t solution of (1) starting from point x at time t (i.e. $X_t = x$) and using the feedback control u. When $t=0$, $E^u_{x,t}$ is denoted by E.

So we get

$$v(x,t) \leq E^{u}_{x,t}\left[\int_t^{t+h} f(X_s,\tilde{u}(X_s,s))\,ds + v(X_{t+h},t+h)\right] ,$$

and the equality holds if $\overline{u} = \hat{u}^t$, which proves (6).

Hamilton–Jacobi–Bellman equation The value function (4) satisfies the following equation

$$\frac{\partial}{\partial t}v(x,t) + \inf_{u\in U}\left[\mathcal{L}^u v(x,t) + f(x,u)\right] = 0\,, \quad v(x,T) = h(x)\,. \tag{8}$$

Proof. For notational convenience we consider the 1–dimensional case. $E^u_{x,t}$ is denoted by E. Using (6)

$$\inf_{u\in\mathcal{U}} E\left[\frac{v(X_{t+h},t+h) - v(x,t)}{h} + \frac{1}{h}\int_t^{t+h} f(X_s,u(X_s,s))\,ds\right] = 0\,. \tag{9}$$

Moreover

$$\begin{aligned} v(X_{t+h},t+h) = v(x,t) &+ v'(x,t)\,(X_{t+h}-x) \\ &+\frac{1}{2}v''(x,t)\,(X_{t+h}-x)^2 + h\,\frac{\partial}{\partial t}v(x,t) + o(h)\,, \end{aligned}$$

so

$$\begin{aligned} E\left[\frac{v(X_{t+h},t+h) - v(x,t)}{h}\right] = v'(x,t)\,E\left[\frac{X_{t+h}-x}{h}\right] \\ +\frac{1}{2}v''(x,t)\,E\left[\frac{(X_{t+h}-x)^2}{h}\right] \\ +\frac{\partial}{\partial t}v(x,t) + o(1)\,. \end{aligned} \tag{10}$$

Equation (1) with initial condition $X_t = x$, leads to

$$\begin{aligned} E\left[\frac{X_{t+h}-x}{h}\right] &= E\frac{1}{h}\int_t^{t+h} b(X_s,u(X_s,s))\,ds + E\frac{1}{h}\int_t^{r}\sigma(X_s)\,dW_s \\ &= E\frac{1}{h}\int_t^{t+h} b(X_s,u(X_s,s))\,ds \\ &\underset{h\to 0}{\to} b(x,u(x,t))\,. \end{aligned} \tag{11}$$

Similarly

$$\begin{aligned}
E\left[\frac{(X_{t+h}-x)^2}{h}\right] &= \frac{1}{h}E\left[\int_t^{t+h} b(X_s,u(X_s,s))\,ds\right]^2 \\
&\quad +\frac{2}{h}E\left[\int_t^{t+h} b(X_s,u(X_s,s))\,ds \int_t^{t+h}\sigma(X_s)\,dW_s\right] \\
&\quad +\frac{1}{h}E\left[\int_t^{t+h}\sigma(X_s)\,dW_s\right]^2 \\
&\underset{h\to 0}{\simeq} \frac{1}{h}E\left[\int_t^{t+h}\sigma(X_s)\,dW_s\right]^2 \\
&= \frac{1}{h}E\int_t^{t+h}\sigma^2(X_s)\,ds \\
&\underset{h\to 0}{\to} \sigma^2(x)\,. \qquad (12)
\end{aligned}$$

¿From (11,12) and (10) we get

$$E\left[\frac{v(X_{t+h},t+h)-v(x,t)}{h}\right]$$
$$\underset{h\to 0}{\simeq} v'(x,t)\,b(x,u(x,t)) + \frac{1}{2}v''(x,t)\,\sigma^2(x) + \frac{\partial}{\partial t}v(x,t)\,.$$

Let $h\to 0$ in (9), so we get

$$\inf_{u\in\mathcal{U}}\left[\frac{\partial}{\partial t}v(x,t) + v'(x,t)\,b(x,u(x,t)) + \frac{1}{2}v''(x,t)\,\sigma^2(x) + f(x,u(x,t))\right] = 0\,.$$

This shows (8). The condition at final time is an easy consequence of (5,4). The preceding argument is made rigorous if we know *a priori* that

$$v\in C^{2,1}(\mathbb{R}^n\times]0,\infty[)\,,$$

but the main question lies in the fact that v is not in general $C^{2,1}$ (cf. [31]).

Verification result Let v be the solution of the HJB equation (8). Any function $\tilde{u}(x,t)$ which satisfies the relation

$$\mathcal{L}^{\tilde{u}(x,t)}\,v(x,t) + f(x,\tilde{u}(x,t)) = \inf_{u\in U}\left[\mathcal{L}^u\,v(x,t)+f(x,u)\right]\,,\quad \forall (x,t)\,, \qquad (13)$$

is an optimal feedback control, i.e.

$$J(\tilde{u}) = \inf_{u\in\mathcal{U}} J(u)\,. \qquad (14)$$

Proof. We need the following lemma (cf. [23] for a precise statement)

Lemma 2 a Feynman–Kac formula. *Let $\Psi(x,s)$ be a solution of the following backward partial differential equation*

$$\frac{\partial}{\partial s}\Psi(x,s) + \mathcal{L}^u\,\Psi(x,s) + M(x,s) = 0\ ,$$

then, for any $T > t$,

$$\Psi(x,t) = E^u_{x,t}\left[\int_t^T M(X_s,s)\,ds + \Psi(X_T,T)\right]\ .$$

Moreover, from (13) and (8) we get

$$\frac{\partial}{\partial t}v(x,t) + \mathcal{L}^{\tilde{u}(x,t)}\,v(x,t) + f(x,\tilde{u}(x,t)) = 0\ ,$$

hence, using lemma 2 we find

$$v(x,t) = E^{\tilde{u}}_{x,t}\left[\int_t^r f(X_s,\tilde{u}(X_s,s))\,ds + v(X_r,r)\right]\ ,\qquad t < r\ ,$$

and for $r = T$ and $t = 0$, this last equality becomes

$$v(x,0) = E\left[\int_0^T f(X_s,\tilde{u}(X_s,s))\,ds + h(X_T)\right]\ .$$

The definitions of J and v lead to (14).

Rigorous statements [7] The main hypothesis is the nondegeneracy of the matrix $a(x) \triangleq \sigma(x)\,\sigma(x)^*$. The case of degenerate diffusion is much more tricky. So we suppose that

$$a(x) \triangleq \sigma(x)\,\sigma(x)^* \geq \sigma_0\,I > 0\ ,\quad \forall x \in \mathbb{R}^n\ . \tag{15}$$

Up to now, $\mathcal{U}$ was a class of Markovian controls (i.e. feedback controls); throughout this section, we use a wider class containing controls which are not necessarily Markovian (i.e. stochastic processes which are not only function of the current state value X_t). We define the class $\mathcal{U}$ of admissible controls as follow ([5])

$$u = \{u(t); 0 \leq t \leq T\} \in \mathcal{U} \quad\Longleftrightarrow\quad \begin{cases} u \in L^2_{\mathcal{F}}(0,T;\mathbb{R}^k)\ , \\ u(t,\omega) \in U,\ t \text{ a.e.},\ \omega \text{ a.s.}\ . \end{cases}$$

[5] the space $L^2_{\mathcal{F}}(0,T;\mathbb{R}^k)$ is defined by

$$z \in L^2_{\mathcal{F}}(0,T;\mathbb{R}^k) \quad\Longleftrightarrow\quad \begin{cases} (t,\omega) \to z(t,\omega) \text{ is measurable,} \\ E\int_0^T |z(t)|^2\,dt < \infty\ , \\ z(t) \in L^2(\Omega,\mathcal{F}_t,P)\ ,\quad t \text{ a.e.} \end{cases}$$

where $\{\mathcal{F}_t\}$ is the filtration associated with the Wiener process w.

We also make the following hypotheses

- U is a closed convex subset of $\mathbb{R}^k$, (16)
- $f(x,u) \geq f_0\,|u|^2 - C_0$, $h(x) \geq -C_0$, $\forall (x,u) \in \mathbb{R}^n \times U$, (17)
- $b : \mathbb{R}^n \times \mathbb{R}^k \to \mathbb{R}^n$, $\sigma : \mathbb{R}^n \to L(\mathbb{R}^n, \mathbb{R}^n)$, (18)

  b, σ are continuously differentiable, the derivatives are bounded,

  $|b(x,u)| \leq \bar{b}\,(1+|x|+|u|)$, $|\sigma(x)| \leq \bar{\sigma}\,(1+|x|)$,
- $f : \mathbb{R}^n \times \mathbb{R}^k \to \mathbb{R}$, $h : \mathbb{R}^n \to \mathbb{R}$, (19)

  f, h are continuously differentiable,

  $|f(x,u)| \leq \bar{f}\,(1+|x|^2+|u|^2)$, $|h(x)| \leq \bar{h}\,(1+|x|^2)$,

  $|\partial f/\partial x_i|$, $|\partial f/\partial u_i| \leq \bar{f}\,(1+|x|+|u|)$, $|\partial h/\partial x_i| \leq \bar{h}\,(1+|x|)$.

Concerning the quasilinear partial differential equation (8), we have the following result

Theorem 3 Bensoussan[7]. *Under assumptions (15—19), there exists one and only one solution v of (8) such that* ([6])

$$v \in L^p(0,T;W^{2,p}_{loc}(\mathbb{R}^n)) , \quad \frac{\partial}{\partial t} v \in L^p(0,T;L^p_{loc}(\mathbb{R}^n)) , \tag{20}$$

for any $2 \leq p < \infty$, and

$$|v(x,t)| \leq C\,(1+|x|^2) , \quad \left|\frac{\partial v}{\partial x_i}\right| \leq C\,(1+|x|) . \tag{21}$$

This solution is explicitly given by (4).

Then there exists $\hat{u} : \mathbb{R}^n \times [0,\infty[\to \mathbb{R}^k$ such that

$$\mathcal{L}^{\hat{u}(x,t)}\, v(x,t) + f(x,\hat{u}(x,t)) = \inf_{u \in U} \left[\mathcal{L}^u\, v(x,t) + f(x,u)\right] .$$

Using the estimates (21), we can prove that

$$|\hat{u}(x,t)| \leq C\,(1+|x|) . \tag{22}$$

This last theorem does not establish that the infimum in the right hand side of (4) is reached. The reason is that the feedback $\hat{u}(x,t)$ is not smooth enough to ensure that the s.d.e. (1) has a strong solution.

With additional assumptions we can prove that the feedback $\hat{u}(x,t)$ is optimal for the original problem (1,2). Indeed

---

[6] The space $L^p_{loc}(\mathbb{R}^n)$ is defined by

$$z \in L^p_{loc}(\mathbb{R}^n) \iff z\phi \in L^p(\mathbb{R}^n),\ \forall \phi \in C_0^\infty(\mathbb{R}^n) ,$$

and the space $W^{2,p}_{loc}(\mathbb{R}^n)$ is defined by

$$z \in W^{2,p}_{loc}(\mathbb{R}^n) \iff z\,,\ \frac{\partial z}{\partial x_i}\,,\ \frac{\partial^2 z}{\partial x_i\, \partial x_j} \in L^p_{loc}(\mathbb{R}^n) .$$

Theorem 4 Bensoussan[7]. *Under the assumptions of Theorem 3, assume that*

- $$\frac{\partial^2 f}{\partial x_i \, \partial u_q}, \; \frac{\partial^2 f}{\partial u_p \, \partial u_q}, \; \frac{\partial^2 b}{\partial x_i \, \partial u_q}, \; \frac{\partial^2 b}{\partial u_p \, \partial u_q} \quad \textit{exist and are continuous, bounded} \tag{23}$$
- $$\left[\frac{\partial^2 f}{\partial u_p \, \partial u_q}\right] \geq f_0 \, I \,, \quad \left|\frac{\partial^2 b}{\partial u_p \, \partial u_q}\right| \leq \frac{b_0}{1+|x|} \,, \quad f_0 > b_0 \sup_{x\in\mathbb{R}^n} \frac{|\nabla v|}{1+|x|} \,. \tag{24}$$

Then $\hat{u}(x,t)$ is an optimal feedback control for the problem (1,2).

2.2 Infinite–horizon problems

The discounted case We consider the stochastic control problem defined by the state equation

$$dX_t = b(X_t, u(X_t))\, dt + \sigma(X_t)\, dW_t \,, \quad t \geq 0 \tag{25}$$

and the cost function

$$J(u) \triangleq E \int_0^\infty e^{-\alpha t} f(X_t, u(X_t))\, dt \,. \tag{26}$$

Note that the feedback control does not depend on t anymore.

Formally, we can take $h \equiv 0$ and replace f by $e^{\alpha t} f$ in (2), we get

$$J_T(u) \triangleq E \int_0^T e^{-\alpha t} f(X_t, u(X_t))\, dt \,,$$

and (8) becomes

$$\frac{\partial}{\partial t} v_\alpha^T(x,t) + \inf_{u\in U} \left[\mathcal{L}^u v_\alpha^T(x) + e^{-\alpha t} f(x,u)\right] = 0 \,, \quad v_\alpha^T(x,T) \equiv 0 \,. \tag{27}$$

Let

$$\tilde{v}_\alpha^T(x,t) \triangleq e^{\alpha t} v_\alpha^T(x,t) \,, \tag{28}$$

so (27) reads

$$\frac{\partial}{\partial t} \tilde{v}_\alpha^T(x,t) + \inf_{u\in U} \left[\mathcal{L}^u \tilde{v}_\alpha^T(x,t) + f(x,u)\right] = \alpha \tilde{v}_\alpha^T(x,t) \,, \quad e^{-\alpha T} \tilde{v}_\alpha^T(x,T) \equiv 0 \,,$$

taking $T \to \infty$ in this last equation yields

$$\frac{\partial}{\partial t} \tilde{v}_\alpha(x,t) + \inf_{u\in U} \left[\mathcal{L}^u \tilde{v}_\alpha(x,t) + f(x,u)\right] = \alpha \tilde{v}_\alpha(x,t) \,, \quad t \geq 0 \,, \tag{29}$$

where $\tilde{v}_\alpha(x,t) = \tilde{v}_\alpha^\infty(x,t)$.

In fact $\tilde{v}_\alpha(x,t)$ does not depend on t, this can be checked using (28) and the definition of $v_\alpha^T(x,t)$, indeed

$$\begin{aligned}\tilde{v}_\alpha(x,t) &= e^{\alpha t}\, v_\alpha^\infty(x,t)\\ &= e^{\alpha t} \min_{u\in\mathcal{U}} E^u_{x,t} \int_t^\infty e^{-\alpha s} f(X_s,u(X_s))\, ds\\ &= \min_{u\in\mathcal{U}} E^u_{x,t} \int_t^\infty e^{-\alpha(s-t)} f(X_s,u(X_s))\, ds\\ &= \min_{u\in\mathcal{U}} E \int_0^\infty e^{-\alpha s} f(X_s,u(X_s))\, ds\end{aligned}$$

this last equality comes from the fact that equation (25) is homogeneous in t. Hence $\tilde{v}_\alpha(x,t)$ is now denoted $\tilde{v}_\alpha(x)$ and (29) can be rewritten as follows

$$\inf_{u\in U} [\mathcal{L}^u\, \tilde{v}_\alpha(x) + f(x,u)] = \alpha\, \tilde{v}_\alpha(x)\ , \quad t \geq 0\ . \tag{30}$$

Equation (30) is the HJB for the infinite–horizon control problem associated with state equation (25) and the discounted cost function (26). The optimal feedback control is given by

$$\hat{u}_\alpha(x) \in \operatorname{Arg}\min_{u\in U} [\mathcal{L}^u\, \tilde{v}_\alpha(x) + f(x,u)]\ .$$

The undiscounted case We consider the diffusion process (25). We want to minimize an *average cost* of the form

$$J(u) \triangleq \liminf_{T\to\infty} \frac{1}{T} E \int_0^T f(X_t,u(t))\, dt\ . \tag{31}$$

Most frequently the *discounted cost function* (26) is used. In many applications, the cost functional (31) is more realistic than (26) because it represents a time average while (26) involves a discount factor which is often difficult to evaluate and not always relevant. In general, (31) implies that the control stabilizes the system. In fact we need to suppose some recurrence and stability conditions (cf. [27]). From the mathematical viewpoint, the discounted problem is easier than the undiscounted one, since the former avoid considering the behavior of the controlled process as t goes to infinity.

For the average case (31), we want to find a pair (v,ρ) where ρ is a constant and v is a smooth function, such that the following HJB equation is satisfied

$$\inf_{u\in U} [\mathcal{L}^u\, v(x) + f(x,u)] = \rho\ . \tag{32}$$

Let (v,ρ) be a solution of (32), then the optimal feedback is given by

$$\hat{u}(x) \in \operatorname{Arg}\min_{u\in U} [\mathcal{L}^u\, v(x) + f(x,u)]\ ,$$

and ρ is the optimal cost, i.e.

$$\rho = J(\hat{u}) = \inf_{u\in U} J(u)\ .$$

The average cost problem can be viewed as a limit model for the discounted problem as $\alpha \to 0$, indeed

$$\alpha \tilde{v}_\alpha(\cdot) \underset{\alpha \to 0}{\to} \rho ,$$
$$\tilde{v}_\alpha(\cdot) - \tilde{v}_\alpha(x_0) \underset{\alpha \to 0}{\to} v(\cdot) ,$$

(in a suitable sense, see [7] for a rigorous proof).

3 A Class of Ergodic Stochastic Control Problems

We present a class of models which derive from a problem of control for semi–active suspension systems. In these models — like in most realistic models — difficulties of the following type are met: the coefficients of the diffusion which we want to control are discontinuous and strongly nonlinear.

In section 3.1 we introduce the class $\mathcal{C}$ of problems. Then, in section 3.2, we consider μ_u the invariant measure associated to a system of the class $\mathcal{C}$ for a given admissible control u, and we prove that μ_u exists and is unique. Finally, in section 3.3, we present the original semi–active suspensions problem.

3.1 The problem

Let us consider the following stochastic system

$$dX_t = b(u(X_t), X_t)\, dt + \begin{pmatrix} 0 \\ \sigma \end{pmatrix} dW_t , \tag{33}$$

where X is a process which takes values in $\mathbb{R}^2$, W is a real standard Wiener process and $\sigma > 0$. b maps $\mathbb{R} \times \mathbb{R}^2$ in $\mathbb{R}^2$ and is defined by

$$b(u,x) \triangleq \begin{pmatrix} b_1(u,x) \\ b_2(u,x) \end{pmatrix} \triangleq \begin{pmatrix} x_2 \\ -u\, x_2 - \gamma_1\, x_1 - \gamma_2 \operatorname{sign}(x_2) \end{pmatrix} , \quad x \triangleq \begin{pmatrix} x_1 \\ x_2 \end{pmatrix} ,$$

where γ_1, γ_2 are strictly positive constants. In (33), u is a feedback control which belongs to the class $\mathcal{U}$ of admissible controls defined by (fix $\underline{u}$, $\overline{u}$ such that $0 < \underline{u} < \overline{u} < \infty$)

$u \in \mathcal{U} \iff$ $u : \mathbb{R}^2 \to [\underline{u}, \overline{u}]$ and there exists a finite number of submanifolds of $\mathbb{R}^2$ with dimension less than or equal to 1 outside of which u is continuous.

We are concerned with an ergodic type control problem, whose cost functional is

$$J(u) \triangleq \lim_{T \to \infty} \frac{1}{T} E \int_0^T f(u(X_t), X_t)\, dt , \quad \forall u \in \mathcal{U} , \tag{34}$$

where the instantaneous cost function f is defined by

$$f(u,x) \triangleq (u\, x_2 + \gamma_1\, x_1 + \gamma_2 \operatorname{sign}(x_2))^2 . \tag{35}$$

$$dX_t = b(u(X_t), X_t)\, dt + \begin{pmatrix} 0 \\ \sigma \end{pmatrix} dW_t$$

$$b(u,x) \triangleq \begin{pmatrix} b_1(u,x) \\ b_2(u,x) \end{pmatrix} \triangleq \begin{pmatrix} x_2 \\ -u\,x_2 - \gamma_1\,x_1 - \gamma_2\,\mathrm{sign}(x_2) \end{pmatrix}, \quad \gamma_1, \gamma_2 > 0$$

$$J(u) \triangleq \lim_{T\to\infty} \frac{1}{T} E \int_0^T f(u(X_t), X_t)\, dt$$

$$f(u,x) \triangleq (u\,x_2 + \gamma_1\,x_1 + \gamma_2\,\mathrm{sign}(x_2))^2$$

Table 1. The class C of ergodic control problems

¿From now on, we denote

$$b^u(x) \triangleq b(u(x), x)\,, \quad f^u(x) \triangleq f(u(x), x)\,, \quad \forall u \in \mathcal{U}\,.$$

The Hamilton–Jacobi–Bellman equation for the ergodic control problem (33) (34) can be formally written as (see section 2)

$$\min_{u\in[\underline{u},\overline{u}]} (\mathcal{L}^u v(\cdot) + f(u,\cdot)) = \rho \quad \text{on } \mathbb{R}^2, \tag{36}$$

where $v : \mathbb{R}^2 \to \mathbb{R}$ is defined up to an additive constant, ρ is a constant and $\mathcal{L}^u$ is the infinitesimal generator associated with (33).

$$\mathcal{L}^u \phi(x) \triangleq b_1^u(x) \frac{\partial\phi(x)}{\partial x_1} + b_2^u(x) \frac{\partial\phi(x)}{\partial x_2} + \frac{\sigma^2}{2} \frac{\partial^2\phi(x)}{\partial^2 x_2}\,. \tag{37}$$

Remark. The arguments presented bellow may be applied to a wider class of problems. Indeed, we can consider a system of the form

$$d\begin{pmatrix} X_t^1 \\ X_t^2 \end{pmatrix} = \begin{pmatrix} b_1(X_t) \\ b_2(u, X_t) \end{pmatrix} dt + \begin{pmatrix} 0 \\ \sigma \end{pmatrix} dW_t\,,$$

where X_t^1 (resp. X_t^2) takes values in $\mathbb{R}^{n_1}$ (resp. $\mathbb{R}^{n_2}$) and W is a standard Wiener process. The main hypotheses are

(i) the discontinuous terms appear only in the "noisy part" of the system, that is $b_1(x)$ is smooth and $\sigma\sigma^* > 0$,
(ii) the system satisfies a stability property (e.g. $E|X_t|^2 \leq C\,,\ \forall t \geq 0$).

Point (i) permits us to use a Girsanov transformation to remove the discontinuous terms.

Remark. In this case the choice of the value of the function "sign" at point 0 is not important. Indeed, in (33) the noise is added to the second component, so we can prove that

$$P(X_t^2 = 0) = 0\,, \quad \forall t\,. \tag{38}$$

Property (38) implies that, if we change the value of sign(0), the (weak) solution of (33) will not be changed. If (38) was false, we should use differential inclusion techniques to give a meaning to the stochastic differential equation (33).

3.2 The invariant probability measure

The cost function (34) can be rewritten as

$$J(u) = \langle f^u , \mu_u \rangle , \quad \forall u \in \mathcal{U} , \tag{39}$$

where μ_u is the invariant probability measure associated with system (33). In this section we present an existence and uniqueness property for μ_u which gives a meaning to expressions (34,39). For the results presented in this section all the details can be found in [13].

Proposition 5. *For any $u \in \mathcal{U}$, the diffusion process (33) admits an invariant probability measure μ_u.*

Proof. We fix $u \in \mathcal{U}$. By means of usual techniques (e.g. [16] th. 9.3 ch. 4), it is sufficient to prove the following properties

(i) There exists a constant C such that

$$E\,|X_t|^2 \leq C\,, \quad \forall t \geq 0\,. \tag{40}$$

(ii) The process X_t solution of (33) has the Feller property, i.e. for any $t \geq 0$ and $\phi \in C_b(\mathbb{R}^2)$, the function

$$\mathbb{R}^2 \ni x \longrightarrow E\phi\,(X_t^x) \tag{41}$$

is continuous, where $\{X_t^x\}$ denotes the diffusion process (33) starting at point x at time 0.

proof of (i) We define

$$V(t) \triangleq E\,\mathcal{V}(X_t)\,, \qquad \mathcal{V}(x) \triangleq \gamma_1\,(x_1)^2 + \varepsilon\, x_1\, x_2 + (x_2)^2\,.$$

There exists $\varepsilon_0 > 0$ such that for any $\varepsilon_0 > \varepsilon > 0$

$$\mathcal{V}(x) \geq \frac{1}{2}\,\left(\gamma_1\,(x_1)^2 + (x_2)^2\right)\,.$$

Hence, it is sufficient to show that $V(t) \leq$ Cte for any $t \geq 0$.

We can check [13] that there exist strictly positive constants ε and δ such that

$$\frac{d}{dt}V(t) \leq -C(\varepsilon,\delta)\,V(t) + \frac{\varepsilon}{2\,\delta} + \sigma^2\,,$$

where $C(\varepsilon,\delta) > 0$, which yields the conclusion.

proof of (ii) In (33), the drift coefficient can be written as

$$b(u,x) = \overline{B}\,x + \begin{pmatrix} 0 \\ -u\,x_2 - \gamma_2\,\mathrm{sign}(x_2) \end{pmatrix}$$
$$\triangleq \begin{pmatrix} 0 & 1 \\ \gamma_1 & 0 \end{pmatrix} \begin{pmatrix} x_1 \\ x_2 \end{pmatrix} + \begin{pmatrix} 0 \\ -u\,x_2 - \gamma_2\,\mathrm{sign}(x_2) \end{pmatrix} .$$

Let

$$\overline{W}_t \triangleq W_t + \int_0^t \psi(X_s^x)\,ds \,,$$
$$\psi(x) \triangleq -\frac{1}{\sigma}\left(u(x)\,x_2 + \gamma_2\,\mathrm{sign}(x_2)\right) \,,$$
$$Z_t^x \triangleq \exp\left(\int_0^t \psi(X_s^x)\,d\overline{W}_s - \frac{1}{2}\int_0^t \psi(X_s^x)^2\,ds\right) . \tag{42}$$

We define a new probability law

$$\left.\frac{d\overline{P}}{dP}\right|_{\mathcal{F}_t} \triangleq (Z_t^x)^{-1} \quad , t \geq 0 \,.$$

X^x satisfies

$$dX_t^x = \overline{B}\,X_t^x\,dt + \begin{pmatrix} 0 \\ \sigma \end{pmatrix} d\overline{W}_t \,, \tag{43}$$

where — from Girsanov's theorem — $\overline{W}_t$ is a real standard Wiener process under the probability law $\overline{P}$.

For any sequence $x_n \to x$, we want to prove that

$$E\phi(X_t^{x_n}) = \overline{E}\left[\phi(X_t^{x_n})\,Z_t^{x_n}\right] \underset{n\to\infty}{\to} E\phi(X_t^x) = \overline{E}\left[\phi(X_t^x)\,Z_t^x\right] \,, \tag{44}$$

where $\overline{E}$ denotes the expectation with respect to $\overline{P}$. We can check that it is sufficient to prove that

$$X_t^{x_n} \underset{n\to\infty}{\to} X_t^x \quad \overline{P}\text{–a.s.} \,, \tag{45}$$
$$Z_t^{x_n} \underset{n\to\infty}{\to} Z_t^x \quad \text{in } \overline{P}\text{–probability.} \tag{46}$$

Under the probability law $\overline{P}$, X_t is the solution of a *linear* stochastic differential system, so (45) is obvious. For (46), we show that

$$\overline{E}\int_0^t \left[u(X_s^{x_n})\,X_s^{x_n,2} - u(X_s^x)\,X_s^{x,2}\right]^2 ds \underset{n\to\infty}{\to} 0 \,, \tag{47}$$
$$\overline{E}\int_0^t \left[\mathrm{sign}(X_s^{x_n,2}) - \mathrm{sign}(X_s^{x,2})\right]^2 ds \underset{n\to\infty}{\to} 0 \,. \tag{48}$$

The difficulty comes from the discontinuity of the functions $\mathrm{sign}(\cdot)$ and $u(\cdot)$, but using the definition of $\mathcal{U}$ we know that theses function are continuous a.e., which, using standard arguments, is enough to conclude.

Proposition 6. *For any $u \in \mathcal{U}$, the diffusion process (33) admits a unique invariant measure μ_u.*

Proof. ¿From now on, we fix $u \in \mathcal{U}$ and we suppose that μ denotes an invariant probability measure of the system (33), and X_t is the solution of this system with μ as initial law (i.e. X_0 has law μ). We also define Z_t by (42) where X^x is replaced by X. It is sufficient to prove that

$$\text{μ has a density $p(x)$ with respect to Lebesgue measure, and $p(x) > 0$ for almost all x.} \tag{49}$$

Indeed (49) implies that if there exist two invariant measures, they are equivalent. So there exits at most one extremal invariant measure, which establishes the proposition.

We first prove the following result

$$\text{Under $\overline{P}$, for any $t > 0$, the law of X_t has a density $\overline{p}(t,x)$ such that $\overline{p}(t,x) > 0$, $\forall x$.} \tag{50}$$

Under $\overline{P}$, consider the system (43) where $d\overline{W}$ is replaced by $v\,dt$ ($v \in L^2(\mathbb{R}^+)$), we get

$$\begin{pmatrix} \dot{x}_1 \\ \dot{x}_2 \end{pmatrix} = \begin{pmatrix} x_2 \\ -\beta\, x_1 \end{pmatrix} + \begin{pmatrix} 0 \\ \sigma \end{pmatrix} v\,, \quad x(0) = x\,. \tag{51}$$

Let $x^{x,v}(t)$ denote the solution of this last equation. We define the reachability set

$$\mathcal{A}(t,x) \triangleq \{x^{x,v}(t)\ ;\ \forall v \in L^2(\mathbb{R}^+)\}\ .$$

(51) can be rewritten as $\dot{x} = A\,x + B\,v$ and the matrix $[B|A\,B]$ has full rank. Hence this system is controllable [29]. So

$$\forall t > 0\,, \quad \forall x \in \mathbb{R}^2\,, \quad \mathcal{A}(t,x) = \mathbb{R}^2\ . \tag{52}$$

Using [33] §3.6.1, we prove that — under $\overline{P}$ — the law of X_t is absolutely continuous with respect to Lebesgue measure and that its density $\overline{p}(t,x)$ is strictly positive for any $t > 0$ and x.

Now we prove (49). For any $\phi \in C_b(\mathbb{R}^2)$

$$\begin{aligned} \langle \mu\,, \phi \rangle &= \overline{E}[\phi(X_t)\,Z_t]\,, \\ &= \overline{E}\left[\phi(X_t)\,\overline{E}[Z_t|X_t]\right]\,, \\ &= \int_{\mathbb{R}^2} \phi(x)\,\overline{E}[Z_t|X_t = x]\,\overline{p}(t,x)\,dx\ . \end{aligned}$$

Since $\overline{E}[Z_t|X_t] > 0$ $\overline{P}$–a.s. and under $\overline{P}$ the law of X_t is equivalent to Lebesgue measure, we get $\overline{E}[Z_t|X_t = x] > 0$ $\forall x$–a.e.. Using (50) and the last inequality, we prove that μ has a density

$$q(x) \triangleq \overline{E}[Z_t|X_t = x]\,\overline{p}(t,x)\,,$$

and that this density is strictly positive for almost all $x \in \mathbb{R}^2$.

3.3 An example: a semi–active suspension system

In this section we present a damping control method for a nonlinear suspension of road vehicle (comprising a spring, a shock absorber, a mass, and taking into account the dry friction, cf. figure 1). The aim is to improve the ride comfort.

Among alternatives to classical suspension systems (passive systems) we distinguish between active and semi–active techniques. An active suspension system consists in force elements in addition to a spring and a damper assembly. Force elements continuously vary the force according to some control law. In general, an active system is costly, complex, and requires an external power source [19]. In contrast, a semi–active system requires no hydraulic power supply, and the implementation of its hardware is simpler and cheaper than a fully active system. A semi–active suspension system acts only on damping or spring laws, so it can only dissipate or store energy.

Here we consider a system with control on the damping law, the forces in the damper are generated by modulating its orifice for fluid flow [2, 37]. We use the simplest model which consists in a one degree–of–freedom model (this model can be represented as a problem of the class $\mathcal{C}$).

The equation of motion for a one degree–of–freedom model is

$$m\ddot{y} + c\dot{y} + k_s y + z = -m\ddot{e} , \tag{53}$$

(cf. figure 1 and table 2 for the exact definition of the terms).

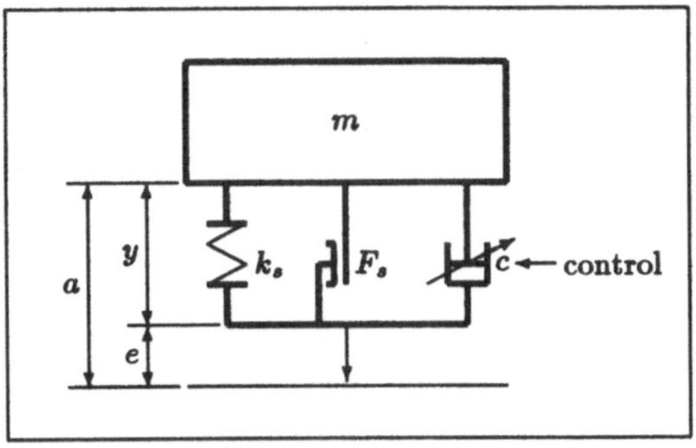

Fig. 1. One degree–of–freedom model.

$\ddot{e}$ denotes the input acceleration. The restoring force $k_s y + z$, has a linear part $k_s y$, and a nonlinear part z which describes the dry friction force [11, 12] defined by

$$\dot{z} + \beta\,(|\dot{y}|z + \dot{y}|z|) = (k - k_s)\dot{y} , \tag{54}$$

where $\beta > 0$ and $k > k_s > 0$. For "large displacements", z degenerate to a Coulomb friction force

$$z = F_s \,\mathrm{sign}(\dot{y}) , \tag{55}$$

a absolute displacement of mass m
y absolute displacement ($y = a - e$)
e stochastic input (surface road acceleration)
m sprung mass
c shock–absorber damping constant (controlled)
k_s spring constant
F_s dry friction constant

Table 2. Notations.

with $F_s = (k - k_s)/2\beta$. The damping force is $c\dot{y}$ where $c > 0$ is the instantaneous damping coefficient (the control is acting on this term).

The general model is described by equations (53,54). The problem is to compute a feedback law $c = c(y, \dot{y})$ such that the solution of the system (53,54) minimizes a criterion — related to the vibration comfort —

$$J(u) \triangleq \lim_{T\to\infty} \frac{1}{T} E \int_0^T |\ddot{a}|^2 \, dt = \lim_{T\to\infty} \frac{1}{T} E \int_0^T |\ddot{y} + \ddot{e}|^2 \, dt \, .$$

This model leads to a control problem for a 3–dimensional diffusion process. If we want to obtain a 2–dimensional problem, we must use the system given by equations (53,55) (i.e. we use a Coulomb force term). We get the following simplified model

$$m\ddot{y} + c\dot{y} + k_s x + F_s \operatorname{sign}(\dot{y}) = -m\ddot{e} \, , \tag{56}$$

$\ddot{e}$ is supposed to be a white Gaussian noise process, $\ddot{e} = -\sigma \, dW/dt$ where W is a standard Wiener process.

Using $u = c/m$, $\gamma_1 = k_s/m$, $\gamma_2 = F_s/m$ and

$$X \triangleq \begin{pmatrix} y \\ \dot{y} \end{pmatrix} ,$$

equation (56) can be rewritten as (33).

We get the following system

$$dX_t = b(u(X_t), X_t) \, dt + \begin{pmatrix} 0 \\ \sigma \end{pmatrix} dW_t \, , \tag{57}$$

where

$$b(u, x) = \begin{pmatrix} b_1(u, x) \\ b_2(u, x) \end{pmatrix} \triangleq \begin{pmatrix} x_2 \\ -(u\, x_2 + \gamma_1 x_1 + \gamma_2 \operatorname{sign}(x_2)) \end{pmatrix} . \tag{58}$$

Hence the instantaneous cost function is

$$f(u, x) \triangleq |\ddot{y} + \ddot{e}|^2 = |u\, x_2 + \gamma_1 x_1 + \gamma_2 \operatorname{sign}(x_2)|^2 \, . \tag{59}$$

4 Numerical Approximation

We use the following procedure: we do not discretize directly the HJB equation but we transform the original ergodic control problem to a control problem for a Markov process in continuous time and finite state space (section 4.1). Then, for the discrete case, we can write a dynamic programming equation (section 4.2); this equation is solved numerically via an iterative algorithm (section 4.3).

We describe the approximation procedure in the case of a diffusion process defined by

$$dX_t = b(u(X_t), X_t)\,dt + \sigma(X_t)\,dW_t\ , \tag{60}$$

and with the following cost function

$$J(u) = \liminf_{T\to\infty} \frac{1}{T} E \int_0^T f(u(X_t), X_t)\,dt\ , \tag{61}$$

where

$$\begin{aligned} b &: \mathbb{R}^k \times \mathbb{R}^n \to \mathbb{R}^n\ ,\\ \sigma &: \qquad \mathbb{R}^n \to \mathbb{R}^n \times \mathbb{R}^d\ ,\\ f &: \mathbb{R}^k \times \mathbb{R}^n \to \mathbb{R}^+\ . \end{aligned}$$

X takes values in $\mathbb{R}^n$ and W in $\mathbb{R}^d$. u belongs to a given class $\mathcal{U}$ of applications from $\mathbb{R}^n$ to $U \subset \mathbb{R}^k$. We suppose that, for any $u \in \mathcal{U}$, the solution X_t of (60) admits a unique invariant probability measure, so the cost function (61) is well defined.

The infinitesimal generator associated with (60) is

$$\mathcal{L}^u\phi(x) \triangleq \sum_{i=1}^n b_i(x,u)\,\frac{\partial\phi(x)}{\partial x_i} + \frac{1}{2}\sum_{i,j=1}^n a_{ij}(x)\,\frac{\partial^2\phi(x)}{\partial x_i\,\partial x_j}\ , \tag{62}$$

where $a(x) \triangleq \sigma(x)\,\sigma(x)^*$. We note $b_i^u(x) = b_i(u,x)$ and $f^u(x) = f(u,x)$.

In section 4.4, we apply this approximation technique to the class $\mathcal{C}$. For this class we also present some convergence results which were proved in [4, 13].

4.1 The finite state space problem

In a first step we approximate the solution X_t of (60) by a controlled Markov process X_t^h in continuous time and discrete (but infinite) state space. In a second step, X_t is approximated by a controlled Markov process $X_t^{h,D}$ in continuous time and finite state space.

$$\frac{\partial \phi(x)}{\partial x_i} \simeq \begin{cases} \dfrac{\phi(x+e_i\,h_i)-\phi(x)}{h_i} & \text{if } b_i^u(x) > 0 \\[2mm] \dfrac{\phi(x)-\phi(x-e_i\,h_i)}{h_i} & \text{if } b_i^u(x) < 0 \end{cases}$$

$$\frac{\partial^2 \phi(x)}{\partial x_i^2} \simeq \frac{\phi(x+e_i\,h_i) - 2\,\phi(x) + \phi(x-e_i\,h_i)}{h_i^2}$$

$$\frac{\partial^2 \phi(x)}{\partial x_i\,\partial x_j} \simeq \begin{cases} \dfrac{2\,\phi(x)+\phi(x+e_i\,h_i+e_j\,h_j)+\phi(x-e_i\,h_i-e_j\,h_j)}{2\,h_i\,h_j} \\ \quad -\dfrac{\phi(x+e_i\,h_i)+\phi(x-e_i\,h_i)+\phi(x+e_j\,h_j)+\phi(x-e_j\,h_j)}{2\,h_i\,h_j} \\ \hfill \text{if } a_{ij}(x) > 0 \\[2mm] -\dfrac{2\,\phi(x)+\phi(x+e_i\,h_i-e_j\,h_j)+\phi(x-e_i\,h_i+e_j\,h_j)}{2\,h_i\,h_j} \\ \quad +\dfrac{\phi(x+e_i\,h_i)+\phi(x-e_i\,h_i)+\phi(x+e_j\,h_j)+\phi(x-e_j\,h_j)}{2\,h_i\,h_j} \\ \hfill \text{if } a_{ij}(x) < 0 \end{cases}$$

$i,j = 1,\ldots,n\,,\ i \neq j\,,$ e_i unit vector in the ith coordinate direction

Table 3. Finite difference schemes.

first step: discrete state space Let h_i (resp. e_i) denote the finite difference interval (resp. the unit vector) in the ith coordinate direction and $h = (h_1, \ldots, h_n)$. We define $\mathbb{R}_h^n$, the h-grid on $\mathbb{R}^n$, by

$$\mathbb{R}_h^n \triangleq \{x \in \mathbb{R}^n\,;\, x_i = n_i\,h_i + h_i/2\,,\ i = 1,\ldots,n,\ n_i \in \mathbf{Z}\}\ .$$

The infinitesimal generator (62) is approximated using finite difference schemes given in table 3. The reason for the choices in the schemes will be explained below.

$\mathcal{L}^u$ is approximated by an infinite dimensional matrix $\mathcal{L}_h^u$ of $\mathbb{R}^{\mathbb{N}} \times \mathbb{R}^{\mathbb{N}}$ given as follows

$$\mathcal{L}^u\phi(x) \simeq \mathcal{L}_h^u\phi(x) \triangleq \sum_{y\in\mathbb{R}_h^n} \mathcal{L}_h^u(x,y)\,\phi(y)\,, \quad \forall x \in \mathbb{R}_h^n$$

the terms $\mathcal{L}_h^u(x,y)$ of this matrix are detailled in table 4.

The matrix $\mathcal{L}_h^u$ has the following property

$$\sum_{y\in\mathbb{R}_h^n} \mathcal{L}_h^u(x,y) = 0\,, \qquad \forall x \in \mathbb{R}_h^n\ .$$

$$\mathcal{L}_h^u(x,x) \triangleq -\sum_i \left(\frac{a_{ii}(x)}{h_i^2} - \frac{1}{2} \sum_{k;k\neq i} \frac{|a_{ik}(x)|}{h_i\, h_k} \right) - \sum_i \frac{|b_i^u(x)|}{h_i}$$

$$\mathcal{L}_h^u(x,x+e_i\,h_i) \triangleq \frac{1}{2} \left(\frac{a_{ii}(x)}{h_i^2} - \sum_{k;k\neq i} \frac{|a_{ik}(x)|}{h_i\, h_k} \right) + \frac{(b_i^u(x))^+}{h_i}$$

$$\mathcal{L}_h^u(x,x-e_i\,h_i) \triangleq \frac{1}{2} \left(\frac{a_{ii}(x)}{h_i^2} - \sum_{k;k\neq i} \frac{|a_{ik}(x)|}{h_i\, h_k} \right) + \frac{(b_i^u(x))^-}{h_i}$$

$$\mathcal{L}_h^u(x,x+e_i\,h_i+e_j\,h_j) \triangleq \mathcal{L}_h^u(x,x-e_i\,h_i-e_j\,h_j) \triangleq \frac{a_{ij}^+(x)}{2\,h_i\,h_j}$$

$$\mathcal{L}_h^u(x,x+e_i\,h_i-e_j\,h_j) \triangleq \mathcal{L}_h^u(x,x-e_i\,h_i+e_j\,h_j) \triangleq \frac{a_{ij}^-(x)}{2\,h_i\,h_j}$$

$$i,j = 1,\ldots,n\,,\ i \neq j$$

Table 4. The discrete infinitesimal generator.

Suppose that

$$a_{ii}(x) - \sum_{j;j\neq i} |a_{ij}(x)| \geq 0\,, \qquad \forall x \in \mathbb{R}_h^n,\ i = 1,\ldots,n\,, \tag{63}$$

then

$$\mathcal{L}_h^u(x,y) \geq 0\,, \qquad \forall x,\, y \in \mathbb{R}_h^n,\ x \neq y\,.$$

Remark. The choice of the finite difference schemes we use (cf. table 3) depends on the sign of the drift coefficients of the diffusion process. The reason for the choice is the following: if (63) is true then $\{\mathcal{L}_h^u(x,y)\,;\ x,\, y \in \mathbb{R}_h^n\}$ can be viewed as the infinitesimal generator of a continuous–time Markov process X_t^h with discrete state space $\mathbb{R}_h^n$ [18]. We will see later why this is important.

So we get a stochastic control problem for a Markov processX_t^h with infinitesimal generator $\mathcal{L}_h^u$, and the following cost function

$$J_h(u) \triangleq \lim_{T\to\infty} E\,\frac{1}{T} \int_0^T f^u(X_t^h)\,dt\,. \tag{64}$$

u belongs to the class $\mathcal{U}_h$ defined by

$$u \in \mathcal{U}_h \quad \Longleftrightarrow \quad u \text{ is an application from } \mathbb{R}_h^n \text{ to } U.$$

second step: finite state space X_t^h has a discrete but infinite state space; if we want to perform computations it is necessary to work on a finite state space. We consider a bounded domain D of $\mathbb{R}^n$. We define a new state space

$$\mathbb{R}^n_{h,D} \triangleq \mathbb{R}^n_h \cap D = \{x^1, x^2, \dots, x^N\}\,, \quad N \triangleq \mathrm{Card}\left(\mathbb{R}^n_{h,D}\right)\,. \tag{65}$$

Because we are working on a bounded domain, we must specify boundary conditions. In practice, D will be chosen large enough so that the process will rarely reach the border. Hence, the choice of the boundary conditions is of little importance, provided that all the states communicate. Example of such conditions (usually reflecting conditions) will be given later for the suspension problem.

So we get an approximation $\mathcal{L}^u_{h,D}$ to $\mathcal{L}^u_h$

$$\mathcal{L}^u_{h,D}\,\phi(x) = \sum_{y \in \mathbb{R}^n_{h,D}} \mathcal{L}^u_{h,D}(x,y)\,\phi(y)\,,$$

$\mathcal{L}^u_{h,D}$ is a $N \times N$–matrix.

Remark. The choices in the finite difference schemes (cf. table 3) imply that

$$\sum_{y \in \mathbb{R}^n_{h,D}} \mathcal{L}^u_{h,D}(x,y) = 0\,, \qquad \forall x \in \mathbb{R}^n_{h,D}\,,$$

moreover, hypothesis (63) implies that

$$\mathcal{L}^u_{h,D}(x,y) \geq 0\,, \qquad \forall x,\, y \in \mathbb{R}^n_{h,D}\,, x \neq y\,.$$

Hence $\mathcal{L}^u_{h,D}$ can be interpreted as the infinitesimal generator of a controlled Markov process $X_t^{h,D}$ in continuous time and finite state space. $X_t^{h,D}$ is described by the following terms

- a sequence $\{\Delta_l^{h,D}\,;\, l \geq 0\}$ where the random variable $\Delta_l^{h,D}$ denotes the elapsed time between the lth and the $(l+1)$th jump,
- a Markov chain $\{\xi_l^{h,D}\,;\, l \geq 0\}$ with state space $\mathbb{R}^n_{h,D}$, $\xi_l^{h,D}$ denotes the state of the process between the lth and the $(l+1)$th jump.

The law of the random variable $\Delta_l^{h,D}$ and the transition probabilities of the Markov chain $\{\xi_l^{h,D}\,;\, l \geq 0\}$ are defined as follows

- the pair $(\Delta_l^{h,D}, \xi_{l+1}^{h,D})$ depends only on $\xi_l^{h,D}$,
- under the conditional probability law $P(\cdot\,|\xi_l^{h,D})$, the random variables $\Delta_l^{h,D}$ and $\xi_{l+1}^{h,D}$ are independant.

And for any $x \in \mathbb{R}^n_{h,D}$, under the conditional probability law $P(\cdot\,|\xi_l^{h,D} = x)$

- the random variable $\Delta_l^{h,D}$ obeys an exponential law of parameter $\delta_l^{h,D}(x)$ where

$$\delta_l^{h,D}(x) \triangleq -\frac{1}{\mathcal{L}^u_{h,D}(x,x)}\,,$$

– the transition probabilities $\{\pi^{h,D}(x,y)\,;\, y \in \mathbb{R}^n_{h,D}\}$ are defined by

$$\pi^{h,D}(x,y) \triangleq -\frac{\mathcal{L}^u_{h,D}(x,y)}{\mathcal{L}^u_{h,D}(x,x)}\,, \quad x \neq y\,.$$

With a suitable choice of boundary conditions (usually reflecting conditions) and with the finite differences schemes we used, we have: $\mathcal{L}^u_{h,D}(x,x) < 0$ for all $x \in \mathbb{R}^n_{h,D}$.

With remark 4.1, the discretized problem can be viewed as a control problem for a Markov process $X^{h,D}_t$ in continuous time, finite state space, and infinitesimal generator $\mathcal{L}^u_{h,D}$. The cost function is

$$J_{h,D}(u) \triangleq \lim_{T\to\infty} E\,\frac{1}{T}\int_0^T f^u(X^{h,D}_t)\,dt\,, \tag{66}$$

and u belongs to a class $\mathcal{U}_{h,D}$ of control defined by

$$u \in \mathcal{U}_{h,D} \quad \Longleftrightarrow \quad u \text{ is an application from } \mathbb{R}^n_{h,D} \text{ to } U.$$

The solution to this problem is given by the dynamic programming equation.

Remark. Let $\mu^{h,D}_u$ be the invariant measure of the process $X^{h,D}_t$. Using $\mu^{h,D}_u$, the cost function (66) can be rewritten as

$$J_{h,D}(u) = \sum_{x\in\mathbb{R}^n_{h,D}} f^u(x)\,\mu^{h,D}_u(x)\,.$$

The measure $\mu^{h,D}_u$ is solution of the following linear system

$$\begin{cases} \displaystyle\sum_{y\in\mathbb{R}^n_{h,D}} \mathcal{L}^u_{h,D}(y,x)\,\mu^{h,D}_u(y) = 0\,, \quad \forall x \in \mathbb{R}^n_{h,D}\,, \\[2ex] \displaystyle\sum_{y\in\mathbb{R}^n_{h,D}} \mu^{h,D}_u(y) = 1\,. \end{cases}$$

4.2 The "discrete" Hamilton–Jacobi–Bellman equation

Associated with the control problem defined in the last section we have the following dynamic programming equation

$$\min_{u\in U}\left[\sum_{y\in\mathbb{R}^n_{h,D}} \mathcal{L}^u_{h,D}(x,y)\,v(y) + f^u(x)\right] = \rho\,, \quad \forall x \in \mathbb{R}^n_{h,D}\,, \tag{67}$$

where ρ is a strictly positive constant and $v : \mathbb{R}^n_{h,D} \to \mathbb{R}$ (i.e. $v \in \mathbb{R}^N$) is defined up to an additive constant.

If (v, ρ) is a solution to (67) then

$$\hat{u}(x) \in \operatorname{Arg}\min_{u\in U} \left[\sum_{y\in \mathbb{R}^n_{h,D}} \mathcal{L}^u_{h,D}(x,y)\, v(y) + f^u(x) \right] \,, \quad x \in \mathbb{R}^n_{h,D} \qquad (68)$$

is an optimal feedback control law, and ρ is the minimal cost

$$\rho = J_{h,D}(\hat{u}) = \min_{u\in\mathcal{U}_{h,D}} J_{h,D}(u) \,.$$

Equation (67) can be viewed as an approximation to the HJB equation (36). Equation (67) gives the solution to the ergodic control problem for the Markov process $X^{h,D}_t$.

4.3 The policy iteration algorithm

In order to solve (67), we use the policy iteration algorithm [15, 22]: suppose that $u^0 \in \mathcal{U}_{h,D}$ — the initial policy — is given. Starting with u^0 we generate a sequence $\{u^j; j \geq 1\}$. The iteration $u^j \to u^{j+1}$ proceeds in two steps (cf. table 5).

1 *compute* (v^j, ρ^j)	we compute $(v^j, \rho^j) \in \mathbb{R}^N \times \mathbb{R}^+$ the solution of the linear system $$\sum_{y\in\mathbb{R}^n_{h,D}} \mathcal{L}^{u^j}_{h,D}(x,y)\, v^j(y) + f^{u^j}(x) = \rho^j \,, \quad \forall x \in \mathbb{R}^n_{h,D}$$
stopping test	$\lvert \rho^{j+1} - \rho^j \rvert \leq \varepsilon$.
2 *compute* u^{j+1}	we solve the N following optimization problems: $\forall x \in \mathbb{R}^2_{h,D}$ $$u^{j+1}(x) \in \operatorname{Arg}\min_{u\in[\underline{u},\overline{u}]} \left(\sum_{y\in\mathbb{R}^2_{h,D}} \mathcal{L}^{h,D}_u(x,y)\, v^j(y) + f^u(x) \right) .$$

Table 5. The policy iteration algorithm, iteration $u^j \to u^{j+1}$.

Remark. The first step of this algorithm leads to a linear system of dimension N. Let $\mathbb{R}^2_{h,D} = \{x^i; i = 1, \ldots, N\}$, then the unknown parameters are

$$v(x^2),\ v(x^3), \ldots,\ v(x^N),\ \rho\,,$$

and we take $v(x^1) = 0$.

Remark. For the second step, the optimization problems are nonlinear and they are solved by means of iterative routines. The nonlinearity comes from the discretization technique we use. Indeed, the choice of finite difference approximation (cf. table 3) depends on u. Instead of the schemes of the table 3, we can use central difference approximation (so that it does not depend on u), in which case the second step becomes explicit because the functions to be optimized are now quadratic in u. On the other hand, with this kind of difference approximation, a certain condition on the parameter h has to be fulfilled (h must be small enough) for the matrix $\mathcal{L}_u^{h,D}$ to be the generator of a Markov process. See [26] p.175–179 for further considerations.

4.4 Application to the class of problem $\mathcal{C}$

The approximation In this example, the discretized state space are $\mathbb{R}_h^2$ and $\mathbb{R}_{h,D}^2$ where $h = (h_1, h_2)$ and D is of the form

$$D = [-\overline{x}_1, \overline{x}_1] \times [-\overline{x}_2, \overline{x}_2] \ ,$$

so

$$\mathbb{R}_{h,D}^2 = \left\{x_1^{(1)}, x_1^{(2)}, \ldots, x_1^{(N_1)}\right\} \times \left\{x_2^{(1)}, x_2^{(2)}, \ldots, x_2^{(N_2)}\right\} \ ,$$

with

$$\begin{aligned} x_1^{(i)} &= -\overline{x}_1 + 2\overline{x}_1 \frac{i-1}{N_1-1} \ , \quad h_1 = \frac{1}{N_1-1} \ , \\ x_2^{(j)} &= -\overline{x}_2 + 2\overline{x}_2 \frac{j-1}{N_2-1} \ , \quad h_2 = \frac{1}{N_2-1} \end{aligned}$$

(cf. figure 2).

The matrix $a(x)$ is degenerate

$$a(x) = \begin{pmatrix} 0 & 0 \\ 0 & \sigma^2 \end{pmatrix} \ .$$

Condition (63) is fulfilled. The finite difference schemes of table 3 are simplified, they are presented on table 6; the terms of the matrix $\mathcal{L}_h^u$ are presented on table 7.

For this example we give explicit boundary conditions. We define

$$\begin{aligned} \Gamma_{h,D} \triangleq & \left\{x_1^{(1)}, x_1^{(N_1)}\right\} \times \left\{x_2^{(1)}, x_2^{(2)}, \ldots, x_2^{(N_2)}\right\} \\ & \cup \left\{x_1^{(1)}, x_1^{(2)}, \ldots, x_1^{(N_1)}\right\} \times \left\{x_2^{(1)}, x_2^{(N_2)}\right\} \ . \end{aligned}$$

$\Gamma_{h,D}$ the set of points on the border. We chose very simple reflecting conditions, we obtain the matrix $\mathcal{L}_{h,D}^u$ described table 8.

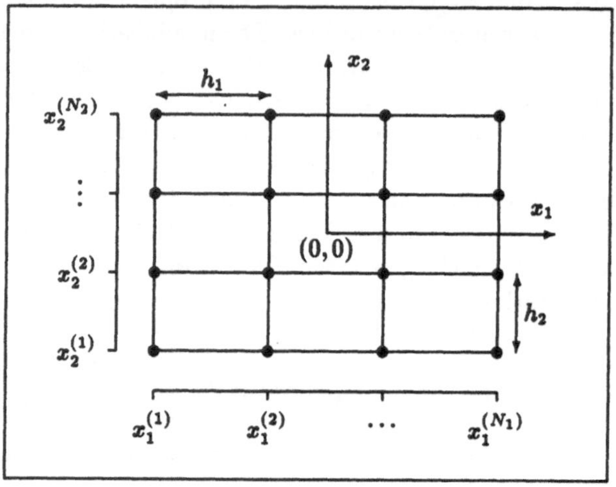

Fig. 2. Discretized state space.

$$\frac{\partial \phi(x)}{\partial x_i} \simeq \begin{cases} \dfrac{\phi(x+e_i\,h_i)-\phi(x)}{h_i} & \text{if } b_i^u(x) > 0 \\ \dfrac{\phi(x)-\phi(x-e_i\,h_i)}{h_i} & \text{if } b_i^u(x) < 0 \end{cases}$$

$$\frac{\partial^2 \phi(x)}{\partial x_2^2} \simeq \frac{\phi(x+e_2\,h_2)-2\,\phi(x)+\phi(x-e_2\,h_2)}{h_2^2}$$

$i = 1, 2$, e_i unit vector in the ith coordinate direction

Table 6. Finite difference schemes (class $\mathcal{C}$).

The convergence results We present two kinds of results. Firstly, consider the discrete HJB equation (67), we can prove that it admits a unique solution and that the policy iteration algorithm converges to this unique solution. Secondly, we can also prove a convergence result for the approximation as the discretization step h tends to 0. These results are presented for the class $\mathcal{C}$.

existence and uniqueness of a solution to the discrete HJB equation We have the following results

Theorem 7. *The HJB equation (67) (with $v(x^1) = 0$) admits a unique solution $(v, \rho) \in \mathbb{R}^N \times \mathbb{R}^+$.*

For the existence part of theorem (7), we use the following

$$\mathcal{L}_h^u(x,x) \triangleq -\frac{\sigma^2}{h_2^2} - \sum_{i=1,2} \frac{|b_i^u(x)|}{h_i}$$

$$\mathcal{L}_h^u(x, x + e_1\, h_1) \triangleq \frac{(b_1^u(x))^+}{h_1}$$

$$\mathcal{L}_h^u(x, x - e_1\, h_1) \triangleq \frac{(b_1^u(x))^-}{h_1}$$

$$\mathcal{L}_h^u(x, x + e_2\, h_2) \triangleq \frac{\sigma^2}{2\, h_2^2} + \frac{(b_2^u(x))^+}{h_2}$$

$$\mathcal{L}_h^u(x, x - e_2\, h_2) \triangleq \frac{\sigma^2}{2\, h_2^2} + \frac{(b_2^u(x))^-}{h_2}$$

Table 7. Discrete infinitesimal generator (class $\mathcal{C}$).

for $x \in \mathbb{R}^2_{h,D} \setminus \Gamma_{h,D}$ et $y \in \mathbb{R}^2_{h,D}$	$\mathcal{L}_{h,D}^u(x,y) = \mathcal{L}_h^u(x,y)$
for $x \in \Gamma_{h,D}$ such that $x_1 = x_1^{(1)}$	$\mathcal{L}_{h,D}^u(x,x) = \mathcal{L}_h^u(x,x)$ $\mathcal{L}_{h,D}^u(x, x + h_1\, e_1) = -\mathcal{L}_h^u(x,x)$
for $x \in \Gamma_{h,D}$ such that $x_1 = x_1^{(N_1)}$	$\mathcal{L}_{h,D}^u(x,x) = \mathcal{L}_h^u(x,x)$ $\mathcal{L}_{h,D}^u(x, x - h_1\, e_1) = -\mathcal{L}_h^u(x,x)$
for $x \in \Gamma_{h,D}$ such that $x_2 = x_2^{(1)}$	$\mathcal{L}_{h,D}^u(x,x) = \mathcal{L}_h^u(x,x)$ $\mathcal{L}_{h,D}^u(x, x + h_2\, e_2) = -\mathcal{L}_h^u(x,x)$
for $x \in \Gamma_{h,D}$ such that $x_2 = x_2^{(N_2)}$	$\mathcal{L}_{h,D}^u(x,x) = \mathcal{L}_h^u(x,x)$ $\mathcal{L}_{h,D}^u(x, x - h_2\, e_2) = -\mathcal{L}_h^u(x,x)$
all other terms are null	

Table 8. Discrete infinitesimal generator $\mathcal{L}_{h,D}^u$ (class $\mathcal{C}$).

Corollary 8. *The policy iteration algorithm converge to an optimal feedback control.*

These results are proved in [4], but one can find the same kind of results in a more general setup in [10].

Approximation: a convergence result We present a convergence result concerning the approximation, when the discretization parameter h tends to 0 and when the domain D tends to $\mathbb{R}^2$ (for the complete proof of this result cf. [13]).

Theorem 9. *We consider two strictly increasing sequences*

$$\{\overline{x}_1^h; h > 0\} \quad \textit{and} \quad \{\overline{x}_2^h; h > 0\}$$

such that $\overline{x}_i^h > 0$ *and* $\overline{x}_i^h \to \infty$ *as* $h \to 0$. *We define*

$$D_h = [-\overline{x}_1^h, \overline{x}_1^h] \times [-\overline{x}_2^h, \overline{x}_2^h] \,.$$

We suppose that

$$\lim_{h\to 0} h\,\delta_h = 0\,, \quad \textit{where} \quad \delta_h \triangleq \textit{radius}(D_h)\,. \tag{69}$$

Then, for any $u \in \mathcal{U}$,

$$J_{h,D_h}(u) \underset{h\to 0}{\to} J(u)\,.$$

Remark. Theorem 7 proves the existence of an optimal feedback control law for the discretized problem. With such a control, we can associate a feedback control law $\hat{u}_h$ for the continuous state space problem, where $\hat{u}_h$ is piecewise constant. Using theorem 9 we can easily conclude that

$$\limsup_{h\to 0} J_{h,D_h}(\hat{u}_h) \leq \inf_{u\in\mathcal{U}} J(u)\,.$$

We would like to prove the stronger result that the sequence $\{\hat{u}_h; h > 0\}$ is a minimizing sequence for the functional J, i.e.

$$J(\hat{u}_h) \to \inf_{u\in\mathcal{U}} J(u)\,, \quad \text{when } h \to 0\,.$$

A numerical example

parameters As an example, we use values which roughly correspond to a suspension system for the seat of a truck: $m = 60$(kg), $k_s = 3500$(N/m), $F_s = 40$(N). These values have already been used in [4]. We also set $\sigma = 0.5$.

We use the following discretization parameters

$$\begin{aligned}
\overline{x}_1 &= y_{max} = -y_{min} = 0.1 \text{ (m)}\,,\\
\overline{x}_2 &= \dot{y}_{max} = -\dot{y}_{min} = 1 \text{ (m/s)}\,,\\
n_1 &= n_2 = 30\,.
\end{aligned}$$

So we get a $30 \times 30 = 900$ points grid.

optimal feedback control[4] The approximated optimal feedback control (68) (plotted on figure 3) is computed using the policy iteration algorithm. The value of the minimal cost is given below.

Now we present suboptimal control laws, for a more general discussion concerning these techniques one can consult [4, 3].

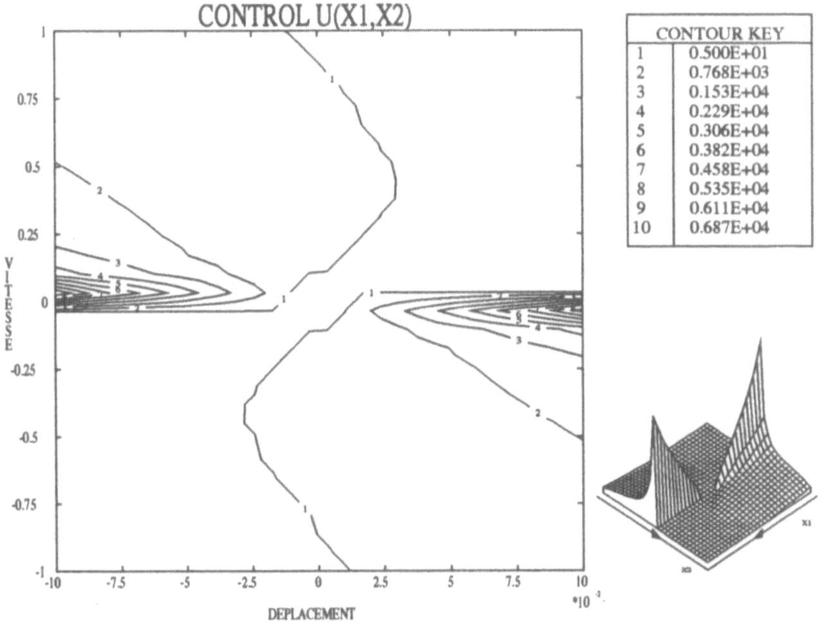

Fig. 3. The optimal feedback control.

suboptimal feedback control #1 ne possibility is to find a feedback control which minimizes the instantaneous cost function (59). We obtain

$$\tilde{u}(x) = \frac{-k_s\, x_1 \mathrm{sign}(x_2) - F_s}{|x_2|} \,.$$

To take into account the constraint $\underline{u} \leq u \leq \overline{u}$, we use the following control law

$$u(x) = (\tilde{u}(x) \vee \underline{u}) \wedge \overline{u} \,,$$

(cf. figure 4) (we take $\underline{u} = 0$ and $\overline{u}$ large).

suboptimal feedback control #2 The previous results lead us to the class of suboptimal feedback controls — parametrized by $\theta \in \mathbb{R}^2$ — of the following form

$$u_\theta(x) \stackrel{\Delta}{=} [(\theta_1 + \theta_2\, x_1\, \mathrm{sign}(x_2)) \vee \underline{u}) \wedge \overline{u} \,, \quad \theta = (\theta_1, \theta_2) \in \mathbb{R}^2 \,. \tag{70}$$

The techniques presented above can also be applied to compute the suboptimal feedback control $u_{\hat{\theta}}$ such that

$$J_{h,D_h}(u_{\hat{\theta}}) = \min_{\theta \in \Theta} J_{h,D_h}(u_\theta) \,,$$

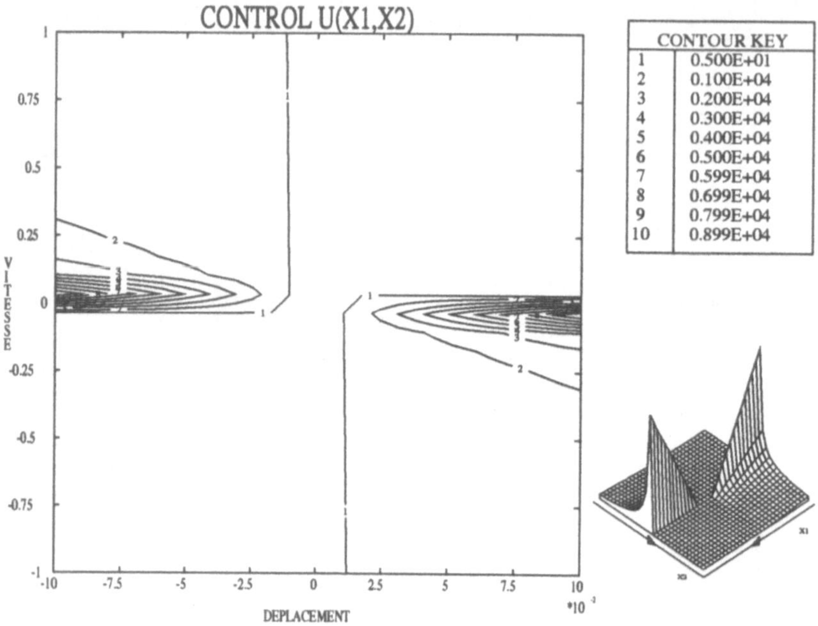

Fig. 4. The suboptimal feedback control #1.

where $\Theta = \{\theta \in \mathbb{R}^2\,;\, u_\theta \in \mathcal{U}\}$. We get

$$\hat{\theta}_1 = 137.2,\ \hat{\theta}_2 = -12130. \tag{71}$$

The control law $u_{\hat{\theta}}(x)$ is plotted on figure 5. A feedback control where the sign of the product $x_1\, x_2$ (i.e. $y\,\dot{y}$) appears has already been proposed in [37].

comparison of the feedback controls Now we compare the three feedback controls presented above to the constant control $u(X) \equiv u_0$. The optimal constant u_0 (i.e. the constant which minimizes the cost) is 188. The different values of the cost are given in the following table

control type	*cost*
constant control	2.93
suboptimal feedback control #1	2.68
suboptimal feedback control #2	2.37
optimal control	2.22

References

1. M. AKIAN, J.P. CHANCELIER, and J.P. QUADRAT. Dynamic programming complexity and application. In *Proceedings of 27th Conference on Decision and Control*, pages 1551–1558, Austin, Texas, December 1988. IEEE.

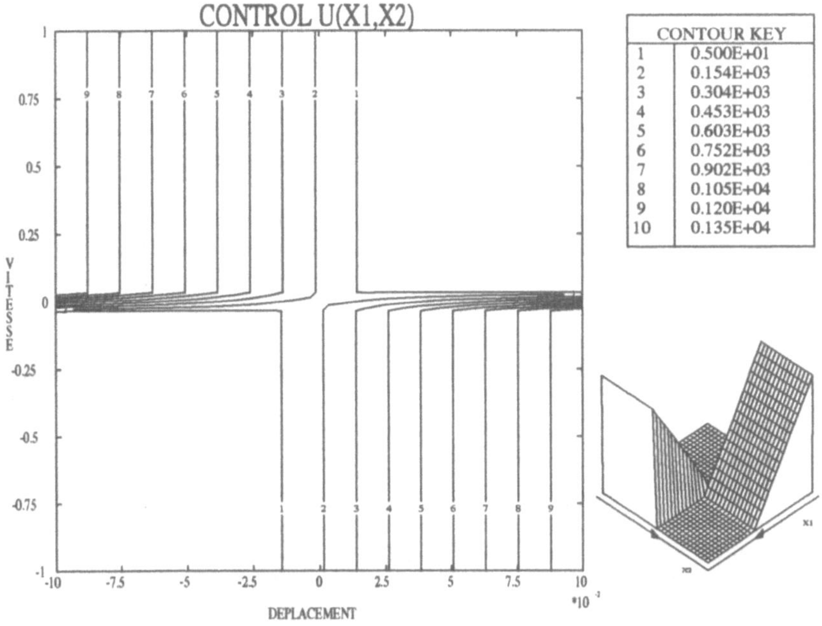

Fig. 5. The suboptimal feedback control #2.

2. J. ALANOLY and S. SANKAR. Semi-active force generators for shock isolation. *Journal of Sound and Vibration*, **126**(1):145–156, 1988.
3. S. BELLIZZI and R. BOUC. Adaptive suboptimal parametric control for nonlinear stochastic systems — Application to semi-active vehicle suspensions. In *Effective Stochastic, P. Krée & W. Wedig (eds.)*, 1989. To appear.
4. S. BELLIZZI, R. BOUC, F. CAMPILLO, and E. PARDOUX. Contrôle optimal semi-actif de suspension de véhicule. In *Analysis and Optimization of Systems, A. Bensoussan and J.L. Lions (eds.)*. INRIA, Antibes, 1988. Lecture Notes in Control and Information Sciences 111, 1988.
5. A. BENSOUSSAN. *Stochastic Control by Functional Analysis Methods*, volume 11 of *Studies in Mathematics and its Applications*. North-Holland, Amsterdam, 1982.
6. A. BENSOUSSAN. Discretization of the Bellman equation and the corresponding stochastic control problem. In *26th Conference on Decision and Control*, pages 2251–2254, Los Angeles, CA, December 1987.
7. A. BENSOUSSAN. *Perturbation Methods in Optimal Control.* John Wiley & Sons, New-York, 1988.
8. A. BENSOUSSAN and J.L. LIONS. *Applications des Inéquations Variationnelles en Contrôle Stochastique*, volume 6 of *Méthodes Mathématiques pour l'Informatique*. Dunod, Paris, 1978.
9. A. BENSOUSSAN and J.L. LIONS. *Contrôle Impulsionnel et Inéquations Quasi Variationnelles*, volume 11 of *Méthodes Mathématiques pour l'Informatique.* Dunod, Paris, 1982.

10. D.P. BERTSEKAS. *Dynamic Programming and Stochastic Control*, volume 123 of *Mathematics in Science and Engineering*. Academic Press, New–York, 1976.
11. R. BOUC. Forced vibration of mechanical system with hysteresis. In *Proceedings of 4th conference ICNO*, Prague, 1967. Résumé.
12. R. BOUC. Modèle mathématique d'hystérésis. *Acustica*, **24**(3):16–25, 1971.
13. F. CAMPILLO, F. LE GLAND, and E. PARDOUX. Approximation d'un problème de contrôle ergodique dégénéré. In *Colloque International Automatique Non Linéaire*, Nantes, 13–17 Juin 1988. CNRS. To appear.
14. F. DELEBECQUE and J.P. QUADRAT. Contribution of stochastic control singular perturbation averaging and team theories to an example of large–scale systems: management of hydropower production. *IEEE Transactions on Automatic Control*, **AC–23**(2):209–221, April 1978.
15. B.T. DOSHI. Continuous time control of Markov processes on an arbitrary state space: average return criterion. *Stochastic Processes and their Applications*, **4**:55–77, 1976.
16. N. ETHIER and T.G. KURTZ. *Markov Processes - Characterization and Convergence*. J. Wiley & Sons, New–York, 1986.
17. W.H. FLEMING and R.W. RISHEL. *Deterministic and Stochastic Optimal Control*, volume 1 of *Applications of Mathematics*. Springer–Verlag, New York, 1975.
18. F. LE GLAND. *Estimation de Paramètres dans les Processus Stochastiques, en Observation Incomplète — Applications à un Problème de Radio–Astronomie*. Thèse de Docteur–Ingénieur, Université de Paris IX – Dauphine, 1981.
19. R.M. GOODDALL and W. KORTUM. Active controls in ground transportation — A review of the state–of–the–art and future potential. *Vehicle systems dynamics*, **12**:225–257, 1983.
20. U.G. HAUSSMANN. *A Stochastic Maximum Principle for Optimal Control of Diffusions*, volume 151 of *Pitman Research Notes in Mathematics Series*. Longman, Harlow, UK, 1986.
21. R.H.W. HOPPE. Multi–grid methods for Hamilton–Jacobi–Bellman equations. *Numerische Mathematik*, **49**:239–254, 1986.
22. R.A. HOWARD. *Dynamic Programming and Markov Processes*. J. Wiley, New–York, 1960.
23. I. KARATZAS and S.E. SHREVE. *Brownian Motion and Stochastic Calculus*, volume 113 of *Graduate Texts in Mathematics*. Srpinger–Verlag, New–York, 1988.
24. N. EL KAROUI. *Les aspects probabilistes du contrôle stochastique. Ecole d'Eté de Probabilité de Saint–Flour IX–1979, P.L. Hennequin (ed.)*, volume 876 of *Lecture Notes in Mathematics*. Springer–Verlag, Berlin, 1981.
25. H.J. KUSHNER. *Introduction to Stochastic Control*. Holt, Rinehart and Winston Inc., New York, 1971.
26. H.J. KUSHNER. *Probability Methods for Approximations in Stochastic Control and for Elliptic Equations*, volume 129 of *Mathematics in Science and Engineering*. Academic Press, New–York, 1977.
27. H.J. KUSHNER. Optimality conditions for the average cost per unit time problem with a diffusion model. *SIAM Journal of Control and Optimization*, **16**(2):330–346, March 1978.
28. H.J. KUSHNER. Numerical methods for stochastic control problems in continuous time. Technical report, Lefschetz Center for Dynamical Systems, 1988.
29. H. KWAKERNAAK and R. SIVAN. *Linear Optimal Control Systems*. John Wiley & Sons, New York, 1972.

30. P.L. LIONS. *Generalized solutions of Hamilton–Jacobi equations*, volume 69 of *Research Notes in Mathematics*. Pitman, Boston, 1982.
31. P.L. LIONS. On the Hamilton–Jacobi–Bellman equations. *Acta Applicandae Mathematicae*, 1:17–41, 1983.
32. P.L. LIONS and B. MERCIER. Approximation numérique des équations de Hamilton–Jacobi–Bellman. *R.A.I.R.O. Numerical Analysis*, 14(4):369–393, 1980.
33. D. MICHEL and E. PARDOUX. An introduction to Malliavin's calculus and some of its applications. To appear.
34. J.P. QUADRAT. Existence de solution et algorithme de résolution numérique, de problèmes de contrôle optimal de diffusion stochastiques dégénérée ou non. *SIAM Journal of Control and Optimization*, 18(2):199–226, March 1980.
35. J.P. QUADRAT. *Sur l'Identification et le Contrôle de Systèmes Dynamiques Stochastiques*. Thèse, Université de Paris IX – Dauphine, 1981.
36. M. ROBIN. Long–term average cost control problems for continuous time Markov processes: a survey. *Acta Applicandae Mathematicae*, 1:281–299, 1983.
37. S. TAKAHASHI, T. KANEKO, and K. TAKAHASHI. A damping force control method which reduce energy to the vehicle body. *JSAE Review*, 8(3):95–98, 1987.

Stochastic Dynamics of Hysteretic Media

Fabio Casciati

Abstract

Classic plasticity theory regards the yielding condition as discontinuity between elastic and plastic phases. This discontinuity leads obvious negative consequences in the mathematical features of the algorithm one uses in solving solid and structural mechanics problems.

Smoothed plasticity models are presently available in the framework of endochronic theory. This contribution discusses in particular three-dimensional tensorial smoothed idealizations of the Prager's model. Multivariate smoothed constitutive laws are also provided at a section level. Some aspect of the stochastic equivalent linearization algorithm which makes direct use of this smoothed model are discussed.

1 Introduction

The design of a structual component is likely supported by linear elastic models of the material constitutive law. Safety assessment, damage calculation or behaviour under extreme events, however, need non linear analyses.

Non linearity requires the system is considered in its three- dimensional extension. It can be discretized into 3D finite elements and a step-by-step integration of the equilibrium or motion equations can be conducted. The inconveniences of such a procedure are not of minor nature: i) a time history of the external action has to be assigned, ii) the computational effort required is still significant even with the recent developments of calculus facilities, iii) the result is relevant to a case-study but is poor for the subsequent engineer's decision making.

The computational effort could be strongly reduced by using smoothed forms of the elasto-plastic constitutive law [1]. This would also open the way to the application of stochastic techniques able to incorporate the variability of the external actions and to give a probabilistic description of the resulting response [2,3].

A further reduction of the computational effort could be obtained by assigning the constitutive law at a section level, rather than as a stress strain relationship. Thus, the three dimensional extension of the structural system is maintained, but the use of 3D finite elements is avoided [4,5]. Unfortunately, one discovers here that no experimental result exists to support any model of the elasto-plastic behaviour under cyclic biassial loads [6].

Section 2 considers the problem of smoothing the plasticity law. Stochastic equivalent linearization is summarized in Section 3. Section 4, finally, approaches the problem of developing a similar method at a structural level. For the appropriate numerical examples the reader is referred to [7] and [5].

2 Smoothed Plasticity

2.1 Univariate Case

Independently of endochronic theory, with which relations were successively stated (see [8]), Bouc proposed in [1] a functional form able to provide a mathematical representation of the hysteresis in a system on condition that it does not depend on time (for this extension see [9]). Under some regularity conditions, one writes:

$$\Psi(t) = a^2 U(t) + f(U(t)) + \int_{t_0}^{t} F(\delta U) d_s \rho(U(s)) \tag{1}$$

In (1) δU is the total variation of U and, for U(t) regular:

$$\delta U = \int_s^t \left| \frac{dU}{dv} \right| dv \tag{2}$$

A model appropriate for mechanical use is found by putting:

$$\left.\begin{array}{c} f(U) = 0, \quad \rho(U) = U \\ F(U) = A\exp(-\beta U), \quad \beta > 0, \quad A > 0 \end{array}\right\} \tag{3}$$

leading to

$$\Psi(t) = a^2 U(t) + \int_{t_0}^{t} A\exp\left[-\beta \int_s^t \left|\frac{dU}{dv}\right| dv\right] \left(\frac{dU}{ds}\right) ds = a^2 U(t) + Z(t) \tag{4}$$

In (4) Z(t), i.e. the integral over the interval (t_0, t), is also the solution of the differential equation

$$\dot{Z} + \beta|\dot{U}|Z = A\dot{U} \tag{5}$$

In the previous equations, Ψ denotes a generalized force, U a generalized displacement and Z an auxiliary variable whose relation with U (see equation 5) is of hysteretic nature. Improved models were provided by Bouc [10] (6), Wen [2] (7) and the author [11] (8), respectively, acting over (5):

$$\dot{Z} + \beta|\dot{U}|Z + \gamma\dot{U}|Z| = A\dot{U} \tag{6}$$

$$\dot{Z} + \beta|\dot{U}||Z|^{n-1}Z + \gamma\dot{U}|Z|^n = A\dot{U} \tag{7}$$

$$\dot{Z} + \beta|\dot{U}||Z|^{n-1}Z + \gamma\dot{U}|Z|^n - \left[\beta_1|\dot{U}||Z|^{n_1-1}Z + \beta_1\dot{U}|Z|^{n_1}\right] = A\dot{U} \qquad n_1 < n \tag{8}$$

Equation (7), for $a^2 = 0$, (i.e: in absence of post-yielding hardening), provides a relationship $\Psi - U$ which tends to be elasto perfectly plastic as n tends to ∞ [2]. Note that the yielding value is obtained for $\dot{Z} = 0$ as

$$\Psi_y = \left[\frac{A}{\beta + \gamma}\right]^{1/n} \tag{9}$$

In this paper attention is focused on $n = 1$ (and hence $n_1 = 0$), i.e. on a smoothed model of the elasto-plastic behaviour. It is easy to show that the energy dissipated in every hysteresis cycle is positive for $\beta + \gamma > 0$ and $\gamma - \beta \leq 0$ (β being positive by assumption). Still assume $a^2 = 0$ and divide U into the elastic, U_e, and plastic, U_p, parts:

$$U_p = U - U_e \tag{10}$$

with

$$U_e = \frac{Z}{A} \tag{11}$$

Introduce now the following positions:

$$\begin{gathered} \zeta = \beta + \gamma > 0 \\ \beta = \nu\zeta \\ \gamma = (1 - \nu)\zeta \end{gathered} \tag{12}$$

from which $\nu > 0$ (since $\beta < 0$) and $1 - 2\nu \leq 0$ (since $\gamma - \beta \leq 0$). From (6), one writes

$$\dot{U}_e + \beta|\dot{U}|U_e + \gamma\dot{U}|U_e| = \dot{U} \tag{13}$$

and, by (10):

$$\beta|\dot{U}|U_e + \gamma\dot{U}|U_e| = \dot{U}_p \tag{14}$$

Multiplying (14) by $Z = AU_e$ (11), one obtains

$$\beta A|\dot{U}|U_e^2 + \gamma A\dot{U}|U_e|U_e = Z\dot{U}_p \tag{15}$$

and (12) leads to

$$\zeta A\left[\nu|\kappa| + (1 - \nu)\kappa\right] = Z\dot{U}_p \geq 0 \tag{16}$$

where $\kappa = \dot{U}|U_e|U_e$. In (15) $\kappa > 0$ implies a positive r.h.s. since ζ and A are positive and $\kappa < 0$ implies a non-negative r.h.s. since $1 - 2\nu \leq 0$. Equation (15) is therefore the analytical form of the Drucker's stability postulate of classic plasticity theory. It ensures the positiveness of the dissipated energy over the cycle.

Equation (15), however, does not completely agree with classic plasticity theory because it violates the so-called complementary rule. Let

$$\Phi(\Psi) = (\Psi - \Psi_y) = 0 \tag{17}$$

be the yielding surface: the complementary rules states that

$$\Phi(\Psi) \cdot \dot{U}_p = 0 \tag{18}$$

In the unloading stage, (18) is satisfied only when $\gamma = \beta$. In this case the unloading occurs along a straight line of slope A, and hence, $\dot{U}_p = 0$. For $\beta > \gamma$, $\dot{U}$ is greater than 0 even when Z decreases (unloading). During the loading stage, (18) is never satisfied since $\dot{U}_p$ is greater than 0 for any β and γ satisfying (12). Nevertheless, since $\dot{U}_p$ increases as Z increases, one can regards (6) as a smoothed elasto-plastic constitutive law. Its main inconvenience is that the hysteresis loops during an unload-reload cycle for which Ψ does not change sign do not close. As a consequence, the work done on the material during this cycle is negative and, hence, the Drucker's stability postulate is violated despite (15) [12]. Of course, the entity of this violation can be reduced by increasing n.

High values of n make obviously cumbersome a step-by-step integration of the non-linear equilibrium/motion equations; also they reduce the accuracy achievable by using stochastic equivalent linearization techniques [13]. Therefore, the author tried to reduce the violation of the Drucker's postulate without increasing the exponent n. This is obtained by adding (8) a second hysteretic item I_2 to the Bouc's constitutive law equation contributed by a single hysteretic device I_1 . This additional item is counterclockwise, as $\beta_1 > 0$ appears with negative sign in (8), and is effective only during the loading ($\gamma_1 = \beta_1$). During the loading phase, therefore, (10) gives rise to "negative" inelastic displacements U_p with associated negative dissipated energy linearly proportional to U^2 for $n_1 = 0$.

The energy dissipated by I_1 is initially unable to conpensate the negative dissipation of I_2, but the rate of dissipating energy is higher for I_1, since its equation is characterized by an exponent n higher than n_1 in I_2. Therefore, in the initial stage of loading the energy ($\int_{t_0}^{t} Z\dot{U}dt$) is the one of an elastic system of stiffness $A + 2\beta_1$ rather than A and, in case of immediate unloading, the dissipated energy comes out to be negative. This inconvenience will be discussed later.

With reference to Fig. 1, during the reloading BC a clockwise loop with the unloading branch AB is produced: it ends when

$$U_{pm} - U_{pu} = U_{pr} \tag{19}$$

and produces a dissipated energy ΔE_1. If the selection of β_1 is made such that the work $\Delta E_1 - \Delta E_2$ (in Fig. 1) done on the material during the unload-reload cycle is non-negative, the Drucker's stability postulate would still be satisfied. For this purpose it is sufficient that, for C appropriately close to the yielding level

$$U_{pm}\Psi_y \leq \int_A^B Z\dot{U}_p dt - \int_B^C (Z\dot{U}_p)_1 dt \tag{20}$$

where $Z\dot{U}_p$ is given by 15 and $(Z\dot{U}_p)_1$ is the analogous term, everywhere negative, derived from (8):

$$\zeta A\left[\nu|\kappa| + (1-\nu)\kappa\right] - \beta_1 A\left[\frac{|\kappa|}{|U_e|} + \frac{\kappa}{|U_e|}\right] = (Z\dot{U}_p)_1 \tag{21}$$

The negativeness of the r.h.s. of (21) defines the position of point C

$$\beta_1\left[|\kappa| + \kappa\right] \geq \zeta\left[\nu|\kappa| + (1-\nu)\kappa\right]|U_e| \tag{22}$$

which for $|U_e| = |U_y| = |\frac{\Psi_y}{A}|$ provides a further constraint on β_1. It is immediate

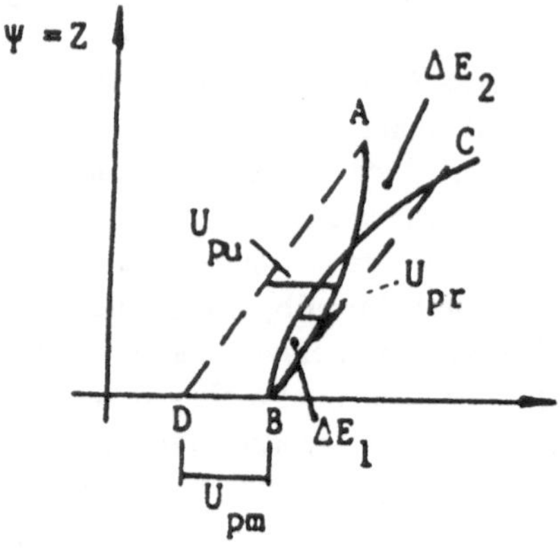

Fig. 1. Drucker's stability postulate is violated everywhen $\Delta E_1 - \Delta E_2 < 0$

to realize that for $AB \equiv AD$ in Fig. 1 (i.e. $U_{pm} = 0$ since $\gamma - \beta = 0$) (22) is the only condition for having the Drucker's stability postulate satisfied in the unload-reload cycle. On the other hand, any unloading curve will cross the loading one, violating the postulate in the load-unload cycle. Similar situations arise for $\gamma - \beta < 0$. In this sense one must interpretate the following theoretical result recently announced by Iwan [14]: "when endochronic models are adopted, local violations of the Drucker's stability postulate cannot be avoided".

Therefore, if these smoothed forms of plasticity are accepted due to their computational advantages, the objective should be the reduction of the Drucker's postulate violations either by increasing n in (7) or, better, by adopting appropriate parameters in the linear form ($n = 1$, $n_1 = 0$) of (8). The experience of [11] suggests $\beta = -2\gamma$ (i.e. $U_{pm} > 0$) in order to reduce load-unload cycle violations and $\beta_1 = 2$ in order to reduce small-amplitude cycle violations: the global result makes $\Delta E_1 - \Delta E_2$ a small negative quantity but not a positive one as Drucker's postulate would require.

About the need of substituting (6) with (8), it is worth noting that comparisons were made for systems dynamically loaded by a stationary stochastic external excitation [5] and no menaningful difference was detected in terms of dissipated energy. However, this is a general result of non-linear random vibration (i.e. for stationary excitation the dissipated energy does not depend on the form of the system constitutive law) and, hence, comparisons under non-stationary external actions are presently in progress.

In the previous models, the hardening can be included by putting

$$\Psi = K_H U + Z \tag{23}$$

instead of $\Psi = Z$ and, for instance for (7),

$$\dot{Z} + \beta|\dot{U}||Z|_{n-1}Z + \gamma\dot{U}|Z|^n = (K - K_{\mathrm{H}})\dot{U} \tag{24}$$

with K denoting the small-cycle stiffness (initial stiffness; it coincides with A in (7)) and $K_{\mathrm{H}} = a^2 = \alpha K (0 < \alpha < 1)$ the large-cycle stiffness (post-yielding stiffness). The ratio α is called the hardenging ratio. Note that (9) in this case becomes

$$\Psi_{\mathrm{y}} = \frac{K_{\mathrm{H}}\Psi_{\mathrm{y}}}{K} + \left[\frac{K - K_{\mathrm{H}}}{\beta + \gamma}\right]^{1/n} = \frac{K(K - K_{\mathrm{H}})^{1/n-1}}{(\beta+\gamma)^{1/n}} \tag{25}$$

Alternatively one can write

$$\Psi = K_{\mathrm{H}}U + (K - K_{\mathrm{H}})Z_{\mathrm{a}} \quad \text{in place of (23)} \tag{26}$$

$$\dot{Z}_{\mathrm{a}} + \beta_{\mathrm{a}}|Z_{\mathrm{a}}|^{n-1}|\dot{U}|Z_{\mathrm{a}} + \gamma_{\mathrm{a}}\dot{U}|Z_{\mathrm{a}}|^n = \dot{U} \quad \text{in place of (24)} \tag{27}$$

$$\Psi_{\mathrm{y}} = \frac{K}{(\beta_{\mathrm{a}} + \gamma_{\mathrm{a}})^{1/n}} \quad \text{in place of (25)} \tag{28}$$

In the report [9] and in the successive papers [3,11,13,15,4,5] (28) is written as

$$\Psi_{\mathrm{y}} = \frac{K - K_{\mathrm{H}}}{(\beta_{\mathrm{a}} + \gamma_{\mathrm{a}})^{1/n}} \tag{29}$$

It corresponds to a different defintion of the yielding level, aiming at defining the amplitude of the loop of the hysteretic item rather than the actual yielding value of plasticity theory. The difference between the values provided by (25) and (29), however, is meaningless for low values of the ratio $\alpha = \frac{K_{\mathrm{H}}}{K}$. These low values characterize the constitutive laws (26) and (27) of potential plastic hinges as in [3,4,5]. By contrast for large values of α (28) should be preferred when comparisons with results of elasto-plastic analyses are pursued.

2.2 Multivariate Case

According to classic plasticity theory (Prager's model), the plastic strain tensor $\varepsilon^p_{ij}(t)$ in any point of an elasto-plastic medium should be computed by integrating along the appropriate path Γ the plastic strain rate tensor $\dot{\varepsilon}^p_{ij}$ [7].

$$\varepsilon^p_{ij}(t) = \int_\Gamma \dot{\varepsilon}^p_{ij}(\tau)d\tau \tag{30}$$

The strain rate tensor is a function of the plastic potential Φ, which, in turn, is a function of the stress tensor σ_{ij} and of the internal variables η_{ij} depending on the previous $\sigma - \varepsilon$ history:

$$\dot{\varepsilon}^p_{ij} = \frac{\partial\Phi(\sigma_{ij}, \eta_{ij})}{\partial\sigma_{ij}}\dot{\lambda} \tag{31}$$

where the function to be derived (i.e. the plastic potential) coincides with the yielding condition Φ for materials with associate flow rule. The $\dot{\lambda}$ is the corresponding plastic multiplier. Equation (31) must be completed by the normality law:

$$\begin{aligned} \dot{\lambda} &\geq 0 \\ \Phi(\sigma_{ij}, \eta_{ij}) &\leq 0 \\ \Phi\dot{\lambda} &= 0 \\ \dot{\lambda}\frac{\partial \Phi}{\partial \sigma_{ij})}\dot{\alpha}_{ij}) &\geq 0 \end{aligned} \tag{32}$$

and by the kinematic hardening rule

$$\eta_{ij} = C\dot{\varepsilon}^{p}_{ij} \tag{33}$$

A classical form of $\Phi(\sigma_{ij}, \eta_{ij})$ is the von Mises model

$$\Phi(\sigma_{ij}, \eta_{ij}) = \|S_{ij} - \eta_{ij}\| - S_{\mathrm{M}} \leq 0 \tag{34}$$

where S_{ij} denotes the deviatoric part of the stress tensor and

$$\|S_{ij} - \eta_{ij}\| = \sqrt{(S_{ij} - \eta_{ij})(S_{ij} - \eta_{ij})} \tag{35}$$

In (34) S_{M} is the yielding constant. Equation (35) leads to write

$$\frac{\partial \Phi}{\partial \sigma_{ij}} = \frac{S_{ij} - \eta_{ij}}{\|S_{ij} - \eta_{ij}\|} \tag{36}$$

Of course, if μ denotes the elastic shear modulus, one has

$$\dot{S}_{ij} = 2\mu\left(\dot{e}_{ij} - \lambda\frac{S_{ij} - \eta_{ij}}{\|S_{ij} - \eta_{ij}\|}\right) \tag{37}$$

where e_{ij} is the deviatoric part of the strain tensor.

For sake of lightining the notation, from now on, any tensor component y_{ij} is denoted by y, the tensorial product of y by x is marked $y \otimes x$ and the position

$$Y = S - \eta \tag{38}$$

is introduced. In [16], on the suggestion of the adviser Bouc, it is noted that

1. by the third constraint of (25) the plastic multiplier is:

$$\dot{\lambda} = 0 \text{ for } \begin{cases} \Phi < 0 \\ \Phi = 0 \quad \frac{\partial \Phi}{\partial \sigma} \otimes \dot{\sigma} < 0 \end{cases} \tag{39}$$

$$\dot{\lambda} > 0 \text{ for } \Phi = 0 \quad \frac{\partial \Phi}{\partial \sigma} \otimes \dot{\sigma} \geq 0 \tag{40}$$

2. the consistency condition during the plastic flow states that

$$\frac{\partial \Phi}{\partial \sigma} \otimes \dot{\sigma} + \frac{\partial \Phi}{\partial \eta} \otimes \dot{\eta} = 0 \tag{41}$$

which ensures

$$Y \otimes \dot{S} = Y \otimes \dot{\eta} \tag{42}$$

By (37), (31) provides, after re-arrangement,

$$2\mu \|Y\| \dot{\lambda} = 2\mu(Y \otimes \dot{e}) - (Y \otimes \dot{S}) = 2\mu(Y \otimes \dot{e}) - (Y \otimes \dot{\eta}) \tag{43}$$

and, in force of (33), during the plastic flow one can write

$$\dot{\lambda} = \frac{2\mu(Y \otimes \dot{e})}{(2\mu + C)\|Y\|} \tag{44}$$

Equations (31,39,40,44) can be summarized in the single expression

$$\dot{\varepsilon}^p = 2\mu \frac{Y \otimes \dot{e}}{(2\mu + C)\|Y\|} \frac{Y}{\|Y\|} H_1(\|Y\|) H_2(\frac{\partial \Phi}{\partial \sigma} \otimes \dot{\sigma}) \tag{45}$$

with the help of two Heaviside functions:

$$\begin{aligned} H_1(x) &= 1 \text{ for } x > S_{\mathrm{M}} \\ H_2(x) &= 1 \text{ for } x > 0 \end{aligned} \tag{46}$$

With the substitution of (46) in (37), the latter equation becomes, (remembering that $\dot{S} = \dot{Y} + C\dot{\varepsilon}^p$)

$$\dot{Y} = 2\mu \left(\dot{e} - \frac{Y \otimes \dot{e}}{\|Y\|} \right) \frac{Y}{\|Y\|} H_1(\|Y\|) H_2(\frac{\partial \Phi}{\partial \sigma} \otimes \dot{\sigma}) \tag{47}$$

Smoothing the plasticity laws means now to smooth the Heaviside function H_1 for which Bouc proposed:

$$H_1(x) = \frac{x^n}{S_{\mathrm{M}}^n} \tag{48}$$

Writing the other Heaviside function in the form

$$H_2(\frac{\partial \Phi}{\partial \sigma} \otimes \dot{\sigma}) = \frac{1 + \operatorname{sign}(\frac{\partial \Phi}{\partial \sigma} \otimes \dot{\sigma})}{2} = \frac{Y \otimes \dot{e} + |Y \otimes \dot{e}|}{2(Y \otimes \dot{e})} \tag{49}$$

one finds the three-dimensional Bouc's model, for $\beta = \frac{\mu}{S_{\mathrm{M}}^n}$ and $A = 2\mu$:

$$\dot{Y} + \beta \|Y\|^{n-2}(Y \otimes \dot{e})Y + \beta \|Y\|^{n-2}|Y \otimes \dot{e}|Y = A\dot{e} \tag{50}$$

Its generalization is

$$\dot{Y} + \beta \|Y\|^{n-2}|Y \otimes \dot{e}|Y + \gamma \|Y\|^{n-2}(Y \otimes \dot{e})Y = A\dot{e} \tag{51}$$

for which it is easy to prove the corresponding of (9):

$$\|Y\| \le \left[\frac{A}{\beta + \gamma} \right]^{1/n} \quad \text{for } \beta + \gamma > 0 \tag{52}$$

In [7] a constitutive equation alternative to (51) is proposed. It is founded on a piece-wise linearization of the yielding surface and the Koiter's hardening rule [4,5]. Its discussion is postponed to Sect.4.

3 Stochastic Equivalent Linearization

Consider an elasto-plastic medium: its behaviour can be regarded as the sum of the elastic response to the loads W and the ont to inelastic strain $\underline{\varepsilon}_p$ at the Gauss points of the finite elements by which the continuum is discretized. Since the response is always of elastic type, the equation of motion can be written:

$$\underline{m}\ddot{\underline{u}} + c\underline{k}\dot{\underline{u}} + \underline{k}\underline{u} - \underline{k}_\varepsilon\underline{\varepsilon}_p - \underline{W} = 0 \tag{53}$$

where the matrices of mass, $\underline{m}$, abd stiffness $\underline{k}$ and $\underline{k}_\varepsilon$, are assembled over the structure. The damping matrix is assumed to be proportional to $\underline{k}$ by the coefficient. Matrix $\underline{k}_\varepsilon$ is evaluated by elastic analysis of the system under an imposed deformation 1 in the single Gauss point. The stress at the Gauss points is

$$\underline{\sigma} = \underline{E}_1\underline{u} + \underline{E}_2\underline{\varepsilon}_p \tag{54}$$

where $\underline{E}_1$ and $\underline{E}_2$ are static matrices to be estimated by elastic analysis: E_{1kj} is the stress in the kth Gauss point due to a value 1 of the jth displacement; E_{2kl} is the stress in the kth Gauss point due to a value 1 of the strain in the lth Gauss point. The constitutive law in the kth Gauss point is then

$$\underline{\varepsilon}_k = \underline{e}_k + e_s\underline{\delta} \tag{55}$$

$$\underline{\varepsilon}_k^p = \underline{e}_k - \underline{e}_k^e \tag{56}$$

$$\underline{\sigma}_k = \underline{S}_k + \sigma_s\underline{\delta} \tag{57}$$

$$\underline{e}_k^e = \frac{\underline{S}_k}{2\mu} \tag{58}$$

$$\sigma_s = (3\Lambda + 2\mu)\varepsilon_s \tag{59}$$

$$\underline{S}_k = (1-\alpha)\underline{K}_k\underline{Z}_k + \alpha\underline{K}_k\underline{e}_k \tag{60}$$

$$\dot{\underline{Z}}_k = \dot{\underline{Z}}_k(\dot{\underline{e}}_k, \underline{Z}) \tag{61}$$

where $\underline{\delta}$ is a vector of Kronecker indexes and Λ is the Lame's elastic constant. Strain, ε, deviatoric strain, e, and isotropic strain, e_s, were already defined in the previous section, as well as the stress contributions.

When $\underline{W}$ is a vector of stochastic processes $\underline{W}'$, (53) is well tackled by stochastic equivalent linearization techniques [17,18,3,7]. At the kth Gauss point, the endochronic constitutive law (61), expressed in terms of the auxiliary variable Z, is linearized and usual matrix algebra [3,7] leads eventually to write

$$\dot{\underline{d}} + \underline{\Omega}\underline{d} = \underline{w} \tag{62}$$

with $\underline{d}^T = \{\underline{u}^T, \dot{\underline{u}}^T, \underline{e}^T\}$. Some elements of the matrix in (62) depends on the linearization coefficients which are "a priori" unknown. The covariance matrix of $\underline{d}$ is the solution of the matrix equation

$$\frac{d\underline{\Sigma}^d}{dt} + \underline{\Omega}\underline{\Sigma}_d + \underline{\Sigma}_d\underline{\Omega}^T = \underline{T} \tag{63}$$

The r.h.s. $\underline{T} = E[\underline{w}\underline{d}^{\mathrm{T}}] + E[\underline{d}\underline{w}^{\mathrm{T}}]$ can be easily computed when the system is excited by a single white noise (f.i. the ground acceleration due to a seismic excitation).

For the problem till now considered the task of the analyst is therefore twofold:

1. to express the structural matrices which form $\underline{\Omega}$ by finite element discretization and successive algebra;
2. to solve (63) by numerical integration, updating at each step the linearization coefficients.

Significant improvements in the solution procedure [19,20,21] are in progress: they are making possible the adoption of this approach also in the analysis of large hysteretic structural systems. Note, in fact, that the size of the system of first order differential equations (62) is contributed by: i) the number of degrees of freedom of the dynamical system multiplied by a factor two (two first order differential equations for each equation of motion, which is a second order differential equation) and ii) by the number of Gauss points multiplied by a factor six. Moreover, each equation of this second set is characterized by several linearization coefficients to be determined step by step.

It is worth noting, finally, that linearization in the meaning of this paper does not denote the substitution of the elasto-plastic behaviour with an elastic one. The hysteresis loops (Fig.2) described by a non linear differential equation, in fact, is substituted by the loop described by a linearized differential equation (Fig.3).

4 Structural Constitutive Law

As usual in structural engineering, after the analysis problem is formulated in the context of solid mechanics, a simplification can be pursued at a structural level. With reference to frames, this can be made by stating the constitutive law either for the single beam section [3,4,5,19] or the whole storey [22]. Both these constitutive laws are of the multivariate type due to the spatial extension of the frame and the bi-directional nature of the external excitation.

In [22] the constitutive law was put in the form

$$\begin{aligned} \dot{z}_x &= \dot{u}_x - \beta|\dot{u}_x||z_x|z_x - \gamma\dot{u}_x z_x^2 - \beta|\dot{u}_y||z_y|z_x - \gamma\dot{u}_y z_x z_y \\ \dot{z}_y &= \dot{u}_y - \beta|\dot{u}_y||z_y|z_y - \gamma|\dot{u}_y|z_y^2 - \beta|\dot{u}_x||z_x|z_y - \gamma\dot{u}_x z_x z_y \end{aligned} \tag{64}$$

where u_x and u_y are the two horizontal displacements of the storey. They are related with the corresponding forces by the auxiliary variables z_x and z_y. The hardening in not included. Equation (64) is quite similar to (51) with n=2. However not all the terms of the product $Y \otimes \dot{e}$ are considered in it. This also occurs in its extension to the continuous case of [23], where some mechanical requirements, however are not satisfied (the relationship is not stated for the deviatoric tensors; the yielding curve is not convex etc...).

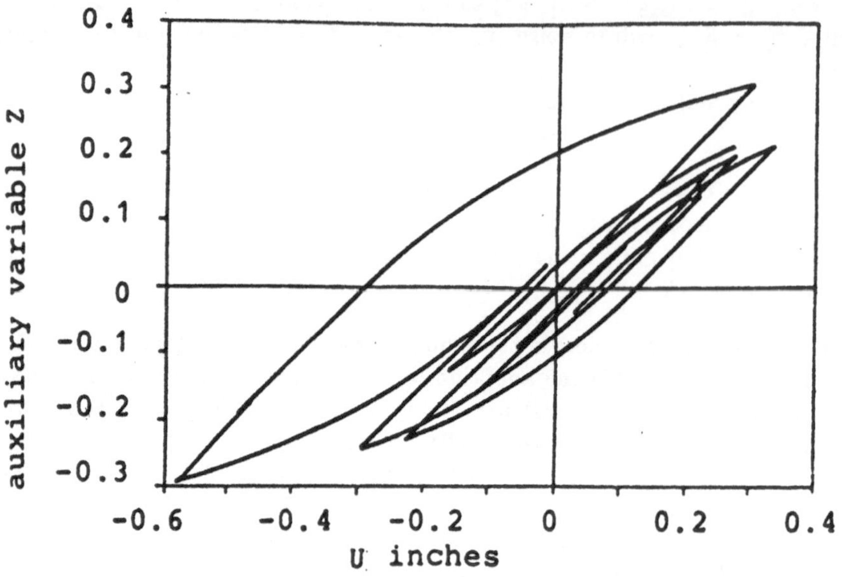

Fig. 2. Hysteresis loop described by (6) under an excitation which is a realization of a stochastic white noise

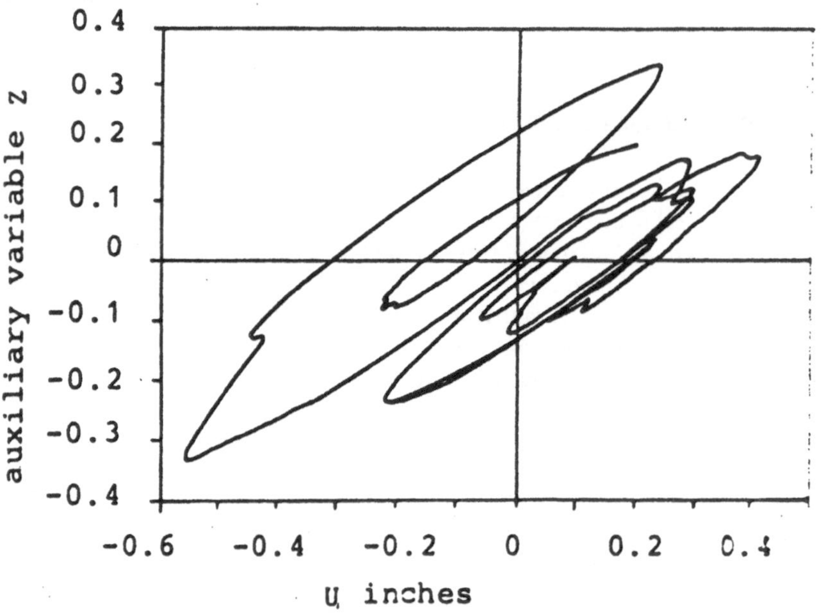

Fig. 3. Hysteretic loop described by the linearized differential equation (6) under the same realization of the stochastic external excitation

In [8] it is emphasized that different results are reached by using the relationship of [3][4] and [5]. There the yielding surface is introduced in a piece-wise linear form. For the hth (h =1,..., n_h) boundary line, forming an angle θ_h with the first axis x, the deformation λ_h is introduced. It is normal to the yielding surface and it plays the role of a plastic multiplier in classic plasticity theory:

$$\begin{bmatrix} u_x \\ u_y \end{bmatrix} = \begin{bmatrix} \sin\theta_h \\ \cos\theta_h \end{bmatrix} \quad \lambda_h = \underline{Q}_h \lambda_h \tag{65}$$

whith implicit sum over h. Its static counterpart is:

$$\Phi_h = \underline{Q}_h^{\mathrm{T}} \begin{bmatrix} q_x \\ q_y \end{bmatrix} \tag{66}$$

The multivariate constitutive law is then introduced directly between Φ_h and λ_h [4][5]:

$$\dot{z}_{\Phi_h} = \dot{\lambda}_h - \beta_h |\dot{\lambda}_h| |z_{\Phi_h}|^{n-1} z_{\Phi_h} - \gamma_h |z_{\Phi_h}|^n \dot{\lambda}_h \tag{67}$$

In this way the normality rule is always respected, the Koiter's hardening rule is introduced and any form of yielding surface can be considered.

Also (67) is a particular form of (51), but it requires a value of the exponent n equal to 1. Since in (64) n is equal to 2, the different results achievable by (64) and (67) emphasized in [8] are now theoretically explained.

The proof that (67) is a particular form of (51) is reached by writing again (51) for n = 1. Making use of (44) then, one finds:

$$\dot{Y} + \beta \frac{2\mu + C}{2\mu} |\dot{\lambda}| Y + \gamma \frac{2\mu + C}{2\mu} \dot{\lambda} Y = A\dot{e} \tag{68}$$

Remember that in (65) u plays the role of e and in (66) q the one of Y. Multipliy both sides of (68) by $\underline{Q}_h^{\mathrm{T}}$, one obtains (67) for n=1 and the appropriate values of β_h, λ_h and A. (Note that (67) is written in the scheme of (26,27), while (68) in that of (23,24)). The only difference is the absolute value of Y in the last term of the l.h.s.. This absolute value is introduced in order to include two symmetric branches of the yielding surface in a single relation. Equation (67) for $n > 1$ is obtained by appropriate modifications in (48). The comparison of (67) with (51) gives rise to the followind remarks:

1. equation (51) depends on the particular shape (von Mises) of the yielding criterion; (67) is independent of it
2. equation (51) holds at a tensorial level: in a single point it provides the relation between the stress and strain tensors. By contrast, (67) was originally proposed for a beams section under bi-axial bending moments. However, (67) also holds at a tensorial level, as well as it can express the relation between storey displacement and force vectors.
3. the hardening rule which characterizes (51) is the classical Prager's rule. The one of (67) is presently the Koiter's hardening rule: may-be some improvements are required to remove this limitation in view of more accurate practical applications.

5 Conclusion

The original objective of stating a multivariate constitutive law in a smoothed form has been reached by two different ways. In the single point of a hysteretic medium, the relationship between the stress and strain tensors is obtained first as an extension of the univariate Bouc's endochronic model. Explicit use of the von Mises's yielding criterion is made and the Prager's hardening rule is adopted.

An alternative model was originally proposed for biaxial bending moments in [4] and [5] and makes use of a piece-wise linear yielding condition. It can also be used to describe the relationship between the stress and strain tensors. In this form, it is shown to be a special case of the previous Pouc's multivariate model, but it leads to and idealization of the constitutive law much more flexible. It can be written for bi-axial bending noments in beam sections, between stress and strain tensors in Gauss points and, also, for displacement and force vectcors in frame storeys Moreover, it is quite independent of the shape of the yielding condition provided the latter one is given in a piecewise linear form. Of course, stating these relationships for beam sections or frame storeys would require experimental checks for which laboratory tests are not presently available [6]. Nevertheless, these models were already adopted with success for the analysis of complex structural systems in [4,5,7,22].

Acknowledgement

The research summarized in this paper has been supported by grants from the Ministry of Public Education (M.P.I.) and from the National Research Council (C.N.R.)

References

1. Bouc R., Modèle Mathématique d'Hysteresis (in French), Acustica, 24, 1971, pp 16-25
2. Wen Y.K., Equivalent Linearization for Hysteretic Systems under Random Excitation, J. Applied Mechanics, 47, 1980, pp. 150-154
3. Casciati F., Faravelli L., Methods of Non-linear Stochastic Dynamics for the Assessment of Structural Fragility, Nuclear Engineering and Design, 90, 1985, pp. 341-356
4. Casciati F., Faravelli L., Stochastic Linearization for 3-D Frames, accepted for publication in J. of Eng. Mech., ASCE
5. Casciati F., Faravelli L., Hysteretic 3-D Frames under Stochastic Excitation, accepted for publication in Res Mechanica
6. Ceradini G., Un legame costitutivo elasto-plastico pluridimensionale per materiali con degradazione (in Italian), in Sandro Dei Poli, A Festschrift for the 70th Birthday, Politecnico di Milano, 1985, pp.195-207
7. Casciati F., Faravelli L., Stochastic Equivalent Linearization for Dynamic Analysis of Continuous Structures, Proc. ASME/SES Meeting on Computational Probabilistic Methods, Berkeley, 1988, pp. 205-210

8. Casciati F., Faravelli L., Endochronic Theory and Nonlinear Stochastic Dynamics of 3D-Frame, ASCE Spec. Conf. on Probabilistic Methods in Civil Engineering, Blacksburg, 1988, pp. 400-403
9. Baber T.T., Wen Y.K., Stochastic Equivalent Linearization for Hysteretic Degrading Multistory Structures, UILU-ENG-80-2001, SRS 471, Univ. of Illinois, 1980
10. Bouc R., Forced Vibrations of a Mechanical System with Hysteresis, Proc. of 4th Conf. on Non-linear Oscillations, Prague, 1967
11. Casciati F., Non-Linear Stochastic Dynamics of Large Structural Systems by Equivalent Linearization, Proc. ICASPS, Vancouver, 1987, pp. 1165-1172
12. Bazant Z.P., Krizek R.J., Shieh C-L., Hysteretic Endochronic Theory for Sand, J. of Eng. Mech., ASCE, 109,1983, pp.1073-1095
13. Casciati F., Faravelli L., Singh M.P., Non-Linear Structural Response and Modeling Uncertainty on System Parameters and Seismic exxitation, Proc. 8th ECEE, Lisbon, 1986, 6.3, pp.41-48
14. Iwan W.D., Private Communication, 1988
15. Casciati F., Faravelli L., Non-Linear Stochastic Dynamics by Equivalent Linearization, in Casciati F. and Faravelli L. (eds.), Methods of Stochastic Structural Mechanics, SEAG, Pavia, 1986, pp.571-586
16. Karray M.A., Etude de l'Efficacité d'un Système d'Isolation à la Base avec Amortissement par Plasticité (in French), Ph.D.Thesis, Univ. d'Aix-Marseille II, 1987.
17. Spanos P., Stochastic Linearization in Structural Dynamics, Appl. Mech. Rev., 34, 1981, pp 1-11
18. Casciati F., Faravelli L., Equivalent Linearization in Non Linear Random Vibration Problems, Proc. Int. Conf. on Vibration Problems in Eng., Xian, Cina, 1986, pp. 986-991
19. Baber T.T., Modal Analysis for Random Vibration of Hysteretic Frames, Earth. Eng. and Struct. Dyn., 14, 1986, pp.841-859
20. Casciati F., Faravelli L., Singh M.P., Stochastic Equivalent Linearization Algorithms and Their Applicability to Hysteretic Systems, accepted for publ. in Meccanica
21. Singh M.P., Maldonado G., Heller R., Faravelli L., Modal Analysis of Nonlinear Hysteretic Structurs for Seismic Motions, in Ziegler F., Schueller G. (eds.), Non Linear Struct. Dynamics in Engineering Systems, Springer Verlag, 1988, pp. 443-454
22. Park Y.J., Wen, Y.K., Ang A.H-S., Random Vibration of Hysteretic Systems under Bi-Dimensional Ground Motion, Earth. EnR. and Struct. Dyn., 14, 1986, pp.543-557
23. Park Y.J., Ang A.H.-S., Seismic Damage Analysis of r/c Nuclear Structures, Proc. 9th SMiRT Conference, Lausanne, 1987, Vol. M, pp. 229-236

Exact Steady-State Solution of FKP Equation in Higher Dimension for a Class of Non Linear Hamiltonian Dissipative Dynamical Systems Excited by Gaussian White Noise

Christian Soize

Abstract

This paper deals with the study of nonlinear stochastic dynamical systems. We have obtained an exact steay-state probability density function for a class of multi-dimensional nonlinear Hamiltonian dissipative dynamical systems excited by Gaussian white noise. The damping can be nonlinear and parametric excitation can be taken into account. When the Hamiltonian function has a radial form, an explicit expression of the Fourier transform of the probability density function is obtained. In this case, for any finite dimension of the system, one can easily calculate any moments of random variables, by using this characteristic function.

1 Introduction

We know that one of the basic tools for the treatment of diffusive Markov processes is the Kolmogorov's forward equation, also called the Fokker-Planck equation (FKP equation). For the mechanical applications this tool allows the study of the response of nonlinear oscillators to stochastic excitation.

1.1 One dimensional nonlinear oscillator

(A) Let us consider the one dimensional nonlinear oscillator of [10]:

$$\ddot{Q}(t) + \xi f(H)\dot{Q}(t) + k(Q(t)) = s\dot{W}_2(t)$$

where: $s > 0$ is the factor intensity of the $\mathbb{R}$-valued normalized gaussian white noise $\dot{W}_2(t)$ on $\mathbb{R}$; $\xi > 0$; $q, \dot{q} \to H(q, \dot{q})$ is the real-valued function defined on $\mathbb{R} \times \mathbb{R}$ such that:

$$H(q, \dot{q}) = \frac{1}{2}\dot{q}^2 + \int_0^q k(x)dx$$

f and k are two real-valued functions on $\mathbb{R}$, with $f(x) > 0$ for all x, such that the initial value problem has a unique solution $\{Q(t), \dot{Q}(t)\}$ which tends asymptotically for $t \to +\infty$, towards a stationary process $\{Q_s(t), \dot{Q}_s(t)\}$. For any fixed t, the steady-state probability density function (p.d.f), denoted by $\rho_{Q,\dot{Q}}(q, \dot{q})$, of the random variable $\{Q_s(t), \dot{Q}_s(t)\}$, is given by:

$$\rho_{Q,\dot{Q}}(q, \dot{q}) = C_0 \exp\left\{-\frac{2\xi}{s^2}\int_0^{H(q,\dot{q})} f(x)\, dx\right\}$$

where C_0 is a positive constant defined by the normalization condition:

$$\int_{\mathbb{R}} \int_{\mathbb{R}} \rho_{Q,\dot{Q}}(q,\dot{q})\, dq\, d\dot{q} = 1$$

Using this result, we deduce the following classical cases:

- For the **linear oscillator**, $\xi > 0, T_0 > 0$:

$$\ddot{Q}(t) + \xi\dot{Q}(t) + T_0 Q(t) = s\dot{W}_2(t),$$

  we have

$$\rho_{Q,\dot{Q}}(q,\dot{q}) = C_0 \exp\left\{-\frac{\xi}{s^2}(\dot{q}^2 + T_0 q^2)\right\}.$$

- For the **nonlinear damping case**, $\varepsilon > 0$:

$$\ddot{Q}(t) - \varepsilon(1 - \dot{Q}(t)^2 - Q(t)^2)\dot{Q}(t) + T_0 Q(t) = s\dot{W}_2(t),$$

  we have:

$$\rho_{Q,\dot{Q}}(q,\dot{q}) = C_0 \exp\left\{\frac{2\varepsilon}{s^2} H(1 - H)\right\} \quad H(q,\dot{q}) = \frac{1}{2}\dot{q}^2 + \frac{1}{2}T_0 q^2$$

- For the **cubic restoring force**, $\xi > 0, T_0 > 0, T_1 > 0$:

$$\ddot{Q}(t) + \xi\dot{Q}(t) + T_0 Q(t) + T_1 Q(t)^3 = s\dot{W}_2(t),$$

  we have:

$$\rho_{Q,\dot{Q}}(q,\dot{q}) = C_0 \exp\left\{-\frac{\xi}{s^2}\left(\dot{q}^2 + T_0 q^2 + \frac{1}{2}T_1 q^4\right)\right\}$$

- For the **bang-bang restoring** force, $\xi > 0, T_0 > 0$:

$$\ddot{Q}(t) + \xi\dot{Q}(t) + T_0 \operatorname{sgn} Q(t) = s\dot{W}_2(t),$$

  we have:

$$\rho_{Q,\dot{Q}}(q,\dot{q}) = C_0 \exp\left\{-\frac{\xi}{s^2}\left(\dot{q}^2 + 2T_0|q|\right)\right\}$$

(B) Recently T.K. Caughey [9] drew a review of the state of the art in the field of the exact steady-state probability density function for one-dimensional nonlinear oscillator to Gaussian white noise excitation:

$$\ddot{Q}(t) + \left[H_y f(H) - \frac{H_{yy}}{H_y}\right]\dot{q}(t) + \frac{H_q}{H_y} = s\dot{W}_2(t)$$

where $y = \frac{1}{2}\dot{q}^2$ and where $H(q,y)$ is now a more general energy integral corresponding to the following conservative oscillator:

$$\ddot{Q} + \frac{H_q}{H_y} = 0$$

and where each subscript q and y denotes a partial derivative. For this case we have:

$$\rho_{Q,\dot{Q}}(q,\dot{q}) = C_0 H_y \exp\left\{-\frac{2}{s^2}\int_0^H f(r)\,dr\right\}$$

(C) If the excitation is not a white noise, but a ̀colored́process, the stochastic averaging method [46, 51, 74, 76, 77] can be used. For one-dimensional nonlinear oscillator, many results have been obtained in the last decade [32, 55, 56, 57, 58, 74].

If the non-white exictation is a ̀coloredśtationary and physically realizable gaussian process, a Markov realization of this process can be built to obtain the FKP equation [40, 42, 43, 44] instead of the averaging stochastic method. But in this case, the dimension of the FKP equation is greater than 2 for a one dimensional nonlinear oscillator, and no exact steady-state p.d.f. can be obtained. Generally, in this case, a numerical method must be used to solve the FKP equation.

1.2 Multi-dimensional nonlinear oscillator

(A) For nonlinear dynamical systems of higher dimension excited by a gaussian white noise, the situation is different and very little is known. The exact steady-state p.d.f. has been obtained for some specific nonlinear oscillator of higher dimension [10,11,13]. For instance, the multi-dimensional nonlinear oscillor of [11] is:

$$\begin{cases} \dot{Q}_j = h_j(P_j) & j \in \{1,\ldots,m\} \\ \dot{P}_j = -\frac{\partial}{\partial Q_j}U(Q) - f(H)\dot{Q}_j + s\dot{W}_{2,j} & j \in \{1,\ldots,m\} \end{cases}$$

where:

$$q \to U(q) : \mathbb{R}^m \to \mathbb{R};$$

$$x \to f(x) : \mathbb{R} \to]0,+\infty[;$$

$$s > 0;$$

$$x \to h_j(x) : \mathbb{R} \to \mathbb{R};$$

$$q,p \to H(q,p) : \mathbb{R}^m \times \mathbb{R}^m \to \mathbb{R}$$

such that:

$$H(q,p) = \sum_{j=1}^{m} \int_0^{p_j} h_j(x)\,dx + U(q).$$

In this case, the p.d.f. of the random variable $\{Q_s(t), P_s(t)\}$ for any fixed t is written as:

$$\rho_s(q,p) = C_0 \exp\left\{-\frac{2}{s^2}\int_0^H f(r)\,dr\right\}$$

(B) In the general case, no exact solution can be obtained and numerical methods must be performed. Unfortunately the numerical solution of the FKP equation in higher dimension is very difficult to build. This is the reason why the FKP method is not used in this situation and approximate methods have been developped. For instance the stochastic linearization method allows the construction of an approximation of some problems of higher dimension [5,12,31,47,67,71,72,81]. Nevertheless, the most general approximate method is, in the present time, the direct numerical simulation of the nonlinear stochastic differential equations [4,23,62,63,64,65,66,71,73].

1.3 Purpose of the present paper

The purpose of the present paper is to study the steady-state p.d.f. for mechanical systems of higher dimension, using the FKP equation method.

We shall consider a nonlinear dynamical system of dimension $m \geq 1$ described in the canonical form:

$$\begin{cases} \dot{Q} = \partial_p H(Q,P) \\ \dot{P} = -\partial_q H(Q,P) - f(H)G\dot{Q} + g(H)S\dot{W}_2 \end{cases}$$

where: $q,p \rightarrow H(q,p) : \mathbb{R}^m \times \mathbb{R}^m \rightarrow \mathbb{R}$ is the Hamiltonian function of the corresponding associated conservative oscillator; G and S are two real $(m \times m)$ matrices; $\dot{W}_2$ is a $\mathbb{R}^m$-valued normalized gaussian white noise; and, f and g are two functions from $\mathbb{R}$ to $\mathbb{R}$.

By assumptions, we shall assume that the equation $\partial_p H(q,p) = \dot{q}$ can be solved locally in p for all $\dot{q}$ in $\mathbb{R}^m$. Let $p = h(q,\dot{q}) \in \mathbb{R}^m$ be the solution which can be built for all $\dot{q}$ in $\mathbb{R}^m$. We denote by $[\partial_{\dot{q}}h]$ and $[\partial_q h]$ the real $(m \times m)$ matrices such that:

$$[\partial_q h]_{jk} = \frac{\partial h_j(q,\dot{q})}{\partial_{q_k}} \quad [\partial_{\dot{q}} h]_{jk} = \frac{\partial h_j(q,\dot{q})}{\partial_{\dot{q}_k}}$$

Let $H(q,\dot{q})$ be the real-valued function on $\mathbb{R}^m \times \mathbb{R}^m$ such that:

$$H(q,\dot{q}) = H(q,h(q,\dot{q})),$$

and $F(q,\dot{q})$ be the $\mathbb{R}^m$-valued function defined on $\mathbb{R}^m \times \mathbb{R}^m$ such that:

$$F(q,\dot{q}) = F(q,h(q,\dot{q})) \quad F(q,p) = \partial_p H(q,p).$$

With these notations, the canonical form is equivalent ot the following second order nonlinear differential equation on $\mathbb{R}^m$:

$$[\partial_{\dot{q}} h]\ddot{Q} + \{[\partial_q h] + f(H)G\} + F(Q,\dot{Q}) = g(H)S\dot{W}_2.$$

We see that for $m \geq 1$, this class of nonlinear dynamical systems, contains the result 1.1.(B) from ref. [9], and the result 1.2.(A) from ref. [11].

We shall note that the nonlinear conservative part is, in fact, a general Hamiltonian formulation for time invariant nonlinear dynamical system, and that the non conservative part has a nonlinear damping and a parametric excitation.

This article contains three parts: 1) In part I (section 2), we recall results about Ito stochastic differential equations and FKP equation, and we set the problem. 2) In part II (section 3), we shall present a result concerning the exact steady-state p.d.f. for the introduced class of nonlinear dynamical systems of higher dimension. 3) In part III (section 4), we shall give some complements for the particular case where the Hamiltonian function has a radial form. In this case, one can explicitly calculate the Fourier transform on $\mathbb{R}^m \times \mathbb{R}^m$ of the p.d.f. (characteristic function).

2 Prerequisites from the theory of nonlinear stochastic differential equations

The aim of this section is to recall certain notions and to state the results that we shall use in the following parts of the present paper.

2.1 Notations and setting the problem

Let $x = (x_1, \dots, x_n)$ be a $\mathbb{R}^n$ vector. In all this paper, we identify the x vector with the $(n \times 1)$ column matrix of its x_j components. The Euclidean space $\mathbb{R}^n$ is equipped with the usual inner product $\langle x, y \rangle = \sum_{j=1}^n x_i y_j$, x and y in $\mathbb{R}^n$ and the associated norm $\|x\| = \langle x, x \rangle^{1/2}$. We denote by $\mathrm{Mat}_{\mathbb{R}}(n, n)$ the set of all the $(n \times n)$ real matrices, we take

$$\|a\| = \left(\sum_{j,k=1}^{n} a_{jk}^2 \right)^{1/2}$$

for $a \in \mathrm{Mat}_{\mathbb{R}}(n, n)$, and a^T denotes the transpose of a. Let

$$W(t) = (W_1(t), \dots, W_n(t))$$

be the $\mathbb{R}^n$-valued normalized Wiener process on $\mathbb{R}^+ = [0, +\infty[$. Its $\mathrm{Mat}_{\mathbb{R}}(n, n)$-valued covariance function is $C_W(t, t') = \min(t, t')I$ where I denotes the unit $(n \times n)$ matrix.

Let $X_0 = (X_{0,1}, \dots, X_{0,n})$ be a $\mathbb{R}^n$-valued second order random variable which is independent from the σ-algebra generated by the random variables $\{W(t), t \geq 0\}$. We denote by $m_{x_0}(dy)$ the probability distribution on $\mathbb{R}^n$ of the random variable X_0. We consider the Ito stochastic differential equation on $\mathbb{R}^n$:

$$dX(t) = b(X(t), t)\, dt + a(X(t), t)\, dW(t), \quad t > 0 \tag{1}$$

with the random initial condition:

$$X(0) = X_0 \text{ a.s,} \tag{2}$$

where $x, t \rightarrow b(x, t) = (b_j(x, t))_j$ and $x, t \rightarrow a(x, t) = (a_{jk}(x, t))_{jk}$ are two mappings defined on $\mathbb{R}^n \times \mathbb{R}^+$ with values respectively in $\mathbb{R}^n$ and $\mathrm{Mat}_{\mathbb{R}}(n, n)$. Concerning the existence and uniqueness of the diffusion process $X(t)$ solution of problems (1)-(2) in the uniform Lipschitz case, we refer to [21,22,28]. But this case is rarely encountered and in the following parts other tools are necessary in order to prove the existence and uniqueness. On the other hand, we are concerned here with the stationary solution of the initial value problem (1)-(2). So it is necessary that all coefficients $b_j(x,t) = b_j(x)$ and $a_{jk}(x,t) = a_{jk}(x)$ are independent of time t. In all this paper we restrict the development to this case, i.e. we consider only stochastic differential equations of the time homogeneous markovian type.

2.2 Existence and uniqueness of solutions defined up to explosion time

Suppose we are given continuous functions $x \to b(x) : \mathbb{R}^n \to \mathbb{R}^n$ and $x \to a(x) : \mathbb{R}^n \to \mathrm{Mat}_{\mathbb{R}}(n,n)$ with the following properties:

P1 For all fixed positive real number $R > 0$ and for all x and y with $\|x\| \leq R$, $\|y\| \leq R$, there is a positive constant K_R such that:

$$\begin{aligned} \|b(x) - b(y)\|^2 + \|a(x) - a(y)\|^2 \leq K_R\|x-y\|^2 \\ \|b(x)\|^2 + \|a(x)\|^2 \leq K_R(1 + \|x\|^2) \end{aligned} \tag{3}$$

P2 Let L be the following differential operator on $\mathbb{R}^n$ such that for any C^2-function $x \to u(x) : \mathbb{R}^n \to \mathbb{R}$ we have:

$$(Lu)(x) = \sum_{j=1}^{n} b_j(x)\frac{\partial u(x)}{\partial x_j} + \frac{1}{2}\sum_{j,k=1}^{n} \sigma_{jk}(x)\frac{\partial^2 u(x)}{\partial x_j \partial x_k} \tag{4}$$

where $\sigma(x) = a(x)a(x)^T \in \mathrm{Mat}_{\mathbb{R}}(n,n)$. We suppose that there exists a function $x,t \to V_1(x,t)$ defined on $\mathbb{R}^n \times \mathbb{R}^+$ with values in $\mathbb{R}^+$, C^2 with respect to x and C^1 with respect to t which satisfies the two conditions:

$$\exists\lambda > 0 : \forall(x,t) \in \mathbb{R}^n \times \mathbb{R}^+,\ \frac{\partial}{\partial t}V_1(x,t) + (LV_1)(x,t) \leq \lambda V_1(x,t) \tag{5}$$

$$\inf_{\|x\|>r,t>0} V_1(x,t) \to +\infty \text{ as } r \to +\infty \tag{6}$$

V1 is known as a Liapounov function.

With all the above assumptions, we have the following results [24,44]:

R1 The initial value problem:

$$\begin{aligned} dX(t) = b(X)\,dt + a(X)\,dW(t),\ t > 0 \\ X(0) = X_0 \text{ a.s.} \end{aligned} \tag{7}$$

has a unique solution $X(t)$ which is a $\mathbb{R}^+$-valued diffusive Markov process and which does not explode. Consequently the stochastic process $X(t)$ is defined a.s. for all $t \geq 0$.

R2 Let B be an arbitrary Borel set of $\mathbb{R}^n$ and $y \in \mathbb{R}^n$. Let Q be the transition probability of the process $X(t)$:

$$B \to Q(y,t,B) = P(X(t) \in B | X(0) = y),\ t > 0.$$

In addition, we assume that the probability distribution $m_{x_0}(dy)$ has a density $\rho_{X_0}(y)$ with respect to dy. Hence, for every $t \in \mathbb{R}^+$, the probability distribution of the random variable $X(t)$ has a density $\rho(t,x)$ with respect to dx:

$$\rho(t,x)\,dx = \int_{\mathbb{R}^n} Q(y,t,dx)\rho_{X_0}(y)\,dy \tag{8}$$

which satisfies the FKP equation on $\mathbb{R}^n$:

$$\frac{\partial}{\partial t}\rho + \sum_{j=1}^{n} \frac{\partial}{\partial x_j}(b_j(x)\rho) - \frac{1}{2}\sum_{j,k=1}^{n} \frac{\partial^2}{\partial x_j \partial x_k}(\sigma_{jk}(x)\rho) = 0 \tag{9}$$

for $t > 0$, with the initial condition:

$$\rho(t,x) \to r_{X_0}(x) \text{ as } t \downarrow 0. \tag{10}$$

When results (R1) and (R2) hold, we shall say that $X(t)$ is a *regular solution* of problem (7).

2.3 Existence of an asymptotic stationary solution and steady-state FKP equation

Suppose that the problem (7) has a regular solution $X(t)$. In addition there exists another C^2-function $x \to V_2(x)$ defined on $\mathbb{R}^n$ with values in $\mathbb{R}^+$ such that:

$$\sup_{\|x\| > \mathbb{R}} LV_2(x) = -C_R, \quad C_R \to +\infty \text{ as } R \to +\infty. \tag{11}$$

With these assumptions we have the following results [24,28,44]:

R3 The regular solution $X(t)$ of problem (7) tends in probability for $t \to +\infty$, to a $\mathbb{R}^n$-valued stationary process $X_S(t)$ and we have:

$$\rho_S(x) = \lim_{t \to +\infty} \rho(t,x), \tag{12}$$

where $\rho_S(x)$ is the p.d.f. on $\mathbb{R}^n$ of the random variable $X_s(t)$ for any fixed t. The steady-state p.d.f. ρ_S satisfies the steady-state FKP equation associated with (9):

$$\sum_{j=1}^{n} \frac{\partial}{\partial x_j}(b_j(x)\rho_S(X)) - \frac{1}{2}\sum_{j,k=1}^{n} \frac{\partial^2}{\partial x_j \partial x_k}(\sigma_{jk}(x)\rho_S(x)) = 0 \tag{13}$$

with the normalization condition:

$$\int_{\mathbb{R}^n} \rho_S(x)\,dx = 1 \tag{14}$$

3 Exact steady-state p.d.f. for a class of nonlinear mechanical system of higher dimension

In this section we establish the exact steady-state p.d.f. for a class of nonlinear dynamical system described by canonical equations.

3.1 Canonical equations of a class of dynamical systems

We consider a nonlinear dynamical system of dimension $m \geq 1$ described with the canonical form.

Let $Q(t) = (Q_1(t), \ldots, Q_m(t))$ be the generalized coordinates and $\dot{Q}(t) = \frac{\partial Q(t)}{dt}$ the generalized velocity. We denote by $P(t) = (P_1(t), \ldots, P_m(t))$ the generalized momentum canonically conjugated from $Q(t)$. The variable P is also called the generalized impulsion and the variables P and Q are the canonical variables. We denote by $p = (p_1, \ldots, p_m)$ and $q = (q_1, \ldots, q_m)$ the variables associated with the $P(t)$ and $Q(t)$ stochastic processes. Let $p, q \to h(p, q)$ be a real function defined on $\mathbb{R}^m \times \mathbb{R}^m$. For $z \in \{p, q\}$ we shall denote by $\partial_z h$ and $\partial_z^2 h$ respectively the $\mathbb{R}^m$-vector and the $(m \times m)$ real matrix such that:

$$(\partial_z h)_j = \frac{\partial h}{\partial z_j} \quad [\partial_z^2 h]_{jk} = \frac{\partial^2 h}{\partial z_j \partial z_k}$$

The nonlinear dynamical system is subjected to a $\mathbb{R}^m$- valued normalized gaussian white noise denoted by $\dot{W}_2(t) = (\dot{W}_{2,1}(t), \ldots, \dot{W}_{2,m}(t))$. It is described by the following generalized nonlinear stochastic differential equation on $\mathbb{R}^n = \mathbb{R}^m \times \mathbb{R}^m$ with $n = 2m$:

$$\begin{aligned} \dot{Q} &= \partial_p H(Q, P) \quad t > 0 \\ \dot{P} &= -\partial_q H(Q, P) + F(Q, P, \dot{W}_2) \end{aligned} \tag{15}$$

with the random initial condition:

$$Q(0) = Q_0 \quad P(0) = P_0 \text{ a.s.} \tag{16}$$

where:

i The derivatives $\dot{Q}$ and $\dot{P}$ with respect to t of the stochastic processes $Q(t)$ and $P(t)$, should be understood in the sense of the theory of generalized stochastic processes [44].

ii The Hamiltonian function of the corresponding associated conservative system (i.e., $F \equiv 0$) is denoted by H. We shall assume that 1) $q, p \to (q, p)$ is a C^2-function on $\mathbb{R}^m \times \mathbb{R}^m$ with values in $\mathbb{R}^+ = [0, +\infty[$, independent of time t; 2) for all p and q in $\mathbb{R}^m$, the hessian matrix $[\partial_p^2 H] \in \mathrm{Mat}_{\mathbb{R}}(m, m)$ is positive-definite; 3) the mapping $p, q \to [\partial_p^2 h]$ is bounded on $\mathbb{R}^m \times \mathbb{R}^m$. Hence we have the following inequalities for all p and q in $\mathbb{R}^m$:

$$0 < \langle [\partial_p^2 H] z, z \rangle \leq C_T \|z\|^2, \forall z \in \mathbb{R}^m, \|z\| > 0 \tag{17}$$

where C_T is a positive real constant which does not depend on p and q. In addition we shall assume that:

$$\inf_{\|q\|^2+\|p\|^2>R^2} H(q,p) \to +\infty \text{ if } R \to +\infty \tag{18}$$

iii The initial condition (Q_0, P_0) is a given $\mathbb{R}^m \times \mathbb{R}^m$-valued second order random variable and its probability distribution $m_{Q_0,P_0}(dq,dp)$ has a density $\rho_{Q_0,P_0}(dq,dp)$ with respect to $dq\,dp$.

iv The non conservative force $F = (F_1, \dots, F_m)$ is expressed in terms of the canonical variables. We consider here the restricted class of systems such that

$$F(Q,P,\dot{W}_2) = -f(H)G\dot{Q} + g(H)S\dot{W}_2 \tag{19}$$

with $\dot{Q} = \partial_p H$ (first equation (15)), and where S and G are two constant real $(m \times m)$ matrices such that $G = S\,S^T$ and $\|S\| > 0$. Hence, denoting by N the rank of S, we have $1 \leq N \leq m$, and G is a positive matrix, but not necessarily a positive definite matrix. Functions f and g are continuous on $\mathbb{R}^+$ with values in $\mathbb{R}^{+*} =]0, +\infty[$, and g is differentiable. In addition, there exists $r_0 > 0$ and real constants

$$\mu > 0, C_0 > 0, C_1 > 0, C_2 > 0, -\infty < \alpha_2 \leq 1, \alpha 1 > \alpha 2 - 1, \alpha 0 < \alpha 1 - \alpha 2 + 1$$

such that:

$$\forall r \geq r0,\ f(r) \geq C_1 r^{\alpha_1} \tag{20}$$

$$\forall r \geq r0,\ C_0 \exp(-\mu r^{\alpha_0}) \leq g^2(r) \leq C_2 r^{\alpha_2} \tag{21}$$

v In order to apply the results (R1) to (R3) of section 2, we introduce the following notations:

$$x = \begin{pmatrix} q \\ p \end{pmatrix} \in \mathbb{R}^n;\ X(t) = \begin{pmatrix} Q(t) \\ P(t) \end{pmatrix} :\ X_0 = \begin{pmatrix} Q_0 \\ P_0 \end{pmatrix} \tag{22}$$

$$b(x) = \begin{pmatrix} \partial_p H \\ -\partial_p H - f(H)G\partial_p H \end{pmatrix} \in \mathbb{R}^n \tag{23}$$

$$a(x) = \begin{pmatrix} 0_{m\times m} & 0_{m\times m} \\ 0_{m\times m} & g(H)S \end{pmatrix} \in \mathrm{Mat}_{\mathbb{R}}(n,n) \tag{24}$$

Hence the initial value problem (15-16) can be written as:

$$\begin{gathered} \dot{X} = b(X) + a(X)\dot{W},\ t > 0 \\ X(0) = X_0 \text{ a.s.} \end{gathered} \tag{25}$$

with

$$\dot{W} = \begin{pmatrix} \dot{W}_1 \\ \dot{W}_2 \end{pmatrix}$$

a $\mathbb{R}^n$-valued normalized gaussian white noise. We know (see ref. [44]) that the generalized stochastic problem (25) is equivalent to the following Ito stochastic differential equation with initial value:

$$\begin{array}{c} dX(t) = b(X(t))\,dt + a(X(t))dW(t),\, t > 0 \\ X(0) = X_0 \text{ a.s.} \end{array} \tag{26}$$

Finally, we shall suppose functions f, g and H to be such that the mappings b and a on $\mathbb{R}^n$, defined by relations (23)-(24), verify the inequalities (3) and that the random variable X_0 is independent from $\{W(t), t \geq 0\}$.

3.2 Existence and uniqueness of a regular solution and FKP equation

Let u and v be two C^1-real functions defined on $\mathbb{R}^m \times \mathbb{R}^m$. We introduce the Poisson bracket notation:

$$[u, v] = \sum_{j=1}^{m} \left(\frac{\partial u}{\partial q_j}\frac{\partial v}{\partial p_j} - \frac{\partial u}{\partial p_j}\frac{\partial v}{\partial q_j} \right) = \langle \partial_q u, \partial_p v\rangle - \langle \partial_p u, \partial_q v\rangle \tag{27}$$

Let L be the differential operator on $\mathbb{R}^m \times \mathbb{R}^m = \mathbb{R}^n$ which is defined by (4), and which is associated with the a and b mappings given by (23) and (24).

Let $q, p \to V(q,p)$ be a C^2-real function on $\mathbb{R}^m \times \mathbb{R}^m$. Hence we have:

$$LV = [V, H] - f(H)\langle S^T\partial_p H, S^T\partial_p V\rangle + \frac{1}{2}g^2(H)\sum_{k=1}^{m}\langle(\partial_p^2 V)S^k, S^k\rangle \tag{28}$$

with $S^k \in \mathbb{R}^m$ such that $S^k_j = S_{jk}$. In the particular case where $V = \varphi(H)$, with $r \to \varphi(r)$ a C^2-real function on $\mathbb{R}$, we obtain:

$$\begin{aligned} LV = {} & \left(\frac{1}{2}g^2(H)\varphi''(H) - f(H)\varphi'(H)\right)\|S^T\partial_p H\|^2 \\ & + \frac{1}{2}g^2(H)\varphi'(H)\sum_{k=1}^{m}\langle(\partial_p^2 H)S^k, S^k\rangle \end{aligned} \tag{29}$$

with $\varphi'(r) = \frac{d}{dr}\varphi(r)$. We observe that in this case we have used the Poisson bracket property: $[\varphi(H), H] = \varphi'(H)[H, H] = 0$ and the usual differential calculus: $\partial_p V = \varphi'(H)\partial_p H$, $\partial_p^2 V = \varphi''(H)\partial_p H \otimes \partial_p H + \varphi'(H)\partial_p^2 H$.

Let us consider the Lyapounov function $V_1(q,p) = H(q,p) + \gamma$ with γ a positive real number. The condition (6) is satisfied in view of (18). Note that $\frac{\partial V_1}{\partial t} = 0$. Using the relation (29) we obtain:

$$LV_1 = -f(H)\|S^T\partial_p H\|^2 + \frac{1}{2}g^2(H)\sum_{k=1}^{m}\langle(\partial_p^2 H)S^k, S^k\rangle$$

It is clear that condition (5) is satisfied for all p and q in any compact subset of $\mathbb{R}^m \times \mathbb{R}^m$ because $q, p \to (LV_1)(q,p)$ is a bounded mapping on every compact subset of $\mathbb{R}^m \times \mathbb{R}^m$. Therefore we can limit the proof of the condition (5) to the case $H \geq H_0 > r_0$ with $H_0 > 1$ and where r_0 is the positive real number introduced in (21). The inequalities (17) yield

$$0 < \sum_{k=1}^{m} \langle (\partial_p^2 H) S^k, S^k \rangle \leq C_T \|S\|^2 \tag{30}$$

and (21) gives $g_2(H) \leq C_2 H$ because $\alpha_2 \leq 1$ and $H \geq H_0 > 1$.

Hence we have for $H \geq H_0$:

$$LV_1 \leq -f(H)\|S^T \partial_p H\|^2 + \frac{1}{2} C_2 C_T \|S\|^2 H$$

and condition (5) is satisfied with $\lambda = \frac{1}{2} C_2 C_T \|S\|^2 > 0$ because $\gamma > 0$ and $-f(H)\|S^T \partial_p H\|^2 \leq 0$. So, we have proved the following result in using (R1) and (R2) of section 2.2:

Result 1: The nonlinear stochastic problem (15) subjected to all conditions defined in (ii), (iii), (iv) and (v), has a unique regular solution. For every $t > 0$ the probability distribution of the random variable $X(t)$ has a density $\rho(t, q, p)$ with respect to $dq\, dp$ which satisfies the FKP equation on $\mathbb{R}^m \times \mathbb{R}^m$:

$$\frac{\partial \rho}{\partial t} + [\rho, H] - \mathrm{div}_p J(\rho) = 0,\ t > 0 \tag{31}$$

with the initial condition:

$$\rho(t, q, p) \to \rho_{Q_0, P_0}(q, p) \text{ as } t \downarrow 0 \tag{32}$$

where $\mathrm{div}_p J = \sum_{j=1}^{m} \frac{\partial}{\partial p_j} J_j$ and where $J(\rho)$ is the $\mathbb{R}^m$-vector:

$$J(\rho) = \rho[f(H) + g(H) g'(H)] G \partial_p H + \frac{1}{2} g^2(H) G \partial_p \rho \tag{33}$$

3.3 Existence and construction of the exact steady-state p.d.f.

Asymptotic stationary solution of system (15)- (16) does not exist with the previous assumption. We must change the assumption on S introduced in (iv).

Result 2: With the assumptions of the result 1, if the rank of S is equal to m (hence G is a positive definite matrix), the unique regular process $X(t)$ tends in probability for $t \to +\infty$ to a $\mathbb{R}^m \times \mathbb{R}^m$-valued stationary process $X_S(t) = \{Q_S(t), P_S(t)\}$. For every t, the steady-state p.d.f. $\rho_S(q, p)$ of the random variable $X_S(t)$ satisfies the steady-state FKP equation on $\mathbb{R}^m \times \mathbb{R}^m$:

$$[\rho_S, H] - \mathrm{div}_p J(\rho_S) = 0 \tag{34}$$

with the normalization condition:

$$\int_{\mathbb{R}^m \times \mathbb{R}^m} \rho_S(q, p)\, dq\, dp = 1 \tag{35}$$

where J is given by (33). Problem (34)- (35) has an exact unique solution which is written as:

$$\rho_S(q,p) = C_N \frac{1}{g^2(H(q,p))} \exp\left(-2\int_0^{H(q,p)} g^{-2}(r) f(r)\, dr\right) \tag{36}$$

where C_N is a positive real constant defined by condition (35).

Proof. a Existence of $X_S(t)$-process is obtained applying result (R3) of section 2.3. The proof of this point is lengthy because several cases must be considered. Hence we shall only give some elements. First of all, one can prove that the assumptions on H allow us to consider only the case $\|q\|^2 + \|p\|^2 > R^2$ with $\|p\| > 0$. Hence $\|S^T \partial_p H\| > 0$ because rank$S = m$ and $\|\partial_p H\| > 0$ in this case. Next we can exhibit a V_2-function the condition (11) to be verified. For instance the case $\alpha_2 < \alpha_1$ is easy to perform. The choice $V_2 = H$ yields as $R \to +\infty$,

$$LV_2 \leq H(-C_1 H^{\alpha_1 - \alpha_2} \|S^T \partial_p H\|^2 + \frac{1}{2} C_2 C_T \|S\|^2) \to -\infty.$$

b The application of (13) and (14) yields (34) and (35) taking into account result 1. Exact steady-state p.d.f. is built by searching it in the form $\rho_S(q,p) = \varphi(H)$. Hence $[\rho_S, H] = 0, J(\rho_S) = \psi G \partial_p H$ with $\psi \equiv (f + gg')\varphi + \frac{1}{2} g^2 j'$ and equation (34) is verified when φ is solution of equation $\psi = 0$. Note that ρ_S is integrable on $\mathbb{R}^m \times \mathbb{R}^m$ because we have (18) and as $H \to +\infty$,

$$\rho_S(q,p) \leq \frac{C_N}{C_0} \exp\left(-H^\gamma \left(\frac{C_1}{C_2} - \mu H^{\alpha_0 - \gamma}\right)\right) \sim \frac{C_N}{C_0} \exp(-H^\gamma)$$

with $\gamma = \alpha_1 - \alpha_2 + 1 > 0$ and $\alpha_0 - \gamma < 0$.

Remarks

1 Conditions (ii) yield that equation $\partial_p H(q,p) = \dot{q}$ can be solved locally in p for $\dot{q}$ in $\mathbb{R}^m$. Let $p = h(q, \dot{q}) \in \mathbb{R}^m$ be the solution. If we assume that h can be built for all $\dot{q}$ in $\mathbb{R}^m$ and that $\dot{q} \to h(q, \dot{q})$ is a continuously differentiable $\mathbb{R}^m$-valued function defined on $\mathbb{R}^m$, then the steady-state p.d.f. $\rho_{Q,\dot{Q}}(q, \dot{q})$ of the random variable $\{Q_S(t), \dot{Q}_S(t)\}$, with respect to $dq\, d\dot{q}$ is given by:

$$\rho_{Q,\dot{Q}}(q, \dot{q}) = \rho_S(q, h(q, \dot{q})) |\det[\partial_{\dot{q}} h(q, \dot{q})]| \tag{37}$$

where $[\partial_{\dot{q}} h] = \frac{\partial h_j}{\partial \dot{q}_k}$ is the Jacobian matrix.

2 It is clear that all the assumptions introduced in (ii) of section 3.1 are sufficient but not necessary. In some cases, it is possible to relaxe some hypotheses. For instance, the C^2-regularity of function H on $\mathbb{R}^m \times \mathbb{R}^m$ can be weakened. In fact, it is necessary for $\partial_p H$ and $\partial_q H$ to be functions and not distributions on $\mathbb{R}^m \times \mathbb{R}^m$.

3.4 Examples

Examples 1: Multi-dimensional extensions of classical cases

We give hereafter a multi-dimensional extension ($m > 1$) of the classical examples given in section 1.1.

Let M and K be two symmetric real ($m \times m$) positive definite matrices.

- For the *linear oscillator*, $\xi > 0$:

$$M\ddot{Q}(t) + \xi\dot{Q}(t) + KQ(t) = s\dot{W}_2(t)$$

we have:

$$\rho_{Q,\dot{Q}}(q,\dot{q}) = C_0 \exp\left\{-\frac{\xi}{s^2}[\langle M\dot{q},\dot{q}\rangle + \langle Kq,q\rangle]\right\}$$

- For the *nonlinear damping case*, $\varepsilon > 0$:

$$M\ddot{Q} - \varepsilon(1 - \langle M\dot{Q},\dot{Q}\rangle - \langle KQ,Q\rangle)\dot{Q} + KQ = s\dot{W}_2$$

we have:

$$\rho_{Q,\dot{Q}}(q,\dot{q}) = C_0 \exp\left\{-\frac{2\varepsilon}{s^2}H(1-H)\right\}$$

$$H(q,\dot{q}) = \frac{1}{2}\langle M\dot{q},\dot{q}\rangle + \frac{1}{2}\langle Kq,q\rangle$$

- For the *cubic restoring force*, $\xi > 0$, $T_0 > 0$, $T_1 > 0$:

$$M\ddot{Q} + \xi\dot{Q}(t) + (T_0 + T_1\langle KQ,Q\rangle)KQ = s\dot{W}_2$$

we have:

$$\rho_{Q,\dot{Q}}(q,\dot{q}) = C_0 \exp\left\{-\frac{\xi}{s^2}\left[\langle M\dot{q},\dot{q}\rangle + T_0\langle Kq,q\rangle + \frac{1}{2}T_1\langle Kq,q\rangle^2\right]\right\}$$

This example is developped in details in the section 4.

- For the *bang-bang restoring force*, $\xi > 0, (T_j > 0, j \in \{1,\ldots,m\})$:

$$M\ddot{Q} + \xi\dot{Q} + T\mathrm{sgn}Q = s\dot{W}_2$$

where $(T\mathrm{sgn}Q)_j = T_j \mathrm{sgn}Q_j \forall j \in \{1,\ldots,m\}$, we have:

$$\rho_{Q,\dot{Q}}(q,\dot{q}) = C_0 \exp\left\{-\frac{\xi}{s^2}\left(\langle M\dot{q},\dot{q}\rangle + 2\sum_{j=1}^{m} T_j|q_j|\right)\right\}$$

Example 2: Particular multi-dimensional nonlinear dynamical system with parametric excitation

Let $M(q)$ be a symmetric real $(m \times m)$ matrix such that for all $q \in \mathbb{R}^m$, $M(q)$ is positive definite and $q \to [M(q)]^{-1}$ is a C^2-function bounded on $\mathbb{R}^m$. Let $U(q)$ be a C^2-potential function on $\mathbb{R}^m$. Hence we have:

$$H(q,p) = \frac{1}{2}\langle M(q)^{-1}p, p\rangle + U(q)$$

The generalized impulsion is $P(t) = [M(Q(t))]\dot{Q}(t)$, and $p = h(q,\dot{q}) = M(q)\dot{q}$. We see that $\partial_{\dot{q}}h] = M(q)$, and:

$$H(q,\dot{q}) = \frac{1}{2}\langle M(q)\dot{q}, \dot{q}\rangle + U(q)$$

For the nonlinear dynamical system on $\mathbb{R}^m$, with nonlinear damping and nonlinear parametric excitation:

$$\frac{d}{dt}\{[M(Q)]\dot{Q}\} + f(H)G\dot{Q} + \partial_q U(Q) = g(H)S\dot{W}_2$$

we have, if $G = SS^T$ and $\text{rank}S = m$:

$$\rho_{Q,\dot{Q}}(q,\dot{q}) = C_0 \frac{|\det M(q)|}{g^2(H(q,\dot{q}))} \exp\left\{-2\int_0^{H(q,\dot{q})} g^{-2}(r)f(r)\,dr\right\}$$

Example 3: Extension of example 2

One can generalize example 2 in considering a more general form of the kinetic energy of the dynamical systems:

$$T(q,\dot{q}) = \frac{1}{2}\langle M(q)\dot{q}, \dot{q}\rangle + \langle B(q), \dot{q}\rangle + C(q)$$

such that $T(q,\dot{q}) \geq 0 \forall q, \dot{q} \in \mathbb{R}^m$, $M(q)$ as in example 2, $B(q) \in \mathbb{R}^m$ and $C(q) \in \mathbb{R}$.

In this case the generalized impulsion is given by $p = \partial_{\dot{q}}T$, i.e. by $p = h(q,\dot{q}) = M(q)\dot{q} + B(q)$, and the Hamiltonian function defined by

$$H = < p, \dot{q} > - l(q,\dot{q})$$

$$l(q,\dot{q}) = T(q,\dot{q}) - U(q)$$

is written as:

$$H(q,p) = \frac{1}{2}\langle M^{-1}(q)(p - B(q)), p - B(q)\rangle$$
$$+ 2\langle M^{-1}(q)(p - B(q)), B(q)\rangle + C(q) + U(q).$$

For this general conservative part, the exact steady-state p.d.f. is given by (36)-(37) if $\text{rank}S = m$, and we have:

$$H(q,\dot{q}) = \frac{1}{2}\langle M(q)\dot{q}, \dot{q}\rangle + 2\langle B(q), \dot{q}\rangle + C(q) + U(q)$$

Example 4: The general multi-dimensional case

Using the remark 1 of section 3.3, the general multi-dimensional case is given by the following nonlinear equation on $\mathbb{R}^m$:

$$[\partial_{\dot{q}}]\ddot{Q} + \{[\partial_q h] + f(H)G\}\dot{Q} + F(Q, \dot{Q}) = g(H)S\dot{W}_2$$

where $[\partial_{\dot{q}} h]$ and $[\partial_q h]$ the matrices:

$$[\partial_{\dot{q}} h]_{jk} = \frac{\partial h_j(q, \dot{q})}{\partial \dot{q}_k}$$

$$[\partial_q h]_{jk} = \frac{\partial h_j(q, \dot{q})}{\partial_{\dot{q}_k}}$$

and where:

$$H(q, \dot{q}) = H(q, h(q, \dot{q}))$$
$$F(q, \dot{q}) = F(q, h(q, \dot{q}))$$
$$F(q, p) = \partial_q H(q, p)$$

For this equation we have, if rank of $S = m$:

$$\rho_{Q,\dot{Q}}(q, \dot{q}) = C_0 \frac{|\det[\partial_{\dot{q}} h]|}{g^2(H(q, \dot{q}))} \exp\left\{-2 \int_0^{H(q,\dot{q})} g^{-2}(r) f(r)\, dr\right\}$$

4 Complements for the radial Hamiltonian case

In this section, we give some complements for the case where the Hamiltonian function H has a radial form.

4.1 Definition of a class of radial Hamiltonian functions

In all this section 4, we shall consider the nonlinear stochastic differential equation (15)-(16) and we shall suppose that all the assumptions of the result 2 are verified. Hence, the steady-state p.d.f. $\rho_S(q, p)$ is given by the relation (36).

Let M and K be two positive definite symmetric real $(m \times m)$ matrices. Taking advantage of the fact that matrices M and K are positive definite, $\dot{q}_1, \dot{q}_2 \rightarrow \langle M\dot{q}_1, \dot{q}_2\rangle$ and $q_1, q_2 \rightarrow \langle Kq_1, q_2\rangle$ are inner products on $\mathbb{R}^m$, and we can write:

$$\begin{aligned} u &= K^{1/2}q, \quad \langle Kq, q\rangle = \langle K^{1/2}q, K^{1/2}q\rangle = \|u\|^2 \\ v &= M^{1/2}\dot{q}, \quad \langle M\dot{q}, \dot{q}\rangle = \langle M^{1/2}\dot{q}, M^{1/2}\dot{q}\rangle = \|v\|^2 \end{aligned} \tag{38}$$

There is a basis $\{\varphi_1, \ldots, \varphi_m\}$, $\varphi_j \in \mathbb{R}^m$, of the Euclidean space $\mathbb{R}^m$ and a normalization of $\{\varphi_j\}_j$, such that:

$$\langle M\varphi_j, \varphi_k\rangle = \delta_{jk}; \quad \langle K\varphi_j, \varphi_k\rangle = \lambda_j \delta_{jk} \tag{39}$$

where δ_{jk} is Kronecker's symbol and where $\lambda_j > 0$ such that

$$0 < \lambda_1 \leq \lambda_2 \leq \ldots \leq \lambda_m$$

Using (39) we get:

$$\sqrt{\det MK} = (\det M)\left(\prod_{j=1}^{m} \lambda_j^{1/2}\right) \tag{40}$$

This particular case is defined by the following kinetic energy and potential function:

$$T(\dot{q}) = \frac{1}{2}\langle M\dot{q},\dot{q}\rangle; \quad U(q) = \Lambda(\frac{1}{2}\langle Kq,q\rangle) \tag{41}$$

where $r \to \Lambda(r)$ is a mapping from $\mathbb{R}^+$ to $\mathbb{R}^+$.

In these conditions, the Hamiltonian function $H(q,p)$ is such that:

$$H(q,p) = \frac{1}{2}\langle M^{-1}p,p\rangle + \Lambda(\frac{1}{2}\langle Kq,q\rangle) \tag{42}$$

and we shall say that H has a radial form.

Let us note that function Λ can be any mapping from $\mathbb{R}^+$ to $\mathbb{R}^+$ such that all the assumptions on H, introduced in section 3, be verified.

The generalized impulsion is given by $p = \partial_{\dot{q}}T = h(q,\dot{q}) = M\dot{q}$. Hence, the steady-state p.d.f. $\rho_{Q,\dot{Q}}(q,\dot{q})$ on $\mathbb{R}^m \times \mathbb{R}^m$ is given by (37) with (36):

$$\rho_{Q,\dot{Q}}(q,\dot{q}) = C_0\frac{1}{g^2(H(q,\dot{q}))}\exp\left(-2\int_0^{H(q,\dot{q})} g^{-2}(y)f(y)\,dy\right), \tag{43}$$

where C_0 is a positive real constant defined by the normalization condition:

$$\int_{\mathbb{R}^m}\int_{\mathbb{R}^m} \rho_{Q,\dot{Q}}(q,\dot{q})dq\,d\dot{q} = 1 \tag{44}$$

and where $q,\dot{q} \to H(q,\dot{q}) : \mathbb{R}^m \times \mathbb{R}^m \to \mathbb{R}^+$ is such that:

$$h(q,\dot{q}) = \frac{1}{2}\langle M\dot{q},\dot{q}\rangle + \Lambda(\frac{1}{2}\langle Kq,q\rangle) \tag{45}$$

4.2 General result for the radial class

With all the preceding assumptions and notations, we have the following result:

Result 3: Let $r1,r2 \to H_0(r1,r2)$ and $r1,r2 \to \mu(r1,r2)$ be two functions from $\mathbb{R}^+ \times \mathbb{R}^+$ to $\mathbb{R}^+$ such that:

$$H_0(r_1,r_2) = \Lambda(\frac{1}{2}r_1^2) + \frac{1}{2}r_2^2 \tag{46}$$

$$\mu(r_1,r_2) = \frac{1}{g^2(H_0(r_1,r_2))}\exp\left(-2\int_0^{H_0(r_1,r_2)} g^{-2}(y)f(y)\,dy\right) \tag{47}$$

Then, we have

$$H(q,\dot{q}) = H_0(\langle Kq,q\rangle^{1/2},\langle M\dot{q},\dot{q}\rangle^{1/2}) \tag{48}$$

$$\rho_{Q,\dot{Q}}(q,\dot{q})dq\,d\dot{q} = \sqrt{\det MK}\frac{\Gamma\left(\frac{m}{2}\right)^2}{4\pi^m}\frac{\mu(\langle Kq,q\rangle^{1/2},\langle M\dot{q},\dot{q}\rangle^{1/2})dq\,d\dot{q}}{\int_0^{+\infty}\int_0^{+\infty} r_1^{m-1}r_2^{m-1}\mu(r_1,r_2)\,dr_1\,dr_2} \tag{49}$$

where $\Gamma(z) = \int_0^{+\infty} t^{z-1}e^{-t}\,dt$ is the gamma function.

Proof. **a** Let $x \to s(x) : \mathbb{R}^m \to \mathbb{R}$ be a function that depends only on $r = \|x\|$. We denote a such function $s(x) = S(\|x\|)$ with $r \to S(r) : \mathbb{R}^+ \to \mathbb{R}$. We have the classical result:

$$\int_{\mathbb{R}^m} s(x)\,dx = \frac{2\pi^{m/2}}{\Gamma(\frac{m}{2})} \int_0^{+\infty} S(r) r^{m-1}\,dr \tag{50}$$

b the equalities (45) and (46) yield (48).
c Taking into account (47) and (48), the p.d.f. $\rho_{Q,\dot{Q}}$ given by (43), is written as:

$$\rho_{Q,\dot{Q}}(q,\dot{q})dq\,d\dot{q} = C_0 \mu(\langle Kq,q\rangle^{1/2}, \langle M\dot{q},\dot{q}\rangle^{1/2})dq\,d\dot{q} \tag{51}$$

The normalization condition (44) yields:

$$C_0^{-1} = \int_{\mathbb{R}^m}\int_{\mathbb{R}^m} \mu(\langle Kq,q\rangle^{1/2}, \langle M\dot{q},\dot{q}\rangle^{1/2})dq\,d\dot{q}$$

Using (38) and applying the theorem on integration with respect to the image of a measure, we can also write:

$$C_0^{-1} = (\det MK)^{-1/2} \int_{\mathbb{R}^m}\int_{\mathbb{R}^m} \mu(\|u\|, \|v\|)\,du\,dv$$

We obtain (49) in applying (50) and by substitution in (51).

4.3 Moment of total energy

For every t, we denote $e_s(t) = H(Q_s(t), \dot{Q}_s(t))$ the $\mathbb{R}^+$-valued random variable which represents the total energy at time t of the dynamical system in its steady-state. The moment of order j of the r.v. $e_s(t)$ is written as:

$$E\{e_s(t)^j\} = \int_{\mathbb{R}^m}\int_{\mathbb{R}^m} H(q,\dot{q})^j \rho_{Q,\dot{Q}}(q,\dot{q})\,dqd\dot{q}$$

Using (38), (49) and (50) we obtain:

$$E\{e_s(t)^j\} = \frac{\int_0^{+\infty}\int_0^{+\infty} H_0(r_1,r_2)^j r_1^{m-1} r_2^{m-1} \mu(r_1,r_2)\,dr_1 dr_2}{\int_0^{+\infty}\int_0^{+\infty} r_1^{m-1} r_2^{m-1} \mu(r_1,r_2)\,dr_1 dr_2} \tag{52}$$

4.4 Characteristic function

For every t, the characterisitic function of $\{Q_s(t), \dot{Q}_s(t)\}$ is the mapping $\alpha,\beta \to \hat{\rho}_{Q,\dot{Q}}(\alpha,\beta)$ from $\mathbb{R}^m \times \mathbb{R}^m$ to $\mathbb{C}$ which is defined by the rule:

$$\begin{aligned} \hat{\rho}_{Q,\dot{Q}}(\alpha,\beta) &= E\{\exp(i\langle\alpha, Q_s(t)\rangle + i\langle\beta, \dot{Q}_s(t)\rangle)\} \\ &= \int_{\mathbb{R}^m}\int_{\mathbb{R}^m} e^{i\langle\alpha,q\rangle + i\langle\beta,\dot{q}\rangle} \rho_{Q,\dot{Q}}(q,\dot{q})\,dqd\dot{q} \end{aligned} \tag{53}$$

Result 4: With the preceding assumptions, the characterisitic function $\hat{\rho}_{Q,\dot{Q}}$ is a real valued function on $\mathbb{R}^m \times \mathbb{R}^m$ such that:

$$\hat{\rho}_{Q,\dot{Q}}(\alpha,\beta) = \Phi(\langle K^{-1}\alpha,\alpha\rangle^{1/2}, \langle M^{-1}\beta,\beta\rangle_{1/2}) \tag{54}$$

with $R_1, R_2 \rightarrow \Phi(R_1, R_2)$ a $\mathbb{R}$-valued function defined on $\mathbb{R}^+ \times \mathbb{R}^+$ which is written as:

$$\Phi(R_1,R_2) = [2^\nu \Gamma(\nu+1)]^2 \times \frac{\int_0^{+\infty}\int_0^{+\infty} R_1^{-\nu} R_2^{-\nu} r_1^{\nu+1} r_2^{\nu+1} J_\nu(r_1 R_1) J_\nu(r_2 R_2) \mu(r_1,r_2)\, dr_1 dr_2}{\int_0^{+\infty}\int_0^{+\infty} r_1^{2\nu+1} r_2^{2\nu+1} \mu(r_1,r_2)\, dr_1 dr_2} \tag{55}$$

with $\nu = \frac{m}{2} - 1$, and J_ν the Bessel function:

$$J_\nu(x) = \frac{\left(\frac{x}{2}\right)^\nu}{\sqrt{\pi}\,\Gamma\left(\nu+\frac{1}{2}\right)} \int_{-1}^{+1} (1-t^2)^{\nu-1/2} \cos(t)\, x\, dt$$

Proof. **a** Let be

$$x \rightarrow s(x) = S(\|x\|) : \mathbb{R}^m \rightarrow \mathbb{C}$$

and

$$r \rightarrow S(r) : \mathbb{R}^m \rightarrow \mathbb{C}.$$

If $s \in L^1(\mathbb{R}^m, \mathbb{C})$, the Fourier transform $\hat{s}(y)$ is defined for any $y \in \mathbb{R}^m$ by:

$$\hat{s}(y) = \hat{S}(\|y\|) = \int_{\mathbb{R}^m} e^{-i\langle x,y\rangle} S(\|x\|)\, dx \tag{56}$$

with $R \rightarrow \hat{S}(R) : \mathbb{R}^+ \rightarrow \mathbb{C}$ the function given by the rule:

$$\hat{S}(R) = 2\pi \left(\frac{2\pi}{R}\right)^{m/2-1} \int_0^{+\infty} r^{m/2} J_{m/2-1}(rR) S(r)\, dr \tag{57}$$

b Using (38)-(49), and because $K^{-1/2}$ and $M^{-1/2}$ are two symmetric real matrices, function $\hat{\rho}_{Q,\dot{Q}}$ which is given by (53), can be written as:

$$\hat{\rho}_{Q,\dot{Q}}(\alpha,\beta) =$$

$$\frac{\Gamma^2(\frac{m}{2}) \int_{\mathbb{R}^m}\int_{\mathbb{R}^m} e^{i\langle K^{-1/2}\alpha,u\rangle + i\langle M^{-1/2}\beta,v\rangle} \mu(\|u\|,\|v\|)\, du\, dv}{4\pi^m \int_0^{+\infty}\int_0^{+\infty} r_1^{m-1} r_2^{m-1} \mu(r_1,r_2)\, dr_1 dr_2}$$

Because μ is a $\mathbb{R}^+$-valued function, we can conjugate the equality (56) and we obtain (54)-(55), using formula (56)-(57) for the u and v variables.

Remark: Let us note that $\nu \neq -1, -2, \ldots$, we have

$$J_\nu(x) \equiv \frac{\left(\frac{x}{2}\right)^\nu}{\Gamma(\nu+1)} \text{ if } x \to 0$$

Hence we deduce from equality (55) that:

$$\Phi(R_1, 0) = 2^\nu \Gamma(\nu+1) \times \frac{\int_0^{+\infty}\int_0^{+\infty} R_1^{-\nu} r_1^{\nu+1} r_2^{2\nu+1} J_\nu(r_1 R_1) \mu(r_1, r_2)\, dr_1 dr_2}{\int_0^{+\infty}\int_0^{+\infty} r_1^{2\nu+1} r_2^{2\nu+1} \mu(r_1, r_2)\, dr_1 dr_2} \tag{58}$$

$$\Phi(0, R_2) = 2^\nu \Gamma(\nu+1) \times \frac{\int_0^{+\infty}\int_0^{+\infty} R_2^{-\nu} r_1^{2\nu+1} r_2^{\nu+1} J_\nu(r_2 R_2) \mu(r_1, r_2)\, dr_1 dr_2}{\int_0^{+\infty}\int_0^{+\infty} r_1^{2\nu+1} r_2^{2\nu+1} \mu(r_1, r_2)\, dr_1 dr_2} \tag{59}$$

Result 5:

a Let us introduce the following notations:

$$\mathcal{Y}(z, z') = \frac{1}{\Gamma(z)\Gamma(z')} \int_0^{+\infty}\int_0^{+\infty} r_1^{2z-1} r_2^{2z'-1} \mu(r_1, r_2)\, dr_1 dr_2 \tag{60}$$

$$\mathcal{A}_{kk'} = \frac{\mathcal{Y}(\nu+1+k, \nu+1+k')}{\mathcal{Y}(\nu+1, \nu+1)} \tag{61}$$

Hence, the function Φ, defined by (55) can be calculated by the series:

$$\Phi(R_1, R_2) = \sum_{k=0}^{+\infty}\sum_{k'=0}^{+\infty} \frac{1}{k!k'!} \left(-\frac{1}{4}R_1^2\right)^k \left(-\frac{1}{4}R_2^2\right)^{k'} \mathcal{A}_{kk'} \tag{62}$$

b If function μ can be written as:

$$\mu(r_1, r_2) = \mu_1(r_1) \times \mu_2(r_2), \tag{63}$$

hence, Φ is given by:

$$\Phi(R_1, R_2) = \Phi_1(R_1) \times \Phi_2(R_2) \tag{64}$$

with, for $j \in \{1, 2\}$:

$$\Phi_j(R_j) = \sum_{k=0}^{+\infty} \frac{1}{k!} \left(-\frac{1}{4}R_j^2\right)^k A_{j,k} \tag{65}$$

$$A_{j,k} = \frac{Y_j(\nu+1+k)}{Y_j(\nu+1)} \tag{66}$$

$$Y_j(z) = \frac{1}{\Gamma(z)} \int_0^{+\infty} r^{2z-1} \mu_j(r)\, dr \tag{67}$$

c If μ_j is given by:

$$\mu_j(r) = ae^{-br^2} \quad a > 0, b > 0 \tag{68}$$

we have:

$$\Phi_j(R_j) = \exp\left(-\frac{1}{4b}R_j^2\right) \tag{69}$$

d If μ_j is given by:

$$\mu_j(r) = ae^{-br^2-\frac{c}{2}r^4} \quad a > 0, b > 0, c > 0 \tag{70}$$

we have:

$$\Phi_j(R_j) = \sum_{k=0}^{+\infty} \frac{1}{k!}\left(-\frac{1}{4b}R_j^2\right)^k A_k \tag{71}$$

with

$$\left.\begin{aligned} A_{k+1} &= A_k Z_k \quad k \geq 0 \\ A_0 = 1; \quad Z_0 &= \frac{Y(\nu+2)}{Y(\nu+1)} \\ Z_{k+1} &= \frac{1}{\gamma(\nu+2+k)}\left[\frac{1}{Z_k} - 1\right], \; k \geq 0 \end{aligned}\right\} \tag{72}$$

and where:

$$\left.\begin{aligned} Y(z) &= \frac{1}{\Gamma(z)}\int_0^{+\infty} r^{2z-1}e^{-r^2-\frac{\gamma}{2}r^4}\,dr \\ \gamma &= \tfrac{c}{b^2} \end{aligned}\right\} \tag{73}$$

Remark (d.1): Z_k which is given by recurence (72), is such that:

$$Z_k = \frac{Y(\nu+1+k+1)}{Y(\nu+1+k)} \tag{74}$$

Remark (d.2): Let $U(a,x)$ be the parabolic cylindrical function:

$$U(a,x) = \frac{1}{\Gamma(\frac{1}{2}+a)}e^{-x^2/4}\int_0^{+\infty} e^{-xR-\frac{1}{2}R^2}R^{a-1/2}\,dR \tag{75}$$

Then, we can write $Y(z)$ given by (73), as:

$$Y(z) = \frac{e^{1/4\gamma}}{2\gamma^{z/2}}U\left(z-\frac{1}{2},\frac{1}{\sqrt{\gamma}}\right) \tag{76}$$

and:

$$Z_0 = \frac{Y(\nu+2)}{Y(\nu+1)} = \frac{1}{\sqrt{\gamma}}\frac{U\left(\frac{m}{2}+\frac{1}{2},\frac{1}{\sqrt{\gamma}}\right)}{U\left(\frac{m}{2}-\frac{1}{2},\frac{1}{\sqrt{\gamma}}\right)} \quad \nu = \frac{m}{2}-1 \tag{77}$$

We have the asymptotic behaviour:

$$m \to +\infty, \; Z_0 \sim \sqrt{\frac{2}{\gamma}}\frac{1}{\sqrt{m}}e^{-\frac{1}{\sqrt{2\gamma m}}} \sim \sqrt{\frac{2}{\gamma}}\frac{1}{\sqrt{m}} \tag{78}$$

Proof. **a** Relation (62) is obtained by substituting into the right hand side of (55), the ascending series of the Bessel function:

$$J_\nu(z) = \left(\frac{1}{2}z\right)^\nu \sum_{k=0}^{+\infty} \frac{\left(-\frac{1}{4}z^2\right)^k}{k!\Gamma(\nu+k+1)} \tag{79}$$

b Relations (64) to (67) are deduced from (60) to (62) in taking into account the expression (63) of μ.

c We can write in this case:

$$Y_j(z) = \frac{a}{2b^z}\frac{1}{\Gamma(z)}\int_0^{+\infty} (br^2)^{z-1} e^{br^2} d(br^2) = \frac{a}{2b^z} \tag{80}$$

where we have used the Euler integral definition of the gamma function. Replacing expression (80) of $Y_j(z)$ into (66) we obtain:

$$A_{j,k} = b^{-k} \tag{81}$$

Equation (69) is deduced from (81) and (65).

d Taking into account (70), $Y_j(z)$, given by (67), can be written as:

$$\left.\begin{aligned} Y_j(z) &= \frac{a}{b^z} Y(z) \\ Y(z) &= \frac{1}{\Gamma(z)}\int_0^{+\infty} r^{2z-1} e^{r^2 - \frac{\gamma}{3}r^4}\, dr \end{aligned}\right\} \tag{82}$$

with $\gamma = cb^{-2}$. Integrating by parts the integral (82), and using the relation $\Gamma(z+1) = z\Gamma(z)$, we obtain the following recurence:

$$Y(z) = Y(z-1) - \gamma z Y(z+1) \tag{83}$$

Putting:

$$A_k = \frac{Y(\nu+1+k)}{Y(\nu+1)} \tag{84}$$

and taking into account (66) and (82) we obtain: $A_{j,k} = b^{-k}A_k$. Then, relation (65) yields relation (71). Let us introduce the variable Z_k defined by (74). Using (84), we see that $A_{k+1} = A_k Z_k$, $A_0 = 1$ and $Z_0 = Y(\nu + 2)/Y(\nu+1)$. Finally the recurence on Z_k is deduced from (83) and (74).

(d.2) Starting from the definition (75) and by the change of variable: $x = \gamma^{-1/2}$,$R = \gamma^{1/2}r^2$ and $a = z - 1/2$ into the integral (75), we obtain (76). The asymptotic expression (78) is deduced from (77) and from the following results:

1. For fixed $x, a > 0$, we have:

$$\left.\begin{aligned} U(a,x) &= \sqrt{\pi}2^{-a/2-1/4}\frac{1}{\Gamma(3/4+a/2)}e^{-\sqrt{a}x + v_1(a)} \\ v_1(a) &\sim \frac{1}{\sqrt{a}} \text{ if } a \to +\infty \end{aligned}\right\} \tag{85}$$

2. For $z \to +\infty$:

$$\frac{\Gamma(z)}{\Gamma(z+1/2)} \sim z^{-1/2} \tag{86}$$

4.5 Example

Let M and K be two symmetric positive definite real $(m \times m)$ matrices. We consider the nonlinear dynamical system on $\mathbb{R}^m$:

$$M\ddot{Q}(t) + \xi\dot{Q}(t) + [T_0 + T_1\langle KQ(t), Q(t)\rangle]KQ(t) = s\dot{W}_2(t) \tag{87}$$

where ξ, s, T_0, T_1 are positive scalars and $\dot{W}_2(t)$ the $\mathbb{R}^m$-valued normalized gaussian white noise. Equation (87) can be seen, for instance, as a finite approximation of order m, of dynamical transverse motions of a nonlinear elastic string, clamped at its ends, subjected to a transverse random loading.

In this case we have $p = M\dot{q}$ and the Hamiltonian function H(q,p) on $\mathbb{R}^m \times \mathbb{R}^m$, associated to the conservative system is such that:

$$H(q,p) = \frac{1}{2}\langle M^{-1}p,p\rangle + T_0\langle Kq,q\rangle + T_1(\frac{1}{2} < Kq,q >)^2 \tag{88}$$

Hence equation (87) can be written with the form (15)-(19):

$$\left\{\begin{array}{c} \dot{Q} = \partial_p H(Q,P) \\ \dot{P} = -\partial_q H(Q,P) + F(Q,P,\dot{W}_2) \\ F(Q,P,\dot{W}_2) = -\xi\dot{Q} + s\dot{W}_2 \end{array}\right\} \tag{89}$$

Then we have $g(H) = s$ and $f(H) = \xi$. Comparing (88) with (42) we see that function $r \to \Lambda(r)$ is such that:

$$\Lambda(r) = T_0 r + T_1 r^2 \tag{90}$$

In this condition we can apply result 3 and result 4. Relations (46) and (47) yield:

$$H_0(r_1,r_2) = \frac{1}{2}T_0 r_1^2 + \frac{1}{4}T_1 r_1^4 + \frac{1}{2}r_2^2 \tag{91}$$

$$\mu(r_1,r_2) = s^{-2}\exp\{-2\xi s^{-2}H_0(r_1,r_2)\} \tag{92}$$

We can also write the relation (92) as follow:

$$\mu(r_1,r_2) = mu_1(r_1) \times \mu_2(r_2) \tag{93}$$

with

$$\mu_1(r_1) = \exp(-\xi s^{-2}T_0 r_1^2 - \frac{1}{2}\xi s_{-2}T_1 r_1^4) \tag{94}$$

$$\mu_2(r_2) = s^{-2}\exp(-\xi s^{-2}r_2^2) \tag{95}$$

This particular form of μ, allows the use of result 5, parts (c) and (d).

The characteristic function $\alpha,\beta \to \hat{\rho}_{Q,\dot{Q}}(\alpha,\beta)$ on $\mathbb{R}^m \times \mathbb{R}^m$ is given by (54). The $\mathbb{R}$-valued function $R_1, R_2 \to \Phi(R_1,R_2)$ on $\mathbb{R}^+ \times \mathbb{R}^+$ is given by (64). Functions Φ_1 and Φ_2 can be calculated by using respectively (71) and (69).

Finally we obtain the following result for the characteristic function:

For any $\alpha,\beta \in \mathbb{R}^m \times \mathbb{R}^m$:

$$\hat{\rho}_{Q,\dot{Q}}(\alpha,\beta) = \Phi(\langle K^{-1}\alpha,\alpha\rangle^{1/2}, \langle M^{-1}\beta,\beta\rangle^{1/2}) \tag{96}$$

with:

$$\Phi(R_1,R_2) = e^{-s^2R_2^2/4\xi} \sum_{k=0}^{+\infty} \frac{1}{k!}\left(-\frac{s^2}{4\xi T_0}R_1^2\right)^k A_k \tag{97}$$

$$\left\{\begin{array}{c} A_{k+1} = A_k Z_k \quad k \geq 0 \\ Z_{k+1} = \dfrac{1}{\gamma(\nu+2+k)}\left[\dfrac{1}{Z_k} - 1\right] \quad k \geq 0 \\ A_0 = 1 \quad Z_0 = \dfrac{Y(\nu+2)}{Y(\nu+1)} \end{array}\right\} \tag{98}$$

$$\nu = \frac{m}{2} - 1 \quad \gamma = \frac{s^2 T_1}{\xi T_0^2} \tag{99}$$

$$Y(z) = \frac{1}{\Gamma(z)}\int_0^{+\infty} r^{2z-1}e^{-r^2-\gamma r^4/2}\,dr \tag{100}$$

$$m \to +\infty,\ Z_0 \sim \sqrt{\frac{2}{\gamma}}\frac{1}{\sqrt{m}} \tag{101}$$

References

1. BATHE K.J. and WILSON E.L., "Numereical Methods in Finite Element Analysis". Prentice-Hall, Inc. Englewood Cliffs, N.J. (1976).
2. BELLMAN R., RICHARDSON J.M., "Closure and preservation of moment properties", J. Math. Anal. Appl., 23, pp. 639-644, (1968).
3. BENSOUSSAN A., LIONS J.L., PAPANICOLAOU G., "Asymptotic analysis for periodic structures", North Holland, Amsterdam, (1978).
4. BERNARD P. et al., "Un algorithme de simulation stochastique par markovianisation approchée. Application à la mécanique aléatoire", J. de Méch. Théo. et Appl., (1984).
5. BOUC R., "Forced vibration of mechanical systems with hysteresis", Proc. 4ème Conf. sur les oscillations non linéaires, Prague, (1967).
6. BOURET R.C., FRISCH U., POUQUET A., "Brownian motion of harmonic oscillator with stochastic frequency", Physica, 65, pp. 303-320, North-Holland, (1973).
7. BRISSAUD A., FRISCH U., "Solving linear stochastic differential equations", J. Math. Phys., Vol. 15, N¡ 4, (1974).
8. BUCY R.S., JOSEPH P.D., "Filtering for stochastic processes with applications to guidance", Inters.Tracks in pure and applied Math., Wiley, (1968).
9. CAUGHEY T.K., "On the response of non linear oscillators to stochastic excitation", Probabilistic engineering mechanics, (1)1, March (1986).
10. CAUGHEY T.K., "Non linear theory of random vibrations", Advances in applied mechanics, vol. 11, pp. 209-253 (1971).
11. CAUGHEY T.K.and F.-Ma, "The steady-state response of a class of dynamical system to stochastic excitation", Journal of Applied Mechanics, Vol. 49, pp. 622-632, September, (1982).
12. CAUGHEY T.K., "Equivalent linearization techniques", J. Acoust. Soc. Amer., Vol. 35, pp. 1706-1711, (1963).
13. CAUGHEY T.K., "Derivation and application of the Fokker-Planck equation to discrete nonlinear dynamic systems subjected to white random excitation", J. of Acous. Soc. Amer. Vol. 35, n¡ 11, (1963).

14. CHEN K.K. and SOONG T.T., "Covariance properties of waves propagating in a random medium", J. Acoust. Soc. Amer., 49, pp. 1639-1642, (1971).
15. CLOUGH R.W. and PENZIEN J., "Dynamic of structures", McGraw-Hill, New York, (1975).
16. CRAMER H. and LEADBETTER M.R., "Stationary and related stochastic processes". John Wiley and Sons, New York, (1967).
17. CRANDALL S.H., "Perturbation technique for random vibration of nonlinear system", J. Acoust. Soc. Amer., Vol 35, n¡ 11, (1963).
18. CRANDALL S.H., "Random vibration of systems with non linear restoring forces", AFOSR, 708, June, (1961).
19. CRANDALL S.H., MARK D.W., "Random vibration in mechanical systems", Academic Press, New York, (1973).
20. DOOB J.L., "Stochastic processes", John Wiley and Sons, New York, (1967).
21. FRIEDMAN A., "Stochastic differential equations and applications", Vol. 1 and 2, Academic Press, New York, (1975).
22. GUIKHMAN L. and SKOROKHOD A.V., "The theory of stochastic processes", Springer Verlag, Berlin, (1979).
23. HARRIS C.J., "Simulation of multivariate nonlinear stochastic system", Int. J. for Num. Meth. in Eng., Vol. 14, pp. 37-50, (1979).
24. HASMINSKII R.Z., "Stochastic stability of differential equations", Sijthoff and Noordhoff, (1980).
25. HASMINSKII R.Z., "A limit theorem for the solutions of differential equations with random right-hand sides", Theory of probability and applications, 11(3), pp. 390- 405, (1966).
26. HILLE E., PHILLIPS R.S., "Functional analysis and semi-groups", Amer. Math. Soc., Providence, Rhode Island, (1957).
27. HORMANDER L., "Hypoelliptic second order differential equations", Acta Math. 119, pp. 147-171, (1967).
28. IKEDA N., WATANABE S., "Stochastic differential equations and diffusion processes", North Holland, (1981).
29. ITO K., "On stochastic differential equations", Memoir. Amer. Math. Soc., 14, (1961).
30. ITO K., McKEAN H.P., "Diffusion processes and their sample paths", Springer-Verlag, Berlin (1965).
31. IWAN W.D., "A generalization of the concept of equivalent linearization", Int. J. Non linear Mechanics, 5, pp. 279-287, Pergamon Press, (1973).
32. IWAN W.D., SPANOS P.T., "Response envelope statistics for nonlinear oscillators with random excitation", J. of Applied Mechanics, 45, (1978).
33. JAZWINSKI A.H., "Stochastic processes and filtering theory", Academic Press, New York (1971).
34. JENKINS G.M., WATT D.G., "Spectral analysis and its applications", Holden Day, San Francisco, (1968).
35. KALMAN R.E., BUCY R.S., "New results in linear filtering and prediction theory", Trans. ASME, (1961).
36. KATO T., "Perturbation theory for linear operators", Springer, N.Y., (1966).
37. KENDALL M.G., STUART A., "The advanced theory of statistics", Griffin, London, (1966).
38. KOZIN F., "On the probability densities of the output of some random systems", J. Appl. Mech., 28, pp. 161-165, (1961).

39. KOZIN F., "A survey of stability of stochastic systems", Automatica Int. J. Control Automation, 5, pp. 95-112, (1969).
40. KREE P., "Markovianization of random vibrations", L. Arnold and P. Kotelenez eds. Stochastic Space time models and limit theorems, pp. 141-162, Reidel, (1985).
41. KREE P., "Diffusion equation for multivalued stochastic differential equations", J. of Functional Analysis, 49, nb 1, (1982).
42. KREE P., "Realization of gaussian random fields", Journal of Math. Phys. 24(11), November, pp. 2573-2580, (1983).
43. KREE P. and SOIZE C., "Markovianization of random oscillators with colored input", Rend. Sem. Math. Univer. Politech. Torino, Editrice Levretto Bella, Torino, (1982).
44. KREE P. and SOIZE C., "Mathematics of random phenomena", Reidel publishing company, Dordrecht, Holland, (1986).
45. KUSHNER H.J., "Stochastic stability and control", Academic Press, N.Y., (1967).
46. LIN Y.K., "Some observations of the stochastic averaging method", Probabilistic engineering mechanics, vol. 1, n¡ 1, March (1986).
47. LIN Y.K., "Probabilistic theory of structural dynamics", Mc Grax-Hill, New York, (1968).
48. LIU S.C., "Solutions of Fokker-Planck equation with applications in nonlinear random vibrations", Bell System Tech. J., 48, pp. 2031-2051, (1969).
49. McKEAN H.P., "Stochastic integrals", Academic Press, New York, (1969).
50. NAHI N.E., "Estimation theory and applications", John Wiley, New York, (1969).
51. PAPANICOLAOU G.C., KOHLER W., "Asymptotic theory of mixing stochastic ordinary differential equations", Communication Pure and Applied Math., 27, (1974).
52. PARDOUX E., PIGNOL M., "Etude de la stabilité d'une EDS bilinéaire à coefficients périodiques", in analysis and optimization of systems, part 2, A. Bensoussan and J.L. Lions Eds., Lecture Notes in Control and Information Sciences, 63, pp. 92-103, Springer-Verlag, (1984).
53. PHILLIPS R.S., SARASON LL., "Elliptic parabolic equations of the second order", J. Math. Mech., 17, pp. 891-917, (1967).
54. ROBERTS J.B., "First passage probability for nonlinear oscillators", J. of the Eng. Mechan. Division EM5, october, (1976).
55. ROBERTS J.B., "Transient response of nonlinear systems to random excitation", J. of Sound and Vibration, 74, pp. 11-29, (1981).
56. ROBERTS J.B., "Energy methods for nonlinear systems with non-white excitation", Proceedings of the IUTAM Symposium on random vibrations and reliability, Frankfurt/Oder, (1982).
57. ROBERTS J.B., "Response of an oscillator with nonlinear damping and a softening spring to non-white random excitation", Probabilisitic engineering mechanics, vol. 1, n¡ 1, march (1986).
58. ROBERTS J.B., "The energy envelope of a randomly excited nonlinear oscillator" JSV, 60(2), pp. 177-185, (1978).
59. ROZANOV Y.A., "Stationary random processes", Holden Day, San Francisco, (1967).
60. SANCHO N.G.F., "On the approximate mement equations of a nonlinear stochastic differential equation", J. Math. Anal. Appl., 29, pp. 384-391, (1970).
61. SCHMIDT G., "Stochastic problems in dynamics", Eds. B.L. Clarkson, Pitman, London, (1977).

62. SHINOZUKA M., WEN Y.K., "Monte Carlo solution of nonlinear vibrations", AIAAJ., vol. 10, n¡ 1, pp. 37-40, (1972).
63. SHINOZUKA M., "Time space domain analysis in the structural reliability assessment", Second Inter. Conf. on Struct. Safety and Reliab., ICOSSAR 77, Munich, (1977).
64. SHINOZUKA M., "Simulation of multivariate and multidimensional random processes", J. Acoust. Soc. Amer., Vol. 49, n¡ 1, part 2, pp. 357-367, (1971).
65. SHINOZUKA M.,, JAN C.M., "Digital simulation of random processes and its applications", J. of Sound and Vibration, 25, pp. 111-128, (1972).
66. SHINOZUKA M., "Monte Carlo solution of structural dynamics", J. Computer and Structures, Vol. 2, pp. 855-874, (1972).
67. SINITSYN I.N., "Methods of statistical linearization", Automation and remote control, vol. 35, pp. 765- 776, (1974).
68. SKOROHOD A.V., "Studies in the theory of random processes", Addisa-Wesley, Reading, Massachusetts, (1965).
69. SOIZE C., DAVID J.M., DESANTI A., "Functional reduction of stochastic fields for studying stationary random vibrations", ONERA, J. Rech. Aerosp., ", (1986).
70. SOIZE C., "Markovianization of sea waves and applications", Journée de mécanique aléatoire, AFREM, Paris, in french, (1984).
71. SOONG T.T., "Random differential equations in science and engineering", Academic Press, New York, (1973).
72. SPANOS P.D.T., "Stochastic linearization in structural dynamics", Applied Mech. Reviews, Vol. 34, nb 1, (1981).
73. SPANOS P.D.T., "Monte Carlo simulations of responses of a non-symmetric dynamic systems to random excitations", J. Computer and Structures, 13, Pergamon Press, (1981).
74. SPANOS P.D.T., "Approximate analysis of random vibration through stochastic averaging", in Proceedings of IUTAM Symposium on random vibrations and reliability, Frankfurt/Oder, (1982).
75. STRAND J.L., "Random ordinary differential equations", J. Differential Equations, 7, pp. 538-553, (1970).
76. STRATONOVICH R.L., "Topics in the theory of random noise", Gordon and Breach, New York, (1963).
77. STRATONOVICH R.L., "A new representation for stochastic integrals", J. SIAM control, Vol. 4, n¡ 2, (1966).
78. STRATONOVICH R.L., "Conditional Markov processes and their application to theory of optimal control", Amer. Elsevier, New York, (1968).
79. VAN KAMPEN N.G., "A cumulant expansion for stochastic linear differential equations I", Physica, 74, pp. 215- 238, North-Holland, (1974).
80. VAN KAMPEN N.G., "Stochastic differential equations", Physics reports, Section C of Physics Letters, 24, nb 3, pp. 171-228, North-Holland, (1976).
81. WEN Y.K., "Equivalent linearization for hysteretic systems under random excitation", Trans. of ASME, 47, pp. 150-154, (1980).
82. WEN Y.K., "Approximate method for nonlinear random vibration", J. of the Eng. Mechan. Div., EM4, (1975).
83. WILCOX R.M., BELLMAN R., "Truncation and preservation of moment properties for Fokker-Planck moment equations", J. Math. Anal. Appl., 32, pp. 532-542, (1970).
84. YOSIDA K., "Functional analysis", Third edition, Springer Verlag, (1971).

Power Spectra of Nonlinear Dynamic Systems - Analysis via Generalized Hermite Polynomials

W. Wedig

Abstract

In this paper we introduce a generalized Hermite analysis for the investigation of nonlinear stochastic systems under additive white noise. In the special case that the joint density distribution of the stationary response processes are known the orthogonality relation of classical Hermite polynomials can be generalized replacing the gaussian weighting function by the given nonnormal distribution. The calculated generalized Hermite polynomials represent an orthogonal expansion basis. Therefore, they can be applied to calculate power spectra and correlation functions of nonlinear dynamic systems. Typical examples are the stochastic Duffing oscillator, snap-through problems, restoring with backlash and bang-bang forms. Convergence properties are numerically verified by means of higher order expansions.

1 Introduction to the Problem

Up to now, the spectral analysis of nonlinear oscillators is performed by equivalent linearization techniques [1,2] which replace the nonlinear characteristic by a linear one in such a way that both oscillators, the linear and the nonlinear one, possess the same square means in the stationary response. Clearly, this method gives only a first approximation with the main property that the resonance frequency is shifted according to the applied nonlinear restoring and excitation intensity. Since the method can not be extended to higher order approximations the obtained results have always to be checked by numerical simulations to get rough error estimates only valid for special parameters of the system. To over- come this unsatisfactory situation we need orthogonal expansions which are convergent in the sense of mild solutions. FIrst hints in this direction give the successful application of Hermite polynomials in recent research. Thus, we find in [3] a convergent analysis of second order moments stability and in [4] an extension to Lyapunov exponent calculations for parameter fluctuations by coloured noise. Nonlinear systems under additive noise are investigated in [5,6] and in [7,8]. Because of the orthogonality the Hermite moments possess the basic advantage that higer order moments are decreasing in appropriate cases. Consequently, the associated infinite moments equations can be truncated just by neglecting the higest terms without any closure condition. To improve this Hermite analysis once more considerably we extend it by generalized Hermite polynomials defined by stationary density distributions of the system response.

This has the advantage that we get higher convergence rates. Negative variances and spectral frequency distributions are avoided also in those highly nonlinear cases where the classical Hermite polynomials are failing.

For introductory examples we look at a single degree of freedom system driven by white noise $\dot{W}_t$ with the intensity σ.

$$\ddot{X}_t + 2D\omega_1\dot{X}_t + \omega_1^2 f(X_t) = \sigma\dot{W}_t, \qquad E[(dW_t)^2] = dt. \tag{1}$$

Here, ω_1 denotes the natural frequency, D is a dimensionless measure of the linear damping term and the function f(x) describes the given nonlinear restoring of the oscillator (1). It is convenient to introduce the normalized displacement and velocity processes U_t and V_t by

$$X_t = U_t\frac{\sigma}{\sqrt{4D\omega_1^3}}, \qquad \dot{X}_t = V_t\frac{\sigma}{\sqrt{4D\omega_1}}, \tag{2}$$

and to go over to Itõ differential equations of the form

$$dU_t = \omega_1 V_t dt, \qquad dV_t = -\omega_1[2DV_t + f(U_t)]dt + 2\sqrt{D\omega_1}dW_t. \tag{3}$$

It is well known [9] that the Fokker-Planck equation, associated to (3) is separable in the stationary case and possesses the closed-form solution

$$p(u,v) = C\exp\left[-\int f(u)du\right]\exp\left(\frac{-v^2}{2}\right), \qquad -\infty < u, v < +\infty \tag{4}$$

Obviously, the velocity process V_t is normally distributed meanwhile U_t has a nonnormal distribution according to the applied nonlinearity f(u).

The constant C is calculable from the normalization condition of the joint distribution (4).

2 Generalized Hermite Polynomials

As already mentioned, we introduce the generalized Hermite polynomials $K_n(u)$ by the orthogonality relation

$$\int_{-\infty}^{+\infty} K_n(u)k_m(u)p(u)du = c_n\delta_{n,m}, \qquad n, m = 0, 1, 2, \ldots \tag{5}$$

Here, the weighting function p(u) follows from the joint density distribution given in (4). It possesses the new normalization constant C_u.

$$p(u) = Cu\exp\left[-\int f(u)du\right], \qquad K_n(u) = \sum_{i=0}^{n} a_{ni}u^i \tag{6}$$

The coefficients a_{ni} of the polynomials $K_n(u)$ are calculated according to the well-known Schmidt-Erhard procedure which reduces the orthogonality relation (5) to

$$\int_{-\infty}^{+\infty} K_n(u)u^m p(u)du = c_n\delta_{n,m}, \qquad (m \leq n),\ a_{nn} = 1 \tag{7}$$

and evaluates this equation by means of the Taylor moments $m_p = E(U_t)$.

$$\sum_{i=0}^{n} a_{ni} m_{i+m} = c_n \delta_{n,m}, \qquad m_p = \int_{-\infty}^{+\infty} u^p p(u) du \tag{8}$$

For $n = 2$ e.g. and the usual normalization by $a_{22} = 1$, we obtain from (8) the following matrix equation

$$\begin{bmatrix} m_0 & m_1 & m_2 \\ m_1 & m_2 & m_3 \\ m_2 & m_3 & m_4 \end{bmatrix} \begin{bmatrix} a_{20} \\ a_{21} \\ 1 \end{bmatrix} = \begin{bmatrix} 0 \\ 0 \\ c_2 \end{bmatrix}, \qquad (a_{22} = 1) \tag{9}$$

by which the remaining coefficients a_{n_i} and the associated normalization constant c_n is determined. Consequently, the generalized Hermite polynomials of any stationary process U_t is completely determined by its Taylor moments. The explicite knowledge of the associated density distribution p(u) is not needed.

The generalized Hermite polynomials $K_n(u)$ gives an orthonormal expansion basis in the sense that all expectations are vanishing except the zeroth one, i.e.

$$E[K_n(U_t)] = 0, \text{ for } n > 0, \qquad E[K_0(U_t)] = 1. \tag{10}$$

To prove this basic property we rewrite (9) in the form [10]

$$\begin{bmatrix} m_0 & m_1 & m_2 \\ m_1 & m_2 & m_3 \\ m_2 & m_3 & m_4 \end{bmatrix} \begin{bmatrix} a_{20} \\ a_{21} \\ a_{22} \end{bmatrix} = \begin{bmatrix} 0 \\ 0 \\ c_2^* \end{bmatrix}, \qquad K_2(u) = \begin{vmatrix} m_0 & m_1 & m_2 \\ m_1 & m_2 & m_3 \\ u^0 & u^1 & u^2 \end{vmatrix}$$

by a suitable choice of the normalization constant c_2. In this case the associated second order polynomial $K_2(u)$ can be represented by a determinant expanded with respect to its last row. Taking the expectation in this form the last row coincides uith the first one so that the determinant is vanishing. This is true for all higher order polynomials except the zeroth one where we simply get $K_0(u) = m_0$ with $m_0 = 1$. Note that the orthonormal property is independent on any choice of the stationary density distribution p(u). In the special case of a normal distribution the equations (8) lead to the well-known coefficients of the classical Hermite polynomials.

3 Stochastic Hermite Equations

Coming back to the differential equations (3) of the nonlinear oscillator (1) we apply the classical Hermite polynomials $H_h(v)$ and the generalized ones $K_k(u)$ to calculate their joint increment according to Itõ's formula.

$$\begin{aligned} & d[K_k(U_t)H_h(V_t)] = \\ & 2D\omega_1 K_k(U_t)\ddot{H}_h(V_t)dt + \omega_1 K_k'(U_t)H_h(V_t)dt + \\ & K_k(U_t)\dot{H}_h(V_t)[-2D\omega_1 V_t dt - \omega_1 f(U_t)dt + 2\sqrt{D\omega_1}dW_t]. \end{aligned} \tag{11}$$

Here, dots and dashes denote derivations with respect to the velocity and displacement processes, respectively. The classical Hermite polynomials $H_h(v)$ possess the recursion formulas

$$\dot{H}_h(v) = hH_{h-1}(v), \qquad vH_h(v) = H_{h+1}(v) + hH_{h-1}(v). \tag{12}$$

Correspondingly, derivatives and products of the generalized Hermite polynomials $K_n(u)$ are expanded by

$$K_k'(u) = \sum_{j=0}^{\infty} d_{kj}k_j(u), \qquad f(u)K_k(u) = \sum_{j=0}^{\infty} f_{kj}k_j(u). \tag{13}$$

The differentiation coefficients d_{kj} and nonminearity coefficients f_{kj} are calculated by means of the orthogonality relation (5).

$$d_{kj} + \frac{1}{c_j}\sum_{l=0}^{k}\sum_{i=0}^{j} l a_{kl} a_{ji} m_{l+i-1}, \qquad (k, j = 0, 1, 2, \ldots) \tag{14}$$

$$f_{kj} + \frac{1}{c_j}\sum_{l=0}^{k}\sum_{i=0}^{j} (l+i) a_{kl} a_{ji} m_{l+i-1}, \qquad b_{kj} = d_{kj} - f_{kj} \tag{15}$$

In addition, it is convenient to introduce the basis coefficients b_{kj} by the difference of d_{kJ} and f_{kj}. Finally, the representations (12) and (13) are inserted into (11) in arriving to the following linear equations for the Hermite state processes $K_k(U_t)$ and $H_h(V_t)$ of the dimensionless displacement and velocity.

$$\begin{aligned} &d[K_k(U_t)H_h(V_t)] = \\ &-2D\omega_1 hK_k(U_t)H_h(V_t)dt \\ &+\omega_1 \sum_{j=0}^{\infty}[d_{kj}K_j(U_t) * H_{h+1}(V_t) + hb_{kj}K_j(U_t)H_{h-1}(V_t)]dt \\ &2\sqrt{D\omega_1} hK_k(U_t)H_{h-1}(V_t)dW_t. \end{aligned} \tag{16}$$

The equations (16) are called stochastic Hermite equations. They are of basic importance for both, the simulation of stochastic systems and the analysis of statistical characteristics.

We start with the one-dimensional investigation of the Hermite moments E_{kh}. Applying the expectation operator in (16) the diffusion term is vanishing and the rest of (16) leads to the following linear equations:

$$\dot{E}_{kh} = -2D\omega_1 hE_{kh} + \omega_1 \sum_{j=0}^{\infty}(d_{kj}E_{jh+1} + hb_{kj}E_{jh-1}) \tag{17}$$

$$E_{kh}(t) = E[K_k(U_t)H_h(V_t)], \qquad k, h = 0, 1, 2, \ldots$$

According to the expansion theorem (10) they possess the stationary solutions $E_{kh} = 0$ for all indices k and h except $E_{00} = 1$. It is worth to note that the moments equations (17) can also be derived via the Fokker-Planck equation associated to the nonlinear oscillator (3).

$$\frac{\partial p}{\partial t} + \frac{\partial}{\partial u}(\omega_1 v p) + \frac{\partial}{\partial v}[-\omega_1(2Dv + f(u))p] - 2D\omega_1 \frac{\partial^2 p}{\partial v^2} = 0 \tag{18}$$

A mild nonstationary solution p(u,v,t) of the diffusion equation (18) can be calculated by the orthogonal expansion

$$p(u,v,t) = \sum_{l=0}^{\infty}\sum_{j=0}^{\infty} e_{lj}(t)\frac{1}{c_l\gamma_j}K_l(u)H_j(v)p(u)p(v) \tag{19}$$

$$p(v) = C_v \exp\left(\frac{-v^2}{2}\right), \qquad \gamma_j = \int_{-\infty}^{+\infty} H_j^2(v)p(v)dv$$

Here, p(v) is the given normal distribution of the stationary velocity process V_t, γ_j are normalization constants of the classical Hermite polynomials $H_h(v)$ and the time-dependent coefficients $e_{lj}(t)$ are the Hermite moments $E_{lj}(t)$ in (17). To proof this we simply apply the orthogonality relations of both Hermite polynomials $K_k(u)$ and $H_h(v)$ to their moments E_{kh}, as follows:

$$\begin{aligned} &E[K_k(U_t)H_h(V_t)] \\ &= \iint_{-\infty}^{+\infty} K_k(u)H_h(v)p(u,v,t)dudv = E_{kh} \\ &= \iint_{-\infty}^{+\infty} K_k(u)H_h(v)\sum_{l=0}^{\infty}\sum_{j=0}^{\infty}\frac{1}{c_l\gamma_j}e_{lj}(t)K_l(u)H_j(v)p(u)p(v)dudv \\ &= e_{kh} \end{aligned} \tag{20}$$

Consequently, if there are already simulation results of the Hermite moments E_{kh}, we know the nonstationary density distribution $p(u,v,t)$, as well, simply by applying the expansion (19).

4 Correlations and Power Spectra

For purposes of two-dimensional investigations we multiply the Hermite equations (16) by $K_1(U_s)$ or by U_s which are equal in case of antisymmetric restoring functions $f(u)$. The time s denotes a fixed parameter with $s \le t$. Subsequently, we take the expectation in arriving to the following correlation equations:

$$R'_{kh}(\tau) = -2D\omega_1 h R_{kh}(\tau) + \omega_1 \sum_{j=0}^{\infty}[d_{kj}R_{jh+1}(\tau) + hb_{kj}R_{jh-1}(\tau)] \tag{21}$$

$$R_{kh}(\tau) = E[K_1(U_s)K_k(U_t)H_h(V_t)], \qquad k,h = 0,1,2,\ldots \ (t = s+\tau)$$

$$\text{IC.: } R_{kh}(0) = 0, \text{ for all } k,h \text{ except } R_{10}(0) = E(U_t^2) = m_2.$$

They are valid in the stationary case under the initial consitions, noted above. Dash denotes derivation with respect to the correlation time $\tau \geq 0$. The special initial conditions of the correlations $R_{kh}(\tau)$ follow from the product expansion of the generalized Hermite polynomials $K_k(u)$.

$$uK_n(u) = \sum_{i=0}^{\infty} p_{ki}K_i(u), \qquad p_{ki} = \frac{1}{c_i}\int_{-\infty}^{+\infty} uK_k(u)k_i(u)p(u)du. \tag{22}$$

Applied to the initial conditions in (21) we obtain

$$R_{kh}(0) = \sum_{i=0}^{\infty} p_{ki}EK_{ih} = p_{k0}E_{0h}, \qquad p_{1v} = \frac{1}{c_0}\int_{-\infty}^{+\infty} u^2p(u)du = m_2 \tag{23}$$

taking into account that all Hermite moments E_{ih} are vanishing in the stationary case, except $E_{00} = 1$.

The solution of the correlation equatlons (21) can directly be simulated by means of the exponential matrix of this linear system. If frequency distributions are of interest we apply the Laplace transformation

$$\mathcal{L}[R'_{kh}(\tau)] = sX_{kh}(s) - R_{kh}(0), \qquad s = j\omega \tag{24}$$

$$S_{kh}(\omega) = X_{kh}(j\omega) + X_{kh}(-j\omega), \qquad k,h = 0,1,\ldots \tag{25}$$

for the determination of the Fourier transform $X_{kh}(j\omega)$. Adding their conjugate complex versions $X_{kh}(-jw)$ we obtain real-valued frequency distributions $S_{kh}(\omega)$ where $S_{kh}(\omega)$ is the power spectrum of U_t.

$$\left[\begin{array}{cc|cccc|c}
s & -\omega_1 d_{10} & 0 & 0 & 0 & 0 & \cdots \\
-\omega_1 b_{01} & (s+2D\omega_1) & -\omega_1 b_{03} & 0 & 0 & 0 & \\
\hline
0 & -\omega_1 d_{30} & s & -\omega_1 d_{32} & 0 & 0 & \\
0 & 0 & -\omega_1 b_{23} & (s+2D\omega_1) & -\omega_1 d_{21} & 0 & \\
0 & 0 & 0 & -2\omega_1 b_{12} & (s+4D\omega_1) & -\omega_1 d_{10} & \\
0 & 0 & 0 & 0 & -3\omega_1 b_{01} & (s+6D\omega) & \\
\hline
\vdots & & & & & &
\end{array}\right]
\left[\begin{array}{c}
X_{10} \\ X_{01} \\ \hline X_{30} \\ X_{21} \\ X_{12} \\ X_{30} \\ \hline \vdots
\end{array}\right] =
\left[\begin{array}{c}
m_2 \\ 0 \\ \hline 0 \\ 0 \\ 0 \\ 0 \\ \hline \vdots
\end{array}\right] \tag{26}$$

Finally, it should be mentioned that the indices of the spectra $X_{kh}(jw)$ are ordered as shown in the matrix equation (26). This corresponds to the situation in linear stochastic systems where higher order spectra or correlations can be calculated by finite equation blocks. However, in nonlinear problems, such expectations are coupled with neighboured blocks so that infinite equations are to be investigated involving the convergence problem.

5 Nondifferentiable Oscillator Restorings

As a first application of the generalized Hermite analysis, derived above, we consider a backlash restoring, i.e. the antisymmetric function f(u) applied in (3) and its associated stationary density (6) are given by

$$f(u) = \begin{cases} u_0 + u & \text{for } |u| \geq u_0 \\ 0 & \text{for } |u| \leq u_0 \end{cases} \qquad p(u) = C_u \begin{cases} \exp(-u^2/2) \\ \exp(-u_0^2/2) \end{cases} \tag{27}$$

Calculating the stationary moments $m_p = E(U_t^p)$ of this density we derive the following recurrence formula:

$$m_{p+2} = (p+1)m_p + \frac{1}{(p+3)(1+B)} u_0^{p+2} \qquad m_0 = 1 \tag{28}$$

$$B = \sqrt{\pi/2}\frac{1 - \Phi\left(u_0/\sqrt{2}\right)}{u_0 \exp\left(-u_0^2/2\right)} \qquad m_1 = 0$$

The recursion is startet with $p = 0$. Φ denotes the error function. The moments m_p are inserted into the algebraic equations (8) in order to calculate the associated polynomial coefficents a_{ni} and normalizatlon constants C_n. For $u_0 = 1$ e.g. we obtain numerical results, as follows:

$$(m_i) = \begin{bmatrix} 1 \\ 0 \\ 1.20 \\ 0 \\ 3.72 \\ \vdots \end{bmatrix} \qquad (c_i) = \begin{bmatrix} 1 \\ 1.20 \\ 2.28 \\ 7.16 \\ 28.3 \\ \vdots \end{bmatrix} \qquad (a_{ni}) = \begin{bmatrix} 1 & 0 & 0 & 0 \cdots \\ 01 & 0 & 0 & \\ -1.20 & 0 & 1 & 0 \\ 0 & -3.10 & 0 & 1 \\ 3.77 & 0 & -6.24 & 0 \\ \vdots & & & \end{bmatrix}$$

$$(d_{kj}) = \begin{bmatrix} 0 & 0 & 0 & 0 & \cdots \\ 1.00 & 0 & 0 & 0 & \\ 0 & 2.00 & 0 & 0 & \\ 0.50 & 0 & 3.00 & 0 & \\ 0 & -0.80 & 0 & 4.00 & \\ \vdots & & & & \end{bmatrix} \qquad (b_{kj}) = \begin{bmatrix} 0 & 0.83 & 0 & 0.07 & 0 \\ 0 & 0 & 1.05 & 0 & -0.0 \\ 0 & 0 & 0 & 0.96 & 0 \\ 0 & 0 & 0 & 0 & 1.01 \\ 0 & 0 & 0 & 0 & 0 \\ \vdots & & & & \end{bmatrix}$$

Subsequently, we perform the summations in (14) and (15) for determining the differentiation and nonlinearity coefficients (d_{kj}) and (b_{kj}).

With these preparations we are ready to evaluate the spectrum equation (26). Figure 1 shows corresponding numerical results for $D = 0.1$, $w_1 = 1$ and three different space lacks $u_0 = 0$, 1.4 and 2. The drawings represent the power spectrum $S_{10}(\omega)$ in dependence of the spectral frequency ω in the range $0 \leq \omega \leq 1.5$. The applied numbers of Hermite polynomials are Na = 5, 7, 9 and 11. As expected, we observe that the resonance of the nonlinear oscillator is shifted to lower frequencies for increasing space lacks u_0. Simultaneously, the power spectrum is raised around the zero frequency. In the upper frequency range, however, the spectral distribution remains almost unchanged. Furthermore, we clearly recognize that the deviations between different approximations become smaller with an increasing expansion order Na. The convergence velocity is the higher the smaller the nonlinearity parameter u_0. For $u_0 = 0$ we obtain the well known result of the linear oscillator.

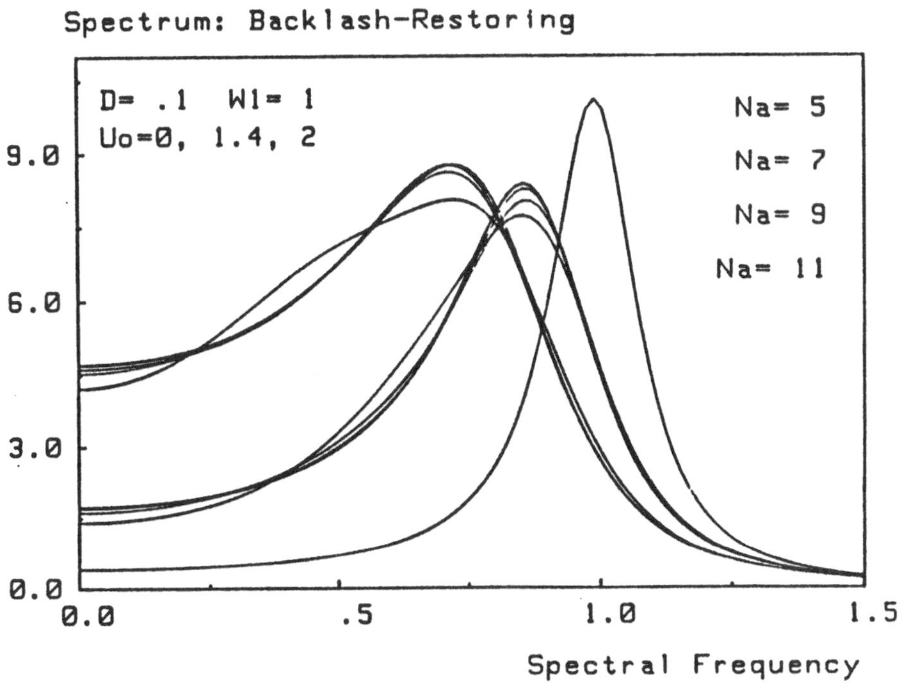

Fig. 1. Oscillator spectra for backlash restorings

We investigate a second form of oscillator restorings. It is given by a bang-bang mechanism. Thus, the antisymmetric characteristic is a sgn function leading to stationary displacement processes U_t which are exponentially distributed.

$$f(u) = \operatorname{sgn}(u), \qquad p(u) = C_u \exp(-|u|),$$

$$m_{p+2} = (p+2)(p+1)m_p, \qquad m_0 = 1,\ m_1 = 0. \tag{29}$$

The associated moments m_p can simply be calculated by means of the recurrence formula, noted above. The following analysis is performed in the same way, as before. Figure 2 shows corresponding results for the power spectra of the stationary displacement processes. The applied natural frequency of the nonlinear oscillator (3) is again $\omega_1 = 1$. The damping values are $D = 0.5$, 0.3 and 0.1. For the high damping $D = 0.5$ the nonlinear bang-bang oscillator possesses spectral distributions like linear low-pass processes. If the damping measure is diminished the bandwidth of the nonlinear spectrum is increasing. For the low damping $D = 0.1$ we observe a resonance distribution. There is a resonance peak below the natural frequency $\omega_1 = 1$. The convergence velocity of the Hermite expansions is removed for decreasing damping of the oscillator.

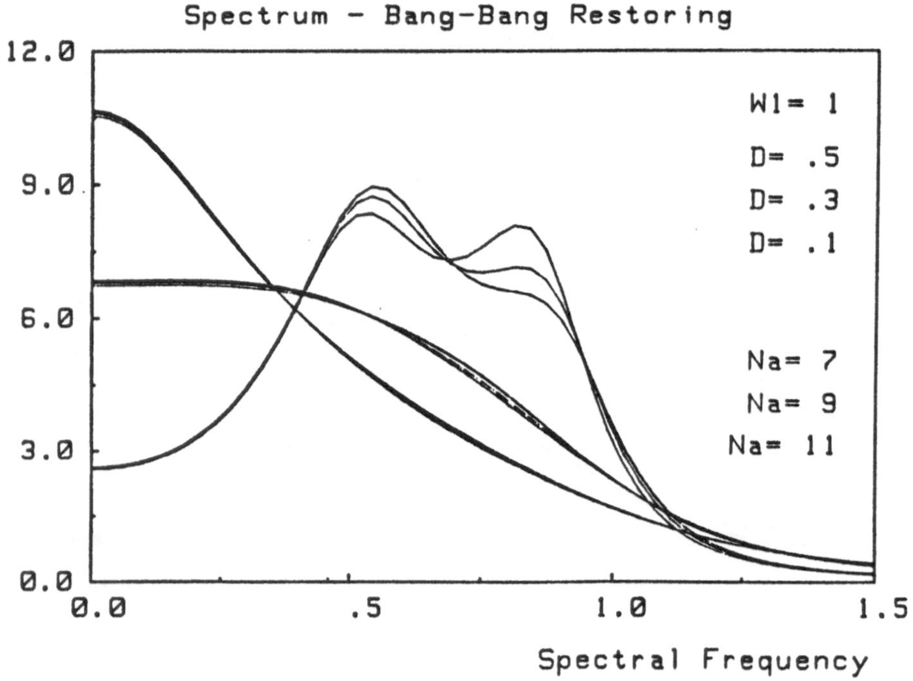

Fig. 2. Oscillator spectra for bang-bang restorings

6 Stochastic Snap-Through and Duffing Problem

Some further examples of interest are given by the Duffing oscillator with a progressive restoring characteristic. If the coefficient of its linear term becomes negative there is a local instability around the zero position leading to the well known snap-through mechanism. Large deviations of the oscillator processes, however, are bounded by damping and cubic terms, both assumed to be positive.

$$\ddot{X}_t + 2\delta\dot{X}_t - \omega_1^2 X_t + \gamma X_t^3 = \sigma\dot{W}_t, \qquad \delta, \gamma > 0$$

$$X_t = U_t \sqrt{\frac{\sigma}{\sqrt{2\delta\gamma}}}, \qquad \dot{X}_t = V_t \frac{\sigma}{\sqrt{4\delta}}. \tag{30}$$

We introduce dimensionless state processes, noted in (30), and go over to corresponding Itõ differential equations.

$$dU_t = \kappa V_t dt, \quad \lambda = \left(\frac{\omega_1}{\kappa}\right)^2, \quad \kappa^2 = \left(\frac{\sigma}{2}\right)\sqrt{\frac{\gamma}{\delta}},$$

$$dV_t = -2\delta V_t dt + \kappa\lambda U_t dt - \kappa U_t^3 dt + 2\sqrt{\delta} dW_t. \tag{31}$$

In the stationary case, the processes U_t and V_t possess a joint two-dimensional density distribution of the following normalized form:

$$p(u,v) = C \exp\left(\frac{-u^4}{4} + \lambda\frac{u^2}{2}\right) \exp\left(\frac{-v^2}{2}\right),$$

$$m_{p+2} = \lambda m_p + (p-1)m_{p-2}, \qquad p = 2, 4, \ldots \tag{32}$$

For the calculation of the displacement moments $m_p = E(U_t^p)$ we derive the recursion formula, noted in (32). To start it we need the initial moments $m_0 = 1$, $m_1 = 0$ and m_2. The latter follows from the integral

$$\int_{-\infty}^{+\infty} u^2 \exp\left(\frac{-u^4}{4} + \lambda\frac{u^2}{2}\right) du = \int_0^{+\infty} \sqrt{z} \exp\left(\frac{-z^2}{2} + \lambda\frac{z}{2}\right) dz$$
$$= \sqrt{2}\left[\Gamma\left(\frac{3}{4}\right) {}_1F_1\left(\frac{3}{4};\frac{1}{2};\frac{\lambda^2}{4}\right) + \lambda\Gamma\left(\frac{5}{4}\right) {}_1F_1\left(\frac{5}{4};\frac{3}{2};\frac{\lambda^2}{4}\right)\right].$$

Here, Γ denotes the gamma function and ${}_1F_1$ is the confluent hypergeometric function [11].

Subsequently, we can perform the same analysis, as shown previously. The nongaussian density p(u) of the dimensionless displacement defines generalized Hermite polynomials $K_n(u)$ the coefficients of which are calculated by the equations (8). Inserted into (14) and (15) we get the associated expansion coefficients d_{kj} and b_{kj}. They determine finally the Hermite equations (16) needed for the correlation and spectral analysis of the nonlinear oscillator (30). In figure 3 we show some typical results representing the displacement spectrum $S_{10}(\omega)$ in the range $0 \le \omega \le 2.5$. The applied expansion orders are $N_a = 7$, 9 and 11, each for two different linear parameters $\lambda = -1$ and $\lambda = 1$, meanwhile the damping value and the cubic nonlinearity are fixed with $\delta = 0.1$ and $\kappa = 1$. For negative $\lambda = -1$ we get the stochastic Duffing oscillator under white noise. Its natural frequency is $\omega_1 = 1$. We recognize clearly the main effect that the linear resonance is shifted to higher frequencies near $\omega = 1.5$ effected by the cubic nonlinearity of the oscillator. For these parameters the Hermite expansions converge rapidly. The drawn spectral distributions coincide graphically for the applied expansion orders N_a. Deviations are only observable if we go back to lower orders $N_a = 5$ or 3. In comparison to these good results the convergence becomes bader for the parameter $\lambda = 1$.

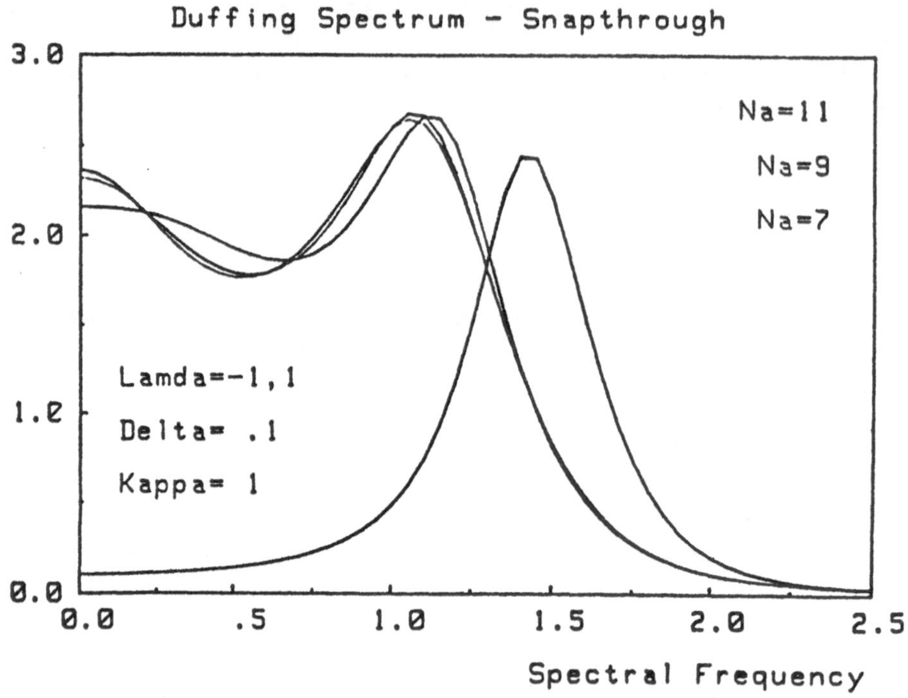

Fig. 3. Duffing and snap-through oscillator spectra

Then, there exists a second resonance at the zero frequency. Also, the main resonance peak is shifted to lower frequency ranges in this case of a snap-through mechanism. To discuss the influence of intensities of the excitation and nonlinearity we go over to the special case that the linear term in (31) is vanishing, i.e. $\lambda = 0$. For this cubic oscillator the spectral densitiy distributions of the displacement process are given in figure 4. Curves with the lower resonance peak are valid for $\kappa = 1.0$, the second one hols for $\kappa = 1.5$. For both cubic parameters the applied expansion order N_a and the damping value are the same, as before. The obtained results correspond clearly to the expected effects. For increasing excitation intensity σ or increasing cubic parameter γ the oscillator resonance is shifted to higher frequency ranges with decreasing spectral amplitudes of the resonance range.

Finally, it is worth noting that the Hermite equations (16) can be once more be simplified in case of differentiable restorings applied in system (31). For this special case we obtain the following reduced form:

$$\begin{aligned} dK_kH_h = {} & 2h\sqrt{2}K_kh_{h-1}dW_t - 2\delta hK_kH_hdt + \kappa d_{k,k-3}K_{k-3}H_{h+1}dt + \\ & h\kappa(\lambda - c_{k,k+1})K_{k+1}H_{h-1}dt - k\kappa K_{k-1}H_{h+1}dt - \\ & k\kappa K_{k+3}H_{h-1}dt \end{aligned} \tag{33}$$

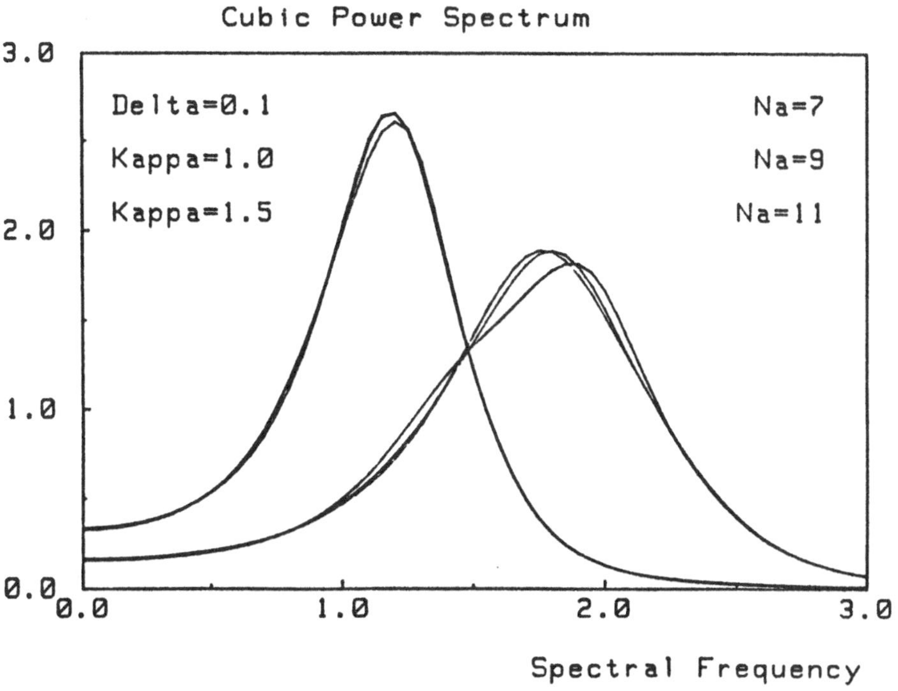

Fig. 4. Power spectra of the cubic oscillator

Here, the coefficients $d_{k,k-3}$ and $c_{k,k+l}$ are determined by

$$d_{k,k-3} = (k-2)a_{k,k-2} - ka_{k-1,k-3}, \qquad c_{k,k+1} = a_{k,k-2} - a_{k+3,k+1}. \tag{34}$$

They follow from the differentiation and multiplication rules of the generalized Hermite polynomials $K_n(u)$. In particular, the following relations hold.

$$\left.\begin{aligned} &K_k'(u) = kK_{k-1} + d_{k,k-3}K_{k-3}, \\ &d_{k,k-3} = c_{k,k-3} \\ &uK_k(u) = K_{k+1} + p_{k,k-1}K_{k-1}, \\ &c_{k,k-1} = k + \lambda p_{k,k-1} \\ &u^3K_k(u) = K_{k+3} + c_{k,k+1}K_{k+1} + c_{k,k-1}K_{k-1} + c_{k,k-3}K_{k-3} \end{aligned}\right\} \tag{35}$$

They can simply be derived by means of the orthogonality condition (5) or in a more direct way by inserting the polynomials $K_n(u)$ and performing a comparison of coefficients. We give only two typical examples for the derivation of multiplication coefficients.

$$uK_k = \sum_{i=0}^{k+1} p_{ki}K_i \;\Rightarrow\; p_{k,k-1}c_{k-1} = \int_{-\infty}^{+\infty} uK_k(u)K_{k-1}(u)p(u)du$$

$$p_{k,k-1}c_{k-1} = \int_{-\infty}^{+\infty} K_k(u)[a_{k-1,k-1}u^k + \cdots + a_{k-1,0}u]p(u)du = c_k$$

$$p_{k,k-2}c_{k-2} = \int_{-\infty}^{+\infty} K_k(u)[a_{k-2,k-2}u^{k-1} + \cdots + a_{k-2,0}u]p(u)du = 0$$

Finally, to obtain the well known multiplication and differentiation rules of the classical Hermite polynomials we just have to replace the corresponding coefficients in (35) by $p_{k,k-1} = k$ and $d_{k,k-3} = 0$.

7 Fast Algorithms, Convergence and Existence

As already mentioned, the spectral analysis of the nonlinear oscillator (3) is based on the linear algebraic equation system (26) which has to be solved for each frequency ω for obtaining frequency distributions. This evaluation can be improved with respect to computing time if only one spectral density, e.g. the displacement spectrum $S_{10}(\omega)$ is of interest. Then it is more efficient to simulate the time-dependent correlation function $R_{10}(\tau)$ by means of exponential matrices and to store the correlation values of a sufficiently large simulation time. Subsequently, we can apply the fast Fourier transform to calculate the associated spectrum. We show a typical correlation behaviour in figure 5 for the data $\omega_1 = \lambda = 0$, $\kappa = 1$ and $\delta = 0.1$. In correspondence to (33) the applied correlation equations are

$$R'_{k,h}(\tau) = -2\delta h R_{k,h}(\tau) + \kappa k R_{k-1,h+1}(\tau) - \kappa h R_{k-3,h-1}(\tau) \qquad (36)$$

$$-\kappa h c_{k,k+1} R_{k+1,h-1}(\tau) + \kappa d_{k,k-3} R_{k-3,h+1}(\tau) \qquad (37)$$

or in matrix form

$$r'(\tau) = Ar(\tau), \qquad r(\tau) = [R_{10}, R_{01} | R_{30}, R_{21}, R_{12}, R_{03} | \ldots]^T \qquad (38)$$

They are simulated with the scan rate $\Delta\tau = 0.01$ for the expanslon orders $N_a = 5$, 7, 9 and 11 under the initial conditions given in (23) and (26). There are several advantages of such a procedure. In the time domain of the correlation function we start with exact initial values at $\tau = 0$ and need only a low expansion order N_a for short correlation times. Moreover, the time simulation is simply performed by multiplications of the correlation vector with the exponential matrix $\exp(A\Delta t)$. As shown in fig. 5, higher order Hermite expansions have to be taken into account for increasing τ parameter and particularly, for small damping values of the oscillator. In case of a vanishing damping, stationary density distributions do not exist. Then, there is no convergence of the applied τ expansion (36) in this extreme case.

It is also worth to note that the generalized Hermite polynomials lead to best expansions in the sense of the physical existence conditions for autocorrelation functions and power spectra.

$$R(0) \geq |R(\tau)|, \qquad S(\omega) \geq 0. \qquad (39)$$

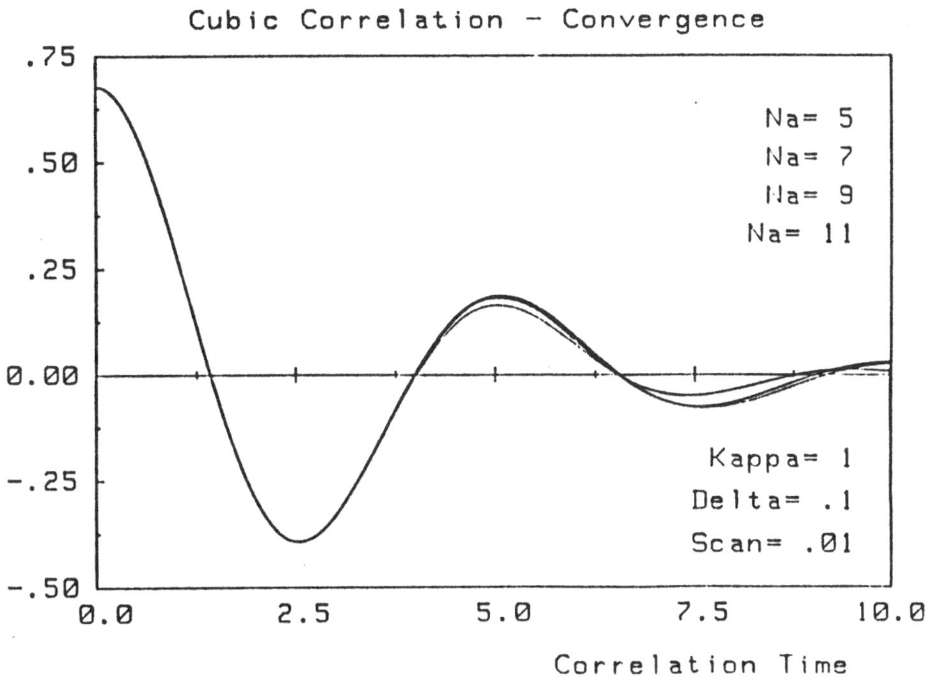

Fig. 5. **Autocorrelation of the cubic oscillator**

Often, they can not be satisfied if simplier expansions are used. For the Duffing oscillator e.g. described by

$$\ddot{X}_t + 2D\omega_1\dot{X}_t + \omega_1^2 X_t + \gamma\omega_1^2 X_t^3 = \sigma\dot{W}_t, \qquad .\omega_1, D, \gamma > 0, \tag{40}$$

$$X_t = U_t\frac{\sigma}{\sqrt{2D\omega_1^3}}, \qquad \dot{X}_t = V_t\frac{\sigma}{\sqrt{2D\omega_1}}, \qquad \gamma_1 = \gamma\frac{\sigma^2}{4D\omega_1^3}$$

we find applications of classical Hermite polynomials already in [6]. Here, they are defined by the joint normal distribution density p(u,v) of the dimensionless displacement and velocity processes U_t and V_t.

$$dU_t = \omega_1 V_t dt,$$

$$p(u,v) = C\exp[-(u^2+v^2)],$$

$$dV_t = \sqrt{2D\omega_1}dW_t - \omega_1(2DV_t + Ut + 2\gamma_1 U_t^3)dt. \tag{41}$$

For the calculation of the stationary moments it is obvious that γ_1 is the corresponding expansion parameter. The numerical results, given in [6], show a monotone convergence property controllable by means of hypergeometric functions. However, if we extend the analysis by classical Hermite polynomials to the investigation of correlation functions or power spectra there exists a second expansion parameter, namely the positive correlation time which reduces rapidly the convergence velocity, in particular for high nonlinearities. Such a result is shown in figure 6 for the system parameters $D = 0.1$, $\omega_1 = 1$, $\sigma = 1$, $\gamma = 0.1$ and the expansion orders $N_a = 1, 3, 5, 7, 9$ and 11. In comparison with the corresponding curves, drawn in figure 3, the convergence rate of the classical Hermite polynomials is much lower. Moreover, we observe negative spectral distributions in the higher frequency ranges. Such a losing of the physical existence is clearly avoided by the generalized Hermite polynomials, applied before.

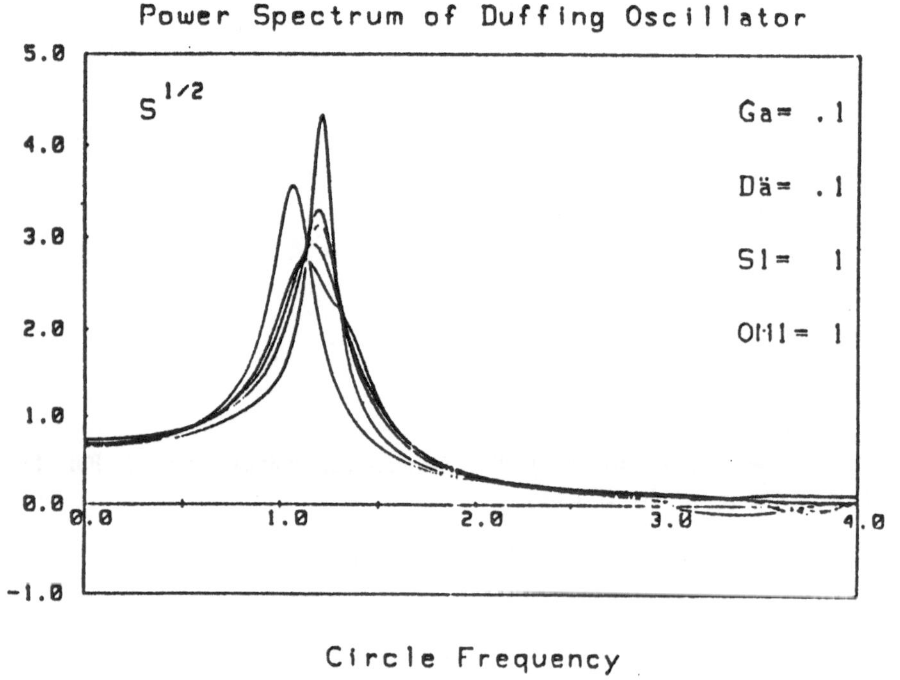

Fig. 6. Dufing spectrum via classical Hermite polynomials

Finally, it is worth to note that the generalized Hermite analysis can easily be extended to multidimensional problems. To show this we consider nonseparable problems of two dimensions where we know the associated joint distribution density p(u,v) of the stationary system processes. It defines a two-dimensional orthogonality relation and Hermite polynomials of the form:

$$\left.\begin{array}{ll} H_{m,n}(u,v) = \sum_{i=0}^{m}\sum_{j=0}^{n} a_{m,n,i,j} u^i v^j, & m,n = 0,1,2,\ldots \\ \int\int_{-\infty}^{+\infty} H_{m,n}(u,v) H_{k,l}(u,v) p(u,v) du dv = c_{m,n}\delta_{m,n,k,l} & \end{array}\right\} \quad (42)$$

Inserting the polynomials into the orthogonality relation we obtain the following linear equation systems for the determination of all polynomial coefficients.

$$\left.\begin{array}{ll} \sum_{i=0}^{m}\sum_{j=0}^{n} a_{mnij} m_{i+k,j+l} = c_{mn}\delta_{mnkl} & (k,l \leq m,n) \\ m_{i,j} = E[U_t^i V_t^j] = \int\int_{-\infty}^{+\infty} u^i v^j p(u,v) du dv \;\; a_{mnmn} = 1 & \end{array}\right\} \quad (43)$$

Here, δ_{mnkl} denotes the Kronecker symbol and m_{kl} are stationary moments of the system processes. The highest coefficient of the two-dimensional polynomial is normed, as usual. The evaluation of (43) leads e.g. to

$$\left[\begin{array}{ccc|cc|c} m_{00} & m_{10} & m_{20} & m_{01} & m_{11} & m_{21} \\ m_{10} & m_{20} & m_{30} & m_{11} & m_{21} & m_{31} \\ m_{20} & m_{30} & m_{40} & m_{21} & m_{31} & m_{41} \\ \hline m_{01} & m_{11} & m_{21} & m_{01} & m_{12} & m_{22} \\ m_{11} & m_{21} & m_{31} & m_{12} & m_{22} & m_{32} \\ \hline m_{21} & m_{31} & m_{41} & m_{22} & m_{32} & m_{42} \end{array}\right] \left[\begin{array}{c} a_{2100} \\ a_{2110} \\ a_{2120} \\ \hline a_{2101} \\ a_{2111} \\ \hline 1 \end{array}\right] = \left[\begin{array}{c} 0 \\ 0 \\ 0 \\ \hline 0 \\ 0 \\ \hline c_{21} \end{array}\right]$$

for the determination of special polynomial

$$H_{21}(u,v) = a_{2100} + a_{2120}u + a_{2101}v + a_{2120}u^2 + a_{2111}uv + u^2v$$

In case that the associated joint density distribution is separable it coincides with the product of $H_2(u)H_1(v)$. Applications of the two-dimensional Hermite polynomials are still lacking.

8 Conclusions

Recently, there is an increasing interest in the application of classical Hermite polynomials for the investigations of moment stability, Lyapunov exponents and stationary characteristics of nonlinear vibration systems. In suited problems the expectations of Hermite polynomials possess the usual advantage of orthogonal expansions; i.e. associated moment equations can be truncated simply by neglecting higher order expectations without any closure condition. It is therefore important to extend to non-gaussian Hermite polynomials and to test convergence properties in more complicated questions.

In the present paper we introduced such generalized Hermite polynomials. They are defined by given non-gaussian distribution densities of stationary system response processes. It is shown that the polynomial moments represent an orthonormal expansion base of associated diffusion equations; i.e. they can be applied for expansions in Fokker-Planck equations to obtain nonstationary solutions of so-called mild forms. The same holds for the expansion of correlation functions or power spectra of nonlinear stochastic systems, investigated here. Typical examples are the stochastic Duffing oscillator, snap-through problems, restoring forces with backlash or bang-bang characteristics.

References

1. T.K. Caughey, Equivalent linearization techniques, J. Acoust. Society America, Vol 35, 1963, p. 1706-1711.
2. F. Kozin, The method of statistical linearization for non linear stochastic vibrations, Nonlinear Stochastic Dynamic Engineering Systems (Iutam Symposium 1987, F. Ziegler, G.I. Schuel ler eds.), Springer-Verlag, Berlin, Heidelberg, 1988, p. 45-56.
3. W. Wedig, Stochastische Schwingungen - Simulation, Schatzung und Stabilität, GAMM-Tagung Dortmund 1986, ZAMM 67, 4, 1987, p. T34-T42 .
4. W. Wedig, Lyapunov Exponent and Rotationszahl der stochastischen Eigenwerttheorie, GAMM-Tagung Wien 1988, to appear in ZAMM 69, 1989, T542.
5. W. Wedig, Mean square stability and spectrum identification of nonlinear stochastic systems, in: Nonlinear Stochastic Dynamic Engineering Systems (IUTAM Symposium 1987, F. Ziegler, G.I. Schueller eds.), Springer-Verlag, Berlin, Heidelberg, 1988, p. 135-152.
6. W. Wedig, Parameter instability and process identification, in: Analysis and Estimation of Stochastic Mechanical Systems (CISM 1987, No 303, W. Schiehlen, W. Wedig, eds.), Springer-Verlag, Wien, New York, 1988, p. 201-242.
7. H. Cramer, Mathematical Methods of Statistics, Princeton University Press, Princeton, 1 961, p. 221-231.
8. S.R. Winterstein, Moment-base Hermite Models of Random Vibration Technical University of Denmark, Lyngby, ISBN 87-87336-78-2.
9. T.K. Caughey, Derivation and application of the Fokker-Planck equation. J. Acoust. Soc. America, Vol. 35, 1963,
10. W. Wedig, Bifurcations in stochastic systems - a generalized Hermite analysis, First European Seminary on Effective Stochastics, Luminy, Marseille, March 1988 (nonpublished).
11. D. Middleton, Statistical Communication Theory, McGraw-Hill, New-York, London, 1960, p. 1071-1079.

Some Remarks Concerning Convergence of Orthogonal Polynomial Expansions

Pierre Bernard

1 Introduction

1- In [9], a numerical computational method for the solutions of the Fokker Planck Kolmogorov equation resting on Hermite polynomial expansion of a probability density is proposed by C.Soize.

Let $\rho_S(x)$ be a probability density on $\mathbb{R}^d$, and $\gamma_d(x)$ be the standard Gaussian density on $\mathbb{R}^d$.

Set $q(x) = \rho_S(x)\gamma_d^{-1}(x)$. Assuming $q \in L^2(\mathbb{R}^d, \gamma dx)$,we have:

$$\int q^2(x)\gamma_d(x)dx < \infty. \tag{1}$$

This is equivalent to:

$$\int \rho_S^2(x)\gamma_d(x)^{-1}dx < \infty, \tag{2}$$

condition that guaranties that q can be expanded in Hermite polynomials with convergence in $L^2(\mathbb{R}_d, \gamma_d dx)$.

But convergence in L^2 space is poor for numerical purposes.

2- In [11], W.Wedig introduced expansions in polynomials orthogonal with respect to the stationary probability density of a non linear oscillator with one degree of freedom excited by a white noise:

$$\ddot{u}_t + \dot{u}_t + f(u) = b_t, \tag{3}$$

In this example, the exact solution is known:

$$p(u,v) = Cexp(-\int(f(u)du))exp(-v^2/2) \tag{4}$$

with some restrictions on the non linear term (antisymetry for example). W.Wedig then uses polynomials orthogonal with respect to the density $p(u)$, named generalized Hermite polynomials, to approximate the power spectrum. There is no convergence result in the paper by W.Wedig, but numerical experience on some examples show a good approximation.

Many applications in random mechanics and other fields use expansions in series of polynomials orthogonal relatively to a kernel ρ. This is the case of Wiener chaos decomposition in Gaussian analysis, one aspect of which is well known [1] : the estimation of a probability density by Gram-Charlier or Edgeworth expansions. In this case, the density is Gaussian.

The pitfall of these methods is the convergence result. The natural convergence is in the $L^2(E_d, \rho dx)$ space; but this does not even imply an almost everywhere convergence. For numerical purposes, one needs point convergence, and, if possible, uniform convergence on the compact sets in E_d.

The goal of this contribution is to give a method available for this problem. It rests on weigthed Sobolev spaces; in the Gaussian case, these spaces are precisely the Gaussian Sobolev spaces introduced by P.Krée and his group in 1974 [2] [3] [6].

Uniform and point convergence for series of orthogonal polynomials, in the case $d = 1$, were studied in the books [7] and [8]. A different method is introduced here, and the results obtained are available in any dimension. In the case of dimension one, they improve some results of these references.
Let ρ be a C^∞ and strictly positive function defined on the Euclidian space E_d.

Lemma 1. *Suppose that ρ has an exponential decay i.e.*

$$\text{[N2]} \qquad \exists \varepsilon_0 > 0 \text{ such that } \int e^{\varepsilon_0 \|x\|} \rho(x) dx < \infty. \tag{5}$$

Then for any given p $(1 < p < \infty)$ the polynomials are dense in $L^p(E_d, \rho dx)$.

Proof:

Since ρ has exponential decay, all polynomials on E_d belong to $L^p = L^p(E_d, \rho dx)$. Since the dual of L^p is $L^{p'}$ with $\frac{1}{p} + \frac{1}{p'} = 1$ we have to show that any $g \in L^{p'}$ satisfying the following conditions vanishes

$$\forall \alpha = (\alpha_1, \ldots, \alpha_d) \in \mathbb{N}^d \quad \int g(x) x^\alpha \rho(x) dx = 0 \tag{6}$$

But Holder inequality combined with N2 gives that $g\rho$ has exponential decay i.e.

$$\exists \varepsilon > 0 \text{ such that } \int e^{\varepsilon \|x\|} |g(x)| \rho(x) dx < \infty. \tag{7}$$

Therefore the following function of $z = (z_j)_1^d = (u_j + iv_j)_1^d$ is defined for $max|u_j|$ small enough and holomorphic

$$\Phi(z) = \int e^{\sum z_j x_j} \rho(x) g(x) dx \tag{8}$$

But (6) means that

$$\left(\frac{\partial}{\partial z}\right)^\alpha \Phi(0) = 0 \text{ for all } \alpha = (\alpha_1, \ldots, \alpha_d) \tag{9}$$

Since Φ is analytic, this means $\Phi = 0$. Hence the Fourier Transform (F.T.) of $g\rho$ vanishes. Since F.T. is bijective, this implies $g\rho = 0$ i.e. $\rho = 0$.

2 Weighted Sobolev Spaces

Let $\rho(x)$ be a probability density satisfying the preceding hypotheses, that is $\rho(x) > 0$ on E_d and $\rho\, C^\infty$. Let p, $1 < p < \infty$, and k a positive integer. $W^{p,k}(E_d, \rho dx)$ is the space of classes of functions $f \in L^p(E_d, \rho dx)$ such that, for any vector of integers, $\alpha = \alpha_1, \cdots, \alpha_d$ satisfying $|\alpha| \le k$, $D^\alpha f \in L^p(E_d, \rho dx)$.

For any bounded open set U in E_d, $W^{p,k}(U, dx)$ is the classical Sobolev space, which contains all classes of functions $g : U \longrightarrow \mathbb{R}$ elements of $L^p(U, dx)$ and such that, for any vector of integers $\alpha = \alpha_1 \cdots \alpha_d$ such that $|\alpha| \le k$, $D^\alpha p \in L^p(U, dx)$. By D^α we mean the derivarive in the distributions sense.

Lemma 2. *For any vector of integers α, any real p, $1 \le p < \infty$, and any bounded open set U in E_d*

$$\rho^{-1/p}\, D^\alpha \rho^{1/p} \quad \textit{is bounded on } U. \tag{10}$$

This results from the calculus of the derivatives and the fact that ρ does not vanish on U.

Let us consider the mapping Γ_p defined for any $f \in L^p(E_d, \rho dx)$, by $\Gamma_p f = f\rho^{1/p}$. It is clear that this is an one to one isometry from $L^p(E_d, \rho dx)$ onto $L^p(E_d, dx)$. For any bounded open set U, denote $|_U$ the restriction to U.

Theorem 3. *Assuming ρ satisfies [N1] and [N2], $\Gamma_p|_U$ is a continuous mapping from $W^{p,s}(E_d, \rho dx)$ into $W^{p,s}(U, dx)$ for any real s and any bounded open set U in E_d.*

Proof :

Let us consider $\Gamma_p : L^p(E_d, \rho dx) \longrightarrow L^p(E_d, dx)$, which is injective. Then take the restriction to U of the images of the elements of $W^{p,s}(E_d, \rho dx)$.

When $s = k$ positive integer , and $|\alpha| \le k$, $D^\alpha(f\rho^{1/p})$ can be computed using Leibniz formula:

$$D^\alpha(f\rho^{1/p}) = \sum_{\substack{\beta \\ \beta \le \alpha}} \binom{\alpha}{\beta} D^{\alpha-\beta} f\, D^\beta \rho^{1/p}$$

which can be written:

$$\sum_{\substack{\beta \\ \beta \le \alpha}} \binom{\alpha}{\beta} D^{\alpha-\beta} f.\rho^{1/p} \,.\, D^\beta \rho^{1/p}.\rho^{-1/p}.$$

In this formula, $D^{\alpha-\beta} f.\rho^{1/p} \in L^p(U, dx)$ by definition of $W^{p,s}(E_d, \rho dx)$ and $D^\beta \rho^{1/p}.\rho^{-1/p}$ is bounded on U by lemma 2. Hence the result is proved in this case. The case s positive real is obtained by interpolation, the case s negative real is obtained by duality.

As a particular case, the local Gaussian Sobolev spaces can be continuously imbedded into the ordinary Sobolev spaces.

When $p = 2$ in the Gaussian case, one can prove, by integration by part, that there is a continuous injection from the Sobolev space $W^{p,s}(E_d, \gamma)$ into $W^{p,s}(E_d, dx)$. The local regularity is the same for both spaces; but for the Gaussian Sobolev, the elements have a growing limitation at infinity.

3 Application to Expansions in Series of Orthogonal Polynomials on E_d

Let $P_\alpha(x)$ be a family of orthogonal polynomials on E_d, deduced by the Gram-Schmidt procedure in $L^2(E_d, \rho dx)$ from the canonical basis of all polynomials on E_d. α denotes a d-dimensional vector of integers.

Hence, any $f \in L^2(E_d, \rho dx)$ can be represented:

$$f = \sum_{\alpha:|\alpha|=0}^{\infty} C_\alpha P_\alpha(x) \tag{11}$$

where the right side converges in $L^2(E_d, \rho dx)$.

If $f \in W^{2,s}(E_d, \rho dx)$, $s > 0$, then f can be expanded as in (11), but the expansion has no reason to converge in $W^{2,s}(E_d, \rho dx)$.

Let us now assume the following property [P]:

[P]: Every $f \in W^{2,s}(E_d, \rho dx)$ can be expanded in orthogonal polynomials in $L^2(E_d, \rho dx)$, with convergence in $W^{2,s}(E_d, \rho dx)$.

The image by Γ_p of the polynomials $P_\alpha(x)$ is an orthogonal family in $L^2(E_d, dx)$, and, from theorem 3 we conclude that $\Gamma_p f|_U$ is the sum of the series

$$\sum C_\alpha \Gamma_p(P_\alpha)|_U$$

in $W^{2,s}(U, dx)$.

By Sobolev injection lemma, $W^{2,s}(U, dx)$ can be continuously imbedded for $s > d/2$ in $C^{s-\frac{d}{2}}(\overline{U})$. Hence, a fortiori in $C^0(\overline{U})$, the space of all continuous functions on $\overline{U}$ endowed with the topology of uniform convergence. Therefore, on every compact set K, $\sum C_\alpha P_\alpha(x)\rho^{1/p}(x)$ is uniformly convergent with sum $f\rho^{1/p}(x)$ and, as $\rho^{1/p}$ and $\rho^{-1/p}$ are bounded on K, $\sum C_\alpha P_\alpha(x)$ converges uniformly to $f(x)$ if $f \in W^{2,s}(E_d, \rho dx)$ with $s > \frac{d}{2}$.

Theorem 4. *Let ρ be a density satisfying [N1] et [N2]. If $f \in W^{2,s}(E_d, \rho dx)$, with $s > \frac{d}{2}$, the expansion in the space $L^2(E_d, \rho dx)$ in orthogonal polynomials $f = \sum_{|\alpha|=0}^{\infty} C_\alpha P_\alpha(x)$, is convergent to $\overline{f}$ uniformly on compact sets ($\overline{f}$ denotes the continuous representant of f).*

4 Application to Uniform Convergence of Hermite and Gram-Charlier Expansions.

Consider the case when $\rho = \gamma$ is a standard Gaussian kernel, and $p = 2$. Let us then prove that property [P] is satisfied.

The Gaussian Sobolev space $W^{2,s}(E_d, \gamma)$ is isometrically isomorphic with $L^2(E_d, \gamma)$ via

$$f \in W^{2,s}(E_d, \gamma) \longrightarrow (I + N)^{s/2} f \in L^2(E_d, \gamma)$$

where N is the operator number of particles defined by:
if $f = \sum C_\alpha P_\alpha$ is the Wiener chaos decomposition of $f \in L^2(E_d, \gamma)$, then

$$Nf = \sum |\alpha| C_\alpha P_\alpha.$$

Hermite polynomials are the eigenfunctions of N and, if $f \in L^2(E_d, \gamma)$,

$$f \in Dom(I+N)^{s/2} \Leftrightarrow f = \sum_\alpha C_\alpha H_\alpha \text{ and } \sum C_\alpha^2 < \infty,\ \sum (1+|\alpha|)^s C_\alpha^2 < \infty.$$

Hence, $(I+N)^{s/2} f = \sum_\alpha (1+|\alpha|)^{s/2} C_\alpha H_\alpha$.

Moreover, $\| f \|_{2,s} = \sum (1+|\alpha|)^s C_\alpha^2$.

Therefore, applying $(I+N)^{-s/2}$, if $f \in W^{2,s}(E_d, \gamma)$, f is the sum of its chaotic expansion (i.e in series of Hermite polynomials) in this space.

Theorem 5. *Let $f \in W^{2,s}(E_d, \gamma)$, with $s > \frac{1}{2}$.*

The expansion of f in Hermite polynomials converges uniformly on every compact set K in E_d.

Proof :

The expansion of f in Hermite polynomials is the expansion of f in chaos, hence it converges to f in $W^{2,s}(E_d, \gamma)$ as a consequence of the preceeding remark. Applying Γ, $\sum_{n=0}^{\infty} C_n H_n \gamma^{1/2}$ converges uniformly on $\mathbb{R}^d$ to $f\gamma^{1/2}$. As $\gamma^{1/2}$ in bounded on every bounded open set U, we obtain the conclusion.

Theorem 6. *Let μ be a probability on E_d, such that $\mu\gamma^{-1} \in W^{2,s}(E_d, dx)$, with $s > \frac{1}{2}$. The Gram-Charlier expansion of μ is uniformly convergent on every compact set in E_d :*

$$\mu = \sum_\alpha (-1)^{|\alpha|} C_\alpha D^\alpha \gamma(x)$$

where α is a vector of integers, $|\alpha|$ its length, $\gamma(x)$ the density of the standard Gaussian probability on E_d and D^α the derivation of order α.

5 Convergence in $L^p(\mathbb{R}^d, \gamma)$ of Chaotic Expansions

The following question arises in a natural way:

every $f \in L^p(\mathbb{R}^d, \gamma)$ has an expansion in chaos converging in the distribution sense. Is this expansion convergent in the L^p space?

The answer is the following, and the proof will be given further:

Theorem 7. *For any $1 < p < \infty$, $p \neq 2$, there exist Gaussian distributions on $\mathbb{R}$ which belong to the space $L^p(\mathbb{R}, \gamma)$ and are not the sum in this space of their expansion in chaos.*

Remark: A proof of this result for $1 < p < 2$ was given by Damien Lamberton [5].

Let us now give a general proof.

5.1 Norm of the Projectors on Chaos

In this section, γ denotes the standard Gaussian measure on $\mathbb{R}$.

$W^{-\infty}(\mathbb{R},\gamma)$ is the space of Watanabe distributions [10] [4]. If A is a linear operator from $L^p(\mathbb{R},\gamma)$ to $L^q(\mathbb{R},\gamma)$, $\|A\|_{p,q}$ denotes its operator norm. If it is unbounded, $\|A\|_{p,q}=\infty$. When $f\in L^p(\mathbb{R},\gamma)$, f defines a distribution on $\mathbb{R}$ as a cylindrical distribution [4], and hence can be expanded in Wiener chaos:

$$f=\sum_{n=0}^{\infty} f_n \tag{12}$$

the series beeing weakly convergent in $W^{-\infty}(\mathbb{R},\gamma)$.
Let π_n be the orthogonal projector on the chaos C_n. Nelson hypercontractivity inequalities imply:

$$\|T_t f\|_p \le \|f\|_q \tag{13}$$

as soon as $e^t \ge \sqrt{\frac{p-1}{q-1}}$.
If $q=2$, the inequality is verified when $e^t\ge\sqrt{p-1}$. Let us apply this result to $\pi_n f$: when $p>2$,

$$\|T_t\pi_n f\|_p \le \|\pi_n f\|_2 \le \|f\|_2 \le \|f\|_p$$

that is $\|\pi_n f\|_p \le e^{nt}\|f\|_p$ as soon as $e^t\ge\sqrt{p-1}$. Therefore:

$$\begin{aligned} \|\pi_n\|_{p,p} &\le (\sqrt{p-1})^n \quad if \;\; p>2 \\ and \;\; \|\pi_n\|_{p,p} &\le \left(\tfrac{1}{\sqrt{p-1}}\right)^n \quad if \;\; 1<p<2. \end{aligned} \tag{14}$$

Let us now assume the following property:
$[P_p]$: For all $f\in L^p(\mathbb{R},\gamma)$, the series 12 is convergent in $L^p(\mathbb{R},\gamma)$.
We know that $[P_2]$ is true. So, we only consider the case $p\neq 2$.

By Banach-Steinhauss theorem, $[P_p]$ implies that there exist a constant A_p such that:

$$\forall n\in\mathbb{N} \quad \|\sum_{k=0}^{n} f_n\|_p \le A_p\|f\|_p;$$

hence a constant C_p independant of n such that:

$$\forall n\in\mathbb{N} \quad \|f_n\|_p \le C_p\|f\|_p. \tag{15}$$

That is to say, the norm $\|\pi_n\|_{p,p}$ of the operator π_n from $L^p(\mathbb{R},\gamma)$ into $L^p(\mathbb{R},\gamma)$ is bounded uniformly in n.

5.2 A Result by Weissler

Let us consider the semi-group $e^{-zN}, z \in \mathbb{C}$, where N is the number of particles operator. Following Weissler [12], we call it Hermite semi-group. When restricted to the domain $z \in \mathbb{R}_+$, it is the Ornstein-Uhlenbeck semi-group.

Theorem 8. *Let* $1 \leq p \leq q \leq \infty$, $Re(z) \geq 0$, $\omega = e^{-z} \neq \pm 1$.
Then, $\|e^{-zN}\|_{p,q} < \infty$ *if and only if:*

$$Re(\frac{1}{1-\omega^2}) \geq \frac{1}{p} \quad Re(\frac{1}{1-\omega^2}) \geq \frac{1}{q'} \quad 1/q + 1/q' = 1 \tag{16}$$

$$[Re(\frac{1}{1-\omega^2}) - \frac{1}{p}][Re(\frac{1}{1-\omega^2}) - \frac{1}{q'}] \geq [Re(\frac{\omega}{1-\omega^2})]^2. \tag{17}$$

Remark.
In the case $z \in \mathbb{R}_+$, put $z = t$, $\omega = e^{-t}$.
Conditions (16) are verified, and condition (17) is equivalent to $\omega^2 \leq \frac{p-1}{q-1}$, which is Nelson's hypercontractivity criterium.

In the case $\omega = e^{-z} = i\alpha$, and $p = q$, we obtain

$$\frac{1}{1+\alpha^2} \geq \frac{1}{p} \qquad \frac{1}{1+\alpha^2} \geq \frac{1}{p'}$$

that is

$$\alpha^2 \leq (p \wedge p') - 1. \tag{18}$$

As, if $p \neq 2$, $p \wedge p' < 2$, we see that for any α such that

$$p \wedge p' - 1 < \alpha^2 < 1 \tag{19}$$

e^{-zN} is unbounded from $L^p(\mathbb{R}, \gamma)$ to $L^p(\mathbb{R}, \gamma)$.

5.3 A Spectral Multiplier

Let us now consider the spectral multiplier P_ω acting as follows on $W^{-\infty}$:
if $f = \sum_{n=0}^{\infty} f_n$ is the expansion of f on Wiener chaos, then $P_\omega f = \sum_{n=0}^{\infty} \omega^n f_n$, in the weak sens. If $|\omega| < 1$, and if the spectral projectors were uniformly bounded from L^p to L^p, then P_ω would be bounded.
Consider $\omega = e^{-z}$, $P_\omega = e^{-zN}$. Then take $\omega = i\alpha$, $p - 1 \leq \alpha^2 < 1$. If $[P_p]$ was true, e^{-zN} would be bounded from L^p to L^p, which contadicts the conclusions of the preceding section. Hence, for $p \neq 2$, $[P_p]$ in not true.

References

1. H. Cramer: Mathematical Methods of Statistics. Princeton University Press (1966)
2. M. Krée: Propriété de Trace en Dimension Infinie d'espaces du type Sobolev. Bull. Soc. Math. France **105**, (1977) 141-163
3. P. Krée: Introduction à la Théorie des Distributions en Dimension Infinie. Bull. Soc. Math. France **46**, (1976) 143-162
4. P. Krée: Dimension Free Stochastic Calculus in the Distribution Sense. Stochastic Analysis, Path Integration and Dynamics. Edited by D.Elworthy and J.C.Zambrini. Pitman Research Notes in Mathematics Series numéro 200 (1989)
5. D. Lamberton (Personal Communication)
6. B. Lascar: Propriétés Locales d'espaces de Sobolev en Dimension Infinie. CRAS (1976), and Communications in Partial Differential Equations I,ch 6,(1976) 561-584, and Séminaire EDP en dimension infinie, IHP (1974-1975)
7. B.M. Levitan, I.S. Sargsjan: Introduction to Spectral Theory. Translations of Math. Monographs. A.M.S. **39** (1975)
8. A.Nikiforov, V. Ouvarov: Eléments de la Théorie des Fonctions Spéciales. Editions MIR (1976)
9. C. Soize: Steady State Solution of Fokker-Planck Equation in Higher Dimension. Publication de la RCP du CNRS de Mécanique Aléatoire (1988)
10. S. Watanabe: Malliavin Calculus in Terms of Generalized Wiener Functions. Theory and Applications of Random Fields. Lecture Notes in Control and Information Sciences, **49** Springer Verlag (1983)
11. W. Wedig: Parameter Identification of Road Spectra and Nonlinear Oscillators. Analysis and Estimation of Stochastic Mechanical Systems. Edited. by W.Schiehlen, W.Wedig. CISM Courses and Lectures, **303** Springer-Verlag (1988) 217-242
12. F.B. Weissler: Two-Point Inequalities, the Hermite Semi-Group, and the Gauss-Weierstrass Semigroup. Jal of Funct Analysis **32** (1979) 102-121

Un Solveur de Wiener Rapide: Résolution des Systèmes de Toeplitz par une Méthode de Gradient Conjugué Préconditionné

P. Fayol

1 Introduction

1.1 Généralités

Les systèmes linéaires de Toeplitz interviennent dans de nombreux problèmes: identification, filtrage, contrôle,... et leur résolution numérique continue de faire l'objet de recherches actives ([15] à [40]). Dans ces travaux, les algorithmes présentés font rarement apparaître l'origine probabiliste du problème sous-jacent et l'interprétation qu'on peut en tirer, souvent avec profit ([41] par exemple).

Par ailleurs, on étudie encore aujourd'hui en théorie des probabilités et en traitement du signal les méthodes de filtrage et de contrôle de Wiener-Kolmogorov ([6], [7], [54]). Les méthodes habituellement utilisées reposent en général sur des résultats de factorisation spectrale et conduisent à des calculs de type algébrique.

Le présent article a deux motivations complémentaires, qui correspondent d'ailleurs aux deux constats évoqués ci-dessus. Abandonnant à la fois les hypothèses liées à la factorisation et le formalisme classique qui conduit à la formule explicite du filtre de Wiener, nous proposons une nouvelle méthodologie pour calculer ce filtre. Comme cette approche nécessite à une certaine étape la résolution d'un système de Toeplitz, nous avons été conduits à proposer un nouvel algorithme de résolution par gradient conjugué préconditionné. Ceci nous amène, au préalable, à passer en revue les algorithmes existants et à présenter dans un cadre fonctionnel les algorithmes "classiques" utilisés usuellement pour résoudre ces systèmes.

Ce travail fait partie d'une thèse de doctorat de mathématiques soutenue par l'auteur en décembre 1989 à l'université Pierre et Marie Curie.

1.2 Le cadre probabiliste et le cadre fonctionnel

Le problème classique de la prédiction, du filtrage ou du contrôle linéaires des systèmes discrets consiste à estimer ou à approcher la valeur présente D_n ou future D_{n+m} d'un processus discret donné, qu'on supposera du second ordre et stationnaire en moyenne d'ordre deux, par une combinaison linéaire des valeurs connues présentes et passées d'un processus X_n du même type. On cherche ainsi à déterminer la suite réelle $(a_k)_{k\geq 0}$ rendant minimale la variance de l'erreur:

$$\min_{(a_k)} E\left(\left|D_{n+m} - \sum_{k\geq 0} a_k X_{n-k}\right|^2\right) \tag{1}$$

Si X et D sont différents, on parle de contrôle. Dans le cas d'un même processus, le filtrage correspond à $m = 0$ et la prédiction à $m > 0$.

Ce problème peut se traiter de deux manières différentes.

La première consiste à développer l'expression contenue dans (1). En tenant compte de la stationnarité des deux processus D_n et X_n, l'optimum se traduit par une équation discrète de type Wiener-Hopf:

$$\sum_{\ell\geq 0} R_{XX}(k-\ell)\, a_\ell = R_{DX}(k+m) \quad k,m \geq 0 \tag{2}$$

où

$$R_{XX}(k) = E(X_n X_{n-k}) = E(X_k X_0)$$

et

$$R_{DX}(k) = E(D_n X_{n-k}) = E(D_k X_0)$$

sont respectivement les coefficients d'auto et d'intercorrélation des deux processus.

La seconde approche utilise la représentation spectrale et permet de traduire le problème initial (1) en un problème de minimisation portant sur des fonctions de la variable complexe. Par exemple, la séquence $R_{XX}(k)\, k \geq 0$ étant de type positif, i.e. :

$$\forall N; \forall k_1, k_2, \ldots, k_N; \forall \underline{\xi} = (\xi_1, \xi_2, \ldots, \xi_N)^t \in \mathbb{R}^N$$

$$\sum_{j,\ell=1}^{N} \xi_j \xi_\ell R_{XX}(k_j - k_\ell) \geq 0$$

Les coefficients d'auto-corrélation sont les moments trigonométriques d'une mesure spectrale positive :

$$R_{XX}(k) = \frac{1}{2\pi}\int_0^{2\pi} e^{ik\theta} d\mu(\theta) \quad d\mu(\theta) \geq 0$$

Il en résulte que la correspondance :

$$L^2_X(\Omega) \ni X_k \longrightarrow e^{ik\theta} \in L^2(d\mu)$$

définit une isométrie bijective entre la fermeture L^2 de l'espace engendré par les variables aléatoires X_k définies sur l'espace probabilisé (Ω, T, P) muni du produit scalaire

$$E(X_k X_\ell) = \int_\Omega X_k(\omega) X_\ell(\omega) dP(\omega)$$

et l'espace $L^2(d\mu)$ défini sur le tore $\mathbb{R}/2\pi\mathbb{Z}$ muni du produit scalaire

$$< f, g >= \frac{1}{2\pi} \int_0^{2\pi} f(e^{i\theta}) \overline{g(e^{i\theta})} d\theta$$

Ainsi :

$$E\big((\Sigma a_m X_m)(\Sigma b_n X_n)\big) = \sum_{m,n} a_m b_n R_{XX}(m-n)$$
$$= \frac{1}{2\pi} \int_0^{2\pi} (\Sigma a_m e^{im\theta})(\overline{\Sigma b_m e^{im\theta}}) d\mu(\theta)$$

dont on déduit :

$$E\left(\left|X_{n+m} - \sum_{k\geq 0} a_k X_{n-k}\right|^2\right) = \frac{1}{2\pi} \int_0^{2\pi} \left|z^{-m} - \sum_{k\geq 0} a_k z^k\right|^2_{z=e^{i\theta}} d\mu(\theta)$$

Cette identité permet donc de traduire le problème initial de prédiction en un problème d'approximation polynomiale L^2 sur le cercle unité.

Le problème du contrôle peut se traiter de la même manière. On définit successivement les densités spectrales directes $h_{XX}(\theta)$ et croisées $h_{DX}(\theta)$ par :

$$d\mu(\theta) = h_{XX}(\theta) d\theta$$

$$E(D_n X_{n-k}) = \frac{1}{2\pi} \int_0^{2\pi} e^{ik\theta} h_{DX}(\theta) d(\theta)$$

On construit ensuite à partir des a_k une fonction (causale)

$$f(z) = \sum_{k\geq 0} a_k z^k$$

qu'on supposera analytique sur le disque unité D et telle que

$$\sum_{k\geq 0} |a_k|^2 < +\infty$$

Il vient alors :

$$E\left(\left|D_{n+m} - \sum_{k\geq 0} a_k X_{n-k}\right|^2\right) = \frac{1}{2\pi} \int_0^{2\pi} \left|f(e^{-i\theta}) - e^{im\theta} \frac{h_{DX}(\theta)}{h_{XX}(\theta)}\right|^2 d\mu(\theta)$$
$$+\text{constante}$$

On voit donc que minimiser le membre de gauche revient à minimiser une fonctionnelle quadratique de forme générale :

$$\int_0^{2\pi} \left|a(\theta) f^*(e^{i\theta}) - b(\theta)\right|^2 d\theta \qquad (3)$$

où $f^*(e^{i\theta})$ est la trace sur le cercle unité $|z| = 1$ de f analytique sur D.

C'est en fait le propre de l'approche de Wiener que d'utiliser cette représentation intégrale pour résoudre le problème initial (1). Qu'il s'agisse d'un système continu ou d'un système discret, cette méthode procède toujours en deux étapes. Dans le cas d'un système discret par exemple, ces deux étapes sont de manière plus précise ([54] chp. 10) :

1. une étape de factorisation spectrale :

$$h_{XX}(\theta) = |G_0(e^{-i\theta})|^2$$

   Pourvu que la condition suivante soit vérifiée :

$$\int_0^{2\pi} \log(h_{XX}(\theta))d\theta > -\infty$$

   on sait que ce problème admet une solution unique

$$G_0(z) = \sum_{k\geq 0} g_k z^k \quad \sum_{k\geq 0} |g_k|^2 < +\infty \quad g_0 \in \mathbb{R}_+$$

   et qui de plus ne s'annule pas sur D ([53] p. 160).

2. une étape de séparation en partie causale $H_+(z)$ et anticausale $H_-(z)$:

$$\begin{aligned} H(z) = \sum_{-\infty}^{+\infty} h_k z^k &= \sum_{k<0} h_k z^k + \sum_{k\geq 0} h_k z^k \\ &= H_+(z) + H_-(z) \end{aligned}$$

   Cette méthode permet la détermination explicite du filtre optimal par la formule désormais classique :

$$f(z) = \frac{H_+(z)}{G_0(z)}$$

   avec :

$$H(e^{-i\theta}) = e^{im\theta} h_{DX}(\theta) / \overline{G_0(e^{-i\theta})}$$

Séduisante sur le plan théorique, cette approche présente dans la pratique de sérieuses limitations. En effet, la recherche de la solution se fait "à la main" via les étapes de factorisation et de séparation et ne peut être en pratique menée à bien que sur des cas simples voire canoniques. D'autre part, elle demande pour les calculs littéraux une représentation particulière des densités spectrales, en particulier sous la forme de fractions rationnelles. Lorsqu'elles sont issues de l'expérience, il en résulte que ces densités spectrales ne peuvent être utilisées telles quelles mais doivent être au préalable lissées aux moindres carrés par des fractions rationnelles.

1.3 Plan de travail

Nous nous intéressons dans cet article à la minimisation de (3) en relaxant cette fois les hypothèses sur les fonctions a et b. Comme la méthode de Wiener s'applique aussi bien aux systèmes continus qu'aux systèmes discrets, le point de départ de notre travail consistera en l'étude et la minimisation d'une fonctionnelle du type :

$$\frac{1}{2}\int_{\mathbb{R}} |a(\omega)f(i\omega) - b(\omega)|^2 d\omega \tag{4}$$

qui constitue l'équivalent de (3) pour les systèmes continus. L'expression $f(i\omega)$ est ici la trace sur l'axe imaginaire u = 0 d'une fonction causale $f(u + i\omega)$ holomorphe dans le demi-plan droit $u > 0$.

Nous proposons dans cet article une méthode constructive et rapide de minimisation de cette fonctionnelle dans la classe de Hardy $\mathcal{H}^2(\mathbb{C}^+)$ dans le cas général où $a \in L^\infty(\mathbb{R}, \mathbb{C})$ et $b \in L^2(\mathbb{R}, \mathbb{C})$ sont données numériquement par des mesures physiques. Aucun mode de représentation particulier de ces fonctions, en particulier rationnel, n'est nécessaire. L'objectif que nous visons est l'obtention d'une méthode systématique, facilement programmable et destinée à être utilisée en "boite noire". A ce titre, une condition nécessaire à satisfaire est qu'elle soit efficace sur le plan numérique.

La première partie du travail est consacrée à l'étude de la fonctionnelle et à la recherche de solutions numériques approchées. Dans un premier temps, nous traitons le problème dans le cas où les fonctions a et b sont connues sur $\mathbb{R}$ tout entier. Par transport de structure, on se ramène au traitement d'un problème sur le disque unité D, similaire à (3), pour lequel nous montrons l'existence et l'unicité d'une solution dans $H^2(D)$. Ainsi la méthode permet-elle de traiter à la fois les systèmes continus et les systèmes discrets.

Dans le cas où les données a et b du problème ne sont connues que sur une bande $(-\omega_0, +\omega_0)$ nous proposons un prolongement simple et consistant qui permet de se ramener au cas précédent.

Les solutions approchées sont obtenues par application de la méthode de Galerkin. Le système linéaire symétrique obtenu est de Toeplitz et traduit une équation de type Wiener-Hopf semblable à celle que l'on obtient naturellement pour un système discret. Les éléments de la matrice et du second membre sont les coefficients de Fourier de fonctions issues des données. Ces éléments sont évalués numériquement à faible coût par application d'un algorithme de Transformation de Fourier Rapide.

La résolution de ce système linéaire permet de déterminer de manière indépendante la réponse en fréquence du filtre causal solution, sa réponse impulsionnelle et de calculer la valeur du reste.

La seconde partie de cet article est consacrée plus particulièrement à la résolution du système linéaire. Nous introduisons d'abord la famille des polynômes de Szegö associée à la matrice de Toeplitz et montrons le lien entre les récurrences de l'algorithme de Levinson- Trench-Zohar et la minimisation de la fonctionnelle dans des sous-espaces de dimensions croissantes.

Nous proposons ensuite de résoudre le système de Toeplitz par la méthode du gradient conjugué, en le préconditionnant par une matrice circulante. Un tel choix nous permet de limiter le coût de la résolution à $O(N \log_2 N)$ opérations par itération du gradient.

L'étude du préconditionnement se porte ensuite sur trois matrices de ce type. Après l'énoncé de conditions suffisantes qui leur assure bien chacune un caractère défini positif, nous illustrons l'efficacité de ces préconditionnements pour les grands systèmes de Toeplitz en montrant la focalisation vers 1 des valeurs propres en moyenne de Césaro.

Nous terminons cet article par un exemple aéronautique qui illustre l'efficacité numérique et la souplesse d'emploi de la méthode proposée.

2 Préliminaires

2.1 Espace $\mathcal{H}^2(\mathbb{C}^+)$

La classe de Hardy $\mathcal{H}^2(\mathbb{C}^+)$ peut se définir de plusieurs manières ([1] à [5]). C'est par exemple l'espace des transformées de Laplace des classes de fonctions de $L^2([0,+\infty[,\mathbb{C})$:

$$f(z) = \int_0^{+\infty} e^{-zt} F(t)dt$$

$$F \in L^2([0,+\infty[,\mathbb{C}) \quad z \in \mathbb{C}^+ = \{z = u + i\omega, u > 0\}$$

Les résultats suivants sont classiques. La classe f admet sur l'axe imaginaire $\{u = 0\}$ une trace $f^* \in L^2(\mathbb{R},\mathbb{C})$ qui est la transformée de Fourier de F :

$$f^*(\omega) = \int_{\infty}^{0} e^{-i\omega t} F(t)dt$$

Muni du produit hermitien induit sur $L^2(\mathbb{R},\mathbb{C})$:

$$((f,g))_{\mathcal{H}^2(\mathbb{C}^+)} = (f^*,g^*) = \int_{\mathbb{R}} f^*(\omega)\overline{g^*(\omega)}d\omega$$

l'espace $\mathcal{H}^2(\mathbb{C}^+)$ est complet et peut s'identifier au sous-espace fermé de $L^2(\mathbb{R},\mathbb{C})$ image de $L^2([0,+\infty[,\mathbb{C})$ par la transformation de Fourier.

2.2 Espace $H^2(D)$

Notons D le disque unité $\{z = x + iy; |z| < 1\}$ de $\mathbb{C}$. L'espace $H^2(D)$ est l'ensemble des fonctions holomorphes définies sur D telles que :

$$\sup_{0 \le r < 1} \frac{1}{2\pi} \int_{-\pi}^{+\pi} |f(re^{i\theta})|^2 d\theta < +\infty$$

Les limites radiales :

$$f^*(e^{i\theta}) = \lim_{r \to 1} f(re^{i\theta}) \text{ dans } L^2([-\pi,+\pi[,\mathbb{C})$$

existent $d\theta$ presque partout et sont dans $L^2([-\pi,+\pi[,\mathbb{C})$. De plus, on a le résultat suivant:

Proposition 1. (cf. [4]) Si f et g sont deux fonctions de $H^2(D)$ dont les limites radiales f^* et g^* coïncident sur un ensemble $\Omega \subset [-\pi, +pi]$ de mesure non nulle, alors f et g sont égales.

L'espace $H^2(D)$ est un espace de Hilbert pour le produit scalaire induit par $L^2([-\pi, +\pi[, \mathbb{C})$:

$$< f, g >= \frac{1}{2\pi} \int_{-\pi}^{+\pi} f^*(e^{i\theta})\overline{g^*(e^{i\theta})}d\theta$$

et peut s'identifier de cette manière à un sous-espace fermé de $L^2([-\pi, +\pi[, \mathbb{C})$. Plus précisement, ce sous-espace, constitué des limites radiales des fonctions de $H^2(D)$ peut être caractérisé de deux façons ([1] chp. 3):

- il est l'adhérence L^2 des polynômes en $e^{i\theta}$,
- il est exactement la classe des fonctions de $L^2([-\pi, +\pi[, \mathbb{C})$ dont les coefficients de Fourier

$$c_n = \int_{-\pi}^{+\pi} e^{-int} f(t)dt$$

sont nuls pour $n < 0$.

Enfin, la fonction $f(z) = \sum_{n\geq 0} a_n z^n$ holomorphe dans D, appartient à $H^2(D)$ si et seulement si :

$$\sum_{n\geq 0} |a_n|^2 < +\infty$$

et dans ce cas :

$$||f||_{H^2(D)} = \left(\sum_{n\geq 0} |a_n|^2 \right)^{1/2}$$

$$a_n = c_n$$

2.3 Passage de $\mathcal{H}^2(\mathbb{C}^+)$ à $H^2(D)$

Soit p_0 un nombre strictement positif, qu'on supposera ici quelconque et dont on détermine la valeur en fonction des applications. Plutôt que de sélectionner la base hilbertienne classique de $\mathcal{H}^2(\mathbb{C}^+)$ obtenue par image des fonctions de Laguerre

$$\Psi_n(t) = \frac{e^{-t/2}}{n!} L_n(t)$$

par la transformation de Fourier, nous développons la solution du problème de minimisation sur la famille $\{f_{p_0,n}; n \geq 0\}$ paramétrée par p_0 et définie par :

$$\begin{aligned} f_{p_0,n} &= \frac{1}{2\sqrt{\pi p_0}} \widehat{\Psi}_n \left(\frac{p}{2p_0} \right) \\ &= \sqrt{\frac{p_0}{\pi}} \frac{1}{p+p_0} \left(\frac{p-p_0}{p+p_0} \right)^n ; \quad \text{Re} p \geq 0 \end{aligned} \tag{5}$$

qui est également dense et orthonormée.

D'autre part, les domaines $\mathbb{C}^+$ et D sont image l'un de l'autre par la transformation conforme

$$\Phi(z) = p = p_0 \frac{1-z}{1+z}$$

qui met également en correspondance l'axe imaginaire $\{u = 0\}$ et le cercle unité $\{|z| = 1\}$. Enfin, l'application continue Ψ :

$$\mathcal{H}^2(\mathbb{C}^+) \ni f \longrightarrow \left(z \to \psi(f)(z) = 2\sqrt{\pi p_0} f \circ \phi(z) \frac{1}{1+z} \right) \in H^2(D)$$

qui transforme $f_{p_0,n}(p)$ en $(-1)^n z^n$ est une isométrie bijective. Via cette transformation conforme, une même fonction $f \in \mathcal{H}^2(\mathbb{C}^+)$ peut donc se lire sous les deux formes suivantes :

$$f(p) = \sum_{n \geq 0} x_n f_{p_0,n}(p); \; x_n \in \mathbb{C}; \; \sum_{n \geq 0} |x_n|^2 < \infty; \;\; \mathrm{Re}\, p \geq 0$$

$$f \circ \Phi(z) = (1+z) \sum_{n \geq 0} a_n z^n = (1+z) g(z); \; |z| \leq 1; \; g \in H^2(D) \qquad (6)$$

$$a_n = \frac{(-1)^n}{2\sqrt{\pi p_0}} x_n$$

Cette transformation nous permet de traduire dans $H^2(D)$ le problème initial dans $\mathcal{H}^2(\mathbb{C}^+)$.

Remarque : En filtrage numérique, on utilise plutôt la transformation $p = p_0(z-1)/(z+1)$, ce qui revient à remplacer z par z^{-1} dans les expressions ci-dessus. Cette transformation porte le nom de transformation algébrique bilinéaire ([6] chp. 6, [7] chp. 2). L'expression $\sum_{n \geq 0} a_n z^{-n}$ est alors "la transformée en z" du signal discret $\{a_n; n \geq 0\}$. Cette transformation constitue un outil fondamental de traitement des signaux échantillonnés, d'analyse et de synthèse des filtres numériques.

3 Recherche de Solutions Approchées

3.1 Le Problème Initial et sa transpotion dans $H^2(D)$

Le problème initial que nous cherchons à résoudre est le suivant :

Trouver $f \in \mathcal{H}^2(\mathbb{C}^+)$, de valeur frontière f^* sur l'axe imaginaire minimisant la fonctionnelle quadratique:

$$(\mathcal{P}_1) \qquad J_1(f) = \frac{1}{2} \int_{\mathbb{R}} |b(\omega) - a(\omega) f^*(\omega)|^2 d\omega$$

sous les hypothèses et les contraintes suivantes: $a \in L^\infty(\mathbb{R}, \mathbb{C})$, $b \in L^2(\mathbb{R}, \mathbb{C})$ avec a,b et f^* vérifiant de plus la relation de conjugaison $h(-\omega) = \overline{h(\omega)}$.

Dans $H^2(D)$, ce problème se traduit comme suit :

Trouver $g \in H^2(D)$ de valeur frontière $g^*(e^{i\theta})$ sur le cercle unité minimisant

$$(\mathcal{P}_2) \qquad J_2(g) = p_0 \int_{-\pi}^{+\pi} \left|\tilde{b}(\theta) - \tilde{a}(\theta) g^*(e^{i\theta})\right|^2 d\theta$$

avec:

$$\bullet\ \tilde{a}(\theta) = a(-i\phi(e^{i\theta})) \in L^\infty([-\pi, +\pi[, \mathbb{C})$$
$$\bullet\ \tilde{b}(\theta) = \frac{b(-i\phi(e^{i\theta}))}{1+e^{i\theta}} \in L^2([-\pi, +\pi[, \mathbb{C})$$
$$\bullet\ \tilde{a}(-\theta) = \overline{\tilde{a}(\theta)};\ \tilde{b}(-\theta) = \overline{\tilde{b}(\theta)};\ g^*(e^{-i\theta}) = \overline{g^*(e^{i\theta})}$$

Preuve 2. Comme

$$\phi(e^{i\theta}) = p_0 \frac{1-e^{i\theta}}{1+e^{i\theta}} = i\omega \text{ et } f^*(\omega) = (1+e^{i\theta}) g^*(e^{i\theta})$$

il vient:

$$\begin{aligned} |b(\omega) - a(\omega) f^*(\omega)|^2 d\omega &= |\tilde{b}(\theta) - \tilde{a}(\theta) g^*(e^{i\theta})|^2 . |1+e^{i\theta}|^2 \phi'(e^{i\theta}) e^{i\theta} d\theta \\ &= -2p_0 |\tilde{b}(\theta) - \tilde{a}(\theta) g^*(e^{i\theta})|^2 d\theta \end{aligned}$$

D'autre part: $-i\phi(\pm\pi) = \mp\infty$.

Enfin, comme $\overline{\phi(e^{i\theta})} = -\phi(e^{i\theta})$ on déduit facilement $\tilde{a}(-\theta) = \overline{\tilde{a}(\theta)}$ et $\tilde{b}(-\theta) = \overline{\tilde{b}(\theta)}$.

Pour simplifier les écritures, nous omettrons le signe tilde sur a et b lorsqu'aucune ambiguïté n'est possible entre par exemple $a(\omega)$ et $a(\theta)$. Toujours pour alléger les notations, nous écrirons indifféremment $f(i\omega)$ pour $f^*(\omega)$ et $g(e^{i\theta})$ pour $g^*(e^{i\theta})$.

Venons en maintenant à l'existence et l'unicité de g dans le cas non dégénéré où a n'est pas la fonction nulle.

Théorème 3. *Si a n'est pas la fonction nulle, le problème d'optimisation*

$$\inf J_2(g)$$

admet une solution et une seule dans $H^2(D)$.

Preuve 4. Quel que soit $h \in H^2(D)$ et $\delta > 0$ la quantité

$$\begin{aligned} \frac{1}{\delta}\left[J_2(g+\delta h) - J_2(g)\right] = \delta p_0 &\int_{-\pi}^{+\pi} |a(\theta)|^2 |h(e^{i\theta})|^2 d\theta \\ -2p_0 &\int_{-\pi}^{+\pi} \left[b(\theta) - a(\theta) g(e^{i\theta})\right] \overline{a(\theta) h(e^{i\theta})} \end{aligned}$$

a pour limite, lorsque δ tend vers zéro, le second terme du membre de droite, que l'on notera $< J_2'(g), h >$ où

$$J_2'(g) = 4\pi p_0 \left[|a(\theta)|^2 g(e^{i\theta}) - b(\theta)\overline{a(\theta)}\right]$$

est la dérivée de Gateaux de J_2 en g .

La convexité de J_2 résulte de celle de la norme L^2. J_2 est même strictement convexe car la dérivée est ici un opérateur strictement monotone. En effet la quantité positive:

$$< J_2'(f) - J_2'(g), f - g >= 2p_0 \int_{-\pi}^{+\pi} |a(\theta)|^2 |f(e^{i\theta}) - g(e^{i\theta})|^2 d\theta$$

ne peut s'annuler que si $f = g$. Si tel est le cas, la fonction $|a(\theta)|^2|f(e^{i\theta}) - g(e^{i\theta})|^2$ est nulle $d\theta$- presque partout. Puisque $a(\theta)$ n'est pas identiquement nulle, il existe une partie $\Omega \subset [-\pi, +\pi[$ de mesure $d\theta$ non nulle sur laquelle $|a(\theta)|^2 > 0$. Les traces sur le cercle unité des deux fonctions f et g de $H^2(D)$ coïncident donc sur Ω. Alors f = g d'après la proposition (1). J_2 étant donc strictement convexe, il existe au plus un minimum.

Pour montrer maintenant l'existence d'un minimum, on considère une suite minimisante $g_n \in H^2(D)$:

$$J_2(g_n) \to \inf_{f \in H^2(D)} J_2(f)$$

Comme J_2 tend vers l'infini lorsque $||g||$ tend vers l'infini, cette suite est bornée. On peut en extraire une sous-suite $g_{\varphi(n)}$ convergeant faiblement vers g dans $L^2([-\pi, +\pi[, \mathbb{C})$. Mais l'espace $H^2(D)$, comme partie fortement fermée et convexe de L^2 est également faiblement fermée. Ainsi $g \in H^2(D)$. Comme

$$J_2(g_{\varphi(n)}) > J_2(g) + < J_2'(g), g_{\varphi(n)} - g >$$

d'après la convexité stricte et

$$< J_2'(g), g_{\varphi(n)} - g > \underset{n \to +\infty}{\longrightarrow} 0$$

par la convergence faible, il vient:

$$\liminf J_2(g_{\varphi(n)}) \geq J_2(g)$$
$$\inf_{f \in H^2(D)} J_2(f) \geq J_2(g)$$

et g est au moins un minimum.

Comme J_2 est Gateaux-différentiable, le minimum g est caractérisé par l'égalité variationnelle

$$< J_2'(g), h >= 0 \qquad \forall h \in H^2(D)$$

Remarque 5. Si on sait que la fonction $|a(\theta)|^2$ est minorée par $m > 0$, J_2 est coercive. Elle vérifie l'inégalité (cf. [52]) :

$$4\pi p_0 \frac{m}{8} \|v - u\|^2 \leq \frac{1}{2} [J_2(u) + J_2(v)] - J_2 \left(\frac{u+v}{2} \right)$$
$$\forall u, v \in H^2(D)$$

Cette formule permet de ne pas faire appel à la compacité faible. Elle permet de plus de montrer la convergence **forte** (et non plus seulement faible) de **toute** suite minimisante vers la solution du problème. Ce point est important pour les algorithmes de calcul.

3.2 Cas des données incomplètes

Venons en maintenant au problème posé par des données incomplètes et que l'on rencontre très souvent en pratique : les fonctions a et b ne sont plus connues que sur une partie stricte $[-\theta_0, +\theta_0]$ de l'intervalle $[-\pi, +\pi]$.

Peut-on dans ces conditions prolonger ces fonctions de manière simple de façon à ce que la solution optimale du problème "tronqué-prolongé" converge vers la solution optimale du problème initial lorsque la borne θ_0 tend vers π ?

Soit θ_n une suite tendant vers π telle que $a(\theta_n)$ soit "défini" (i.e. : a continue dans un voisinage de θ_n) pour tout n. On pose $\Omega_n = [-\theta_n, +\theta_n]$ et $\Omega_n^c = [-\pi, +\pi] - \Omega_n$. Un prolongement possible, et que nous choisirons, est le suivant :

- a et b sont inchangées sur Ω_n
- on prolonge respectivement b par zéro et a par $a(\theta_n)$ sur Ω

Les hypothèses d'existence et d'unicité dans la proposition précédente restent satisfaites et on définit de cette manière une suite g_n d'optima dans $H^2(D)$.

A ce choix de prolongement est assortie une condition suffisante sur le comportement à l'infini de la fonction a pour que g_n tende vers g.

Proposition 6. Si la suite numérique de terme général

$$\delta_n = \left\{ \sup_{\Omega_n^c} |a(\theta)| \right\} / |a(\theta_n)|^2$$

est bornée à partir d'un certain rang, alors:

$$\lim_{n \to +\infty} \int_{-\pi}^{+\pi} |a(g_n - g)|^2 d\theta = 0$$

Preuve 7. Les condition de stationnarité de J_2 s'écrivent pour g et g_n respectivement:

$$\forall h \in H^2(D) \quad \int_{\Omega_n \cup \Omega_n^c} (ag - b)\overline{ah} d\theta = 0$$

$$\forall h \in H^2(D) \quad \int_{\Omega_n} (ag_n - b)\overline{ah}d\theta + |a(\theta_n)|^2 \int_{\Omega_n^c} g_n \overline{h} d\theta = 0$$

Par soustraction et pour $h = g_n - g$:

$$\int_{\Omega_n} |a(g_n - g)|^2 d\theta + |a(\theta_n)|^2 \int_{\Omega_n^c} |g_n - g|^2 d\theta$$
$$= \int_{\Omega_n^c} (ag - b)\overline{a(g_n - g)}d\theta - |a(\theta_n)|^2 \int_{\Omega_n^c} \overline{g(g_n - g)}d\theta$$

Posons pour simplifier :

$$X_n^2 = \int_{\Omega_n} |a(g_n - g)|^2 d\theta$$
$$Y_n^2 = \int_{\Omega_n^c} |g_n - g|^2 d\theta$$
$$A_n^2 = \int_{\Omega_n^c} |ag - b|^2 d\theta$$
$$B_n^2 = \int_{\Omega_n^c} |g|^2 d\theta$$

De l'égalité précédente, on déduit par l'inégalité de Schwarz la majoration:

$$X_n^2 + |a(\theta_n)|^2 Y_n^2 \leq \left(\sup_{\Omega_n^c} |a(\theta)| \right) .A_n Y_n + |a(\theta_n)|^2 B_n Y_n \tag{7}$$

Mais pour $n \geq N$ $\left\{ \sup_{\Omega_n^c} |a(\theta)| \right\} / |a(\theta_n)|^2$ est borné par S. Ainsi

$$Y_n \leq S.A_n + B_n \tag{8}$$

Comme A_n et B_n tendent vers zéro (appliquer pour cela le théorème de convergence dominée de Lebesgue), il en est de même pour Y_n par (8) et par suite pour X_n par (7). Enfin:

$$\int_{\Omega_n^c} |a(g_n - g)|^2 d\theta \leq \|a\|_\infty^2 Y_n^2$$

d'où finalement :

$$\int_{-\pi}^{+\pi} |a(g_n - g)|^2 d\theta \leq X_n^2 + \|a\|_\infty^2 Y_n^2 \underset{n \to +\infty}{\longrightarrow} 0$$

Remarque 8. La condition imposée dans la proposition couvre le cas usuel suivant:

$$|a(\theta)| \text{ est minorée par } m > 0 \text{ pour } \theta > \theta_0$$

3.3 Approximation interne à l'ordre N

Après s'être assuré que nous avions affaire à un problème bien posé, nous cherchons maintenant des solutions approchées de g. Nous utilisons pour cela la méthode de Galerkin. On détermine une approximation g_N de g en minimisant J_2 sur des sous-espaces $\mathcal{V}_N \subset H^2(D)$ de dimension finie; la suite $(\mathcal{V}_N)_{N\geq 0}$ étant croissante ($\mathcal{V}_N \subset \mathcal{V}_{N+1}$) et dense car:

$$\left(\overline{\bigcup_{N\geq 0} \mathcal{V}_N} = H^2(D)\right)$$

La solution $g_N \in \mathcal{V}_N$ est alors caractérisée par la condition $J_2'(g_N) \in \mathcal{V}_N^{\perp}$.

Comme l'espace des polynômes est dense dans $H^2(D)$, nous choisissons de manière naturelle comme espace d'approximation à l'ordre N, l'espace P_N des fonctions polynômes de degré inférieur ou égal à N. La solution approchée dans P_N qui s'écrit :

$$g_N(z) = \sum_{k=0}^{N} x_k z^k$$

est caractérisée par la condition de stationnarité de J_2:

$$\forall h \in P_N \quad \frac{1}{2\pi}\int_{-\pi}^{+\pi} \left[|a(\theta)|^2 g_N(e^{i\theta}) - b(\theta)\overline{a(\theta)}\right] \overline{h(e^{i\theta})}d\theta = 0 \tag{9}$$

Appliquée successivement à $h(z) = z^p;\ 0 \leq p \leq N$, elle conduit au système linéaire:

$$\left.\begin{aligned} \forall 0 \leq p \leq N \ \sum_{k=0}^{N} \left[\tfrac{1}{2\pi}\textstyle\int_{-\pi}^{+\pi} |a(\theta)|^2 e^{i(k-p)\theta} d\theta\right] x_k \\ = \tfrac{1}{2\pi}\textstyle\int_{-\pi}^{+\pi} b(\theta)\overline{a(\theta)} e^{-ip\theta} d\theta \end{aligned}\right\} \tag{10}$$

On pose successivement :

- $\alpha_n = \frac{1}{2\pi}\int_{-\pi}^{+\pi} |a(\theta)|^2 e^{-in\theta}d\theta$ le n[ième] coefficient de Fourier de la fonction 2π-périodique $\theta \rightarrow |a(\theta)|^2$. α_n est réel et $\alpha_{-n} = \alpha_n$ car la fonction est paire.
- $\beta_n = \frac{1}{2\pi}\int_{-\pi}^{+\pi} b(\theta)\overline{a(\theta)}e^{-in\theta}d\theta$ le n[ième] coefficient de Fourier de la fonction 2π-périodique $\theta \rightarrow b(\theta)\overline{a(\theta)}$. β_n est réel en vertu des propriétés de conjugaison de a et b.

Les égalités (10) se réécrivent donc sous la forme d'un système linéaire symétrique de Toeplitz:

$$T_N.\underline{x}_N = \underline{\beta}_N \tag{11}$$

avec $(T_N)_{ij} = \alpha_{|i-j|}$; ${}^t\underline{x}_N = (x_0, \ldots, x_N)$; ${}^t\underline{\beta}_N = (\beta_0, \ldots, \beta_N)$.

Comme T_N et $\underline{\beta}_N$ sont à coefficients réels, les coefficients x_k sont réels. La relation de conjugaison $g_N(e^{-i\theta}) = \overline{g_N(e^{i\theta})}$ est donc vérifiée.

Le système linéaire (11) se lit comme une équation discrète de type Wiener-Hopf qui aurait été tronquée à l'ordre N:

$$\sum_{k=0}^{N} \alpha_{n-k}.x_k = \beta_n \text{ pour } 0 \leq n \leq N$$

Le choix d'une transformation conforme et des fonctions de Laguerre permet donc de réécrire le problème initial $\mathcal{P}_1$ dans le formalisme des systèmes discrets et par voie de conséquence nous permet d'utiliser un certain nombre d'outils déjà mis en place pour traiter ces systèmes.

Le système linéaire auquel nous aboutissons est intéressant à double titre :

- la matrice est de Toeplitz,
- ses éléments ainsi que le second membre sont des coefficients de Fourier.

Sur le plan numérique, la structure de la matrice permet de limiter l'encombrement mémoire au seul stockage de la première ligne (ou de la première colonne) et permet d'utiliser des algorithmes spécifiques aux systèmes de Toeplitz dont les performances sont nettement supérieures aux méthodes classiques. Ce point sera plus amplement détaillé dans la quatrième partie de cet article.

La nature des coefficients de la matrice et du second membre permet d'autre part de les calculer à faible coût en utilisant un algorithme de Transformation de Fourier Rapide. Ces deux caractéristiques laissent envisager la possibilité de traiter facilement des problèmes de grande taille (par exemple $N \geq 2^{10}$) pour un coût informatique raisonnable.

On remarquera enfin que ce problème ne fait intervenir que les moments trigonométriques des données : α_k pour $|a(\theta)|^2$, β_k pour $b(\theta)\overline{a(\theta)}$. En ce qui concerne plus particulièrement les premiers, la donnée de la mesure positive $d\mu(\theta) = |a(\theta)|^2 d\theta$ détermine *tous* ses moments $\alpha_k \quad k \geq 0$. Le problème inverse, de détermination d'une mesure positive $d\mu(\theta)$ obéissant à une suite complète ou limitée à un rang donné de coefficients α_k tels que

$$\alpha_k = \frac{1}{2\pi} \int_{-\pi}^{+\pi} e^{ik\theta} d\mu(\theta)$$

constitue le problème classique des moments trigonométriques dont l'étude reste attachée aux noms de Krein, Carathéodory, Féjer, Toeplitz, F. et M. Riescz ([9] à [13] par exemple).

Nous introduisons maintenant un nouveau produit scalaire, dont l'intérêt apparaitra par la suite.

3.4 Un produit scalaire sur $H^2(D)$

La forme quadratique associée à la matrice T_N est définie positive car:

$$\sum_{i=0}^{N} \sum_{j=0}^{N} x_i x_j \alpha_{|i-j|} = \frac{1}{2\pi} \int_{-\pi}^{+\pi} |a(\theta)|^2 \left| \sum_{k=0}^{N} x_k e^{ik\theta} \right|^2 d\theta \qquad (12)$$

Associons au vecteur $\underline{c} = {}^t(c_0, \dots, c_N) \in \mathbb{R}^{N+1}$ le polynôme trigonométrique de degré N

$$C(z) = \sum_{k=0}^{N} c_k e^{ik\theta} = \sum_{k=0}^{N} c_k z^k \quad |z| = 1$$

On définit naturellement un produit scalaire induit sur P_N par

$$[C, D] = (\underline{c}, T_N \underline{d}) \text{ où } D(z) = \sum_{k=0}^{N} d_k z^k \tag{13}$$

D'après (12):

$$[C, D] = \frac{1}{2\pi} \int_{-\pi}^{+\pi} |a(\theta)|^2 C(e^{i\theta}) \overline{D(e^{i\theta})} d\theta$$

Cette dernière définition s'étend à C et D quelconques dans $H^2(D)$. Cette forme hermitienne est définie positive par le même argument qui a montré que J_2 était strictement convexe. Ce produit scalaire nous sera très utile dans la suite de l'exposé car il nous permettra de justifier simplement un certain nombre de résultats par des considérations de nature géométrique ou fonctionnelle. En particulier, l'inégalité

$$\forall C \in H^2(D)$$

$$[C, C] \leq \left\{ \sup_{[-\pi,+\pi]} |a(\theta)|^2 \right\} \frac{1}{2\pi} \int_{-\pi}^{+\pi} |C(e^{i\theta})|^2 d\theta = \|a\|_\infty^2 . < C, C > \tag{14}$$

montre que la nouvelle norme induit sur $H^2(D)$ une topologie moins fine que la norme usuelle. Nous montrons plus bas que g_N tend vers g au sens de cette norme.

3.5 Polynômes de Szegö

Pour ce nouveau produit scalaire dans $H^2(D)$, la base canonique $1, z, \dots, z^N, \dots$ n'est bien évidemment plus orthogonale. On peut néanmoins construire une telle base $S_0(z), S_1(z), \dots, S_N(z), \dots$, à partir de la base canonique par le procédé d'orthogonalisation de Gram-Schmidt.

Comme à chaque étape k du procédé, les espaces engendrés sont les mêmes:

$$\text{Vect}\{1, z, \dots, z^k\} = \text{Vect}\{S_0(z), S1(z), \dots, S_k(z)\}$$

le polynôme cherché $S_k(z)$ vérifie les k relations

$$[S_k(z), z^\ell] = 0 \quad 0 \leq \ell \leq k-1$$

ce qui s'écrit encore $(T_k \underline{\sigma}_k, \underline{e}_\ell) = 0$ en notant par $\underline{\sigma}_k$ le vecteur de $\mathbb{R}^{k+1}$ associé au polynôme $S_k(z), \underline{e}_\ell$ le $(\ell+1)^{\text{ième}}$ vecteur de la base canonique de $\mathbb{R}^{k+1}$ et T_k la sous-matrice d'ordre k+1

$$(T_k)_{ij} = \alpha_{|i-j|} \quad 0 \leq i, j \leq k$$

Cette dernière relation nous dit que le vecteur $T_k\underline{\sigma}_k$ n'a de composante non nulle que sur $\underline{e}_k$. **On choisit** le polynôme $S_k(z)$ correspondant au choix particulier $T_k\underline{\sigma}_k = \underline{e}_k$. Celui-ci peut être normé à l'aide de son coefficient de plus haut degré s_k d'après la relation:

$$0 < [S_k(z), S_k(z)] = (\underline{\sigma}_k, T_k\underline{\sigma}_k) = (\underline{\sigma}_k, \underline{e}_k) = s_k$$

Les polynômes orthonormés $S_k(z)/\sqrt{s_k}$ ainsi définis sont les polynômes de Szegö associés à la mesure positive $d\mu(\theta) = |a(\theta)|^2 d\theta$. On se réfèrera à [13], [14] et [15] pour la théorie relative à ces polynômes et leurs applications.

D'après (14), la base canonique, qui est dense dans $H^2(D)$ pour la norme usuelle $<.,.>$ l'est également pour $[.,.]$. Par construction, il en est de même pour l'ensemble des polynômes de Szegö, qui, en tant que système orthonormé, constitue donc une base hilbertienne de $H^2(D)$ muni de cette nouvelle norme. Relativement à cette base, toute fonction f donnée de $H^2(D)$ se développe donc sous la forme :

$$f = \sum_{k=0}^{\infty} \frac{[f, S_k]}{s_k} S_k \tag{15}$$

la série étant normalement convergente :

$$\sum_{k=0}^{\infty} \frac{\|[f, S_k]\|^2}{s_k} < +\infty$$

Nous précisons maintenant le lien entre la solution optimale g du problème $\mathcal{P}_2$ et la solution approchée g_N obtenue dans P_N et nous proposons une formule pour les restes $J_2(g)$ et $J_2(g_N)$.

3.6 Relation entre g et g_N

La condition (9) de stationnarité pour la solution approchée peut se réécrire sous la forme :

$$[g_N, S_k] = <b, aS_k> \quad 0 \le k \le N \tag{16}$$

Elle est a fortiori vérifiée par la solution g du problème continu, soit :

$$[g, S_k] = <b, aS_k> \quad 0 \le k \le N \tag{17}$$

Par soustraction: $[g - g_N, S_k] = 0 \quad 0 \le k \le N$ soit:

$$g = g_N + r \quad r \in H^2(D) \ominus P_N \tag{18}$$

Le reste r appartient à l'orthogonal de P_N dans $H^2(D)$, l'orthogonalité étant définie par $[.,.]$ et non par le produit scalaire usuel $<.,.>$.

D'après (15) et (17), g peut s'exprimer en fonction des données du problème sous la forme de la série normalement convergente :

$$g = \sum_{k=0}^{+\infty} \frac{<b, aS_k>}{s_k} S_k \tag{19}$$

En ce qui concerne la solution approchée g_N, comme

$$g_N \in \text{Vect}\{1, z, \ldots, z^N\} = \text{Vect}\{S_0(z), S1(z), \ldots, S_N(z)\}$$

on en déduit de même:

$$g_N = \sum_{k=0}^{N} \frac{<b, aS_k>}{s_k} S_k \tag{20}$$

La somme

$$[g, g] = \sum_{k \geq 0} \frac{|<b, aS_k>|^2}{s_k}$$

est la limite de la suite positive croissante de terme général

$$[g_N, g_N] = \sum_{0 \leq k \leq N} \frac{|<b, aS_k>|^2}{s_k}$$

D'après (18), il vient par Pythagore:

$$[g, g] = [g_N, g_N] + [g - g_N, g - g_N] \tag{21}$$

ce qui montre en définitive que g_N tend fortement vers g dans $H^2(D)$ muni de $[.,.]$.

3.7 Cas où $|a(\theta)|^2 \geq m > 0$

On a dans ce cas :

$$\forall C \in H^2(D) \qquad m<C, C> \leq [C, C] \leq \|a\|_\infty^2 <C, C>$$

Les deux normes sont équivalentes et g_N tend fortement vers g dans $H_2(D)$ muni de l'une quelconque d'entre elles. Ce résultat est cohérent avec la remarque du théorème 3. Dans ce cas particulier, b/a reste dans $L^2([-\pi, +\pi[, \mathbb{C})$ et on peut réécrire g sous la forme

$$g = \sum_{k=0}^{\infty} \frac{[b/a, S_k]}{s_k} S_k \tag{22}$$

La classe b/a est la solution "minimum-minimorum" du problème de minimisation dans L^2 car elle annule le reste $J_2(b/a)$. Dans le cas particulier où $b/a\,(\theta)$ est effectivement la trace sur le cercle unité d'une fonction $b/a\,(z)$ de $H^2(D)$, l'égalité (22) dit simplement que cette fonction est la solution du problème posé. Mais en général $b/a\,(\theta)$ n'a aucune raison d'être dans L^2, ni surtout de se prolonger naturellement sur D en une fonction de $H^2(D)$. En particulier, si on suppose $b/a\,(\theta)$ dans L^2 et décomposée en série de Fourier sous la forme

$$b/a\,(\theta) = \sum_{k \in \mathbb{Z}} \gamma_k e^{ik\theta}$$

la solution g du problème $\mathcal{P}_2$ **n'est pas** la solution intuitive $\sum_{k \geq 0} \gamma_k e^{ik\theta}$!

3.8 Expression du reste

L'expression développée de J_2 est pour f quelconque dans $H^2(D)$:

$$J_2(f) = 2\pi p_0[<b,b> - <af,b> - <b-af,af>]$$

La solution optimale g et la solution approchée g_N vérifient les deux relations $< b-af, af >= 0$ et $< b, af >= [f,f]$. Il vient donc:

$$J_2(g) = J_2(0) - 2\pi p_0[g,g]$$

et:

$$J_2(g_N) = J_2(0) - 2\pi p_0[g_N, g_N]$$

soit encore, en revenant au problème initial $\mathcal{P}_1$:

$$J_1(f) = J_1(0) - 2\pi p_0 \sum_{k\geq 0} \frac{|<\tilde{b}, \tilde{a}S_k>|^2}{s_k}$$

$$J_1(f_N) = J_1(0) - 2\pi p_0 \sum_{0\leq k\leq N} \frac{|<\tilde{b}, \tilde{a}S_k>|^2}{s_k} \tag{23}$$

D'autres expressions sont possibles. Par exemple:

$$J_1(f) = J_1(0) - 2\pi p_0 \sum_{k\geq 0} \beta_k^\infty x_k^\infty$$

$$J_1(f_N) = J_1(0) - 2\pi p_0 \sum_{0\leq k\leq N} \beta_k^N x_k^N \tag{24}$$

Ces relations découlent en effet directement des égalités:

$$[g_N, g_N] = (\underline{x}_N, T_N \underline{x}_N) = (\underline{x}_N, \underline{\beta}_N) = \sum_{k=0}^{N} \beta_k^N x_k^N$$

dont on déduit au passage les relations de monotonie:

$$N \leq M \Rightarrow \sum_{k=0}^{N} \beta_k^N x_k^N = [g_N, g_N] \leq [g_M, g_M] = \beta_k^M x_k^M$$

La convergence de la série $\sum_{k\geq 0} \beta_k^\infty x_k^\infty$ résulte quant à elle des deux inégalités :

$$\sum_{k\geq 0} |x_k^\infty|^2 = <g,g> \; < +\infty$$

$$\sum_{k\geq 0} |\beta_k^\infty|^2 \leq \sum_{k\in\mathbb{Z}} |\beta_k^\infty|^2 = <b\bar{a}, b\bar{a}> \leq \|a\|_\infty^2 <b,b> \; < +\infty$$

et de l'application de l'inégalité de Schwarz.

Les relations (23) n'utilisent que les données a et b, et ne font pas intervenir f et f_N. A l'inverse, les relations (24) font intervenir par leurs coefficients de Fourier à la fois la solution et les données.

3.9 Réponse impulsionnelle

Connaissant $g_N(z) = \sum_{k=0}^{N} x_k^N z^k$, nous pouvons déterminer directement la réponse impulsionnelle $F_N(t)$ associée à $f_N(p)$ et définie par :

$$f_N(p) = \int_0^\infty e^{-pt} F_N(t)dt; \quad p = u + i\omega \quad u \leq 0$$

En effet, d'après (5) et (6):

$$f_N(p) = \frac{2p_0}{p+p_0} \sum_{k=0}^{N} (-1)^k x_k^N \left(\frac{p-p_0}{p+p_0}\right)^k$$

d'où :

$$\left.\begin{array}{ll} F_N(t) = 2p_0 e^{-p_0 t} \sum_{k=0}^{N} (-1)^k x_k^N \frac{L_k(2p_0 t)}{k!} & \text{si } t \geq 0 \\ F_N(t) = 0 & \text{si } t < 0 \end{array}\right\} \quad (25)$$

avec L_k le polynôme de Laguerre d'ordre k.

La réponse en fréquence $f_N(i\omega)$ et la réponse impulsionnelle $F_N(t)$ peuvent ainsi s'évaluer séparément et avec une précision numérique arbitraire donnée. Ce ne serait pas le cas si on passait de l'une à l'autre par une transformation de Fourier Discrète d'ordre $N_F = 2^q$ fixé, pour laquelle les pas de discrétisation en fréquence et en temps sont liés par la relation $N_F.\Delta t.\Delta f = 1$.

4 Résolution par des Méthodes de Directions Conjuguées

4.1 Introduction

Les systèmes de Toeplitz apparaissent dans de nombreuses applications en traitement du signal (prédiction et filtrage linéaires, factorisation spectrale, filtrage de Wiener,...) et en particulier dans l'étude des systèmes et processus discrets.

Des impératifs de rapidité pour les problèmes d'estimation en temps réel ainsi que les grandes dimensions des systèmes que l'on est amené fréquemment à résoudre ont conduit très tôt à tirer le meilleur parti possible de la structure particulière et riche des matrices de Toeplitz pour résoudre les systèmes d'équations correspondants en moins de $O(N^3)$ opérations, comme c'est le cas pour les méthodes générales de Gauss et de Choleski.

Le premier algorithme spécifique de résolution des systèmes de Toeplitz, proposé par Levinson [16], a par la suite été amélioré et étendu, pour la résolution des systèmes symétriques ([18],[19]), l'inversion des matrices de Toeplitz [20] ou de Toeplitz par blocs [23]. La stabilité de ces algorithmes "classiques", qui demandent $O(N^2)$ opérations, a été étudiée entre autres par Cybenko [24].

Plus récemment ([26] à [40]), des algorithmes de résolution plus performants en $O(N(\log_2 N)^2)$ opérations ont été obtenus par diverses approches qui reposent néanmoins toutes sur l'utilisation de l'algorithme de Transformation de Fourier Rapide (TFR) pour évaluer convolutions, produits de polynômes ou inverser les matrices circulantes. Brent, Gustavson et Yun [30] passent ainsi par le calcul d'approximants de Padé et utilisent un algorithme rapide de recherche du

PGCD de deux polynômes. Kumar [31] combine cet algorithme avec l'approche de Trench. Bitmead et Anderson [36] utilisent la notion de "rang de déplacement" introduite par Kailath et Morf ([33],[34]) pour résoudre des systèmes de Toeplitz par blocs en $p^2 O(N(\log_2 N)^2)$ opérations.

Une autre approche consiste à utiliser la formule de Gohberg et Semencul ([37],[38]) de factorisation de l'inverse d'une matrice de Toeplitz :

$$T^{-1} = L_1 L_1^t - L_2 L_2^t$$

Comme les matrices triangulaires inférieures L_1 et L_2 sont de Toeplitz, leur plongement dans des matrices circulantes de plus grande taille permet d'accélérer le calcul du second membre par utilisation de la TFR. Combinée à des procédures rapides de calcul des termes de L_1 et L_2, cette approche conduit à un algorithme également en $O(N(\log_2 N)^2)$ opérations et un encombrement mémoire en $O(N)$.

L'étude de la stabilité de ces algorithmes qualifiés de "superfast" n'est pas terminée et quelques zones d'ombre subsistent encore [25].

Tous les algorithmes qui précèdent font partie des méthodes directes. Pour être compétitive, voire meilleure que ces dernières, une méthodes itérative devra d'une part demander peu d'opérations par itération et d'autre part un nombre restreint d'itérations pour garantir un niveau de précision acceptable. Ce dernier critère conduit naturellement à rechercher de bons préconditionnements pour les matrices de Toeplitz. Dans la suite de l'exposé, nous nous limiterons à la méthode des directions conjuguées en considérant deux cas. Dans un premier temps, les directions conjuguées sont celles générées par le procédé d'orthonormalisation de Gram-Schmidt à partir de la base canonique. Nous montrons que l'algorithme qui en résulte est identique à celui de Zohar [18], ce qui permet d'en donner une interprétation géométrique. Nous considérons ensuite dans un deuxième temps la méthode du gradient conjugé, préconditionné par des matrices circulantes.

Nous montrons d'abord qu'avec de telles matrices, le nombre d'opérations par itération peut être limité à $O(N \log_2 N)$ opérations par utilisation de la TFR. Nous nous penchons ensuite sur trois préconditionnements circulants particuliers dont nous étudions quelques propriétés. Nous terminons cette quatrième partie en montrant l'efficacité de ces trois préconditionnements pour de grandes valeurs de N, qui se traduit par la focalisation des valeurs propres vers 1 en moyenne de Césaro.

4.2 Une méthode de directions conjuguées

Nous revenons maintenant à l'expression de la solution approchée g_N du problème de minimisation. La formule (20), appliquée à un ordre courant $k < N$ montre que les solutions optimales $g_k = \mathrm{Arg}_{g \in P_k} \min J_2(g)$ sont liées par une relation récurrente :

$$g_{k+1} = g_k + \delta_{k+1} S_{k+1} \tag{26}$$

dans laquelle $\delta_{k+1} = \langle b, aS_{k+1} \rangle / s_{k+1} = [g_{k+1}, S_{k+1}] / s_{k+1}$ est déterminé pour que g_{k+1} soit le minimum de J_2 sur P_{k+1}.δ_{k+1} est également la valeur de δ rendant J_2 minimale sur la droite $g_k + \delta S_{k+1}$.

A partir de la base canonique, nous avons par ailleurs construit un système de vecteurs non nuls $S_k(z)$ indépendants car mutuellement orthogonaux pour $[.,.]$. Ces vecteurs de P_N traduisent des directions associées $\sigma_k \in \mathbb{R}^{N+1}$ mutuellement conjuguées par rapport à T_N.

La relation (26) traduite sur $\mathbb{R}^{N+1}$ en:

$$\underline{x}_{k+1} = \underline{x}_k + \delta_{k+1}\underline{\sigma}_{k+1} \tag{27}$$

n'est rien d'autre que l'application d'une méthode de directions conjuguées pour trouver la solution $\underline{x}_N$ du système linéaire (11). Ici les directions $\underline{\sigma}_k$ sont déterminées par orthogonalisation de Gram-Schmidt à partir de la base canonique de $\mathbb{R}^{N+1}$.

Traduite sur $\mathbb{R}^{N+1}$, la relation (26) montre également que **si** $\underline{x}_k$ est solution du système $T_k\underline{x}_k = \underline{\beta}_k$, la méthode des directions conjuguées génère à l'étape suivante $\underline{x}_{k+1}$ solution de $T_{k+1}\underline{x}_{k+1} = \underline{\beta}_{k+1}$. Au bout de N étapes, on évalue donc la solution $\underline{x}_N$ cherchée si on part de la bonne valeur initiale x_0, c'est-à-dire de $x_0 = \beta_0/\alpha_0$, qui est la solution évidente de $T_0.x_0 = \beta_0$.

Connaissant g_k, l'évaluation de g_{k+1} demande le calcul de la nouvelle direction S_{k+1} puis celui de δ_{k+1}.

Calcul de S_{k+1} $S_{k+1}(z)$ est généré par la relation de récurrence de Levinson [41] :

$$\frac{S_{k+1}}{s_{k+1}}(z) = \frac{zS_k(z)}{z_k} + \gamma_k \frac{E_k^0(z)}{s_k} \tag{28}$$

$E_k^0(z)$ est défini fonctionnellement comme le polynôme d'évaluation en zéro sur P_k:

$$\forall p(z) \in P_k \quad p(0) = [p(z), E_k^0(z)]$$

On montre [41] qu'il peut être obtenu à partir de S_k par symétrie miroir entre les coefficients de plus haut et de plus bas degrés :

$$E_k^0(z) = z^k\overline{S_k(z^{-1})} = \check{S}_k(z)$$

de (28) on en déduit alors facilement

$$\frac{E_{k+1}^0}{s_{k+1}}(z) = \frac{E_k^0(z)}{s_k} + \gamma_k \frac{zS_k(z)}{s_k} \tag{29}$$

En physique, les paramètres $\{\gamma_k\}_{k\geq 0}$ sont appelés coefficients de réflexion car (28) et (29) modélisent respectivement le comportement en transmission et en réflexion d'une onde se propageant dans un milieu stratifié [42]. En mathématiques, ces mêmes coefficients portent le nom de paramètres de Schur. Ils interviennent dans des problèmes fondamentaux d'analyse harmonique ([42],[43],[44]).

Puisque $S_{k+1}(z)/s_{k+1}$ et $zS_k(z)/s_k$ ont le même coefficient de plus haut degré, en l'occurence 1, la différence

$$D_k(z) = S_{k+1}(z)/s_{k+1} - zS_k(z)/s_k$$

appartient à P_k.

Dans cet espace, D_k est orthogonal au sous-espace de dimension k constitué des polynômes de la forme $zp_{k-1}(z)$ avec

$$p_{k-1} \in P_{k-1}$$

car

$$[D_k(z), zp_{k-1}(z)] = -[zS_k(z), zp_{k-1}(z)]/s_k = -[S_k, p_{k-1}]/s_k = 0$$

Mais E_k^0 a la même propriété car

$$[E_k^0(z), zp_{k-1}(z)] = (zp_{k-1}(z))(0) = 0$$

D_k et E_k^0 sont donc proportionnels. La condition

$$S_{k+1} \in P_k^{\perp}$$

donne alors :

$$\gamma_k = -[zS_k(z)/s_k, E_k^0(z)] \tag{30}$$

Si on réécrit maintenant (28) sous la forme

$$\frac{S_{k+1}}{s_{k+1}}(z) - \gamma_k \frac{E_k^0}{s_k}(z) = \frac{zS_k(z)}{s_k}$$

les deux polynômes de gauche sont orthogonaux. Comme

$$[E_k^0, E_k^0] = E_k^0(0) = s_k$$

et

$$[zS_k(z), zS_k(z)] = [S_k, S_k] = s_k$$

en prenant la norme des deux côtés :

$$\frac{1}{s_{k+1}} + \gamma_k^2 \frac{1}{s_k} = \frac{1}{s_k}$$

ce qui permet de calculer s_{k+1} :

$$s_{k+1} = \frac{s_k}{1-\gamma_k^2} \quad (|\gamma_k| < 1) \tag{31}$$

On détermine donc S_{k+1} par les trois étapes successives :

$$(30) \rightarrow (31) \rightarrow (28)$$

Le coefficient γ_k peut également s'exprimer en fonction des éléments de la matrice T_N. En utilisant (13) et la définition de E_k^0, on trouve facilement:

$$\gamma_k = -\sum_{\ell=0}^{k} \alpha_{\ell+1}.\sigma_k^{\ell} \tag{32}$$

Calcul de δ_{k+1} De même que γ_k, l'expression donnant δ_{k+1} peut être transformée:

$$\delta_{k+1} = \frac{< b\bar{a}, S_{k+1} >}{s_{k+1}} = \frac{\left(\underline{\beta}_N, \underline{\sigma}_{k+1}\right)}{s_{k+1}} = \frac{\left(\underline{\beta}_N - T_k \underline{x}_k, \underline{\sigma}_{k+1}\right)}{s_{k+1}}$$

car:

$$(T_k \underline{x}_k, \underline{\sigma}_{k+1}) = [g_k, S_{k+1}] = 0$$

Comme par ailleurs $T_k \underline{x}_k = \underline{\beta}_k$ il vient finalement :

$$\begin{aligned} \delta_{k+1} &= \frac{1}{s_{k+1}}(\beta_{k+1} - \alpha_{k+1} x_k^0 - \alpha_k x_k^1 \ldots - \alpha_1 x_k^k) s_{k+1} \\ &= \beta_{k+1} - \sum_{\ell+1}^{k} \alpha_{\ell+1} x_k^{k-\ell} \end{aligned} \tag{33}$$

On pose enfin:

$$\underline{a}_k = {}^t(\alpha_1 \alpha_2 \ldots \alpha_k) \in \mathbb{R}^k \quad \underline{y}_k = \underline{x}_{k-1} \in \mathbb{R}^k$$

et $\underline{e}_k \in \mathbb{R}^k$ associé à $e_k(z)$ défini par:

$$E_k^0(z)/s_k = 1 + z e_k(z)$$

alors (33), (32), (27), (29) et (31) se réécrivent comme suit:

$$\delta_k = (\beta_k - (\underline{y}_k, \check{\underline{a}}_k)) S_k$$

$$\gamma_k = -s_k(\alpha_{k+1} + (\underline{e}_k, \check{\underline{a}}_k))$$

$$\underline{y}_{k+1} = \begin{bmatrix} \underline{y}_k + \delta_k \check{\underline{e}}_k \\ \delta_k \end{bmatrix}$$

$$\underline{e}_{k+1} = \begin{bmatrix} \underline{e}_k + \gamma_k \check{\underline{e}}_k \\ \gamma_k \end{bmatrix}$$

$$s_{k+1} = \frac{s_k}{1 - \gamma_k^2}$$

avec les données initiales:

$$e_1 = \gamma_0 = -\alpha_1/\alpha_0; \quad y_1 = \beta_0/\alpha_0; \quad s_0 = 1/\alpha_0$$

C'est l'algorithme classique de résolution des systèmes de Toeplitz symétriques [18]. Cet algorithme demande $2N^2$ opérations (1 opération = 1 addition + 1 multiplication) pour résoudre un système $N \times N$.

Appliqué à la résolution de notre problème de minimisation, on remarque que cet algorithme peut se définir à la fois comme une méthode directe et comme une méthode itérative car la valeur $\underline{y}_k$ donnée à chaque étape fournit la solution optimale dans P_k. Il n'y a pas besoin d'attendre les N étapes de l'algorithme pour déjà obtenir une bonne solution approchée.

Nous venons de montrer que cet algorithme, obtenu jusqu'ici par des voies purement algébriques, s'interprête en fait géométriquement comme une méthode de directions conjuguées. Les directions conjuguées sont ici définies une fois pour toutes par la donnée de la matrice T_N. En particulier, elles ne font pas intervenir les approximations successives $\underline{x}_k$ de la solution, ni les restes associés. Dans ce contexte, la méthode du gradient conjugué semble intéressante car elle permet de construire de nouvelles directions de descente pour J_2 en fonction des restes successivement calculés.

4.3 Le gradient conjugué préconditionné ([45] chap.1, [46] chap.8)

Principe et étapes élémentaires La méthode du gradient conjugué, préconditionné par la matrice C_N consiste à résoudre le système symétrique $T_N\underline{x} = \beta_N$ en minimisant la fonction

$$f(\underline{y}) = \frac{1}{2}(\underline{y}, \tilde{T}_N\underline{y}) - (\tilde{\underline{\beta}}_N, \underline{y})$$

où

$$\tilde{T}_N = C_N^{-1/2} T_N C_N^{-1/2} \quad \tilde{\underline{\beta}}_N = C_N^{-1/2}\underline{\beta}_N \quad \underline{y} = C_N^{-1/2}\underline{x}$$

et tous les vecteurs sont dans $\mathbb{R}^N$.

L'algorithme est défini par l'exécution à chaque itération des six calculs suivants:

1. $$\alpha_k = \frac{(\underline{z}^k, \underline{r}^k)}{(\underline{p}^k, T_N\underline{p}^k)} = \frac{(\underline{z}^k, \underline{z}^k)}{(\underline{p}^k, T_N\underline{p}^k)}$$

2. $$\underline{x}^{k+1} = \underline{x}^k + \alpha_k\underline{p}^k$$

3. $$\underline{r}^{k+1} = \underline{r}^k - \alpha_k T_N\underline{p}^k$$

4. $$\underline{z}^{k+1} = C_N^{-1}\underline{r}^{k+1}$$

5. $$\beta_k = \frac{(\underline{z}^{k+1}, \underline{r}^{k+1})}{(\underline{z}^k, \underline{r}^k)} = \frac{(\underline{z}^{k+1}, \underline{z}^{k+1})}{(\underline{z}^k, \underline{z}^k)}$$

6. $$\underline{p}^{k+1} = \underline{z}^{k+1} + \beta_k\underline{p}^k$$

et par les conditions initiales: $\underline{x}^0$ arbitraire (si T_N est régulière), $\underline{r}^0 = \underline{\beta}_N - T_N\underline{x}^0$, $\underline{p}^0 = \underline{z}^0 = C_N^{-1}\underline{r}^0$.

Les coefficients α_k et β_k sont déterminés pour que d'une part:

$$\underline{y}^{k+1} = \mathrm{Arg} \min_{\underline{y} = \underline{y}^k + \alpha C_N^{1/2}\underline{p}^k} f(\underline{y}(\alpha))$$

et pour que d'autre part $(\underline{p}^{k+1}, T_N \underline{p}^{\ell}) = 0 \quad 0 \leq \ell \leq k$.

Les restes successifs $\underline{r}^k$ sont C_N orthogonaux (i.e.: $(\underline{r}^k, C_N \underline{r}^{\ell}) = 0$ si $k \neq \ell$) et engendrent le même espace que les directions de descente $\underline{p}$:

$$\text{Vect}\{\underline{p}^0, \underline{p}^1, \ldots, \underline{p}^k\} = \text{Vect}\{\underline{r}^0, \underline{r}^1, \ldots, \underline{r}^k\}$$

Dans ces conditions, on obtient $\underline{r}^N = 0$ et par conséquent la solution $\underline{x}$ en au plus N itérations, du moins en théorie.

La matrice C_N de préconditionnement doit être symétrique et définie positive, proche de T_N pour que la méthode converge vite vers une "bonne solution" approchée et facile à inverser pour qu'à l'étape 4 $\underline{z}^{k+1}$ soit facilement calculable.

Préconditionnements étudiés Les matrices de préconditionnement que nous étudions par la suite sont des matrices circulantes vers la droite, i.e. du type:

$$C_N = \begin{bmatrix} c_0 & c_1 & \cdots & & c_{N-1} \\ c_{N-1} & c_0 & c_1 & \cdots & c_{N-2} \\ & & & & \\ & & & & c_1 \\ c_1 & & & c_{N-1} & c_0 \end{bmatrix}$$

Ce type de préconditionnement contribue de manière significative à l'efficacité globale de l'algorithme, qui s'exerce à deux niveaux :

- au niveau du nombre d'opérations par itération : en tirant parti des structures particulières de T_N et C_N nous montrons que chaque itération peut être calculée en $O(N \log_2 N)$ opérations,
- au niveau du nombre d'itérations : nous montrons que les trois matrices circulantes de préconditionnement étudiées plus bas sont "asymptotiquement équivalentes" à la matrice de Toeplitz T_N lorsque $N \to +\infty$. Le taux de convergence de la méthode du gradient, qui dépend du spectre de $C_N^{-1} T_N$ sera d'autant meilleur que N sera grand.

Mise en oeuvre de l'algorithme Nous rappelons d'abord brièvement quelques propriétés classiques des matrices circulantes.

Les valeurs propres $\lambda_m, m = 0, 1, \ldots, N-1$ de C_N et la première ligne $c0, c1, \ldots, c_{N-1}$ de la matrice forment une paire de série de Fourier discrète:

$$c_n \underset{N}{\longleftrightarrow} \lambda_m$$

c'est-à-dire que :

$$c_n = \frac{1}{N} \sum_{m=0}^{N-1} \lambda_m W_N^{-mn} \text{ (TFD direct)}$$

$$\lambda_m = \sum_{n=0}^{N-1} c_n W_N^{mn} \quad \text{(TFD inverse)}$$

avec $W_N = \exp(2i\pi/N)$ racine N^{ième} de l'unité.

Les vecteurs propres associés

$$\underline{V}_m = {}^t(1, W_N^m, W_N^{2m}, \ldots, W_N^{(N-1)m}) \quad 0 \leq m \leq N-1$$

ne dépendent pas des éléments de la matrice et forment une base orthogonale de $\mathbb{C}^N$ puisque la matrice C_N est symétrique réelle. On notera enfin que la somme, le produit et l'inverse de matrices circulantes sont également des matrices circulantes.

L'utilisation systématique de l'algorithme de TFR pour l'évaluation des transformées de Fourier discrètes va permettre de réduite considérablement le nombre d'opérations effectuées à chaque itération de l'algorithme de gradient conjugué. Le gain va porter principalement sur le calcul de $T_N \underline{p}_k$, qui en général constitue l'étape coûteuse de l'algorithme. On utilise pour cela une approche très voisine de celle qui permet de résoudre rapidement le système circulant $C_N \underline{Z} = \underline{r}^k$.

Résolution du système circulant $C_N \underline{z} = \underline{r}^k$.

On décompose les deux vecteurs $\underline{R} = {}^t(R_0, \ldots, R_m, \ldots, R_{N-1})$ et $\underline{Z} = {}^t(Z_0, \ldots, Z_m, \ldots, Z_{N-1})$ sur la base des vecteurs propres de C_N:

$$\underline{R} = \sum_{n=0}^{N-1} r_n \underline{V}_n \text{ et } \underline{Z} = \sum_{n=0}^{N-1} z_n \underline{V}_n$$

Ces deux égalités se lisent comme des transformées de Fourier discrètes. On a donc :

$$R_m \underset{N}{\longleftrightarrow} r_n \text{ et } Z_m \underset{N}{\longleftrightarrow} z_n$$

D'autre part, la relation $\underline{R} = C_N \underline{Z} = \sum_{n=0}^{N-1} \lambda_n z_n \underline{V}_n$ permet de calculer z_n connaissant r_n:

$$z_n = \frac{r_n}{\lambda_n}$$

La résolution du système circulant se décompose donc en 3 étapes :

1. TFD directe des composantes du vecteur $\underline{R}$
2. division du résultat par λ_n
3. TFD inverse du quotient obtenu.

Calcul du produit $\underline{S} = T_N \underline{P}$

On prolonge la matrice de Toeplitz T_N en une matrice circulante $\tilde{C}$ de dimension 2N dont la première ligne est la séquence:

$$\alpha_0, \alpha_1, \ldots, \alpha_{N-1}, x, \alpha_{N-1}, \ldots, \alpha_1$$

avec x quelconque. On définit ensuite le vecteur $\tilde{P}$ de dimension 2N obtenu en complétant P par N zéros. On retrouve alors les composantes du vecteur $T_N \underline{P}$ dans les N premières composantes du vecteur $\tilde{C}\underline{\tilde{P}}$, lequel aura été calculé en utilisant les trois étapes du paragraphe précédent.

Nombre d'opérations par itération Le calcul des étapes 3 et 4 se réduit donc à $O(N \log_2 N)$ opérations en utilisant un algorithme de TFR pour évaluer les différentes transformées de Fourier discrètes. Ceci fixe l'ordre de grandeur du nombre d'opérations par itération du gradient car les autres étapes n'en demandent que N. Sur la base de calcul suivante :

- 1 opération = 1 addition + 1 multiplication
- 1 opération complexe = 4 opérations réelles

une analyse plus détaillée conduit à $6N \log_2 N$ multiplications réelles et $9N \log N$ additions réelles. Quant à l'encombrement mémoire, il reste proportionnel à N.

4.4 Quelques préconditionnements circulants

Le préconditionnement de Strang ([49],[50]) Cette matrice circulante S_N, qui est définie par

$$s_i = \alpha_i \text{ pour } 0 \leq i \leq \frac{N}{2} \text{ et } s_i = \alpha_{N-i} \text{ pour } \frac{N}{2} + 1 \leq i \leq N - 1 \qquad (34)$$

garde les mêmes sur et sous-diagonales que T_N jusqu'à $\frac{N}{2}$ et complète le reste de la matrice par périodicité.

On remarque que ce préconditionneur ne tient pas compte de tous les a_i mais seulement des $\frac{N}{2} + 1$ premiers.

Dans le cas que nous traitons ici, les coefficients α_i ne sont pas quelconques mais sont les moments trigonométriques d'une fonction $|a(\theta)|^2$ donnée a priori. Ceci nous permet de montrer le résultat suivant :

Proposition 9. On suppose que la fonction $\theta \to |a(\theta)|^2$ est minorée par $m > 0$ et que $|\alpha_n| \leq \frac{K}{n^2}$. Alors S_N est définie positive pour une dimension N assez grande, fonction de K et de m.

Preuve 10. La condition sur α_n assure la convergence uniforme sur $[0, 2\pi]$ de la somme de Fourier

$$\sigma_N(\theta) = \sum_{n=-N}^{n=+N} \alpha_n e^{in\theta} \text{ vers } |a(\theta)|^2$$

Comme:

$$\left||a(\theta)|^2 - \sigma_N(\theta)\right| \leq 2 \sum_{k=N+1}^{+\infty} |\alpha_k| \leq 2K \sum_{k=N+1}^{+\infty} \frac{1}{K_2} \leq \frac{2K}{N}$$

on en déduit la minoration:

$$\sigma_N(\theta) \geq m - \frac{2K}{N}$$

D'autre part, les valeurs propres $\lambda^S_{N,k}$ de S_N sont égales à la transformée de Fourier discrète inverse de la première ligne de la matrice:

$$\lambda^S_{N,k} = \alpha_0 + \sum_{n=1}^{N/2-1} \alpha_n \left(W_N^{kn} + W_N^{-kn}\right) + \alpha_{N/2} W_N^{kN/2} = (-1)^k \alpha_{N/2} + \sigma_{N/2-1}(\theta_N^k)$$

avec θ_N^k défini par $W_N^k = e^{i\theta_N^k}$.

Ainsi

$$\begin{aligned} \lambda^S_{N,k} &\geq m - \frac{2K}{\frac{N}{2}-1} - \frac{K}{\left(\frac{N}{2}\right)^2} \\ &\geq m - \frac{2K}{\frac{N}{2}-1} - \frac{K}{\left(\frac{N}{2}-1\right)^2} \\ &= \frac{m}{\left(\frac{N}{2}-1\right)^2}\left\{X^2 - 2AX - A\right\} \end{aligned}$$

avec $A = \frac{K}{m}$ et $X = \frac{N}{2} - 1$.

Pour assurer $\lambda^S_{N,k} > 0$ pour tout $0 \leq k \leq N-1$ il suffit alors de choisir $N > 2(X_0 + 1)$ où $X_0 = A + \sqrt{A + A^2}$ est la plus grande racine du trinôme.

Remarque 11. La condition sur α_n est en particulier satisfaite dans le cas où $|a(\theta)|^2$ est affine, de pente p_k, par morceaux. Dans ce cas, si on note $V' = \sum_k |p_k - p_{k-1}|$ la variation totale de la fonction dérivée (étagée) de $|a(\theta)|^2$ on a $|\alpha_n| \leq \frac{V'}{2\pi n^2}$.

Le préconditionnement de T.F. Chan [50] La valeur du rayon spectral $\rho(C_N^{-1}(C_N - T_N))$ est un des paramètres mesurant l'efficacité d'un préconditionnement. Comme pour toute norme matricielle:

$$\rho(C_N^{-1}(C_N - T_N)) \leq |||C_N^{-1}(C_N - T_N)|||$$

T.F. Chan cherche la matrice circulante C_N minimisant la norme de Hilbert-Schmidt normalisée:

$$|C_N - T_N|^2_{HS} = \frac{1}{N}\sum_{i,j}^{N-1} |c_{ij} - t_{ij}|^2 = \frac{1}{N}\text{Trace}\left\{{}^t(C_N - T_N)(C_N - T_N)\right\} \tag{35}$$

dont la solution est :

$$c_i = \frac{1}{N}\left\{i\alpha_{N-i} + (N-i)\alpha_i\right\} \quad 0 \leq i \leq N-1 \tag{36}$$

et le minimum donné par :

$$|C_N - T_N|^2_{HS} = \sum_{i=0}^{N-1} \frac{i(N-i)}{N^2}.(\alpha_i - \alpha_{N-i})$$

Cette fois-ci, tous les éléments de la matrice de Toeplitz sont pris en compte dans le préconditionnement.

En ce qui concerne le caractère défini positif de C_N, on a, par rapport à S_N le résultat plus fort:

Proposition 12. La matrice circulante C_N définie en (36) associée à une matrice de Toeplitz symétrique définie positive quelconque T_N est définie positive quel que soit N.

Preuve 13. Supposons dans un premier temps que les éléments T_N soient les moments trigonométiques d'une fonction donnée $|a(\theta)|^2$:

$$\alpha_k = \frac{1}{2\pi}\int_0^{2\pi} |a(\theta)|^2 e^{ik\theta} d\theta; \quad 0 \leq k \leq N-1$$

Les valeurs propres de C_N sont:

$$\begin{aligned}\lambda_{N,m}^C &= \sum_{k=0}^{N-1} \frac{k\alpha_{N-k} + (N-k)\alpha k}{N} W_N^{km} \\ &= \frac{1}{N}\left\{N\alpha_0 + \sum_{k=1}^{N-1}(N-k)\alpha_k(W_N^{km} + W_N^{-km})\right\}\end{aligned}$$

On pose $W_N^m = \exp(i\theta_N^m)$. Les sommes partielles de Fourier

$$\sigma_k(\theta_N^m) = \sum_{\ell=-k}^{\ell=+k} \alpha_\ell W_N^{m\ell} = \alpha_0 + \sum_{\ell=1}^{k} \alpha_k(W_N^{m\ell} + W_N^{-m\ell})$$

permettent de réécrire $\lambda_{N,m}^C$ sous la forme de la moyenne de Césaro:

$$\lambda_{N,m}^C = \frac{\sigma_0 + \sigma_1 + \ldots + \sigma_{N-1}}{N}(\theta_N^m) \tag{37}$$

Par ailleurs

$$\begin{aligned}\sigma_k(x) &= \frac{1}{2\pi}\int_0^{2\pi} |a(\theta)|^2 \left\{1 + 2\cos(x-\theta) + \ldots + 2\cos k(x-\theta)\right\} d\theta \\ &= \frac{1}{2\pi}\int_0^{2\pi} |a(\theta)|^2 \frac{\sin\left(k+\frac{1}{2}\right)(x-\theta)}{\sin\left(\frac{x-\theta}{2}\right)} d\theta\end{aligned}$$

et

$$\sum_{k=0}^{N-1} \sin\left(k+\frac{1}{2}\right)(x-\theta) = \frac{\sin^2 \frac{N}{2}(x-\theta)}{\sin\frac{(x-\theta)}{2}}$$

En remplaçant dans (37) il vient donc :

$$\lambda_{N,m}^C = \frac{1}{2N\pi}\int_0^{2\pi} |a(\theta)|^2 \left[\frac{\sin N\frac{(\theta_N^m-\theta)}{2}}{\sin\left(\frac{\theta_N^m-\theta}{2}\right)}\right]^2 d\theta > 0 \tag{38}$$

Passons maintenant au cas général. T_N est une matrice de Toeplitz quelconque symétrique et définie positive dont la première ligne est constituée des éléments α_k $0 \leq k \leq N-1$. Le problème de trouver l'ensemble $\mathcal{M}_N$ des mesures positives $d\mu(\theta)$ telles que:

$$\alpha_k = \frac{1}{2\pi}\int_0^{2\pi} e^{ik\theta} d\mu(\theta) \qquad 0 \leq k \leq N-1$$

constitue le problème des moments trigonométriques tronqué à l'ordre $N-1$. On en connait une solution particulière ([7] et [41]) :

$$d\mu(\theta) = \frac{s_{N-1}}{|S_{N-1}(e^{i\theta})|^2} d\theta$$

Dans cette formule, $\frac{S_{N-1}}{\sqrt{s_{N-1}}}(z)$ désigne le polynôme de Szegö de degré $N-1$ tel que défini aux paragraphes 3.5 et 4.2. On a $s_{N-1} = [S_{N-1}, S_{N-1}] > 0$. On montre de plus qu'un tel polynôme ne peut s'annuler sur le cercle unitOAé.

On se ramène donc au cas précédent en remplaçant $|a(\theta)|^2$ par $\frac{s_{N-1}}{|S_{N-1}(e^{i\theta})|^2}$.

Les deux matrices de préconditionnement S_N et C_N peuvent être utilisées directement sur tout système de Toeplitz symétrique et défini positif puisqu'on les assemble directement à l'aide des éléments donnés α_k de la matrice T_N. La fonction $|a(\theta)|^2$ n'a pas besoin, pour ces deux matrices, d'être connue.

Dans notre cas particulier, la fonction $|a(\theta)|^2$ dont les α_k sont les N premiers moments est une donnée du problème. Nous allons utiliser cette fonction pour construire un autre préconditionnement circulant.

Un autre préconditionnement circulant de Toeplitz Le préconditionnement que nous proposons se déduit naturellement de l'expression intégrale de la forme quadratique donnée par T_N:

$$(\underline{x}, T_N\underline{x}) = \frac{1}{2\pi}\int_0^{2\pi} |a(\theta)|^2 \left|\sum_{k=0}^{N-1} x_k e^{ik\theta}\right|^2 d\theta$$

Supposons $|a(\theta)|^2$ suffisamment régulière pour être intégrable au sens de Riemann, par exemple affine par morceaux. On approche l'intégrale par la somme partielle:

$$\begin{aligned}\frac{1}{N}\sum_{\ell=0}^{N-1} |a(\theta_N^\ell)|^2 \left|\sum_{k=0}^{N-1} x_k W_N^{k\ell}\right|^2 &= \sum_{k,k'=0}^{N-1} x_k x_{k'} \left\{\frac{1}{N}\sum_{\ell=0}^{N-1} |a(\theta_N^\ell)|^2 W_N^{\ell(k-k')}\right\} \\ &= \sum_{k,k'=0}^{N-1} x_k x_{k'} \tilde{\alpha}_{k'-k}\end{aligned}$$

où $\tilde{\alpha}_k = \frac{1}{N}\sum_{\ell=0}^{N-1} |a(\theta_N^\ell)|^2 W_N^{\ell k}$ est la TFD directe de la séquence $|a(\theta_N^\ell)|^2$.

La parité de la fonction $|a(\theta)|^2$:

$$\left|a\left(\theta_N^{\frac{N}{2}+\ell}\right)\right|^2 = \left|a\left(\theta_N^{\frac{N}{2}-\ell}\right)\right|^2 \quad 0 \leq \ell \leq \frac{N}{2}$$

se transporte par Fourier sur $\tilde{\alpha}_k$:

$$\tilde{\alpha}_{\frac{N}{2}+\ell} = \tilde{\alpha}_{\frac{N}{2}-\ell} \quad 0 \leq \ell \leq \frac{N}{2}$$

La matrice de Toeplitz obtenue par discrétisation de l'intégrale est dans ce cas très particulier une matrice circulante, notée F_N et dont la première ligne est précisément:

$$\tilde{\alpha}_0\tilde{\alpha}_1 \dots \tilde{\alpha}_{\frac{N}{2}-1}\tilde{\alpha}_{\frac{N}{2}}\tilde{\alpha}_{\frac{N}{2}-1} \dots \tilde{\alpha}_2\tilde{\alpha}_1 \tag{39}$$

La forme quadratique associée est positive :

$$(\underline{x}, F_N\underline{x}) = \frac{1}{N}\sum_{\ell=0}^{N-1} |a(\theta_N^\ell)|^2 \left|\sum_{k=0}^{N-1} x_k W_N^{k\ell}\right|^2 \geq 0$$

Elle est même définie positive si on suppose $|a(\theta)|^2$ strictement positive, par exemple minorée par $m > 0$. Dans ce cas:

$$(\underline{x}, F_N\underline{x}) = 0 \iff \forall \ell = 0, 1, \dots, N-1 \qquad \sum_{k=0}^{N-1} x_k W_N^{k\ell} = 0$$

dont on déduit $x_k = 0 \quad 0 \leq k \leq N-1$ par transformée de Fourier inverse. Les valeurs propres de F_N sont immédiates:

$$\lambda_{N,\ell}^F = |a(\theta_N^\ell)|^2 \qquad 0 \leq \ell \leq N-1 \tag{40}$$

4.5 Efficacité asymptotique des préconditionnements circulants

Généralités Nous étudions maintenant l'"efficacité" de ces trois matrices de préconditionnement pour résoudre le système (11) de matrice T_N pour de grandes valeurs de N. Notons génériquement par A_N une quelconque de ces trois matrices circulantes (S_N, C_N ou F_N).

On mesure cette efficacité asymptotique par la convergence, dans un sens qui reste à préciser, de $A_N^{-1}T_N$ vers l'identité lorsque N tend vers l'infini. Dans ce cas, on s'attend donc à un comportement général des valeurs propres de $A_N^{-1}T_N$ à converger vers 1 lorsque N tend vers l'infini.

Les premiers tests numériques de Strang ([49] cité par [50]) sur l'efficacité de S_N mettent effectivement en évidence une convergence des valeurs propres de $S_N^{-1}T_N$ vers 1 à l'infini, à l'exception toutefois de la plus petite et de la plus grande d'entre elles. Pour étayer ces observations on peut d'ailleurs montrer (cité par [50]) que dans le cas particulier où T_N est définie par $\alpha_k = t^{|k|}$ $|t| < 1$, T_N ne possède que cinq valeurs propres distinctes:

$$\begin{array}{ll} \frac{1}{1+t} & \text{simple} \\ \frac{1}{1+t^{N/2}} & \text{ordre } \frac{N}{2} - 2 \\ 1 & \text{double} \\ \frac{1}{1-t^{N/2}} & \text{ordre } \frac{N}{2} - 2 \\ \frac{1}{1-t} & \text{simple} \end{array}$$

Ces valeurs propres tendent vers 1 lorsque N tend vers l'infini, à l'exception des deux valeurs extrêmes $\frac{1}{1+t}$ et $\frac{1}{1-t}$.

On sait d'autre part que l'efficacité de la méthode du gradient conjugué préconditionné est directement liée à la distribution des valeurs propres de $A_N^{-1}T_N$. La convergence est grandement améliorée lorsque ces valeurs propres se regroupent par paquets autour d'un nombre limité de valeurs [47]. A la limite, si $A_N^{-1}T_N$ ne possède que $p < N$ valeurs propres distinctes, on sait que la méthode du gradient converge en p étapes seulement; ceci est particulièrement intéressant lorsque N est grand (quelques milliers) et que p est modéré (quelques dizaines).

En ce sens, l'exemple précédent montre clairement l'intérêt que peut présenter S_N pour la méthode du gradient conjugué. Dans ce cas particulier, l'algorithme converge théoriquement en moins de cinq étapes et son efficacité augmente avec N alors que le paramètre de conditionnement

$$\text{Cond}(S_N^{-1}T_N) = \frac{1+t}{1-t}$$

lui, reste constant !

En ce qui concerne maintenant la convergence du spectre de $A_N^{-1}T_N$ vers 1, l'exemple précédent souligne également que les deux valeurs propres extrêmes et “isolées” ont un poids de plus en plus faible avec N par rapport à l'ensemble de toutes les autres. Ainsi, *en moyenne*, le spectre de $A_N^{-1}T_N$ tend vers 1 lorsque N tend vers l'infini. [51].

C'est cette idée que nous exploitons maintenant.

En désignant par $\Lambda_{N,k}^{A}$ la k+1ième valeur propre de la matrice $A_N^{-1}T_N$ (avec $A_N = S_N, C_N$ ou F_N) nous allons montrer dans un premier temps la convergence en moyenne des valeurs propres vers 1, i.e.:

$$\lim_{N\to+\infty} \frac{1}{N} \left(\sum_{k=0}^{N-1} \Lambda_{N,k}^{A} \right) = 1 \tag{41}$$

Nous étendrons ensuite ce résultat à toute puissance entière des valeurs propres:

$$\forall p \in \mathbb{N} \quad \lim_{N\to+\infty} \frac{1}{N} \left(\sum_{k=0}^{N-1} \left(\Lambda_{N,k}^{A}\right)^p \right) = 1 \tag{42}$$

pour généraliser enfin ce résultat à toute fonction continue:

$$\lim_{N\to+\infty}\frac{1}{N}\left(\sum_{k=0}^{N-1} f\left(\Lambda_{N,k}^{A}\right)\right)=f(1) \tag{43}$$

Pour cela, nous ferons bien entendu un certain nombre d'hypothèses sur S_N, C_N et F_N, qui en fait ne se révèlent pas vraiment restrictives pour les applications pratiques.

Nous aurons alors démontré le

Théorème 14. *Soit T_N une matrice de Toeplitz symétique définie positive de dimension N de terme général $(T_N)_{ij} = a_{|i-j|}$ formé à partir des N premiers moments trigonométriques*

$$\alpha_k=\frac{1}{2\pi}\int_0^{2\pi}|a(\theta)|^2 e^{ik\theta}d\theta;\quad 0\le k\le N-1$$

d'une fonction 2π-périodique $\theta \to |a(\theta)|^2$, majorée par M et minorée par $m > 0$. On construit à partir de T_N les trois matrices circulantes S_N, C_N et F_N définies respectivement en (34), (36) et (39) et notées génériquement par A_N.

On désigne par r_A et s_A les bornes finies de l'ensemble de toutes les valeurs propres de $\Lambda_{N,k}^{A}$ de $A_N^{-1}T_N$ $\quad N \geq 0$.

Alors on a :

$$\lim_{N\to+\infty}\frac{1}{N}\sum_{k=0}^{N-1} f\left(\Lambda_{N,k}^{A}\right)=f(1)$$

pour toute fonction f continue sur $[r_A, s_A]$ sous les conditions:

- $\alpha_k = O\left(\frac{1}{k^2}\right)$ *à l'infini pour S_N*
- $\alpha_k = o\left(\frac{1}{\sqrt{k}}\right)$ *à l'infini pour C_N*
- *la fonction $\theta \to |a(\theta)|^2$ est continue sur $[0, 2\pi]$ et $\alpha_k = o\left(\frac{1}{\sqrt{k}}\right)$ à l'infini pour F_N.*

Choix d'une norme adaptée L'étude des propriétés asymptotiques de $A_n^{-1}T_N$ demande le choix d'une norme matricielle adaptée au problème.

Outre la norme classique $\|A\| = \max_{(\underline{x},\underline{x})\le 1}(A\underline{x}, A\underline{x})^{1/2}$, nous utiliserons également la norme de Hilbert-Schmidt normalisée, définie en (35) et qui peut s'exprimer de différentes manières:

$$|A|_{HS}^2=\frac{1}{N}\mathrm{Trace}(A^*A)=\frac{1}{N}\sum_{i,j=0}^{N-1}|A_{ij}|^2=\frac{1}{N}\sum_{k=0}^{N-1}(A\underline{u}_k, A\underline{u}_k) \tag{44}$$

$\{u_i \quad 0 \leq i \leq N-1\}$ désigne une base orthonormée quelconque pour le produit hermitien canonique de $\mathbb{C}^N$. Nous pouvons prendre en particulier celle formée à partir des vecteurs propres d'une matrice circulante :

$$\underline{u}_k=\frac{1}{\sqrt{N}}\,{}^t(1,(\overline{W}_N)^k,(\overline{W}_N)^{2k},\ldots,(\overline{W}_N)^{(N-1)k});\quad 0\le k\le N-1$$

Cette norme de Hilbert-Schmidt semble bien adaptée au problème. En effet, pour une matrice symétrique :

$$|A|_{HS}^2 = \frac{1}{N}\mathrm{Trace}(A^2) = \frac{1}{N}\left(\sum_{k=0}^{N-1}(\lambda_{N,k})^2\right) \tag{45}$$

de sorte que pour $p = 2$, la relation (42) que l'on cherche à démontrer se traduit directement par:

$$\lim_{N\to+\infty}\left|A_N^{-1}T_N - I_N\right|_{HS}^2 = 1$$

De plus, la normalisation assure $|I_N|_{HS} = 1$ indépendamment de N.

Nous rappelons enfin sans démonstration quatre propriétés classiques de ces deux normes, qui nous serons utiles pour la suite :

$$\left|\frac{1}{N}\mathrm{Trace}(A)\right| \leq |A|_{HS} \tag{46}$$

$$|AB|_{HS} \leq \|A\|.|B|_{HS} \tag{47}$$

$$\text{A symétrique} \Rightarrow \|A\| = \max_{\underline{x}, A\underline{x}\leq 1}|(\underline{x}, A\underline{x})| = \max_i|\lambda_i(A)| \tag{48}$$

$$|(\underline{x}, A\underline{x})| \geq m(\underline{x}, \underline{x}) \Rightarrow \|A^{-1}\| \leq \frac{1}{m} \tag{49}$$

Convergence en moyenne du spectre

Cas général Pour montrer (41) considérons une matrice circulante symétrique inversible A_N. On a d'une part:

$$\begin{aligned}
\mathrm{Trace}(A_N^{-1}T_N) &= \sum_{k=0}^{N-1}(\underline{u}_k, A_N^{-1}T_N\underline{u}_k) \\
&= \sum_{k=0}^{N-1}(A_N^{-1}\underline{u}_k, T_N\underline{u}_k) \\
&= \sum_{k=0}^{N-1}\frac{(\underline{u}_k, T_N\underline{u}_k)}{\lambda_{N,k}^A}
\end{aligned}$$

et d'autre part :

$$\begin{aligned}
(\underline{u}_k, T_N\underline{u}_k) &= \frac{1}{2\pi N}\int_0^{2\pi}|a(\theta)|^2\left|\sum_{m=0}^{N-1}\overline{W}_N^{mk}e^{im\theta}\right|^2 d\theta \\
&= \frac{1}{2\pi N}\int_0^{2\pi}|a(\theta)|^2\left|\sum_{m=0}^{N-1}e^{im(\theta-\theta_N^k)}\right|^2 d\theta \\
&= \frac{1}{2\pi N}\int_0^{2\pi}|a(\theta)|^2\left\{\frac{\sin\left(N\frac{\theta-\theta_N^k}{2}\right)}{\sin\left(\frac{\theta-\theta_N^k}{2}\right)}\right\}^2 d\theta
\end{aligned}$$

soit encore :

$$(\underline{u}_k, T_N \underline{u}_k) = \lambda_{N,k}^C$$

d'après (37).

Il vient donc:

$$\frac{\mathrm{Trace}(A_N^{-1} T_N)}{N} = \frac{1}{N} \left(\sum_{k=0}^{N-1} \frac{\lambda_{N,k}^C}{\lambda_{N,k}^A} \right) \tag{50}$$

Le cas $A_N = C_N$ Le choix particulier $A_N = C_N$, matrice de préconditionnement de T.F. Chan, conduit à l'identité:

$$\frac{\mathrm{Trace}(C_N^{-1} T_N)}{N} = \frac{1}{N} \left(\sum_{k=0}^{N-1} \Lambda_{N,k}^C \right) = 1 \quad \forall N \geq 1 \tag{51}$$

ceci constitue un résultat plus intéressant que:

$$\frac{\mathrm{Trace}(C_N)}{N} = \frac{\mathrm{Trace}(T_N)}{N} = \alpha_0 = \frac{1}{2\pi} \int_0^{2\pi} |a(\theta)|^2 d\theta \tag{52}$$

Le cas $A_N = F_N$ Pour traiter le cas $A_N = F_N$ nous imposons la condition supplémentaire suivante:

$$\text{la fonction } \theta \to |a(\theta)|^2 \text{ est continue sur } [0, 2\pi] \tag{53}$$

En vertu du théorème de Féjer de convergence des séries de Fourier en moyenne de Césaro, la fonction

$$\mathrm{Ces}_N(u) = \frac{\sigma_0 + \sigma_1 + \ldots + \sigma_{N-1}}{N}(u) = \frac{1}{2\pi N} \int_0^{2\pi} |a(\theta)|^2 \left\{ \frac{\sin N \frac{\theta - u}{2}}{\sin \frac{\theta - u}{2}} \right\}^2 d\theta$$

converge uniformément sur $[0, 2\pi]$ vers la fonction continue $|a(\theta)|^2$. Il n'est pas nécessaire pour ce type de convergence qu'elle soit en plus à variation bornée. Par ailleurs, $\mathrm{Ces}_N(\theta_N^k) = \lambda_{N,k}^k$ et $|a(\theta_n^k)|^2 = \lambda_{N,k}^F$. On en déduit que pour tout $\varepsilon > 0$ donné, il existe $N_0(\varepsilon)$ tel que l'on ait:

$$|\lambda_{N,k}^C - \lambda_{N,k}^F| < \varepsilon \tag{54}$$

pour tout $N \geq N_0(\varepsilon)$, et ce, indépendamment de $k = 0, 1, \ldots, N-1$.

Les valeurs propres de C_N et F_N sont donc uniformément proches l'une de l'autre pour de grandes valeurs de N.

En supposant de plus que $|a(\theta)|^2$ est minorée par $m > 0$, pour que F_N soit définie positive, on obtient d'après (50) et (54)

$$\left| \frac{\mathrm{Trace}(F_N^{-1} T_N)}{N} - 1 \right| \leq \frac{1}{N} \left(\sum_{k=0}^{N-1} \frac{|\lambda_{N,k}^C - \lambda_{N,k}^F|}{\lambda_{N,k}^F} \right) \leq \frac{\varepsilon}{m} \text{ pour } N \geq N_0(\varepsilon)$$

d'où

$$\lim_{N \to +\infty} \frac{\mathrm{Trace}(F_N^{-1} T_N)}{N} = \lim_{N \to +\infty} \frac{1}{N} \left(\sum_{k=0}^{N-1} \Lambda_{N,k}^F \right) = 1 \tag{55}$$

résultat voisin de (51).

Convergence en moyenne de tous les moments Pour montrer maintenant (42) en sa forme équivalente:

$$\lim_{N\to+\infty}\frac{1}{N}\text{Trace}\left\{(A_N^{-1}T_N)^p\right\} = 1 \quad p \in \mathbb{N}$$

on procèdera en trois étapes : majoration, borne uniforme, convergence.

Majoration D'après (46):

$$\left|\frac{1}{N}Trace\left\{(A_N^{-1}T_N)^p - I_N\right\}\right| \le \left|(A_N^{-1}T_N)^p - I_N\right|_{HS}$$

On a :

$$(A_N^{-1}T_N)^p - I_N = (I_N + (A_N^{-1}T_N - I_N))^p - I_N = \sum_{k=1}^{p} C_p^k (A_N^{-1}T_N - I_N)^k$$

En appliquant l'inégalité (47) aux puissances de $A_N^{-1}T_N - I_N$, il vient:

$$\begin{aligned}
\left|(A_N^{-1}T_N)^p - I_N\right|_{HS} &\le \left\{\sum_{k=1}^{p} C_p^k \|A_N^{-1}T_N - I_N\|^{k-1}\right\} |A_N^{-1}T_N - I_N|_{HS} \\
&\le \left\{\sum_{k=1}^{p} C_p^k \|A_N^{-1}\|.\|A_N^{-1}T_N - I_N\|^{k-1}\right\}.|T_N - A_N|_{HS} \\
&= K(p,N).|T_N - A_N|_{HS}
\end{aligned}$$

Soit finalement:

$$\left|\frac{1}{N}\text{Trace}\left\{(A_N^{-1}T_N)^p\right\} - 1\right| \le K(p,N).|T_N - A_N|_{HS}$$

Pour conclure, il reste à trouver une borne uniforme en N pour $K(p,N)$, du moins à partir d'un certain rang N_0, et à montrer que pour $A_N = S_N$, C_N et F_N la quantité $|T_N - A_N|_{HS}$ tend vers zéro à l'infini.

Borne uniforme en N pour K(p, N) Il suffit de trouver un majorant uniforme pour $\|T_N\|$ et un autre pour $\|A_N^{-1}\|$. Pour ce faire, nous n'aurons besoin dans un premier temps que de l'hypothèse suivante: la fonction $\theta \to |a(\theta)|^2$ est bornée (par M) et minorée par $m > 0$. Elle n'est pas nécessairement continue.

Pour commencer, on a de manière évidente :

$$\frac{(\underline{x}, T_N\underline{x})}{(\underline{x},\underline{x})} = \frac{\int_0^{2\pi} |a(\theta)|^2 |P_N(e^{i\theta})|^2 d\theta}{\int_0^{2\pi} |P_N(e^{i\theta})|^2 d\theta} \le M$$

dont on tire $\|T_N\| \le M$ d'après (48).

La majoration de $\|A_N^{-1}\|$ se traite au cas par cas:

a) cas $A_N = F_N$

$$\frac{(\underline{x}, F_N\underline{x})}{(\underline{x},\underline{x})} \geq \min_{0\leq k\leq N-1} \lambda^F_{N,k} \geq \min_{[0,2\pi]} |a(\theta)|^2 \geq m$$

d'où $\|F_N^{-1}\| \leq \frac{1}{m}$ d'après (49).

b) cas $A_N = C_N$

$$\frac{(\underline{x}, C_N\underline{x})}{(\underline{x},\underline{x})} \geq \min_{0\leq k\leq N-1} \lambda^C_{N,k}$$

mais, d'après (38):

$$\lambda^C_{N,k} \geq m\frac{1}{2\pi N}\int_0^{2\pi}\left\{\frac{sin N\frac{\theta-\theta^k_N}{2}}{sin\ \frac{\theta-\theta^k_N}{2}}\right\}^2 d\theta = m.I_{N,k}$$

La fonction sous l'intégrale étant 2π-périodique, la quantité $I_{N,k}$ s'écrit encore:

$$I_{N,k} = \frac{1}{2\pi N}\int_{\theta^k_N}^{\theta^k_N+2\pi}\left\{\frac{\sin N\frac{(\theta^k_N-\theta)}{2}}{\sin\ \frac{\theta^k_N-\theta}{2}}\right\}^2 d\theta = \frac{1}{2\pi N}\int_0^{2\pi}\left\{\frac{\sin N\frac{u}{2}}{\sin\frac{u}{2}}\right\}^2 du = 1$$

De même que pour FN, on en déduit :

$$\frac{(\underline{x}, C_N\underline{x})}{(\underline{x},\underline{x})} \geq m$$

d'où

$$\|C_N^{-1}\| \leq \frac{1}{m}$$

c) cas $A_N = S_N$ On rajoute ici la condition supplémentaire $\alpha_n = O\left(\frac{1}{n^2}\right)$, de manière à se retrouver dans les hypothèses de la proposition (9) nous assurant que S_N est définie positive pour N assez grand. Dans ces conditions, il existe un rang N_0 tel que

$$\forall N \geq N_0 \quad \forall k = 0, 1, \ldots, N-1 \quad \lambda^S_{N,k} \geq \frac{m}{2}$$

On en déduit

$$\|S_N^{-1}\| \leq \frac{2}{m} \text{ pour } N \geq N_0$$

En résumé:

$$\|T_N\| \leq M \text{ et } \|A_N^{-1}\| \leq \frac{2}{m} \text{ pour } N \geq N_0$$

Ainsi, pour $N \geq N_0$:

$$K(p,N) \leq \frac{2}{m}\sum_{k=1}^{p} C^k_p\left(1+\frac{2M}{m}\right)^{k-1} \leq \frac{2}{m}\frac{\left\{1+\sum_{k=1}^{p} C^k_p\left(1+\frac{2M}{m}\right)^k\right\}}{1+\frac{2M}{m}}$$

soit encore

$$K(p,N) \leq \frac{2}{m}\frac{\left(2+\frac{2M}{m}\right)^p}{1+\frac{2M}{m}} \leq \frac{\left(2\left(1+\frac{M}{m}\right)\right)^p}{M}$$

Il reste à montrer que $|T_N - A_N|_{HS}$ tend vers zéro à l'infini. C'est l'objet de la proposition suivante.

Convergence

Proposition 15. Pour une matrice circulante A_N d'un des trois types C_N, S_N ou F_N on a:

$$\lim_{N\to+\infty} |T_N - A_N|_{HS} = 0$$

sous la condition supplémentaire commune

$$\lim_{k\to+\infty} k|\alpha_k|^2 = 0$$

Preuve 16. Prenons d'abord le cas de la matrice S_N de Strang. En évaluant la norme par les éléments respectifs des deux matrices S_N et T_N, il vient, compte tenu de la définition (34) de S_N:

$$\begin{aligned}|T_N - S_N|_{HS}^2 &= \frac{2}{N} \sum_{k=1}^{\frac{N}{2}-1} k|\alpha_k - \alpha_{N-k}|^2 \\ &\leq \frac{4}{N} \sum_{k=1}^{\frac{N}{2}-1} k|\alpha_k|^2 + \frac{4}{N}\left(\frac{N}{2}-1\right) \sum_{k=1}^{\frac{N}{2}-1} |\alpha_{N-k}|^2\end{aligned}$$

soit

$$\frac{1}{2}|T_N - S_N|_{HS}^2 \leq \frac{\sum_{k=1}^{\frac{N}{2}-1} k|\alpha_k|^2}{\frac{N}{2}-1} + \sum_{k=\frac{N}{2}+1}^{N-1} |\alpha_k|^2$$

Le premier terme est une moyenne de Césaro. Il tend vers zéro à l'infini d'après notre hypothèse de convergence vers zéro de la suite $k|\alpha_k|^2$. Le second terme tend également vers zéro à l'infini. C'est en effet la différence $R_{N/2+1} - R_N$ des restes de la série convergente $\Sigma|\alpha_k|^2$ dont la somme est

$$\sum_{k=0}^{\infty} |\alpha_k|^2 = \frac{1}{2}\left\{\left(\frac{1}{2\pi}\int_0^{2\pi} |a(\theta)|^2 d\theta\right)^2 + \frac{1}{2\pi}\int_0^{2\pi} |a(\theta)|^4 d\theta\right\}$$

Ainsi:

$$\lim_{N\to+\infty} |T_N - S_N|_{HS} = 0$$

Le traitement du cas $A_N = C_N$ devient alors immédiat. Il découle du cas précédent car $|T_N - C_N|_{HS} \leq |T_N - S_N|_{HS}$ en vertu de la définition de C_N.

Enfin, pour montrer que $\lim_{N\to+\infty} |T_N - F_N|_{HS} = 0$, on décompose cette quantité en deux morceaux: $|T_N - F_N|_{HS} \leq |T_N - C_N|_{HS} + |C_N - F_N|_{HS}$ et il reste alors à montrer que:

$$\lim_{N\to+\infty} |C_N - F_N|_{HS} = 0$$

Or, d'après la définition de cette norme (44):

$$|C_N - F_N|_{HS}^2 = \frac{1}{N}\left(\sum_{k=0}^{N-1} |\lambda_{N,k}^C - \lambda_{N,k}^F|^2\right)$$

et on sait d'autre part que les valeurs propres de C_N et F_N sont uniformément rapprochées pour N assez grand.

Donc $|C_N - F_N|_{HS} < \varepsilon$ si (54) est vérifiée pour $N \geq N_0(\varepsilon)$.

Nous venons donc de montrer que :

$$\forall p \in \mathbb{N} \quad \lim_{N\to+\infty} \frac{1}{N} \sum_{k=0}^{N-1} \left(\Lambda_{N,k}^{A}\right)^{p} = 1$$

Pour cela, plusieurs conditions ont été données:

1. la fonction $\theta \to |a(\theta)|^2$ est minorée par $m > 0$ et bornée;
2. elle est continue sur $[0, 2\pi]$;
3. ses coefficients de Fourier α_n sont des $o(\frac{1}{\sqrt{n}})$ à l'infini;
4. ses coefficients de Fourier α_n sont des $O(\frac{1}{n^3})$ à l'infini.

Les conditions suffisantes qui assurent le résultat sont:

- pour S_N : 1 et 4
- pour C_N : 1 et 3
- pour F_N : 1, 2 et 3.

Généralisation Le résultat précédent se transpose par combinaison linéaires à tout polynôme P

$$\lim_{N\to+\infty} \frac{1}{N} \left\{ \sum_{k=0}^{N-1} P\left(\Lambda_{N,k}^{A}\right) \right\} = P(1)$$

En notant r_A et s_A les bornes des valeurs propres, i.e.:

$$\forall N \quad \forall k = 0, 1, \ldots, N-1 \quad r_A \leq \Lambda_{N,k}^{A} \leq s_A$$

il s'étend ensuite sans difficulté à toute fonction f continue sur $[r_A, s_A]$ à l'aide du Théorème de Stone-Weierstrass. En effet :

$$\begin{aligned}\left| \frac{1}{N} \left\{ \sum_{k=0}^{N-1} f\left(\Lambda_{N,k}^{A}\right) \right\} - f(1) \right| &\leq \left| \frac{1}{N} \sum_{k=0}^{N-1} \left\{ f\left(\Lambda_{N,k}^{A}\right) - P_\ell\left(\Lambda_{N,k}^{A}\right) \right\} \right| \\ &\quad + \left| \frac{1}{N} \sum_{k=0}^{N-1} P_\ell\left(\Lambda_{N,k}^{A}\right) - P_\ell(1) \right| \\ &\quad + |P_\ell(1) - f(1)|\end{aligned}$$

Soit $\varepsilon > 0$ arbitraire donné.

La convergence uniforme de P_ℓ vers f sur $[r_A, s_A]$ permet de rendre le premier terme inférieur à $\frac{\varepsilon}{3}$ pour $\ell \geq \ell_0(\varepsilon)$ indépendant de N et k. Par la convergence simple, le dernier terme est rendu inférieur à $\frac{\varepsilon}{3}$ pour $\ell \geq \ell_1(\varepsilon)$. $\ell = \max(\ell_0(\varepsilon), \ell_1(\varepsilon))$ étant alors fixé, il reste à ajuster N pour rendre le second terme moindre que $\frac{\varepsilon}{3}$ et le total inférieur à ε.

Ceci termine la démonstration du théorème 14.

Remarques

- pour C_N et F_N les bornes r et s des valeurs propres sont respectivement m et M. Elles peuvent être prises égales respectivement à et pour S_N si N est assez grand.
- comme déjà mentionné précédemment, les conditions sur α_k ainsi que la continuité sont remplies pour $|a(\theta)|^2$ affine par morceaux.
- le résultat précédent peut être appliqué à la fonction Log.

La somme des valeurs propres se transforme en produit, ce qui conduit à:

$$\lim_{N\to+\infty} \frac{\log|\det(A_N^{-1}T_N)|}{N} = 0$$

soit :

$$\lim_{N\to+\infty} \frac{\log|\det(T_N)|}{N} = \lim_{N\to+\infty} \frac{\log|\det(A_N)|}{N}$$

résultat valable pour $A_N = S_N$, C_N et F_N. En particulier dans ce dernier cas, le second membre s'exprime plus simplement :

$$\begin{aligned}
\lim_{N\to+\infty} \frac{\log|\det(F_N)|}{N} &= \lim_{N\to+\infty} \left(\frac{1}{N}\sum_{k=0}^{N-1} \log|\lambda_{N,k}^F|\right) \\
&= \lim_{N\to+\infty} \left\{\frac{1}{2\pi}\sum_{k=0}^{N-1} \log\left(|a(\theta_N^k)|^2\right).\frac{2\pi}{N}\right\} \\
&= \frac{1}{2\pi}\int_0^{2\pi} \log(|a(\theta)|^2)d\theta
\end{aligned}$$

et fait apparaitre l'entropie

$$\frac{-1}{2\pi}\int_0^{2\pi} \log(|a(\theta)|^2)d\theta$$

L'égalité classique qui en résulte :

$$\lim_{N\to+\infty} \frac{\log|\det T_N|}{N} = \frac{1}{2\pi}\int_0^{2\pi} \log(|a(\theta)|^2)d\theta$$

est due à Szegö [13].

5 Discussion sur un Exemple Aéronautique du Choix des Paramètres de la Méthode

Le but de cette partie est d'expliquer, à la lumière d'un exemple concret comment, en pratique, mettre en oeuvre la méthode qui vient d'être présentée, et de manière plus précise comment choisir de manière adéquate l'ensemble des paramètres qui définissent la solution.

5.1 Exemple traité

La méthode de minimisation qui vient d'être présentée a été appliquée à un problème de contrôle actif du comportement d'un avion dans la turbulence atmosphérique. Il s'agit de déterminer le filtre linéaire dont l'entrée est une mesure de la vitesse verticale de la turbulence en un point de l'avion et dont la sortie est l'angle de braquage de la gouverne active destinée à assurer le contrôle de l'accélération au niveau du poste pilote.

Ce problème conduit à la recherche d'une fonction causale $f \in \mathcal{H}^2(\mathbb{C}^+)$ rendant minimale la fonctionnelle :

$$J_1(f) = \int_{\mathbb{R}} |b(\omega) - a(\omega)f(i\omega)|^2 d\omega$$

où $a(\omega)$ et $b(\omega)$ sont respectivement les fonctions de transfert accélérométrique de la gouverne et de la turbulence (fig. 1). Ces données n'étaient accessibles que jusqu'à une fréquence de coupure $f_c = 15$ Hz. L'extension à $\mathbb{R}$ de ces données a été réalisée conformément au prolongement proposé au 3.2. Sur ces fonctions de transfert, on met facilement en évidence par les pics de résonance les fréquences des quatres premiers modes mécaniques issues du couplage entre l'écoulement aérodynamique et la structure de l'avion.

La fonction $a(\omega)$ est partout strictement positive, ce qui assure le caractère défini positif de la matrice de Toeplitz et des matrices de préconditionnement. Néanmoins, l'exemple traité n'est pas sur le plan de la convergence de la méthode du gradient un exemple facile car le rapport $|a(\omega)|_{\max}/|a(\omega)|_{\min}$ est assez élevé (environ 75), ce qui entraine un mauvais conditionnement du système linéaire à résoudre. On se trouve donc là dans un cas peu favorable pour l'application des méthodes itératives sans préconditionnement préalable.

5.2 La solution de référence

Sur la figure 2, la solution approchée à un ordre N = 1024 a été obtenue par application de l'algorithme de Zohar, qui rappelons le est une méthode directe qui demande $2N^2$ opérations. On remarquera que la solution tend rapidement vers zéro au-delà de la fréquence de coupure et que la méthode est par ailleurs capable de restituer des solutions présentant des résonances marquées. Les petites oscillations encore perceptibles sur la phase sont dues à l'erreur liée à l'approximation du problème continu par un problème discret d'ordre N. La réponse impulsionnelle est calculée sur 7 secondes. Rappelons que la forme analytique (25) lui permet d'être évaluée en n'importe quelle valeur, indépendamment de la réponse en fréquence.

Cette solution nous servira de référence. Nous la comparerons aux solutions obtenues en faisant varier un quelconque des paramètres qui définissent cette solution.

Sur la figure 3, la comparaison de la valeur crête du reste avec celle de la fonction $a(\omega)$ témoigne de l'efficacité du contrôle.

On a fait figurer sur la planche suivante les coefficients α_n de la matrice, le second membre β_n ainsi que les coefficients solution α_n. Bien que les α_n

tendent assez vite vers zéro, on se convainc assez facilement que la matrice correspondante n'est pas à diagonale dominante.

Nous discutons maintenant du choix et des variations des paramètres de calcul sur la solution.

5.3 Influence de la fréquence de coupure

L'influence de la fréquence de coupure f_c sur la solution est résumée sur les figures 5 et 6. On observe sur la phase un comportement (**caractéristique**) en point de rebroussement à cette fréquence. Nous n'avons pas d'éléments d'explication à apporter à ce phénomène.

Le choix de la fréquence de coupure permet de sélectionner en fonction du problème à traiter, les modes qu'il convient de prendre en compte dans la solution.

5.4 Choix du paramètre p_o

Dans toutes les applications que nous avons traitées, nous avons relié le paramètre p0 défini en II.3 à la fréquence de coupure f_c par la relation $p_0 = 2\pi f_c$. L'examen du comportement des fonctions de base $f_{p_0,n}$ définies en II.3 nous montre en effet que ce choix permet d'obtenir le maximum d'oscillations de ces fonctions dans la bande utile de fréquence $[-f_c, +f_c]$. Nous pensons que ce critère d'oscillation est un bon critère de qualité pour l'approximation. Le choix qui en résulte pour p_0, même s'il n'est pas optimal dans certains cas particuliers, a toujours donné des résultats satisfaisants.

5.5 Influence de l'ordre N de l'approximation

L'évolution de la solution en fonction de l'ordre N de l'approximation est illustrée sur les figures 7 et 8. On retrouve encore ici que la convergence des solutions approchées vers la solution exacte est beaucoup plus lente sur la phase que sur le module. Ce défaut de convergence se traduit par des oscillations dont la fréquence et l'amplitude respectivement croit et décroit lorsque N augmente.

Les dernières figures (9 à 13) sont relatives à la résolution par la méthode du gradient conjugué des systèmes linéaires de Toeplitz obtenus à partir de la séquence des coefficients α_n construite pour traiter ce problème de contrôle.

5.6 Influence du nombre d'itérations du gradient

L'historique de la convergence des solutions générées aux itérations successives est résumé sur les figures 9 à 11. La solution retenue pour l'initialisation de l'algorithme est celle que l'on obtient en assimilant la matrice à sa diagonale soit $x_i = \beta_i/\alpha_0$. La taille du système résolu est ici fixée à 1024 et la matrice de préconditionnement utilisée est celle de Strang.

Nous rappelons qu'à partir des vecteurs successifs $\underline{x}_k$ générés par cette méthode itérative on construit facilement des approximations successives de la solution $f(p)$. Les figures 10 et 11 reproduisent respectivement l'évolution du module et de la phase de la réponse en fréquence $f(i\omega)$.

On retrouve encore ici la même lenteur de convergence sur la phase. La comparaison avec la solution de référence obtenue par la méthode directe de Zohar (fig. 2) permet d'affirmer sur ce système linéaire une convergence satisfaisante de la méthode du gradient conjugué préconditionné en un peu plus de 25 itérations.

5.7 Choix du préconditionneur

Les figures 12 et 13 illustrent quant à elles quelques propriétés de convergence de l'algorithme, avec et sans préconditionnement de la matrice. On remarque d'abord que la méthode du gradient conjugué converge mal sans préconditionnement pour de grandes valeurs de N (≥ 256) et que dans ce cas l'introduction d'un préconditionnement peut se révéler déterminante pour la vitesse de convergence. La convergence est de type linéaire sur l'exemple considéré. Elle devient moins "chahutée", plus régulière, lorsque la taille du système augmente.

Lorsqu'on désire obtenir de très petites valeurs du résidu, le préconditionnement de Chan et celui que nous proposons présentent un intérêt certain. Ils permettent d'obtenir des vitesses de convergence à la fois bonnes et peu sensibles à la taille du système.

Le comportement du préconditionneur de Strang et quant à lui assez paradoxal. Franchement mauvais pour de petites valeurs de N, où il vaut mieux ne pas préconditionner que de l'utiliser, il devient très performant pour $N = 1024$. Il convient néanmoins de rester prudent pour généraliser ces conclusions partielles obtenues avec une séquence donnée de coefficients α_n.

References

1. P.L. Düren, Theory of H^p spaces, Academic Press, New-York et Londres 1970.
2. P. Koosis, Introduction to H^p spaces, Cambridge University Press, Cambridge, MA, 1980.
3. W. Rudin, Real and Complex Analysis, Mac Graw-Hill, 1974.
4. W. Rudin, Analyse Réelle et Complexe, Masson.
5. P. Krée et C. Soize, Mécanique Aléatoire, Dunod, 1983.
6. M. Labarrère, J.P. Krief, B. Gimonet, Le Filtrage et ses Applications, Cepadues Editions, 1980.
7. A. Papoulis, Signal Analysis, New-York, Mac Graw-Hill, 1977.
8. N.I. Akhiezer, The classical moment problem, Hafner, New-York, 1965.
9. C. Carathéodory, Uber den Variabilittsbereich der Koeffizienten von Potenzreihen, die gegebene Werte nicht annehmen, Math. Ann. 64 (1907), pp. 95- 115.
10. C. Carathéodory, Uber den Variabilittsbereich der Fourier'schen Konstanten von positiven harmonischen Funktionen, Rend. Circ. Mat. Palermo 32 (1911), pp. 193-217.

11. C. Carathéodory et L. Féjer, Uber den Zusammenhan der Extremen von harmonischen Fonktionen mit ihren Koeffizienten und über den Picard-Landau'schen Satz, Rend. Circ. Mat. Palermo 32 (1911), pp. 218-239.
12. O. Toeplitz, Uber die Fourier'sche Entwickelung positiver Funktionen, Rend. Circ. Mat. Palermo 32 (1911), pp. 191-192.
13. U. Grenager et G. Szegö, Toeplitz forms and their applications, Univ. of California Press, Berkeley, 1958.
14. G. Szegö, Orthogonal Polynomials, Colloquium Publications, Vol. 23, Troisième Edition (Amer. Math. Soc.), New-York, 1967.
15. J.H. Justice, The Szegö recursion relation and inverses of positive definite Toeplitz matrices, SIAM J. Math. Anal., Vol. 5, pp. 503-508, Mai 1974.
16. N. Levinson, The Wiener RMS error criterion in Filter design and prediction, J. Math. Phys., Vol. 25, numéro 4, pp. 261-278, 1947.
17. P. Delsarte, Y. Genin, Y. Kamp, A generalisation of the Levinson algorithm for hermitian Toeplitz matrices with any rank profile, IEEE Trans. Acoust. Speech Signal Processing, vol. 33, numéro 4, pp. 964-971, Août 1985.
18. S. Zohar, The solution of a Toeplitz set of linear equations, J. Ass. Comput. Mach., Vol. 21, numéro 2, pp. 272-276, Avril 1974.
19. S. Zohar, Fortran subroutines for the solutions of Toeplitz sets of linear equations, IEEE Trans. Acoust. Speech. Signal Processing. vol. ASSP-27, numéro 6, Février 1979.
20. W.F. Trench, Weighting coefficients for the prediction of stationary time series from the finite past, SIAM J. Appl. Math., Vol. 15, numéro 6, pp. 1502-1510, Novembre 1967.
21. S. Zohar, Toeplitz matrix inversion : the algorithm of W.F. Trench, J. Ass. Comp. Mach, Vol. 16, numéro 4, pp. 592-601, Octobre 1969.
22. J.H. Justice, An algorithm for inverting positive definite Toeplitz matrices, SIAM J. Appl. Math. 23, pp. 289-291, 1972.
23. H. Akaike, Block Toeplitz matrix inversion, SIAM J. Appl. Math., Vol. 24, pp. 234-241, Mars 1973.
24. G. Cybenko, The numerical stability of the Levinson-Durbin algorithm for Toeplitz systems of equations, SIAM J. Sci. Stat. Comp., Vol. 1, pp. 303-319, 1980.
25. J.R. Bunch, Stability of methods for solving Toeplitz systems of equations, SIAM J. Sci. Stat. Comp., Vol. 6, pp. 349-364, 1985.
26. J.R. Jain, An efficient algorithm for a large Toeplitz set of linear equations, IEEE Trans. Acoust. Speech. Signal Processing Vol. ASSP-27, pp. 612-615, 1979.
27. A.K. Jain, Fast inversion of banded Toeplitz matrices by circular decompositions, IEEE Trans. Acoust. Speech. Signal Processing vol. ASSP-26, numéro 2, pp. 121-126, Avril 1978.
28. B. Dickinson, Efficient solution of linear equations with banded Toeplitz matrices, IEEE Trans. Acoust. Speech. Signal Processing Vol. ASSP-27, numéro 4, pp. 421-423, Août 1979.
29. F.G. Gustavson et D.Y.Y. Yun, Fast algorithm for rational Hermite approximations and solution of Toeplitz systems, IEEE Trans. Circuits Syst., Vol. 26, pp. 750-755, 1979.
30. R.P. Brent, F.G. Gustavson et D.Y.Y. Yun, Fast solution of Toeplitz systems of equations and computation of Padé approximants, Journal of Algorithms, vol. 1, pp. 259-295, 1980.
31. R. Kumar, Fast algorithm for Toeplitz equations, IEEE Trans. Acoust. Speech. Signal Processing, Vol. ASSP-33, numéro 1, pp. 254-267, Février 1985.

32. B. Frielander, M. Morf, T. Kailath et L. Ljung, Extended Levinson and Chandrasekhar equations for general discrete-time linear estimation problem, IEEE Trans. Automatic Control, Vol. 23, pp. 653-659, Août 1978.
33. T. Kailath, S.Y. Kung et M. Morf, Displacement rank of matrices and linear equations, J. Math. Anal. Appl., vol. 68, numéro 2, pp. 395-407, Avril 1979.
34. B. Frielander, M. Morf, T. Kailath et L. Ljung, New inversion formulas for matrices classified in terms of their distance from Toeplitz matrices, Linear Algebra Applications, Vol. 27, pp. 31-60, Octobre 1979.
35. R.R. Bitmead, S.Y. Kung, B.D.O. Anderson et T. Kailath, Greatest common divisors via generalized Sylvester and Bezout matrices, IEEE Trans. Automatic Control, Vol. 23, numéro 6, pp. 1043-1047, Décembre 1978.
36. R.R. Bitmead et B.D.O. Anderson, Asymptotically fast solution of Toeplitz and related systems of equations, Linear Algebra Applications, Vol. 34, pp. 103-116, 1980.
37. I.C. Gohberg et A.A. Semencul, On the inversion of finite Toeplitz matrices and their continuous analogs, Math. Issled, numéro 2, pp. 201-233, 1972 (en russe).
38. I.C. Gohberg et I.A. Feldman, Convolution equations and projection methods for their solutions, Amer. Math. Soc., Providence, RI, 1974.
39. S.Y. Kung et Y.H. Hu, A Highly Concurrent Algorithm and Pipelined Architecture for Solving Toeplitz Systems, IEEE Trans. Acous. Speech. Signal Processing, Vol. 31, numéro 1, Février 1983.
40. J.M. Delosme, S.C. Eisenstadt et J.R. Massé, Toeplitz solvers and vector processing, Actes du 11ème Colloque GRETSI, Nice 1-5 Juin 1987.
41. H.J. Landau, Maximum Entropy and the Moment Problem, Bulletin of the American Mathematical Society, pp. 45-77, 1987.
42. A.M. Burckstein et T. Kailath, Inverse scattering for discrete transmission-line models, SIAM Review, Vol. 7, pp. 1332-1349, 1986.
43. T. Kailath, A.M. Burckstein et D. Morgan, Fast matrix factorizations via discrete transmission lines, Linear Algebra Applications, Vol. 75, pp. 1-25, 1986.
44. J.P. Burg, Maximum entropy spectral analysis, A new analysis technique for time series data et the relationship between maximum entropy spectra and maximum likehood spectra dans Modern Spectrum Analysis (DG Childers ed.), IEEE Press, New-York, 1978.
45. O. Axelsson et V.A. Barker, Finite Element Solution of Boundary Value Problems. Theory and Computation, Academic Press, 1984.
46. P. Lascaux et R. Théodor, Analyse numérique matricielle appliquée à l'art de l'ingénieur, Tome 2, Masson, 1987.
47. O. Axelsson et G. Lindskog, On the Rate of Convergence of the Preconditioned Conjugate Gradient Method, Numerische Mathematik, Springer Verlag, Vol. 48, pp. 499-523, 1986.
48. L. Andersson, SSOR preconditionning of Toeplitz matrices, Thesis, Chalmers University of Technology, Göteborg, Sweden, 1976.
49. G. Strang, A proposal for Toeplitz matrix calculations, Studies in Applied Mathematics, Vol. 74, pp. 171-176, 1986.
50. T.F. Chan, An optimal preconditionner for Toeplitz systems, SIAM J. of Scientific and Statistical Computing, Vol. 9, numéro 4, pp. 766-771, 1988.
51. R.M. Gray, On the asymptotic eigenvalue distribution of Toeplitz matrices, IEEE Trans. of Information Theory, vol. 18, pp. 725-730, 1972.
52. P. Faurre, Cours d'Analyse Numérique. Notes d'Optimisation ; Ecole Polytechnique. Edition 1982.

53. J.L. Doob, Stochastic Processes, John Wiley & Sons, New-York, 1953.
54. M.B. Priestley, Spectral analysis and time series, Academic Press London, 1981.

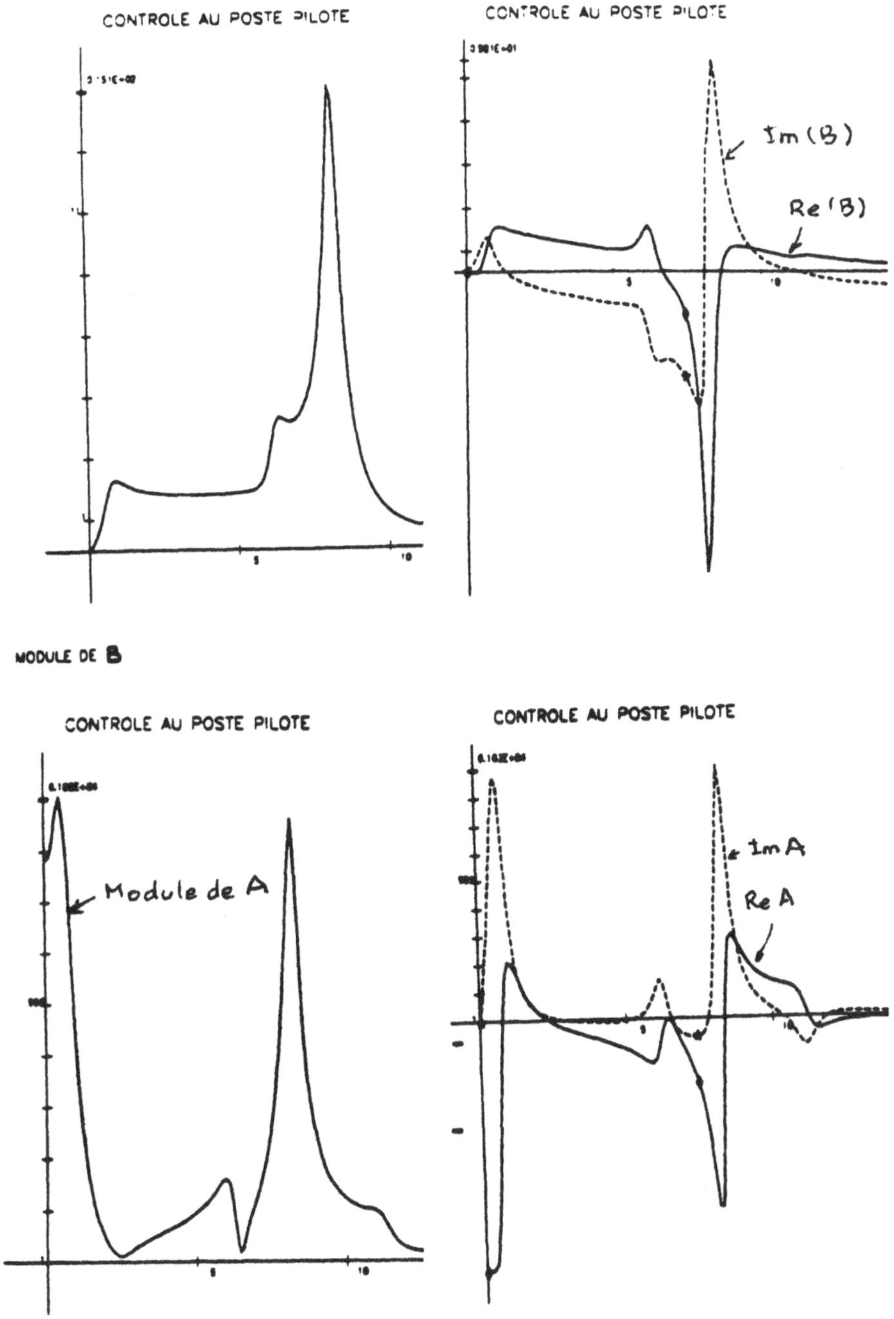

Fig. 1. Les données du problème.

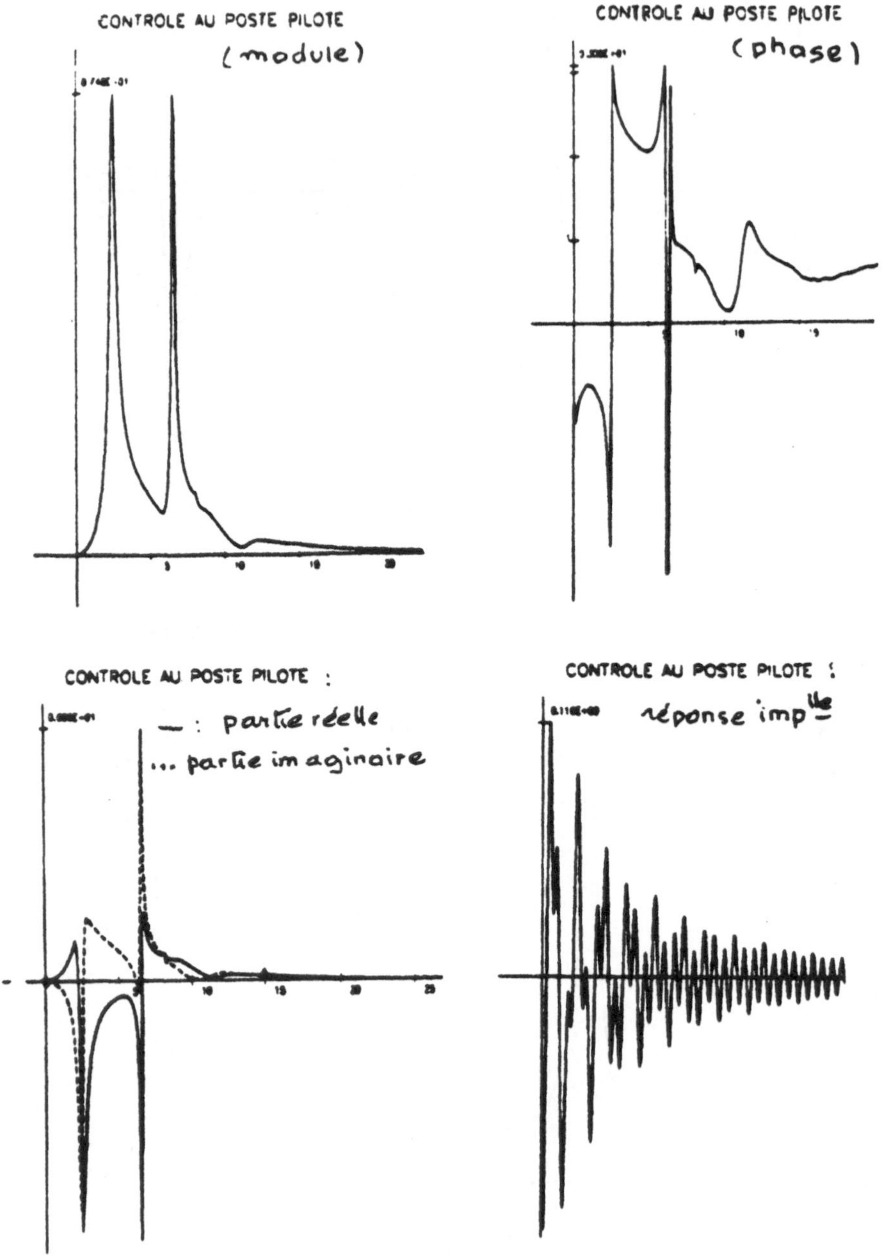

Fig. 2. La solution du problème de minimisation.

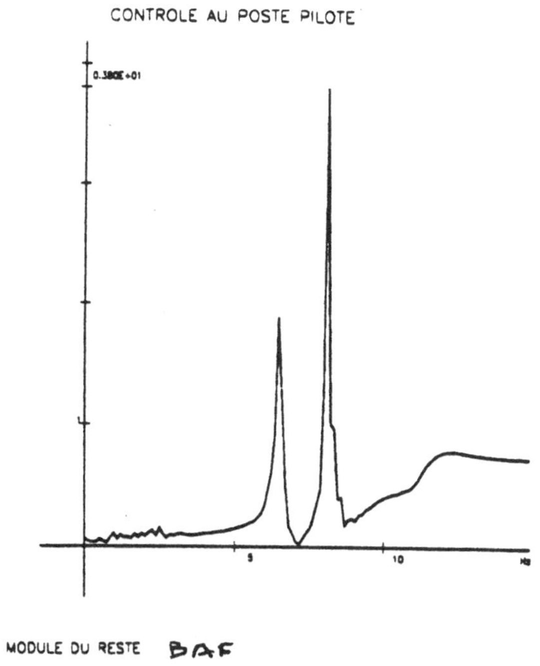

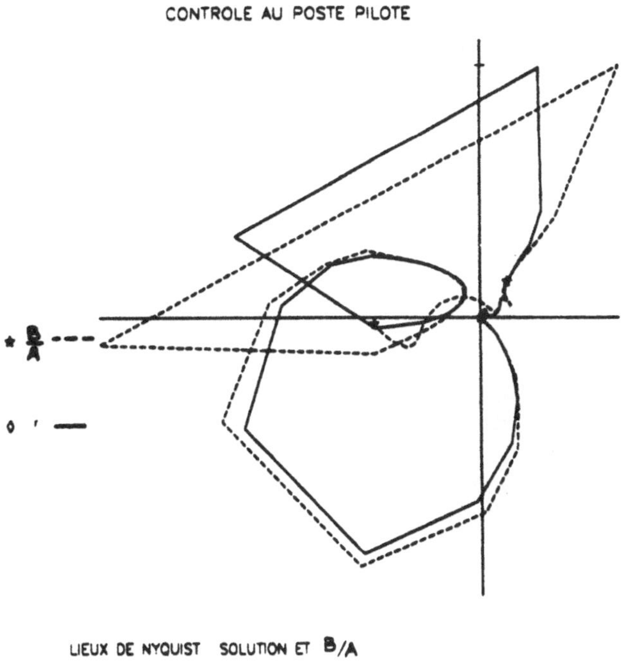

Fig. 3. Diagramme du reste B-A.F et lieux de Nyquist.

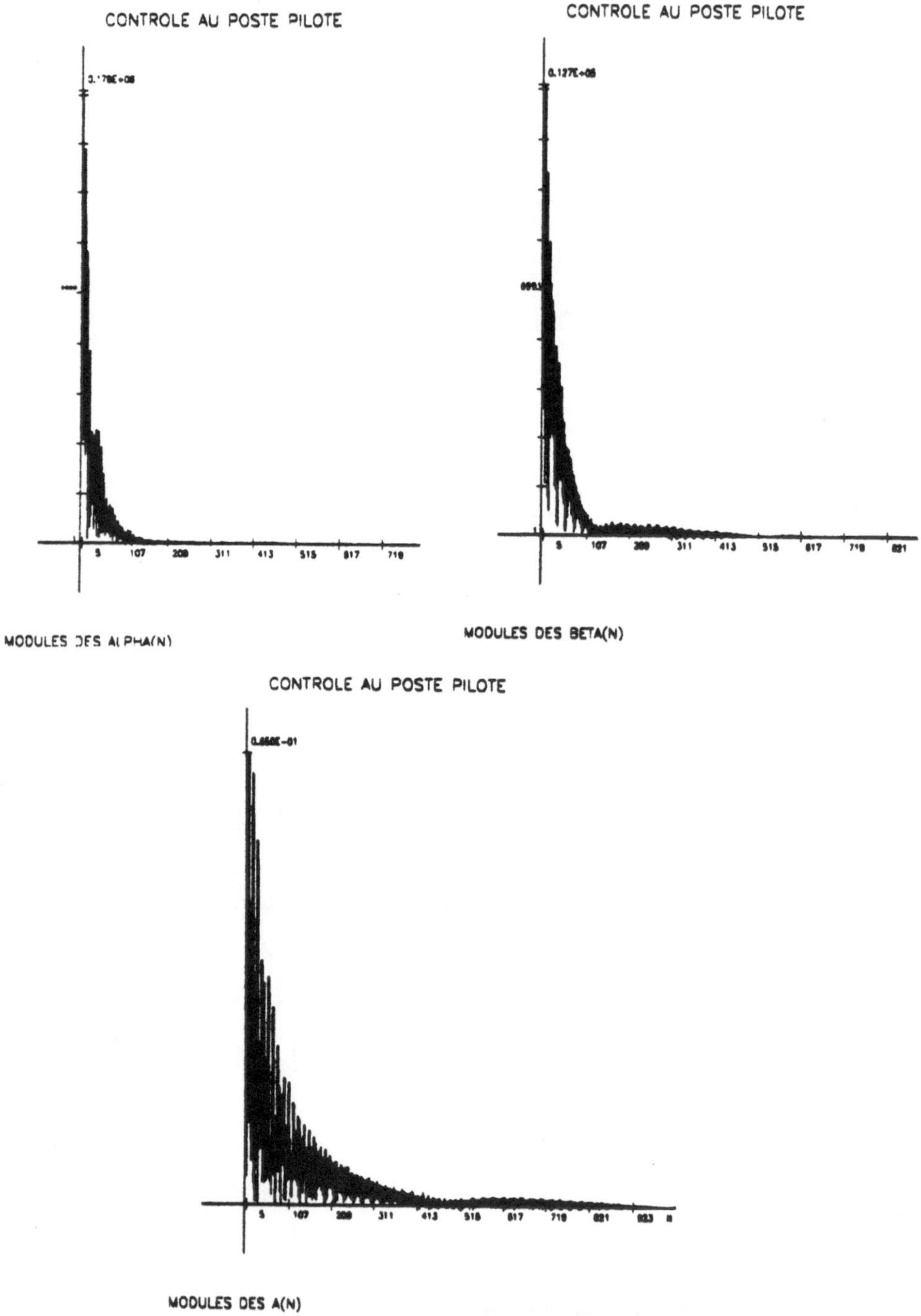

Fig. 4. Diagramme décrivant les éléments de la matrice, le second membre et la solution du sytème linéaire.

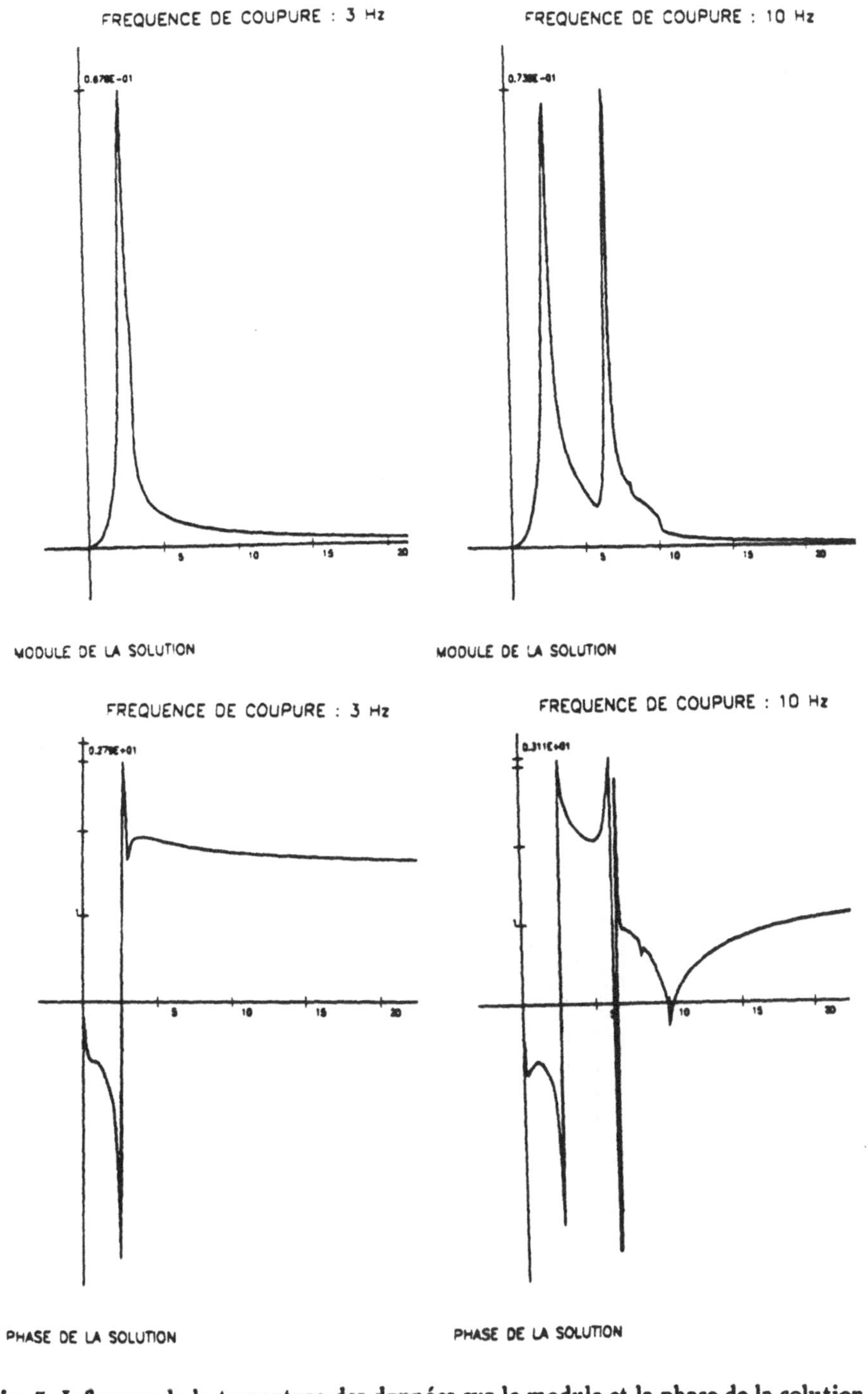

Fig. 5. Influence de la troncature des données sur le module et la phase de la solution.

FREQUENCE DE COUPURE : 3 Hz

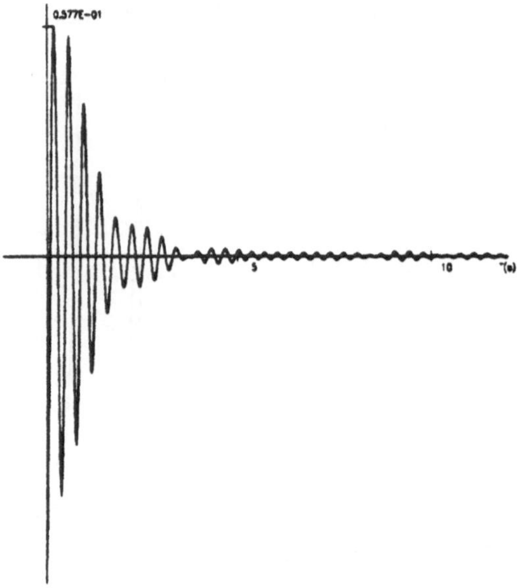

REPONSE IMPULSIONNELLE

FREQUENCE DE COUPURE : 10 Hz

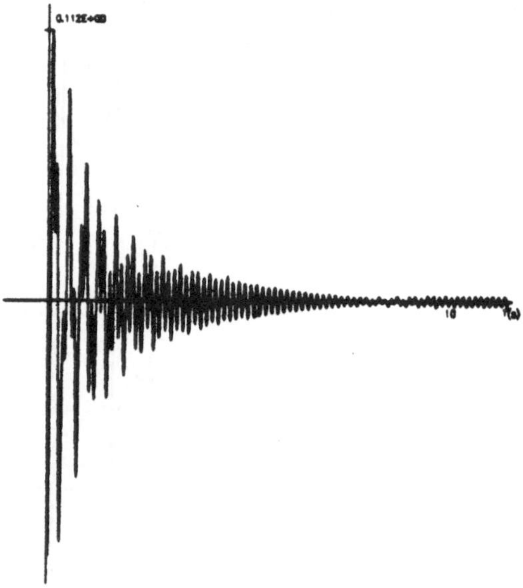

REPONSE IMPULSIONNELLE

Fig. 6. **Influence de la troncature des données sur la réponse impulsionnelle de la solution.**

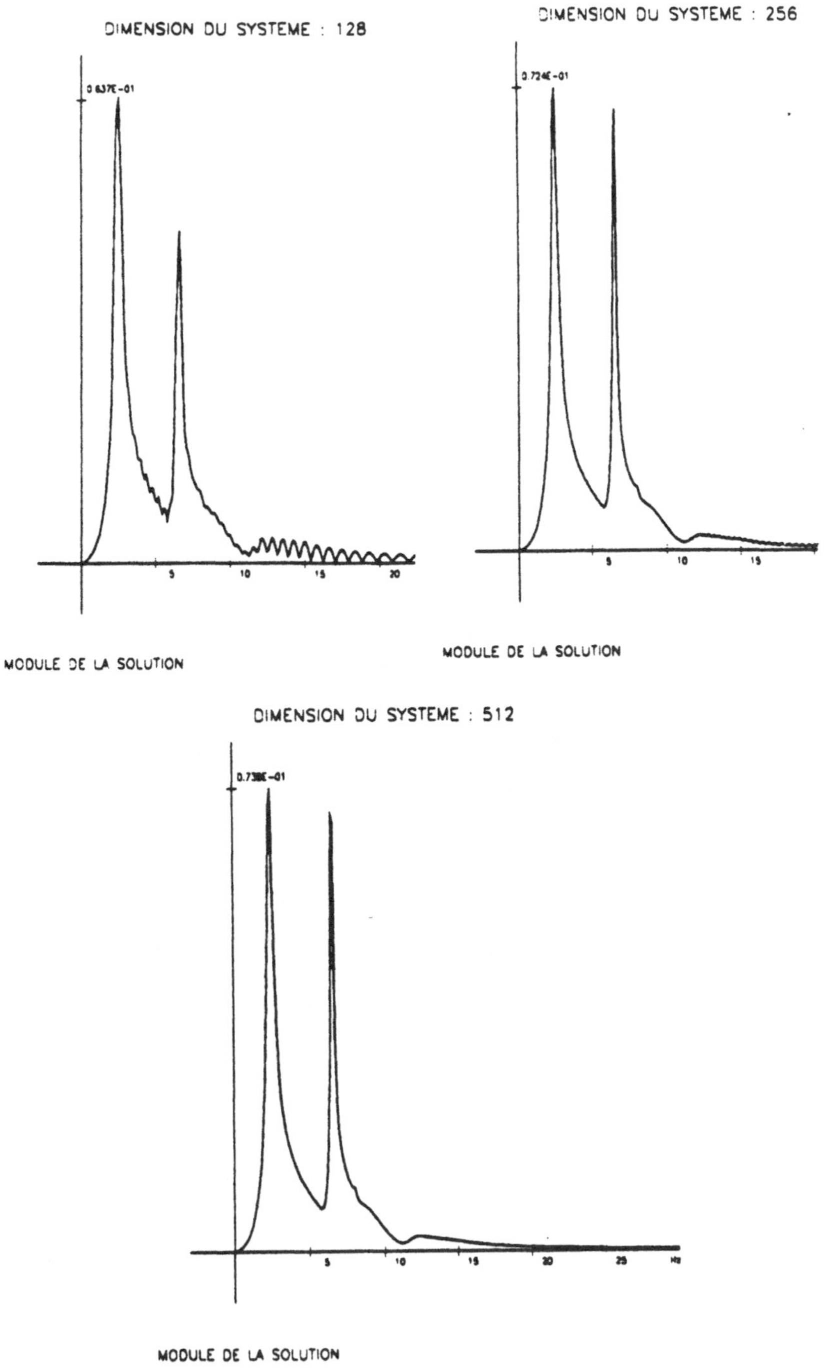

Fig. 7. Diagrammes montrant l'influence de la taille du système linéaire sur le module de la solution.

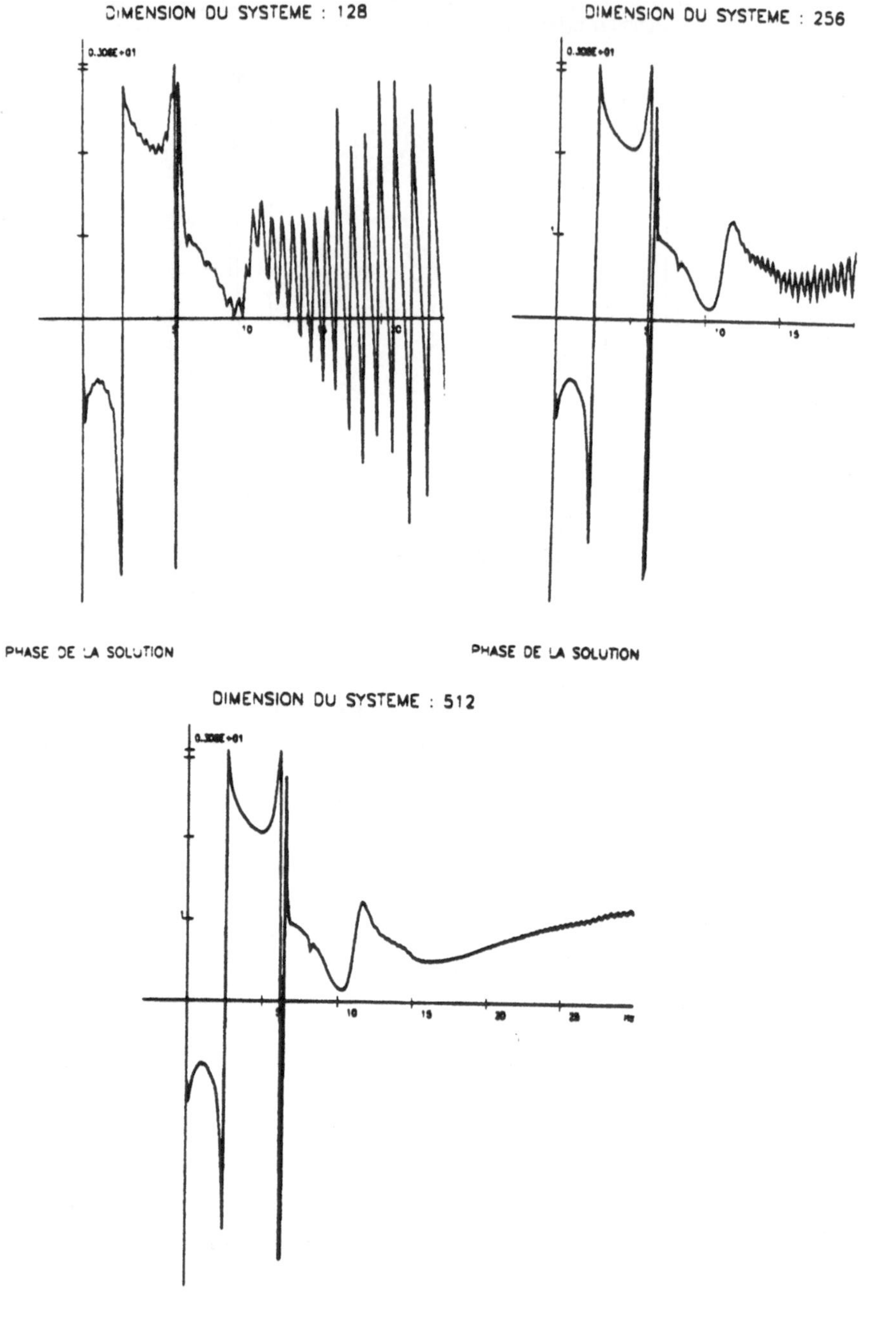

Fig. 8. Diagrammes montrant l'influence de la taille du système linéaire sur la phase de la solution.

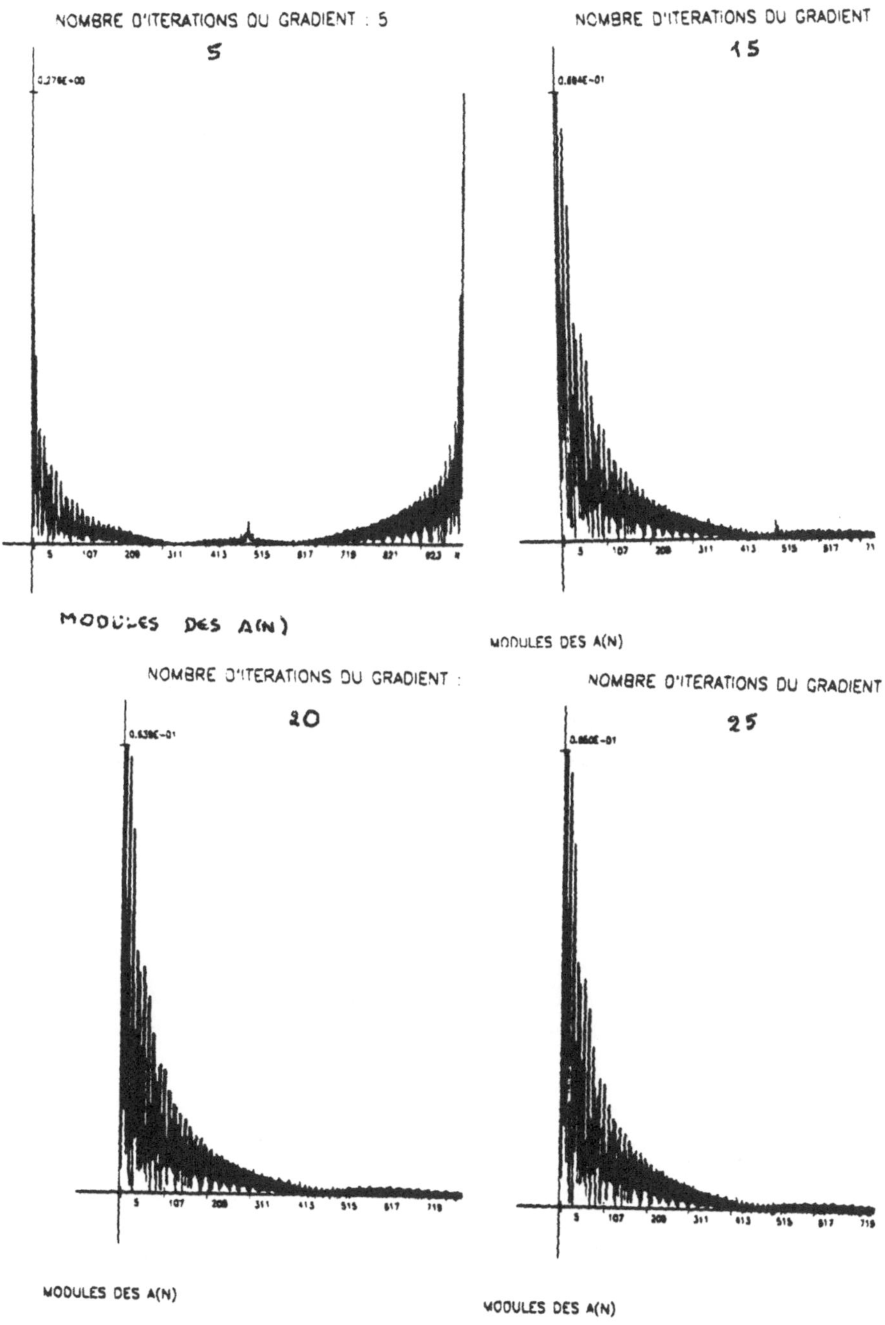

Fig. 9. Diagrammes donnant les modules de la solution du système linéaire à différentes itérations du gradient.

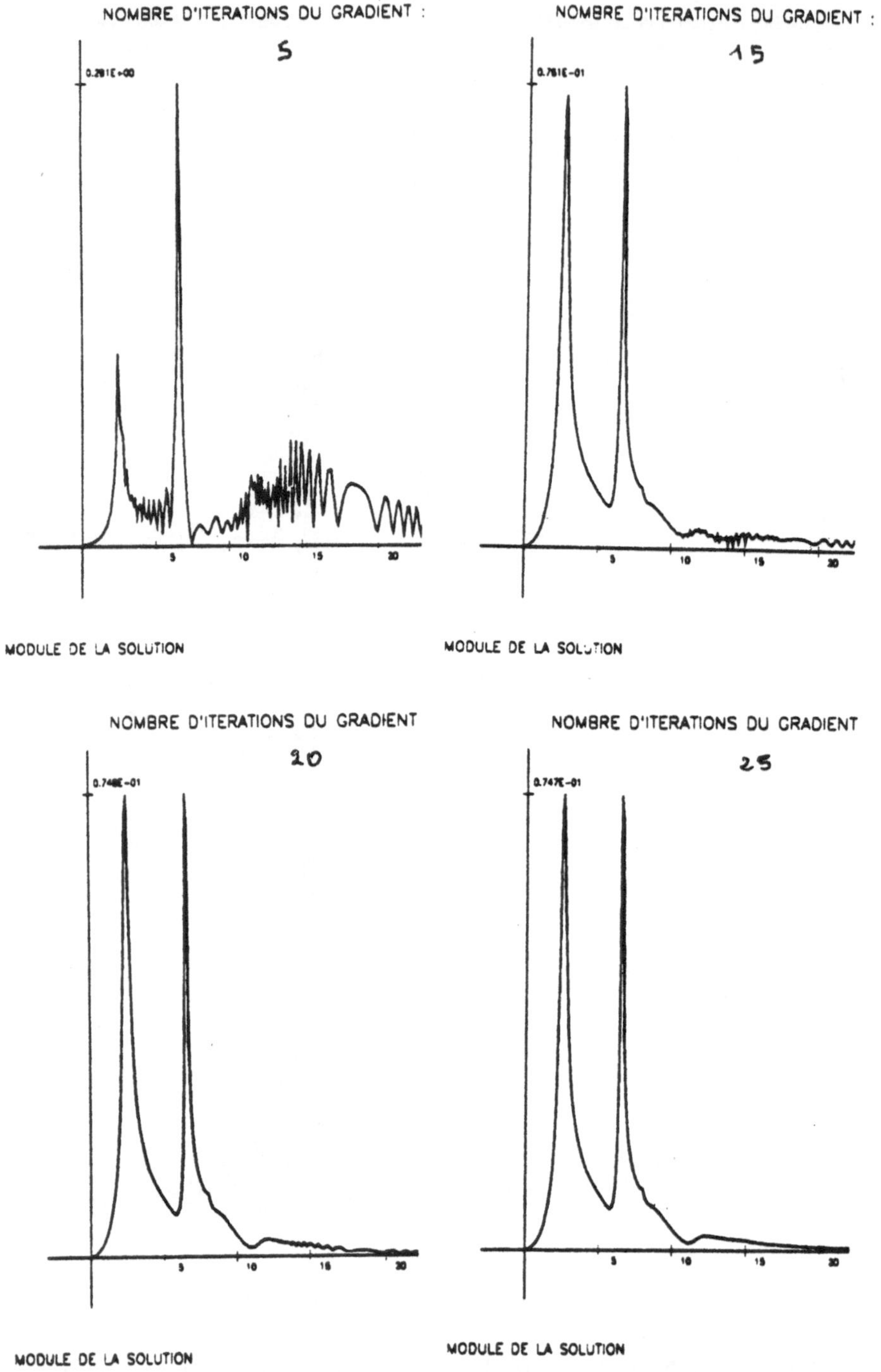

Fig. 10. Diagrammes donnant le module de la solution à différentes itérations du gradient.

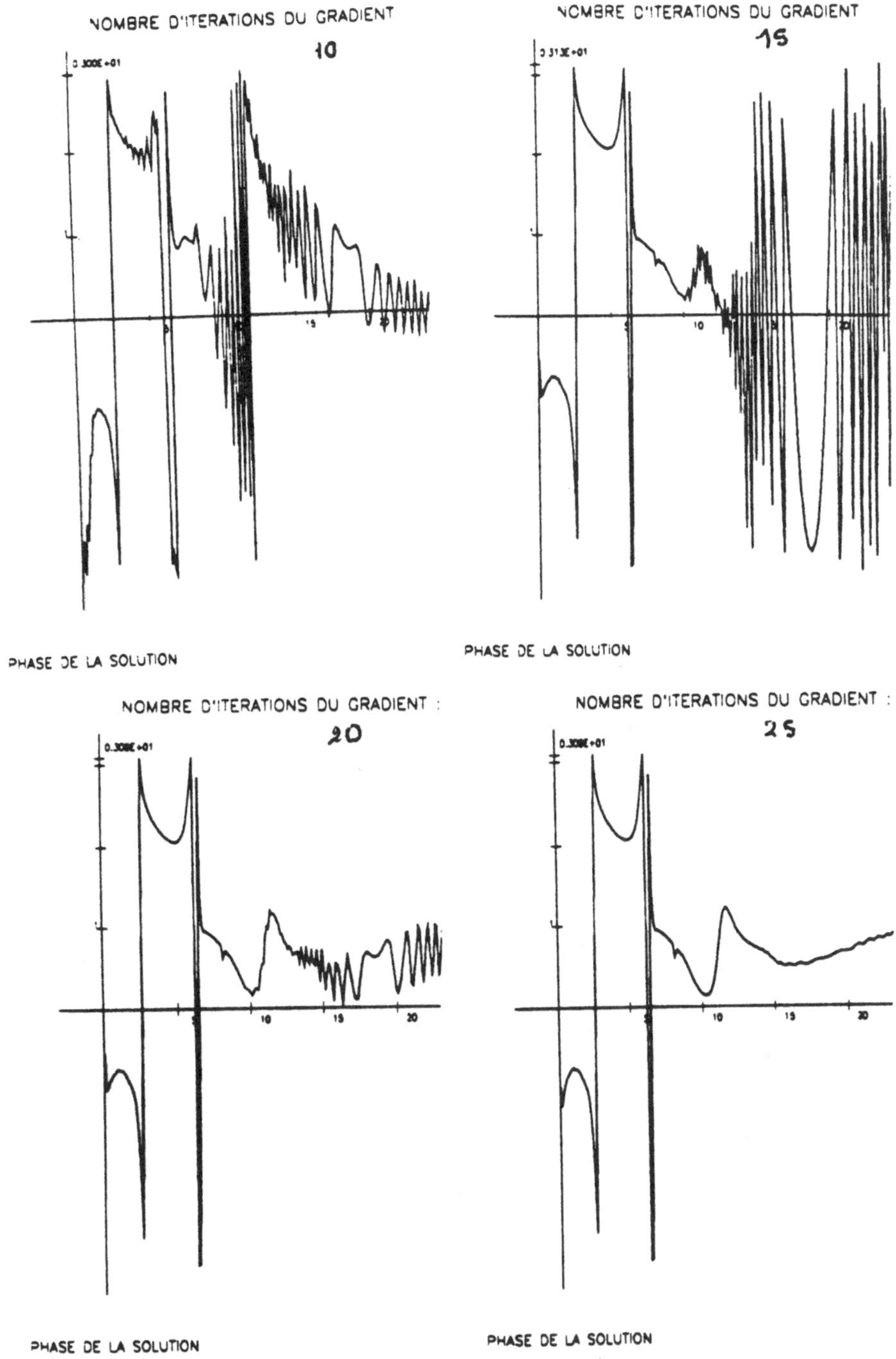

Fig. 11. Diagrammes donnant la phase de la solution à différentes itérations du gradient.

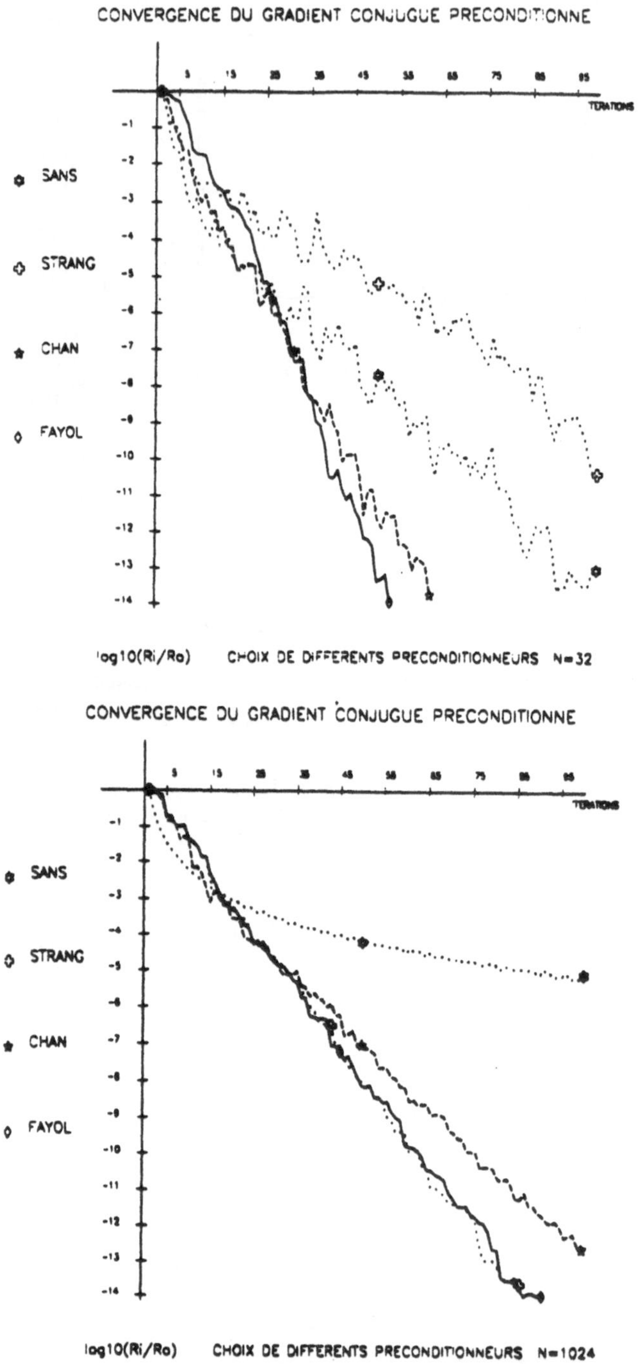

Fig. 12. Diagrammes montrant l'efficacité comparée des préconditionnements circulants en fonction de la taille du système.

CONVERGENCE DU GRADIENT CONJUGUE PRECONDITIONNE

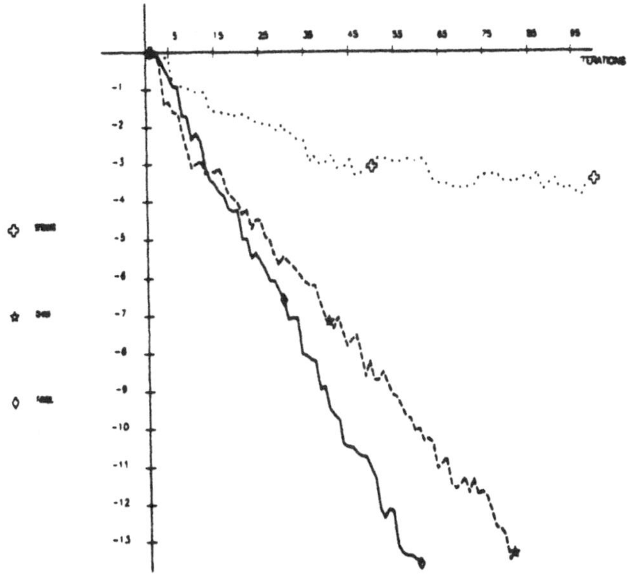

log10(Ri/Ro) CHOIX DE DIFFERENTS PRECONDITIONNEURS N=64

CONVERGENCE DU GRADIENT CONJUGUE PRECONDITIONNE

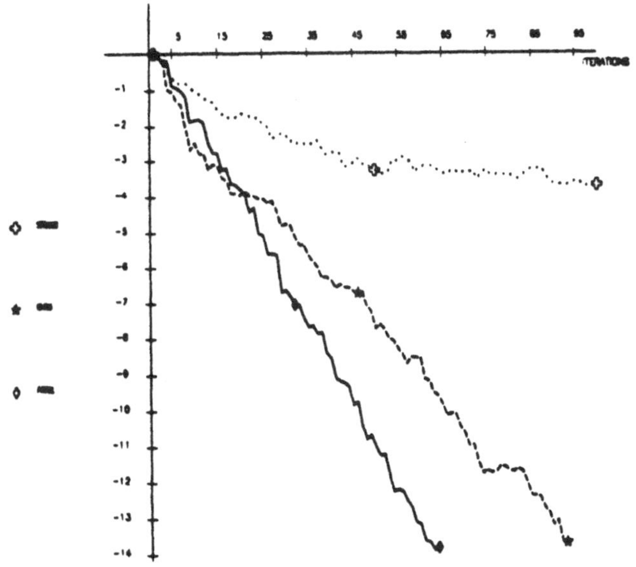

log10(Ri/Ro) CHOIX DE DIFFERENTS PRECONDITIONNEURS N=128

Fig. 13. Diagrammes montrant l'efficacité comparée des préconditionnements circulants en fonction de la taille du système (suite).